Alicia Casals and Anibal T. de Almeida (Eds)

Experimental Robotics V

The Fifth International Symposium
Barcelona, Catalonia, June 15-18, 1997

Springer

Series Advisory Board

A. Bensoussan · M.J. Grimble · P. Kokotovic · H. Kwakernaak
J.L. Massey · Y.Z. Tsypkin

Editors

Professor Alicia Casals
Universitat Politecnica de Cartalunya, Pau Gargallo n.5, 08028 Barcelona, Spain

Professor Anibal T. de Almeida
ISR, Dep. Eng. Electronica, University of Coimbra, POLO II, 3030 Coimbra, Portugal

ISBN 3-540-76218-3 Springer-Verlag Berlin Heidelberg New York

British Library Cataloguing in Publication Data
Experimental robotics V : the Fifth International
  Symposium, Barcelona, Catalonia, June 15-18, 1997. -
  (Lecture notes in control and information sciences ; 232)
  1.Robotics - Congresses
  I.Casals, Alicia II.Almeida, Anibal T. de III.International
  Symposium on Experimental Robotics (5th :1997 : Barcelona,
  Spain)
  629.8'92
  ISBN 3540762183

Apart from any fair dealing for the purposes of research or private study, or criticism or review, as permitted under the Copyright, Designs and Patents Act 1988, this publication may only be reproduced, stored or transmitted, in any form or by any means, with the prior permission in writing of the publishers, or in the case of reprographic reproduction in accordance with the terms of licences issued by the Copyright Licensing Agency. Enquiries concerning reproduction outside those terms should be sent to the publishers.

© Springer-Verlag London Limited 1998
Printed in Great Britain

The use of registered names, trademarks, etc. in this publication does not imply, even in the absence of a specific statement, that such names are exempt from the relevant laws and regulations and therefore free for general use.

The publisher makes no representation, express or implied, with regard to the accuracy of the information contained in this book and cannot accept any legal responsibility or liability for any errors or omissions that may be made.

Typesetting: Camera ready by contributors
Printed and bound at the Athenæum Press Ltd, Gateshead
69/3830-543210 Printed on acid-free paper

PREFACE

Experimental Robotics V - The Fifth International Symposium on Experimental Robotics (ISER) was held at the Universitat Politècnica de Catalunya (UPC), Barcelona, from June 15 to 18, 1997. This meeting was the latest in a series of symposia designed to bring together researchers from institutions around the world that are at the forefront of experimental robotics research. The main objective of these symposia is to pool experience in the various fields of robotics, concentrating not only on theoretical and technological questions, but also on their experimental validation.

The experimental robotics symposia are held every two years rotating through North America, Europe and Asia. The first Symposium of the series was organized by V. Hayward and O. Khatib, in Montreal, Canada in June 1989. The second one, organized by R. Chatila and G. Hirzinger was held in Toulouse, France, in June 1991. The third, organized by T. Yoshikawa and F. Miyazaki, was held in Kyoto, Japan, in 1993, and the 1995 meeting was organized by O. Kathib and K. Salisbury, and held in Stanford, USA.

The International Program Committee for the 1997 meeting comprised:

Anibal T.de Almeida	University of Coimbra, Portugal
Alícia Casals	Universitat Politècnica de Catalunya, Spain
Raja Chatila	LAAS/CNRS France
John Craig	Adept Technology, Inc., U.S.A
Paolo Dario	Schuola Superiore Sta. Anna, Italy
Joris De Schutter	Katholieke Universiteit Leuven, Belgium
Vincent Hayward	McGill University, Canada
Gerhard Hirzinger	DLR, Germany
Oussama Kathib	Stanford University, U.S.A
Jean-Pierre Merlet	INRIA, France
Fumio Miyazaki	Osaka University, Japan
Yoshihiko Nakamura	Tokyo University, Japan
Kenneth Salisbury	MIT, USA
James Trevelyan	University of Western Australia
Tsuneo Yoshikawa	Kyoto University, Japan

As occurred with the previous ISER meetings, the publication of the symposium preprints has been followed by publication in this Lecture-Notes Series of the revised papers and general conclusions of the sessions. This time, the compilation of the experimental results in the video proceedings will provide a further source of information about the advances in experimental robotics.

The symposium was attended by participants from thirteen countries who introduced the studies selected by the International Program Committee (IPC). These experimental studies covered the design, perception, control, planning and robotic applications in areas such as mobile robots, space, underwater, medicine and civil engineering. The standard presentations in one-theme sessions were enriched by four keynotes speeches, one of which opened the proceedings each day. The first one, by Pradeep Koshla, was entitled: Emerging Paradigms and Technologies in Robotics; the second, given by Gerd Hirzinger was: On the way towards a new Robot Generation; the third, presented by Hirochika Inoue, was entitled: Evolution of Experimental Robotics; and the last one, by Georges Giralt, was entitled: Novel Perspectives and Challenges for Robotics in Europe. The quality of these presentations stimulated numerous questions and comments from the participants. The small number of participants and the quality of the experimental research presented enabled the goal of the symposium to be attained: that is, to present the most advanced research works in experimental robotics, in an atmosphere that was conducive to discussion and the exchange of ideas.

Despite a certain feeling among the general public that robotics is antisocial, the fields of application of this technological research and development area show that robots can contribute greatly to a better quality of life. The new techniques emerging in the surgical field make minimally invasive surgery possible in some delicate interventions improving both surgical procedures and results. The advances in rehabilitation robotics open up possibilities for the greater independence of disabled people. Very risky tasks, or tasks which humans can not perform, can be solved by robots or by robots cooperating with humans. Successes in space, underwater, construction, medicine, hazardous manufacturing tasks, the nuclear industry and mining applications show the realities which already exist. The success of this event and the interest it aroused confirmed the appropiateness of its continuing in the future. Therefore, the IPC decided to entrust the sixth edition of ISER to Peter Coorke and Ray Jarvis, who will organize it for 1999 in Cairns, Australia.

On behalf of the International Program Committee we would like to thank the UPC, for hosting the symposium and especially its Rector Jaume Pagès for his kind welcoming address. We would also like to thank Antoni Giró, President of the Catalan Research Agency for his warm reception in the wonderful setting of the Palau de la Generalitat. We also wish to thank the CICYT (Spanish Research Agency), CIRIT (Catalan Research Agency), SCT-IEC (Catalan Society of Technology filial of the Institute for Catalan Studies) and the ETSECCP (School of Civil Engineering of Barcelona) for their contributions to the Symposium.

Our special and warm thanks to the group of people of the Department of Automatic Control and Computer Engineering of UPC who did a tremendous job of organization and support before and during the symposium. Professor Josep Amat provided many good ideas and experiences. Josep Fernandez, Jesús Galceran, Antoni Grau, Pere Marès, and Rita Planas gave their best in support of the Symposium. The administrative support of Rosa Sánchez was paramount to the smooth running of the conference.

Alícia Casals and Anibal T. de Almeida
Barcelona, Catalonia (Spain), September 1997

List of Participants

1. Ahmadi, M.
   Dept of Mechanical Engineering
   McGill University, Montréal
   CANADA

2. Amat, J.
   Departament d'ESAII
   Universitat Politècnica de
   Catalunya, Barcelona
   SPAIN

3. Andersen, N.A.
   Department of Automation
   Technical University of Denmark,
   Lyngby
   DENMARK

4. Araújo, R.
   Institute of Systems and Robotics
   Electrical Engineering Dept.
   University of Coimbra, Coimbra
   PORTUGAL

5. Avello, A.
   Centro de Estudios e
   Investigaciones Tecnológicas
   San Sebastián
   SPAIN

6. Beltran-Escavy, J.
   Dept of Machinery-Engineering
   University of Tokyo, Tokyo
   JAPAN

7. Benavent, M.R.
   Universitat Politècnica de
   València, València
   SPAIN

8. Bidaud, Ph.
   Laboratoire de Robotique de Paris,
   Vélizy
   FRANCE

9. Bilodeau, G.
   Dept of Mechanical Engineering
   McGill University, Montréal
   CANADA

10. Briones, L.
    Centro de Estudios e
    Investigaciones Tecnológicas
    San Sebastián
    SPAIN

11. Butterfass, J.
    DLR, Institute for Robotics and
    System Dynamics, Wessling
    GERMANY

12. Caccavale, F.
    Dipt. Informatica e Sistemistica
    Univ. Studi di Napoli Federico II,
    Napoli
    ITALY

13. Canfield, S.L.
    Dept. of Mechanical Engin.
    Tennessee Tech. University
    Cookeville, TN
    USA

14. Chatila, R.
    LAAS-CNRS
    Toulouse
    FRANCE

15. Corke, P.
    CSIRO - Division of
    Manufacturing Technology
    Kenmore
    AUSTRALIA

16. Cugini, U.
    Dipt. di Ingegneria Industriale
    Università di Parma, Parma
    ITALY

17. Dario, P.
    ARTS Lab.
    Scuola Superiore Sant'Anna, Pisa
    ITALY

18. Davies, B.
    Mechatronics in Medicine Group.
    Dept. of Mechanical Engineering.
    Imperial College, London
    UNITED KINGDOM

19. De Schutter, J.
    Dept. of Mechanical Engineering
    Katolieke Universiteit Leuven,
    Haverlee
    BELGIUM

20. Durrant-Whyte, H.F.
    Dept. of Mechanical and
    Mechatronic Engineering
    University of Sydney, Sydney
    AUSTRALIA

21. Ellis, R.E.
    Department of Computing and
    Information Science
    Queen's University at Kingston
    CANADA

22. Featherstone, R.
    Dep of Engineering Science
    Oxford University, Oxford
    UNITED KINGDOM

23. Fernández, S.
    IKERLAN S. Coop.
    Mondragón
    SPAIN

24. Foulon, G.
    LAAS-CNRS
    Toulouse
    FRANCE

25. Giralt, G.
    LAAS-CNRS
    Toulouse
    FRANCE

26. Hait, A.
    LAAS-CNRS
    Toulouse
    FRANCE

27. Hanebeck, U.D.
    Institute of Automatic Control
    Engineering (LSR)
    Tech. University of Munich
    München
    GERMANY

28. Hayward, V.
    Dept. of Electrical Engineering
    McGill University, Montréal
    CANADA

29. Hirzinger, G.
    DLR, Institute for Robotics and
    System Dynamics
    Wessling
    GERMANY

30. Hosoda, K.
    Dept. of Adaptive Machine
    Systems
    Osaka University, Osaka
    JAPAN

31. Howe, R.
    Division of Engineering and
    Applied Science
    Harvard University
    Cambridge, MA
    USA

32. Inoue, H.
    Dept. of Mechano-Informatics
    The University of Tokyo
    Tokyo
    JAPAN

33. Jaouni, H.
    LAAS-CNRS
    Toulouse
    FRANCE

34. Jarvis, R.
    Intelligent Robotics Research Centre
    Monash University, Victoria
    AUSTRALIA

35. Kaneko, M.
    Industrial and Systems Engineering
    Hiroshima University
    Higashi-Hiroshima
    JAPAN

36. Kapellos, K.
    INRIA
    Sophia Antipolis
    FRANCE

37. Khatib, M.
    LAAS-CNRS
    Toulouse
    FRANCE

38. Khatib, O.
    Robotics Lab
    Dept. of Computer Science
    Stanford University
    Stanford, CA
    USA

39. Khosla, P.
    Institute for Complex Eng. Systems
    Carnegie Mellon University
    Pittsburgh, PA
    USA

40. Konno, A.
    Dept. of Mechano-Informatics
    University of Tokyo, Tokyo
    JAPAN

41. Lahdhiri, T.
    Intelligent Manufact. Syst. Center
    University of Windsor, Windsor
    CANADA

42. Laschi, C.
    ARTS Lab.
    Scuola Superiore Sant'Anna, Pisa
    ITALY

43. Laugier, C.
    INRIA Rhône-Alpes, GRAVIR
    Montbonnot Saint Martin
    FRANCE

44. Liu, J.
    Dept. of Computing Studies
    Baptist University
    Kowloon Tong, Hong-Kong
    CHINA

45. Lloyd, J.E.
    Dept. of Computer Science
    Univ. of British Columbia, VC
    CANADA

46. Maeyama, S.
    Intelligent Robot Laboratory
    University of Tsukuba, Tsukuba,
    JAPAN

47. Martinoli, A.
    Microcomputing Laboratory
    Swiss Federal Institute of Technology, Lausanne
    SWITZERLAND

48. Merlet, J.-P.
    INRIA
    Sophia Antipolis
    FRANCE

49. Mitsuishi, M.
    Dept. of Engineering Synthesis
    University of Tokyo, Tokyo
    JAPAN

50. Miyazaki, F.
Department of Mechanical Engineering
Osaka University, Osaka
JAPAN

51. Morizono, T.
Department of Robotics
Ritsumeikan University
Kusatsu, Shiga
JAPAN

52. Mujika, J.
IKERLAN S. Coop.
Mondragón
SPAIN

53. Nakamura, Y.
Dept. of Mechano-Informatics
University of Tokyo, Tokyo
JAPAN

54. Nakawaki, D.
Dept. of Mechanical Engineering
Osaka University, Osaka
JAPAN

55. Prisco, G.M.
PERCRO
Scuola Superiore Sant'Anna, Pisa
ITALY

56. Renaud, M.
LAAS-CNRS
Toulouse
FRANCE

57. Roβmann, Th.
Tech. Universität München
München
GERMANY

58. Rus, D.
Dartmouth College
Dept. of Computer Science
Hanover, NH
USA

59. Salcudean, S.E.
Dept. of Electrical Engineering
University of British Columbia
Vancouver
CANADA

60. Salisbury, J.K.
AI Lab & Me Dept, MIT
Cambridge, MA
USA

61. Sonck, S.
Robotics Lab. ,Computer Science
Oxford University, Oxford
UK

62. Troisfontaine, N.
Lab. de Robotique de Paris
Vélizy
FRANCE

63. Tsakiris, D.P.
INRIA
Sophia Antipolis
FRANCE

64. Yoshikawa, T.
Dept. of Mechanical Engineering
Kyoto University, Kyoto
JAPAN

65. Yuta, S.
Intelligent Robot Laboratory
University of Tsukuba
Tsukuba Ibaraki
JAPAN

66. Zelinsky, A.
Dept of Systems & Engineering
Australian National University
Canberra
AUSTRALIA

CONTENTS

Author Index   xvi

Keynote   1

"Towards a new Robot Generation", *Hirzinger, G.; Arbter, K.; Brunner, B. and Koeppe, R.* 3

"Towards Evolution of Experimental Robotics", *Inoue, H.* 22

1 Dexterous Manipulation   33

"Experimental Approach on Enveloping Grasp for Column Objects", *Kaneko, M.; Thaiprasert, N. and Tsuji, T.* 35

"DLR's Multisensory Articulated Hand", *Hirzinger, G.; Butterfaβ, J.; Knoch, S. and Liu, H.* 47

"Mechanical Design and Control of a High-Bandwidth Shape Memory Alloy Tactile Displays", *Wellman, P.S.; Peine, W. J.; Favalora, G. and Howe, R.D.* 56

"Toward Dexterous Gaits and Hands", *Leveroni, S.; Shah, V. and Salisbury, K.* 67

"Dexterous Manipulations of the Humanoid Robot *Saika*", *Konno, A.; Nishiwaki, K.; Furukawa, R.; Tada, M.; Nagashima, K.; Inaba, M. and Inoue, H.* 79

2 Dynamics and Control   91

"Experiments of Spatial Impedance Control", *Caccavale, F.; Natale, C.; Siciliano, B. and Villani, L.* 93

"Optimal Control Based Skill Development System for the Kip", *Nakawaki, D.; Joo, S. and Miyazaki, F.* 105

"Toward Virtual Sports with High Speed Motion", *Morizono, T. and Kawamura, S.* 116

"A General Contact Model for Dynamically-Decoupled Force/Motion Control", *Featherstone, R.; Sonck, S. and Khatib, O.* 128

"Control of a Rover-Mounted Manipulator", *Foulon, G.; Fourquet, J.-Y. and Renaud, M.* 140

3 Haptic Devices 153

"Module-Based Architecture of World Model for Haptic Virtual Reality", *Yoshikawa, T. and Ueda, H.* 155

"Haptic Augmented Simulation Supporting Teaching Skill to Robots", *Bordegoni, M.; Cugini, U. and Rizzi, C.* 167

"Interactive Visual and Force Rendering of Human-Knee Dynamics", *Ellis, R.E.; Zion, P. and Tso, C.Y.* 173

4 Mobile Robot Navigation 183

"Long Distance Outdoor Navigation of an Autonomous Mobile Robot by Playback of Perceived Route Map", *Maeyama, S.; Ohya, A. and Yuta, S.* 185

"Etherbot - An Autonomous Mobile Robot on a Local Area Network Radio Tether", *Jarvis, R.* 195

"Automatic Mountain Detection and Pose Estimation for Teleoperation of Lunar Rovers", *Cozman, F. and Krotkov, E.* 207

"A Landmark-based Motion Planner for Rough Terrain Navigation", *Hait, A.; Siméon, T. and Taïx, M.* 216

5 Heavy Robotic Systems 227

"Evaluation of Impedance and Teleoperation Control of a Hydraulic Mini-Excavator", *Salcudean, S.E.; Tafazoli, S.; Hashtrudi-Zaad, K.; Lawrence, P.D. and Reboulet, C.* 229

"Control of Load Sway in Enhanced Container Handling Cranes", *Dissanayake, M.W.M.G.; Coates, J.W.R.; Rye, D.C.; Durrant-Whyte, H.F. and Louda, M.* 241

"The Design of Ultra-High Integrity Navigation Systems for Large Autonomous Vehicles", *Durrant-Whyte, H.F.; Nebot, E.; Scheding, S. Sukkarieh, S. and Clark, S.* 252

"Modeling and Control of a 3500 Tonne Mining Robot", *Corke, P.I.; Winstanley, G.J. and Roberts J.M.* 262

6 Non Holonomic Vehicles 275

"Autonomous Maneuvers of a Nonholonomic Vehicle", *Paromtchik, I.E.; Garnier, Ph. and Laugier, C.* 277

"How to implement dynamic paths", *Khatib, M.; Jaouni, H.; Chatila, R. and Laumond, J.-P.* 289

"From Paths to Trajectories for Multi-body Mobile Robots", *Lamiraux, F. and Laumond, J.-P.* 301

7 Legged Locomotion 311

"Preliminary Experiments with an Actively Tuned Passive Dynamic Running Robot", *Ahmadi, M. and Buehler, M.* 313

"ROBICEN: A Pneumatic Climbing Robot for Inspection of Pipes and Tanks", *Serna, M.A.; Avello, A.; Briones, L. and Bustamante, P.* 325

"Control of an Eight Legged Pipe Crawling Robot", *Roßmann, Th. and Pfeiffer, F.* 335

8 Sensor Data Fusion 347

"Autonomous Vehicle Interaction with In-door Environments", *Henriksen, L.; Ravn, O. and Andersen, N.A.* 349

"An Experimental System for Automated Paper Recycling", *Faibish, S.; Bacakoglu, H. and Goldenberg, A.A.* 361

"Positioning of the Mobile Robot LiAS With Line Segments Extracted from 2D Range Finder Data using Total Least Squares", *Vandorpe, J.; Van Brussel, H.; De Schutter, J.; Xu, H. and Moreas, R.* 373

"Mobile Robot Localization Based on Efficient Processing of Sensor Data and Set-theoretic State Estimation", *Hanebeck, U.D. and Schmidt, G.* 385

9 Modeling and Design 397

"i.ARES Manipulation Subsystem", *Fernández, S.; Mayora, K.; Basurko, J.; Gómez-Elvira, J.; García, R.; González, C. and Selaya, J.* 399

"Modeling of Nonlinear Friction in Complex Mechanisms Using Spectral Analysis", *Popovic, M.R. and Goldenberg, A.A.* 410

"Development of the Carpal Robotic Wrist", *Canfield, S.L. and Reinholtz, C.F.* 423

"First experiments with MIPS 1 (Mini In-Parallel Positioning System)", *Merlet, J.-P.* 435

10 Robots in Surgery 443

"FREEDOM-7: A High Fidelity Seven Axis Haptic Device with Application to Surgical Training", *Hayward, V.; Gregorio, P.; Astley, O.; Greenish, S.; Doyon, M.; Lessard, L.; McDougall, J.; Sinclair; I; Boelen, S.; Chen, X., Demers, J.-G.; Poulin, J.; Benguigui, I.; Almey, N.; Makuc, B. and Zhang, X.* 445

"Tele-micro-surgery: analysis and tele-micro-blood-vessel suturing experiment", *Mitsuishi, M.; Watanabe, H.; Kubota, H.; Iizuka, Y. and Hashizume, H.* 457

"Active Forceps for Endoscopic Surgery", *Nakamura, Y.; Onuma, K.; Kawakami, H. and Nakamura, T.* 471

"Synergistic Robots in Surgery-Surgeons and Robots Working Co-Operatively", *Davies, B.* 481

"Control Experiments on two SMA based micro-actuators", *Troisfontaine, N.; Bidaud, Ph. and Dario, P.* 490

11 Actuation Control 501

"Optimal Nonlinear Position Tracking Control of a Two-Link Flexible-Joint Robot Manipulator", *Lahdhiri, T. and ElMaraghy, H.A.* 503

"Motion Control of Tendon Driven Robotic Fingers Actuated with DC Torque Motors: Analysis and Experiments", *Prisco, G.M.; Madonna, D. and Bergamasco, M.* 515

"Experiments on a High Performance Hydraulic Manipulator Joint: Modelling for Control", *Bilodeau, G. and Papadopoulos, E.* 532

12 Sensor-Based Control   545

"Adaptive Visual Servoing for Various Kinds of Robot Systems", *Hosoda, K. and Asada, M.* 547

"Underwater Hidrojet Explorer Camera Controlled by Vision", *Amat, J.; Aranda, J. and Villà, R.* 559

"Experiments in Real-Time Vision-Based Point Stabilization of a Nonholonomic Mobile Manipulator", *Tsakiris, D.P.; Kapellos, K.; Samson, C.; Rives, P. and Borrelly, J.-J.* 570

"Distributed Control of a Free-floating Underwater Manipulation System", *Kapellos, K.; Simon, D.; Granier, S. and Rigaud, V.* 582

13 Cooperative Multirobots   595

"Towards a Reliable Set-Up for Bio-Inspired Collective Experiments with Real Robots", *Martinoli, A.; Franzi, E. and Matthey, O.* 597

"Experiments in Realising Cooperation between Autonomous Mobile Robots", *Jung, D; Cheng, G. and Zelinsky, A.* 609

"Self-reconfigurable Robots for Navigation and Manipulation", *Kotay, K. and Rus, D.* 621

"Human-Robot Interface System with Robot Group Control for Multiple Mobile Robot Systems", *Beltran-Escavy, J.; Arai, T.; Nakamura, A.; Kakita, S. and Ota, J.* 633

14 Learning & Skill Acquisition   645

"Exploration-Based Path-Learning by a Mobile Robot on an Unknown World", *Araújo, R. and Almeida, A.T.* 647

"An Anthropomorphic Model of Sensory-Motor Co-ordination of Manipulation for Robots", *Laschi, C.; Taddeucci; D. and Dario, P.* 659

"Extracting Robotic Part-mating Programs from Operator Interaction with a Simulated Environment", *Lloyd, J.E. and Pai, D.K.* 675

"Modeling and Learning Robot Manipulation Strategies", *Liu, J.; Tang, Y.Y. and Khatib, O.* 687

Author Index

Ahmadi, M., 313
Almeida, A.T., 645
Almey, N., 443
Amat, J., 557
Andersen, N.A., 347
Arai, T., 631
Aranda, J., 557
Araújo, R., 645
Arbter, K., 3
Asada, M., 545
Astley, O., 443
Avello, A., 323

Bacakoglu, H., 359
Basurko, J., 397
Beltran-Escavy, J., 631
Benguigui, I., 443
Bergamasco, M., 513
Bidaud, Ph., 488
Bilodeau, G., 530
Boelen, S., 443
Bordegoni, M., 167
Borrelly, J.-J., 568
Briones, L., 323
Brunner, B., 3
Buehler, M., 313
Bustamante, P., 323
Butterfaß, J., 47

Caccavale, F., 93
Canfield, S.L., 421
Chatila, R., 289
Chen, X., 443
Cheng, G., 607
Clark, S., 252
Coates, J.W.R., 241
Corke, P.I., 262
Cozman, F., 207
Cugini, U., 167

Dario, P., 488, 657

Davies, B., 479
De Schutter, J., 371
Demers, J.-G., 443
Dissanayake, M.W.M.G., 241
Doyon, M., 443
Durrant-Whyte, H.F., 241, 252

Ellis, R.E., 173
ElMaraghy, H.A., 501

Faibish, S., 359
Favalora, G., 56
Featherstone, R., 128
Fernández, S., 397
Foulon, G., 140
Fourquet, J.-Y., 140
Franzi, E., 595
Furukawa, R., 79

García, R., 397
Garnier, Ph., 277
Goldenberg, A.A., 359, 408
Gómez-Elvira, J., 397
González, C., 397
Granier, S., 580
Greenish, S., 443
Gregorio, P., 443

Hait, A., 216
Hanebeck, U.D., 383
Hashizume, H., 455
Hashtrudi-Zaad, K., 229
Hayward, V., 443
Henriksen, L., 347
Hirzinger, G., 3, 47
Hosoda, K., 545
Howe, R.D., 56

Iizuka, Y., 455
Inaba, M., 79
Inoue, H., 22, 79

Jaouni, H., 289
Jarvis, R., 195
Joo, S., 105
Jung, D., 607

Kakita, S., 631
Kaneko, M., 35
Kapellos, K., 568, 580
Kawakami, H., 469
Kawamura, S., 116
Khatib, M., 289
Khatib, O., 128, 685
Knoch, S., 47
Koeppe, R., 3
Konno, A., 79
Kotay, K., 619
Krotkov, E., 207
Kubota, H., 455

Lahdhiri, T., 501
Lamiraux, F., 301
Laschi, C., 657
Laugier, C., 277
Laumond, J.-P., 289, 301
Lawrence, P.D., 229
Lessard, L., 443
Leveroni, S., 67
Liu, H., 47
Liu, J., 685
Lloyd, J.E., 673
Louda, M.A., 241

McDougall, J., 443
Madonna, D., 513
Maeyama, S., 185
Makuc, B., 443
Martinoli, A., 595
Matthey, O., 595
Mayora, K., 397
Merlet, J.-P., 433
Mitsuishi, M., 455
Miyazaki, F., 105
Moreas, R., 371
Morizono, T., 116

Nagashima, K., 79
Nakamura, A., 631

Nakamura, T., 469
Nakamura, Y., 469
Nakawaki, D., 105
Natale, C., 93
Nebot, E., 252
Nishiwaki, K., 79

Onuma, K., 469
Ohya, A., 185
Ota, J., 631

Pai, D.K., 673
Papadopoulos, E., 530
Paromtchik, I.E., 277
Peine, W.J., 56
Pfeiffer, F., 333
Popovic, M.R., 408
Poulin, J., 443
Prisco, G.M., 513

Ravn, O., 347
Reboulet, C., 229
Reinholtz, C.F., 421
Renaud, M., 140
Rigaud, V., 580
Rives, P., 568
Rizzi, C., 167
Roβmann, Th., 333
Roberts, J.M., 262
Rus, D., 619
Rye, D.C., 241

Salcudean, S.E., 229
Salisbury, K., 67
Samson, C., 568
Scheding, S., 252
Schmidt, G., 383
Selaya, J., 397
Serna, M.A., 323
Shah, V., 67
Siciliano, B., 91
Siméon, T., 216
Simon, D., 580
Sinclair, I., 443
Sonck, S., 128
Sukkarieh, S., 252

Tada, M., 79
Taddeucci, D., 657
Tafazoli, S., 229
Taïx, M., 216
Tang, Y.Y., 685
Thaiprasert, N., 35
Troisfontaine, N., 488
Tsakiris, D.P., 568
Tso, C.Y., 173
Tsuji, T., 35

Ueda, H., 155

Van Brussel, H., 371
Vandorpe, J., 371

Villà, R., 557
Villani, L., 91
Watanabe, H., 455
Wellman, P.S., 56
Winstanley, G.J., 262

Xu, H., 371

Yoshikawa, T., 155
Yuta, S., 185

Zelinsky, A., 607
Zhang, X., 443
Zion, P., 173

Keynotes

The evolution, trends and perspectives of experimental robotics are the main issues of the keynotes. This analysis addresses topics such as real world understanding, human robot cooperation, advanced teleoperation, haptic devices, biologically inspired systems, intelligent systems with tight coupling between perception and action, advanced sensors, sensor data fusion and registration, dexterity, and micro-robotics among others. New concepts and paradigms lead to advanced methodologies and procedures to develop more powerful and intelligent robotic systems.

Hirzinger describes through the DLR's experiences in space robotics, some relevant aspects of the current state of the art of: advanced sensor technologies, mechatronics devices, advanced teleoperation control, interfaces with bilateral control, 3D imaging and virtual environment. Associated with these technologies and methodologies other aspects such as world modeling, skill transfer and learning and neural networks are considered, in relation to the applications developed at DLR.

Inoue analyses experimental robotics evolution through the Japanese research program on intelligent robotics. Biological inspired systems and humanoid robots shows real new trends in robotics research comprising the above mentioned technologies. The program focuses the research towards emerging paradigms of machine intelligence through real world interaction, world and behavior understanding and human-robot interaction and cooperation among other related issues.

Towards a new Robot Generation

G. Hirzinger, K. Arbter, B. Brunner, R. Koeppe

DLR

DLR (German Aerospace Research Establishment),
Institute of Robotics and System Dynamics
Oberpfaffenhofen, D-82234 Wessling/Germany
Phone +49 8153 28-2401, Fax: +49 8153 28-1134
email: Gerd.Hirzinger@dlr.de

Abstract: Key items in the development of a new smart robot generation are explained by hand of DLR's recent activities in robotics research. These items are the design of multisensory gripper and articulated hands systems, ultra-lightweight links and joint drive systems with integrated joint torque control, learning and self-improvement of the dynamical behaviour, modelling the environment using sensorfusion, and new sensor-based off-line programming techniques based on teaching by showing in a virtual environment.

1 Introduction

In the past there has been kind of a very general disappointment about the fairly slow progress in robotics compared to human performance - despite of many years of robotics research involving a large number of scientists and engineers. Robots today in nearly all applications are still purely position controlled devices with may be some static sensing, but still far away from the human arm's performance with its amazingly low *own weight against load* ratio and its online sensory feedback capabilities involving mainly vision and tactile information, actuated by force-torque-controlled muscles.

Space robotics (and service robotics in general) might become a major driver for a new robot generation. The experience we made with ROTEX, the first remotely controlled robot in space, has strongly underlined this. As has been outlined in different papers [1], ROTEX flew with spacelab mission D2 in April 93, performed several prototype tasks (assembly, catching a floating object etc.) in a variety of operational „telerobotic" modes, e.g. on-line teleoperation on board and from ground (via operators and pure machine intelligence) as well as off-line programming on ground. Key technologies for the big success of this experiment have been

- the multisensory gripper technology, which worked perfectly during the mission (redundant force-torque sensing, 9 laser range finders, tactile sensing, stereo TV).

- local (*shared autonomy*) sensory feedback, refining gross commands autonomously
- powerful delay-compensating 3D-stereo-graphic simulation (predictive simulation), which included the robot's sensory behaviour.

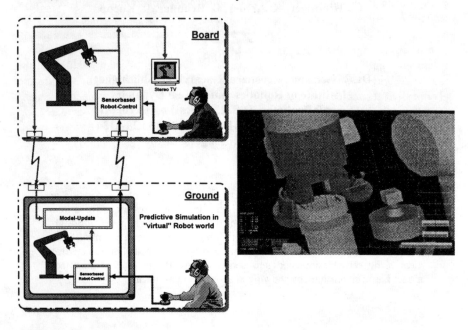

Fig. 1 ROTEX telerobotic control   Fig. 2 Predictive Simulation of sensory perception in the ground station

With this background we are now developing a new light-weight robot generation with manipulative skills and multisensory perception (for space and terrestrial service applications) from bottom up in a unified and integrated, mechatronics approach. Its main features are

- Ultra light carbon fibre grid structures for the links, with structurally optimized integration of torque-controlled actuators, and with all electronics integrated into the arm.
- Multisensory grippers and articulated hands.
- Learning and self-improvement for internal and external sensor control loops.
- Model-based, real-time 3D vision and environment modelling (sensor fusion).
- Implicit, task-level-programming, a sensor-based off-line programming technique (strongly related to learning by showing sensor-based elemental moves).

Let us describe these features in more detail in the subsequent sections.

2 Our mechatronics approach to robot arm and hand design

2.1 General remarks

Our approach in designing *multisensory light-weight robots* and *is an integrated, mechatronics one*. The new sensor and actuator generation developed in the last years does not only show up a high degree of electronic and processor integration, but also a fully modular hard - and software structure. Analogous signal conditioning, power supply and digital pre-processing are typical subsystem modules of this kind. The 20 kHz power supply line connecting all sensor and actuator systems in a galvanically decoupled way, and high-speed (optical) serial data bus systems (SERCOS or CAN) are typical features of our multisensory and multi-actuator concept.

2.2 DLR's first light weight robot
The fully modular, highly integrated hardware design

Ultra-light-weight mechanisms are indispensable for future space as well as terrestrial service robots [4]. Our light-weight robot (Fig. 3, Fig. 4) was designed from scratch in a modular way, each module consisting of a joint, the (carbon fiber grid) link-structure-element and the embedded electronics [5]. Each joint electronic module is of the same type and consists of a power inverter, the joint controller module and the joint-torque-sensor.

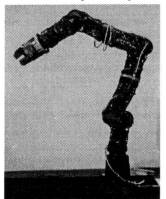
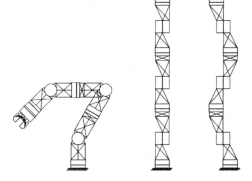

Fig. 3 The 7 axis version of DLR's light-weight-robot with integrated electronics

Fig. 4 Kinematic structure and two different assemblies. The right type is optimally foldable for stowing in space applications

The design-philosophy of this light-weight-robot [5] was to achieve a type of manipulator similar to the kinematic redundancy of human arm , i.e. with seven degrees of freedom, a load to weight ratio of at least 1:2 (industrial robots ≈ 1:20), a total system-weight of about 15 kg, , no bulky wiring on the robot (and no electronics cabinet as it comes with every industrial robot), and a high dynamic performance. As all modern robot control approaches are based on commanding joint torques, we have developed an inductive torque-measurement system that may be seen as an integral part of our double-planetary gearing system (Fig. 5). Its core elements are a six-spokes wheel forming a rotational spring and a differential position measurement sensor (13 bit resolution now with 1 kHz bandwidth!).

Fig. 5 Our 13 bit inductive joint torque sensor (middle, left), an integral part of the gearing (right)

Since each joint has its own control integrated, only a few power and information lines (using a fiber-optical ring) have to be distributed between the joints and the robot-controller.

Dynamic feedback control
From the control point of view [16], [17] the DLR light weight robot belongs to the category of flexible joint robots due to the structure of the gear box and the integrated torque sensor [9]. The dynamic model can be established by applying Lagrange's Equation. For the decoupling of the manipulator dynamics this model is transformed into a new coordinate system in which the joint torque is treated as a state variable instead of the motor position. This leads to the so-called singular perturbation formulation of the robot dynamics. As a result, the fast motion corresponds to the joint-torque loop and the slow motion corresponds to the dynamic path concerned with the link position. On the higher levels, particularly interesting control results so far have been achieved with a hybrid learning approach, it is based on a full inverse dynamic model providing torque control; but as any model never will be perfect, the remaining uncertainties are learnt via backpropagation neural nets (Fig. 6). A first impressive demonstration of this type was learning zero-torque control, i.e. pure gravity compensation so that the arm was just able to sustain itself against gravity, but reacted softly to any external force at any link without additional force sensing [4].

Robot control architecture
Our robot control system is based on VME bus boards, the 4MBaud optical high-speed bus SERCOS and the industrial standard Real-time Multitasking Operating System VxWorks with its powerful development framework. The basic software structure idea is a data flow driven intertask communication approach. That means the necessary software modules (tasks), e.g. path interpolator, coordinate transformation, and drive controller are synchronized via the message passing mechanism supported by VxWorks.

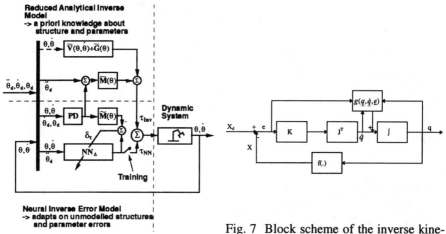

Fig. 6 „Hybrid" learning control scheme applied to gravity compensation of the light weight robot.

Fig. 7 Block scheme of the inverse kinematics algorithm with the *enhanced* Jacobian transpose; function g is chosen via Liapunov theory to optimize diverse criteria (e.g. joint limit avoidance).

The architectural design is based on a model of four layers connected by a global database. Data representation and access are organized by „object oriented programming techniques".

Singularity treatment, i.e. when the Jacobian looses full rank, has always been a problem in robotics. Singular robust Jacobian by damped least squares tends to lead to insufficient behavior within singular regions. We have developed algorithms that make efficient use of redundant motions in singular configurations [8] even with non-redundant robots (using symbolic expressions for the inverse Jacobian) and as an alternative, an extension of the so-called transpose Jacobian approach which we call *enhanced* transpose Jacobian (Fig. 7).

2.3 Sensors and sensorbased 3D-interfaces

A number of essential improvements have been achieved since ROTEX. The patented 6 dof optoelectronic measurement system [28], used in the compliant force torque sensor (Fig. 13) as well as in the teleoperational control balls of ROTEX, has already been optimized a few years ago for use in the low-cost 6 dof controller Space Mouse LOGITECH's MAGELLAN (Fig. 8 or Fig. 9).

It has become the only European 3D computer input device that sells on the market. Several thousand installations in drawing offices (especially of the car manufacturers) make 3D-CAD-constructions faster and more creative (Fig. 10). Moreover, this technology moves back to its origins; companies in medical industry like ZEISS are guiding their surgical microscopes (Fig. 11) by integrating our device into the handles, and the first robot manufacturers have integrated Space Mouse into their programming panels, like STÄUBLI and KUKA (the leading German manufacturer), thus, realizing our very early vision on how to move and program robots in 6 dof intuitively (Fig. 12).

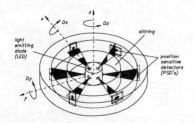

Fig. 8 The patented opto-electronic measuring system

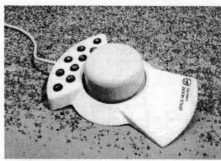

Fig. 9 The Space Mouse

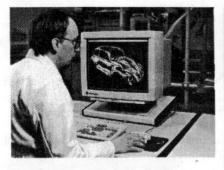

Fig. 10 Space Mouse is going to become a standard interface in 3D-graphics

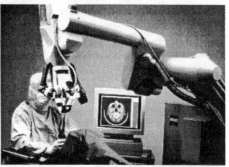

Fig. 11 Guiding surgical microscopes via Space Mouse integrated in the handle

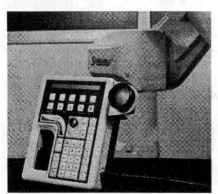

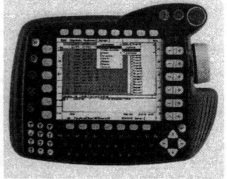

Fig. 12 The Space Mouse is used now by different robot manufacturers in their control panels (STÄUBLI left, KUKA right)

We are meanwhile using the optimization experience of the SPACE MOUSE for the design of a low-cost compliant force-torque wrist sensor, which might lead to kind of a breakthrough with respect to the use of force-torque sensing with industrial robots (e.g. realizing „soft servos").

Fig. 13 The compliant force-torque sensor (left)
and the stiff force-torque-sensor (right)

The stiff strain gauge based force-torque sensors which are perfectly temperature stable show up cycle times of 1 msec (Fig. 13). The medium range (3-40 cm) laser distance sensor based on triangulation as integrated in the ROTEX gripper has been augmented by a tiny motor drive, thus, generating a rotating (or oscillating) miniaturized 2D-Scanner without any mirrors (Fig. 14).

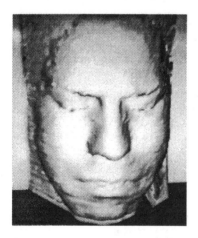 

Fig. 14 Scanning a human face with a laser range finder

Fig. 15 Our 4 finger hand with its 12 actuators and 112 sensors integrates 1000 mechanical and 1500 electronic components

2.4 A four-fingered articulated hand

The design of our four-fingered articulated hand was based on the goal to integrate all actuators (3for each finger) in the fingers or at least in the palm. This became possible by uniformly using our „artificial muscles". For more details see [2]. The fingers are position-force controlled (impedance control), they are gravitiy compensated and they are prevented from colliding by appropriate collision avoidance algorithms.

3 High-level robot programming and information processing

The robots we are developing are supposed to have multisensory (especially visual) capabilities and, thus, should not only be able to on-line react on sensory information, but also to learn about their environment, to update world models as well as improve their dynamic behavior. In the sequel we outline our approaches to telerobotics, visually guided behavior and learning.

3.1 Advances in telerobotics: Task-directed sensor-based tele-programming

After ROTEX we have focused our work in telerobotics on the design of a high-level task-directed robot programming system, which may be characterized as **learning by showing in a virtual environment** [3] and which is applicable to the programming of terrestrial robots as well. The goal was to develop a unified concept for

- a flexible, highly interactive, **on-line programmable teleoperation station** as well as
- an **off-line programming tool**, which includes all the sensor-based control features as tested already in ROTEX, but in addition provides the possibility to program a robot system on an **implicit, task-oriented level**.

A non-specialist user - e.g. a payload expert - should be able to remotely control the robot system in case of internal servicing in a space station (i.e. in a well-defined environment). However, for external servicing (e.g. the repair of a defect satellite) high interactivity between man and machine is requested.

To fulfill the requirements of both application fields, we have developed a 2in2-layer-model, which represents the programming hierarchy from the executive to the planning level.

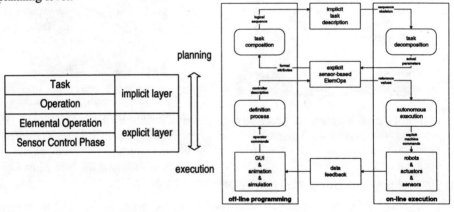

Fig. 16   2in2-layer-model   Fig. 17   Task-directed sensor-based programming

Sensor controlled phases
On the lowest programming and execution level our **tele-sensor-programming** (TSP) concept [6] consists of so-called *SensorPhases*, as partially verified in the

local feedback loops of ROTEX. They guarantee the local autonomy at the remote machine's side. TSP involves *teaching by showing* the reference situation, i.e. by storing the nominal sensory patterns in a virtual environment and generating reactions on deviations. Each SensorPhase is described by
- a controller function, which maps the deviations in the sensor space into appropriate control commands.
- a **state recognition component**, which detects the end conditions and decides with respect to the success or failure of a SensorPhase execution
- the **constraint frame information**, which supports the controller function with task frame information to interpret the sensor data correctly (shared control)
- a **sensor fusion** algorithm, if sensor values of different types have to be transformed into a common reference system (e.g. vision and distance sensors) [7].

Elemental operations

The explicit programming layer is completed by the Elemental Operation (*ElemOp*) level. It *integrates the sensor control facilities with position and end-effector control*. According to the constraint frame concept, the non-sensor-controlled degrees of freedom (dof) of the cartesian space are position controlled

- in case of *teleoperation* directly with a telecommand device like the SpaceMouse
- in case of *off-line programming* by deriving the position commands from the selected task. Each object, which can be handled, includes a relative approach position, determined off-line by moving the end-effector in the simulation and storing the geometrical relationship between the object's reference frame and the tool center point, including the relevant sensory patterns.

A model-based *on-line collision detection* algorithm supervises all the robot activities, it is based on a discrete workspace representation and a distance map expansion. For global transfer motions a *path planning* algorithm avoids collisions and singularities [24].

Operations

Whereas the SensorPhase and ElemOp levels require the robot expert, the implicit, task-directed level provides a powerful man-machine-interface for the non-specialist user. We divide the implicit layer into the Operation and the Task level.

An Operation is characterized by a sequence of ElemOps, which hides the robot-dependent actions. Only for the specification of an Operation the robot expert is necessary, because he is able to build the ElemOp sequence. For the user of an Operation the manipulator is *fully transparent*, i.e. not visible.

Tasks

Whereas the Operation level represents the subtask layer, specifying complete robot tasks must be possible in a task-directed programming system. A *Task* is described by a consistent sequence of Operations. To generate a Task, we use the VR-environment as described above. All the Operations, activated by selecting the desired objects or places, are recorded with the respective object or place description.

Our task-directed programming system with its VR-environment provides a man-machine-interface at a very high level i.e. without any detailed system knowledge, especially w.r.t. the implicit layer. To edit all four levels as well as to apply the SensorPhase and ElemOp level for teleoperation, a sophisticated graphical user interface based on the OSF/Motif standard has been developed (
Fig. 19 bottom, screen down on the left).

Fig. 18 VR-environment with the ROTEX-workcell and the Universal Handling Box, to handle drawers and doors and peg-in-hole-tasks

Fig. 19 DLR's new telerobotic station

3.2 Robotics goes WWW - Application of Telerobotic Concepts to the Internet using VRML and JAVA

New chances towards standardization in telerobotics arise with VRML 2.0, the newly defined Virtual Reality Modeling Language. In addition to the description of geometry and appearance it defines means of animations and interactions. Portable Java scripts allow programming of nearly any desired behavior of objects. Even network connections of objects to remote computers are possible. Our new telerobotic station is fully remotely programmable via VRML now.

A commercial application which gains increasing interest is **teleservicing** of industrial robots. Teleservicing through computer networks could dramatically reduce costs and simultaneously improve customer support in this field. Teleoperation in this case mainly requires remote execution of off-line generated robot programs. VRML will be used as an on-line visualization tool for the entire state of the remote real robot even during fast motions. Using VRML compared to a video feedback channel has the advantage of consuming less bandwith (about 5kByte/s compared to 384 kByte/s) and the possibility to show the scenery from arbitrary points of view.

3.3 Robot vision

The focus of our work is on real-time image post-processing for finding the *pose* of objects in order to enable the robot to grasp them, or to get their *shape* to perform object recognition and world model update. We can only give a few examples here.
In general we apply model-based approaches for visual servoing as well as techniques which do not need calibration or models. These latter techniques are e.g.

preferably applied in the task-level programming environment, when we have to train the mappings from non-nominal multisensory patterns into motion commands. Analytical methods based on the Jacobian are then applied in the same way as neural nets (multilayer perceptrons). Results are comparable.

Model-based vision for robot servoing

The motivation for model-based vision approaches is to robustly analyze complex images with high performance. We assume that model information is available in the form of a polyhedral description of the 3-D object geometry, enhanced by information about circular features. The research reported here is devoted to *multisensory eye-in-hand systems*. The sensors considered are one or more video cameras and optical range sensors that yield an array of 3-D surface point measurements.

The structure of the model-based multisensory tracking algorithm [10] is illustrated in Fig. 20. The task is to track the relative pose x of the target. Image feature vectors ^{c_i}f along with the measurements ^{s_j}m of the range sensing devices are the input to a least squares procedure that computes an estimate x of current pose and its covariance Σ_{xx}. Next, in order to compensate delay and to prepare feature extraction a prediction $\hat{x}$ and its covariance are computed based on assumptions on the object motion, which is used to generate feature values $^{c_i}\hat{f}$ expected in the next processing cycle. Thus, image processing can be limited to small regions of interest, potential occlusions can be predicted by hidden line removal. This structure is equivalent to that proposed by Dickmanns et al. [18] for monocular tracking.

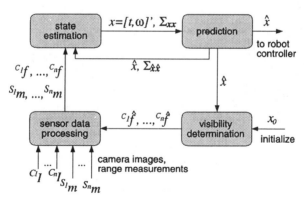

Fig. 20 Processing steps in model-based multisensory 3-D tracking

Based on the prediction $\hat{x}$ expected features values are generated by projection of the model wire frame and subsequent removal of hidden lines. As the computing time required by estimation is low (about 1 msec), *complete analytic hidden-line elimination for non-convex objects* can be realized in one sensor cycle with our recursive pose estimation. Thus, not only self-occlusion can be handled but also occlusions of the target by other objects. Fig. 21 shows some examples of monocular tracking.

 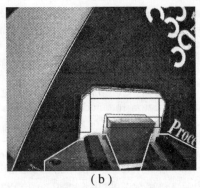

(a) (b)

Fig. 21 Monocular real-time 3-D tracking: (a) The wire frame of the object is projected into the image at the estimated pose (green). The good match indicates high accuracy. Minor occlusions do not affect accuracy. (b) In addition, a model of the gripper is included. The regions for feature search (red) are automatically selected by prediction and hidden-line removal. Thus, mutual occlusion of different objects can be handled.

Model-based pose estimation by registration
Novel algorithms [11] have been developed for registering a 3-D model to one or more of two-dimensional images without a-priori knowledge of correspondences. The method - a generalization of the iterative closest point algorithm (ICP) - handles full six degrees of freedom, without explicit 3-D reconstruction.

Fig. 22 An example of convergence of the registration algorithm. The initial rotational displacement in ZYX-Euler angles is {66.5°, 49.5°, 32.0°}

Experiments show that complex, three-dimensional CAD-models can be registered efficiently and accurately to images, even if image features are incomplete, fragmented and noisy (Fig. 22).

Shape from shading by neural networks
The problem of shape from shading is to infer the surface shape of an object based solely on the intensity values of the image. We proposed a new solution [12] of the shape from shading problem based on the optimization framework. We used a multilayer perceptron to parameterize the object surface.

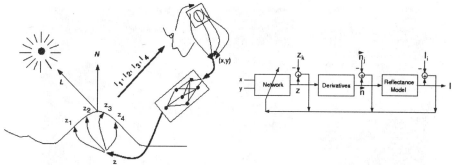

Fig. 23 Shape from shading: A learning scheme minimizes the intensity error.

Fig. 24 A unified training framework

The weights of the network are updated so that the error between the given image intensity and the generated one is minimized. Fig. 23 and illustrate Fig. 24 the mechanism with which the shape is recovered from shading.

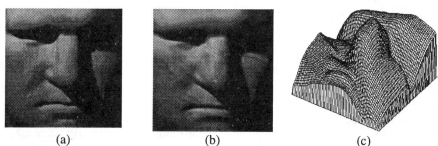

Fig. 25 Surface recovery of the Agrippa statue. (a) Input image; (b) Learned image; (c) Recovered 3D surface.

Typical applications

Automated laparoscope guidance in minimally invasive surgery

Unlike open surgery, where vision and action is concentrated at one person, in laparoscopic surgery the vision part is split up to two persons. A camera assistant has to move the endoscope to the surgeon's area of interest But a human normally cannot do this job without tremor and fatigue. A robot automatically servoing the scope might be superior. We developed and patented a visual tracking system for mono- or stereo-laparoscopes, which is robust, simple and operates in real-time [13]. We use color-coded instruments. The coding color is chosen different from the colors typical for laparoscopic images. With color image segmentation (Fig. 26), the color-coded instrument can be perfectly located in the image.

This approach is very robust against occlusions, smoke, and image saturation. The control loop is designed such that the robot's motion is very smooth. Such a system, was successfully tested in the „Klinikum rechts der Isar" Munich hospital, on animals as well as on humans (Fig. 27). It was found that the surgeon's concentration onto the surgery is massively supported by this technology.

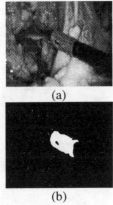

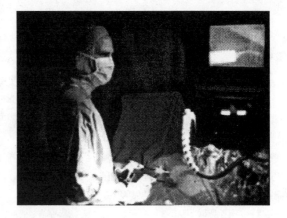

Fig. 26 (a) Laparoscopic image out of the abdomen with a color-marked instrument (b) post-processed segmentation

Fig. 27 Minimally Invasive Surgery using an Autonomous Robot Camera Assistant, for the first time tested on humans in September 1996 in the Munich hospital „Klinikum rechts der Isar"

Model-based visual servoing in satellite capture

The capturing of a target satellite by a free-floating repair robot is intended to be automatically servoed by machine vision. As the geometry of the target satellite is known, a model-based vision approach is pursued, which also includes the well-known dynamics of a rigid body tumbling under zero gravity. Specifically, the prediction module of Fig. 20 is implemented as a Kalman filter that takes x and Σ_{xx} as an input (observation) and continuously estimates the parameters of the dynamic equations. From these estimates both a smoothed version of the object pose and a prediction are computed in each cycle.

3.4 Learning, skill-transfer and self-improvement

Learning and adaptation are the core paradigms of intelligent control concepts which enable to increase skilled manipulation and to achieve higher levels of autonomy (Fig. 28). Data approximation and representation techniques like artificial neural networks enhance or may replace conventional model based approaches.

We use various kinds of neural network architectures for robot perception and manipulation. The multi-layer perceptron with sigmoid transfer functions is capable of solving non linearly separable classification problems. Another universal approximator is the Radial Basis Function network which uses Gaussian transfer functions. Furthermore multi-layer networks may be structured to represent a fuzzy controller with adjustable parameters representing location and shape of the fuzzy sets and the weights of the truth value of rules. Recurrent networks have feedback connections within the network and are therefore dynamic systems themselves. Another important architecture is the self-organizing map, such as the Kohonen Feature map, which can be used to build discrete representations of data at optimal dimensions.

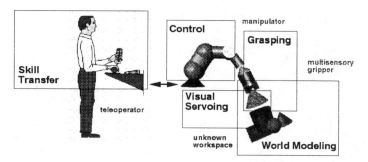

Fig. 28 Areas for learning and adaptation in advanced robotics

World modelling using Kohonen's feature map

Building a geometric description of a robot's workspace via unordered „clouds" of multisensory data out of the robot gripper is of crucial importance for robot autonomy.

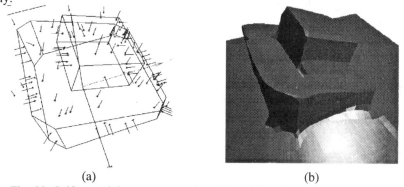

(a) (b)

Fig. 29 Self-organizing reconstruction of an object in the robot's work cell.
Part (a) shows the wireframe representation of the object and the sampled 110 point and normal vector tuples,
(b) the final surface after 3000 steps and a total time of 9 seconds.

To solve these kind of problems a surface reconstruction algorithm, based on Kohonen's self organizing feature map was developed and successfully implemented [14]. The two-dimensional array of neurons in Kohonen's algorithm may be considered as a discrete parametric surface description. To incorporate different types of surface information new training equations were developed for the Kohonen network. Successful surface reconstruction was performed for completely unordered *data clouds* of different surface information (Fig. 29). This method was used as a core for a complete world perception system for *Sensor-in-Hand* robots.

Shape from shading and visual servoing by neural networks

Application of multilayer perceptrons was outlined in Section 3.3.

Learning compliant motions by task demonstration

(a) (b)

Fig. 30 Demonstrating a compliant motion task in a natural way:
(a) by haptic interaction in a telerobotic environment and
(b) by direct manipulation using a teach device.

Neural networks are capable of representing nonlinear compliant motions. The training data preferably consist of the measured forces (input) and the human arm's motion (output). If sensorimotion data are directly recorded from a human demonstrating a task, the correspondence problem has to be solved, since the human executes an action u(t), due to a perception S(t-τ). In [19] we show that the correspondence problem is solved by approximating the function

$$u(t) = f[S(t), \dot{x}(t)]$$

with a neural network, taking the velocity $\dot{x}(t)$ as an additional input to the network.

With recent advances in virtual reality technology, haptics, and human interface systems, demonstrating compliant motion for execution on a robot can be performed by the sole process of human demonstration without the use of the actual robot system. We investigated two approaches: Interaction with a virtual environment using haptic interfaces [19] (Fig. 30 (a)) and direct manipulation of the object using a teach device to measure force and motion of the human (Fig. 30 (b)) [20].

Improving the accuracy of standard position controlled industrial robots

High speed path errors due to the inertia of the robot mechanism can be minimized by independent linear feedforward learning controllers since they are almost decoupled and proportional to the accelerations. Neural networks may be used in addition to reduce the remaining effects, mainly the nonlinear couplings (see Fig. 31 (a) for a simple case). Common approaches for learning control focus on the improvement of fixed paths, thus, allowing accurate repeated execution of the desired trajectory, e.g. [21]. The advantage of our method comes into play if the path differs from the one used for training the controller, as it is valid for on-line planned paths which are modified in relation to unexpected sensor data.

We use one network for each joint. The inputs are the positions of joints 2 to 5, and additional inputs, representing the time series of the desired accelerations of those joints which affect the motion of the actual joint. Multilayer perceptrons with sigmoid activation functions and two hidden layers are used [23]. The training is performed with the Extended Kalman Filter algorithm mentioned in Section 3.3. The learning system was tested for high speed movements reducing path errors to 20 % and less in the first trial (Fig. 31 (c), [15]). Fig. 31 (b) shows a contour following task of a robot with conventional positional interface. First, the linear controllers are trained. Then, the neural networks are learned using training data from similar paths without contact. When applied to the untrained case of contact, the error of the desired force is reduced by 50% within the first execution of the path.

Standard industrial robots can be compensated by a simple Cartesian model of the robot dynamics. The trajectory „a" in Fig. 31 (d) shows the distortion of a high-speed motion path of a KUKA-Robot. Trajectory „b" shows the compensated path.

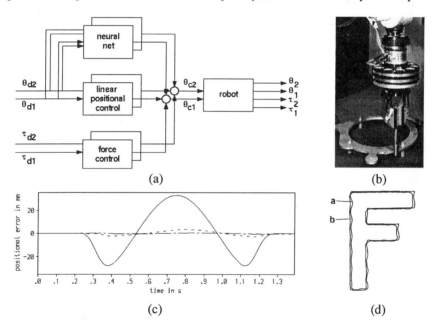

Fig. 31 Learning control structure (a) and contour following task experiment (b). Reduction of path errors to 20% (c). Distorted trajectory due to dynamic effects and compensated trajectory (d).

4 Resume

For us an interesting observation is that not only robotics research has made progress in the past, but that robot manufacturers are now accepting and integrating research results that have been achieved over the last twenty years. The coming years will bring fast sensory interfaces and feedback, more light-weight design and adaptability. Service robotics - be it for space or for terrestrial applications - is going to demonstrate what industrial robots should be capable of in the future.

References

[1] G. Hirzinger, B. Brunner, J. Dietrich, J. Heindl, „ROTEX - The First Remotely Controlled Robot in Space", IEEE Int. Conference on Robotics and Automation, San Diego, California, May 8-13, 1994

[2] G. Hirzinger, J. Butterfass, S. Knoch, H. Liu, DLR's multisensory articulated hand, ISER'95 Fifth International Symposium on Experimental Robotics, Barcelona, Catalonia, June 15-18, 1997

[3] B. Brunner, J. Heindl, G. Hirzinger, and K. Landzettel, „Telerobotic systems using virtual environment display with visual and force display functions", Proc. IEEE International Conference on Robotics and Automation, Atlanta/Georgia, 1993.

[4] G. Hirzinger, A. Baader, R. Koeppe, M. Schedl, "Towards a new generation of multisensory light-weight robots with learning capabilities", Prod. IFAC 1993 World Congress, Sydney, Australia, 1993.

[5] B. Gombert, G. Hirzinger, G. Plank, M. Schedl, „Modular concepts for a new generation of light weight robots", Proc. IEEE Int. Conf. on Industrial Elektronics, Instrumentation and Control (IECON), Bologna, Italy, 1994

[6] B. Brunner, K. Landzettel, B.M. Steinmetz, and G. Hirzinger, „Tele Sensor Programming - A task-directed programming approach for sensor-based space robots", Proc. ICAR'95 7th International Conference on Advanced Robotics, Sant Feliu de Guixols, Catalonia, Spain, 1995.

[7] G. Grunwald and G.D. Hager, „Towards Task-Directed Planning of Cooperating Sensors", Proc. SPIE OE/Technology: Sensor Fusion V, Conf. 1828,, Boston, Mass., 1992.

[8] V. Senft, G. Hirzinger, „Redundant Motions of Non Redundant Robots - A New Approach to Singularity Treatment", 1995 IEEE Int. Conf. on Robotics and Automation ICRA'95, Nagoya, Japan, vol 2, pp. 1553-1558, 1995

[9] J. Shi, Y. Lu, „Chatter Free Variable Structure Perturbarion Estimator on the Torque Control of Flexible Robot Joints with Disturbance and Parametric Uncertainties", Proc. IEEE Conf. on Industrial Electronics, Control and Instrumentation, IECON'96, Taipei, 1996

[10] P. Wunsch, G. Hirzinger, „Real-time visual tracking of 3D-objects with dynamic handling of occlusion", IEEE Int. Conf. on Robotics and Automation, Albuquerque, NM, 1997

[11] P. Wunsch, G. Hirzinger, „Registration of CAD-models to images by iterative inverse perspective matching", ICPR'96 Int. Conf. on Pattern Recognition, Vienna, Austria, 1996

[12] G.Q. Wei, G. Hirzinger, „Learning shape from shading by a multilayer network", IEEE Transactions on Neural Networks, vol. 7, pp. 985-995, 1996

[13] G.Q. Wei, K. Arbter, G. Hirzinger, „Real-time visual servoing for laparoscopic surgery", IEEE Engineering in Medicine and Biology, vol. 16, 1997

[14] A. Baader, G. Hirzinger, „A self-organizing algorithm for multisensory surface reconstruction", Proc. IROS'94 IEEE Int. Conf. on Intelligent Robots and Systems, München, 1994

[15] F. Lange, G. Hirzinger, „Learning of a controller for non-recurring fast movements", Advanced Robotics, pp. 229-244, 1996

[16] M.W. Walker and R.P.C. Paul, „Resolved-Acceleration Control of Mechanical Manipulators", IEEE Trans. Automatic Control, vol. 25, pp. 468-474, 1980

[17] O. Khatib, „A Unified Approach for Motion and Force Control of Robot Manipulators: The Operational Space Formulation", IEEE J. Robotics and Automation, vol. 3, pp.43-53, 1987.

[18] E. D. Dickmanns and V. Graefe. Dynamic monocular machine vision. *Machine Vision and Applications*, 1:223-240,1988.

[19] R. Koeppe, G. Hirzinger, „Learning Compliant Motions by Task-Demonstration in Virtual Environments", Proc. of the Fourth International Symposium on Experimental Robotics, ISER'95, Stanford, CA, pp. 188-193, 1995.

[20] R. Koeppe, A. Breidenbach, and G. Hirzinger, "Skill representation and acquisition of compliant motions using a teach device," presented at Proc. IEEE/RSJ Int. Conference on Intelligent Robots and Systems IROS'96, Osaka, Japan, 1996.

[21] F. Arai, L. Rong, and T. Fukuda. Asymptotic convergence of feedback error learning method and improvement of learning speed. In *IEEE Int. Conf. on Intelligent Robots and Systems (IROS)*, Yokohama, Japan, July 26-30 1993

[22] G. Hirzinger, "Leichtbau, Geschicklichkeit und multisensorielle Autonomie - Schlüsseltechnologien für künftige Raumfahrt-Roboter wie für terrestrische Service-Roboter," presented at Proc. AMS'96 12. Fachgespräch Autonome Mobile Systeme, Munich, Germany, 1996.

[23] F. Lange and G. Hirzinger, "Application of multilayer perceptrons to decouple the dynamical behaviour of robot links," presented at Proc. International Conference on Artificial Neural Networks, ICANN'95, Paris, France, 1995.

[24] E. Ralli and G. Hirzinger, "Fast path planning for 6-dof robot manipulators in static or partially changing environments," presented at Proc. ICAR'95 7th International Conference on Advanced Robotics, Sant Feliu de Guixols, Catalonia, Spain, 1995.

Towards Evolution of Experimental Robotics

Hirochika Inoue
Department of Mechano-Informatics
The University of Tokyo

ABSTRACT: This paper introduces Japanese research program for experimental robotics. Part one describes an ongoing inter-university research program on intelligent robotics. The three year research program, supported by a grant from the Japanese Ministry of Education, aims to study mechanisms for the emergence of robotic intelligence and autonomy. Adopting a tightly-coupled sensor-action approach, this program will tackle the theoretical foundations of machine intelligence, and experimentally study autonomous systems which perceive and behave in the real world. Part two discusses future directions for robotics, including issues, approaches and applications. Part three presents the author's view on a paradigm change for robotics research, and sketches out his group's R&D project, which aims at a break-through to blaze a trail for the robotics of the future.

Part I: Inter-university Research Program on Intelligent Robotics [1,2]

1. Introduction

Part one introduces a research program on intelligent robotics which started in 1995 for three years as a grant from the Japanese Ministry of Education. The title of the program is "Research on Emergent Mechanisms for Machine Intelligence --- A Tightly-Coupled Perception-Motion Behavior Approach" (Principal Investigator H. Inoue, Univ. of Tokyo). This program aims to study mechanisms for the emergence of robotic intelligence and autonomy. Intelligence which emerges from the physical interaction between a machine and the world is essential for real robots. From this viewpoint, adopting a tightly coupled sensor-action approach, this project will work on the theoretical foundations of machine intelligence, and experimentally study autonomous systems which recognize and behave in the real world.

The research program described here was selected as one of the priority areas of Grant-in-Aid Program for Scientific Research, designed to order to focus effort on the challenges of intelligence and autonomy for advanced mechanical systems. It is structured as an inter-university research program, consisting of 21 prescribed research themes and 27 subscribed topics. So far, 50 laboratories from 25 universities has been involved into this program.

2. Research Issues

Four major issues are chosen to structure the research. They are; (A) physical understanding of dexterity, (B) real world understanding through tightly-coupled sensing and action, (C) human-robot cooperative systems, and (D) biologically inspired autonomous systems. All of these issues require a tight coupling of perception with action, and each issue closely relates with others. Issues A and B are fundamentals for machine intelligence that should be coupled each other, and issues C and D are integration oriented. The research program consists framework research topics, which defines the domain and approach, and supplemented research topics which enhance the planned approach.

2.1 Physical understanding of dexterity [3]

A hand is sometimes recognized as a physical agency of the brain. Dexterity of the hand is one of the characteristic features that distinguishes humans from apes. Such intelligence, acquired through the long history of biological development is considered to be implicit intelligence that emerges from the physical interaction without consciousness. This manipulation skill emerges from the interactive process between a hand and objects/environment. The principle of the skill seems to be hidden inside the kinematic nature of interactive processes.

This issue aims to explore the physical principles underlying the dexterity of human/robot motions in execution of sophisticated tasks from the viewpoint of dynamics and control theory. It also aims to discover implementable algorithms based on the physical principles, which can function in robotic arms and hands and execute tasks with sufficient dexterity even if they are physically interacting with objects in the environment.

2.2 Real world understanding through sensing-action coupling [4]

So far, the phases of recognize-plan-act in robotic behavior are connected in a simple loop, if at all. However, recent advances in computing performance provide new possibilities for constructing perception-action coupling. As the speed of visual processing becomes higher, robots can perform very sophisticated perception and perception-action coupling. For instance, based on the partial results of recognition, other perception agents for the lost or lacking information can be invoked reactively. If those cycles for perception and sensing action become very fast, the recognition mechanism can configured into more sophisticated constructs for fast, flexible, efficient or robust perception.

The approach to the problem is to tightly couple sensing and behavior so that behaviors are improved by sensing and better sensing is performed by proper behaviors. Main research themes are the following: 1) Interpretation of sensor information by sensing and behavior, 2) Sensing mechanisms for immediate adaptive behavior, 3) Action control for a given task based on sensing and a world model, 4) Adaptation mechanisms for sensor-behavior systems.

2.3 Intelligence and autonomy for human-robot cooperation [5]

Modern human life is supported by many kinds of machines. However the relationship between those machines and human users is not natural enough. Machines often force humans to adapt to them rather than the machines adapting to humans. Thus, the human centered design of human-machine interaction process is strongly required. The purpose of this issue is the study of intelligence and autonomy that is required to provide new cooperative relationships between human and machines.

To realize novel human-machine cooperation systems, the following three research topics will be investigated in terms of human-robot cooperation intelligence. 1) Research on machine intelligence to cooperate with humans: Multiple robot environments based on agent programming will be realized to supports cooperative works between humans. Also a mechanism, sensor, controller and man-machine interface for mobile robot with TV camera will be constructed. 2) Research on machine intelligence to support human activities: A function to grasp most of the objects in a room by tracking and updating them, plus a function to express emotional information based on facial expression and gesture of the robot will be realized. 3) Research on machine intelligence to realize cooperative maneuvering: After modeling of human skill, a cooperative maneuvering system will be designed.

2.4 Biologically inspired approaches to autonomous systems [6]
Biological systems embody a sort of intelligence that enables flexible adaptation to changes in the environment, and robust pursuit of goals, and thus they keep themselves alive. In this topic, through the study of biological intelligence, we will understand the mechanism of intelligence and autonomy of the living things and attempt to build real mechanical systems of such nature. This will uncover ways to create reliable, robust and flexible machines that behave intelligently in the real environment.

The purpose of this research group is to develop methodologies to realize intelligent robots that achieve high autonomy as in creatures, adaptability in different environments, and sociability. Creatures are to be provided with innate abilities and intelligence for themselves to be alive in a real environment. They can adapt themselves to the variations of their environment flexibly and robustly. They can accomplish their wishes to survive. When we consider the design of intelligent machines that work in the real environment, the creatures should provide us with useful knowledge. We take such insights from intelligence of creatures and make it clear why such intelligence can emerge from them. Finally, we develop a methodology to realize intelligence and autonomy of the machine system.

3. Approach and Guiding Principles
The aim of this program is summed up by the two keywords: perception-motion coupling, and emergent machine intelligence. The objective of this program is to investigate the mechanism or algorithm that gives rise to machine intelligence. For that purpose, we chose a tightly-coupled perception-motion behavior approach. A machine is a physical entity that behaves in the real world. Our research interest is focussed onto some sort of intelligence, which is inherently required for robust, efficient, quick interactions between the machine and the real environment. This project aims to find the principles of such mechanical intelligence and demonstrate them by building physical entities.

3.1 A new research framework

So far, usual research of robot intelligence is divided into components such as motion control, environment recognition, task planning, and so on. However in this program we take a perception-motion coupling approach for creating intelligent systems. The approach will be shifted from functional division to multi-function integration. For instance; from the control of hands to the understanding of manipulation skill; from recognition without action to real world understanding by general use of perception and action. Hence this study of system architecture for emerging intelligence in robotic behavior, learning, memorization, and autonomy.

3.2 Interaction with the outer world

Robot intelligence must cope with various noise, uncertainty, change, and unsteadiness in the real world. And, intelligence for reaction or symbol grounding is much more important than classical symbol based intelligence. Not only perception but also action can be equally recognized as interaction process with outer world. The essence of real world intelligence can be found in the interaction process between a machine and the environment.

3.3 Tight coupling of perception with action

Recent progress of computer hardware is dramatically changing the prerequisites for robot intelligence research, that is, real time vision, massive memory, enormous computing power by massive parallel computers, and so on. Very fast and very flexible coupling of perception process and action has now become available. Better perception might be supported by proper active sensing, on the other hand, behavior must be supported by high and multi-modal sensor integration. The recursive structure of perception and action hints at interesting principles for intelligence for mechanical systems.

3.4 Infrastructures for experimental robotics [7,8]

In robotics research, the development of experimental devices or tools are also very important, because the process of building robots is also very effective approach for understanding robotic principle. However, as proper platforms for robotic experiment are not available, many researchers build similar but not the same experimental devices. If a standard platform like computer hardware were available, software libraries will accumulated and help to accelerate the speed of research. For this purpose, in our research program, we are planning to develop common research tools for distribution. Our current plan is: real time tracking vision based upon correlation technique, another high performance general color vision system in DSP based multi-processor architecture, compact low cost legged robot, servo-module and manipulator system design, and so on.

In the first year of this program, a standard legged robot was designed and built by Hirose's group, and about ten units are now working as common research tools. Another accomplishment is the general purpose real-time vision unit. It employs a DSP chip (the C40) as the processing unit and provides a parallel/pipeline system architecture. This system was developed by Shirai's group and Fujitsu. This research program supports the initial cost for development of both legged robot and vision units, and opens to the way to the development of general-purpose, common research tools for the sensing-action coupling approach.

4. Remarks for part one

The essential nature of intelligence for real mechanical systems can be found in the process of various interactions with real environment. From this viewpoint, the inter-university research program on intelligent robotics sets out to understand the emergent intelligence of mechanical systems, and to create real robots that behave intelligently in the real world based on a tightly-coupled perception-motion approach. One of the important motivations for forming this program was to refresh the field and the approach of robotics research. Currently, the program seems to encourage approaches based upon a tight coupling of sensing and action. We expect that the program will open up new topics, new attempts, new approaches for machine intelligence in the robot-world interaction process.

Part II: Challenges to Address

1. Introduction

Let me start this part by giving my personal observation on the current situation and surroundings of the robotics research in general. At present, nobody considers robotics a very new field. Actually robotics has become mature and the field is almost established in some sense. The number of researchers is increasing, conferences are held very frequently, and, a large number of research papers is produced every year. But research funds are steady or declining because robotics seems not to attract strong general attention any more. The enthusiasm for robotics research decades ago has become calm. On the other hand, in related areas, several novel approaches for systems are growing very rapidly. They are, for instance, micro machines, virtual reality, A-Life, brain technology, new media technology, and so on. And those areas are very active at using robotics as concrete examples or applications for explaining their own approach. Is robotics diverging, and becoming a collection of convenient examples for fresh areas? Or, can robotics change its structure as it comes to include those newly emerging methodologies? Competing with such new vivid research topics, can robotics survive? If so, then how should we envision future robotics? It is a good time to think about what robotics should be, about how we should redefine its structure. As researchers of decades ago started their research from fundamental questions or hopes, we need to reconstruct future robotics upon the results created in past three decades. We really need a paradigm shift for robotics research to refresh it again.

2. Humanoid

Creating a humanoid, whose shape and function is similar to those of human, is considered a grand challenge for integrating mechanical engineering with information technology. By integrating all the state of the art in robotics, can we build an artifact as similar as possible to the human shape and size? It is important to make this a realistic challenge, because making a real attempt clearly reveals what aspects of technology are near the target, and what aspects are still far away. The first step of such a challenge would be the real development of humanoid hardware by integrating a sensor integrated robot head, two arms, five fingered hands, torso, and, two legs and feet. Optimistically, recent advance of electronics suggests the possibility for the aspects of information technology in humanoid. However, pessimistically, the field of mechanisms, actuators, distributed sensors, and skins are not advanced yet enough. Generally, a humanoid that can walk as humans do has been considered very difficult to realize.

In December 1996, Honda R&D Co. Ltd. announced the success of development for a prototype humanoid robot that can walk, climb stairs up and down by very smooth autonomous mode, and also perform simple manipulations by teleoperation mode. It is a good looking humanoid of DC motor driven, battery powered, and complete stand-alone packaged. Its height is 180cm and weights 210kg including battery. Technical detail of the robot is not opened yet, but the demonstration of its walking seems very stable, robust and powerful enough. In some sense, Honda's prototype showed a proof that the technology for building a walking and working humanoid really exists. This robot suggests us to dash into challenges for building another smarter one, if we wish. At the same time, it also provides us with new opportunity for the experimental study of various aspects of intelligence, psychology, communication and behavior of the mechano-informatic systems. Now, a humanoid is not even a target for grand challenge of intelligent systems, but it evolved into a general purpose intelligent machine that recognize and behave in real world, for use in further experiments.

3. R-Cube [10]

R-Cube (R^3) is the acronym of Real-time Remote Robotics. It is defined as advanced remote control technology which enables us to operate robots from a remote cockpit through information network. R-Cube application can expand wider, some of the examples are rescue activity in disasters, R-Cube assisted social activity for elderly or handicapped people working from cockpit at home, and so on. The concept of R-Cube is proposed by a planning committee of MITI Japan for preparing a new national project. Although its technical contents are not clearly defined yet, the report was published as a book that describe some images of R-Cube technology as the story of R-Cubes. The R-Cube story can be read as a SF story, however if we attempt to read it as a specification for new robot application, it includes very interesting aspects to be challenged. R-Cube technology is not a simple remote control, rather it must be strongly supported by very advanced intelligent autonomous functions. The key for making R-Cube technology successful is research on highly advanced real-time intelligent autonomy. Today, the term "robot' is used too much and thus the term is losing its freshness. Thus, it is important that we introduce such new goals not with a name like "xxx-type robot" but with a completely new name, like "R-Cube". In such a way, robots must evolve into many species for future prosperity.

4. Remote Brained Approach [11]

A robot consists of its brain and body. The body is a machine that behaves. The brain is a computer that controls its body. A remote brained robot is constructed by separating the computer from the body, with information between them passed by radio link. For example, Inaba built a whole body type robot with human or animal shape by using wireless servo units for radio-control hobby models. A TV camera on the body is also linked by wireless signal link. The computer receives the video image from the wireless TV camera, performs its image processing, decides its motion plan, and sends back the appropriate motion command via wireless channels. The remote brained approach has the following merits.

(1) Mechanism and software are interfaced through a standard input/output specification, and concurrent development of mechanism and software are made possible.
(2) Motion of whole body is physically free from controller devices, as the body and computer are connected by wireless link instead of real wires.
(3) Wireless connection give no constraint and has no effect on the dynamic properties of robot motion;
(4) As computer and body are connected by radio link, you can use any kind of computers as controllers, from super computer to micro CPU as you wish.

Those properties provide especially convenient experimental environment for the research of brain technology, software, planning/control/learning of robotic behavior, and research of system architecture too. Remote brained approach is particularly good for software research which requires real-world interaction.

5. Network Robotics

There exist various proposals for ubiquitous or network oriented intelligent system architecture. They are, for example, ubiquitous autonomy by ETL, robotic room by RCAST at the Univ. of Tokyo, intelligent distributed agents by Kyoto Univ., intelligent room by MIT, and so on. In those proposals, room, building, and urban environment are constructed as an intelligent system complex where the intelligent machine elements of various levels are mixed and connected via high speed information network. Suppose current home appliances or personal computers evolve to be intelligent, network oriented, and sensor/actuator equipped, and that those machines are eventually connected with messenger robots, cleaning robots, and secretary/assistant robots via advanced information network, and thus highly advanced information environment can be realized. In order to make the new environment more natural, flexible, and human-centered, the following key issues must be solved. Each autonomous systems must be able to guess and understand the intention of humans and other robots, and perform cooperative behaviors. There must be communication in the multi-robot community. Robots must themselves be soft, and must deal with soft objects safely in order to realize physical human-robot symbiosis.

Part III: Micro-mechatronics and Soft-mechanics

1. Introduction

Last year, the author and his group had a chance to begin a new grand project for future robotics that integrates micro-mechatronics with soft-mechanics toward bio-mimetic machines. It is sponsored by Japan Society of Promotion Science as a new research grant referred as Research to the Future Program. In this project we intend to explore micro-mechatronics and soft-mechanics as the key technologies for highly advanced flexible machines of the future. Micro-mechatronics opens novel approach to integrate micro-structure, sensors and processors toward highly compound functional elements. Soft-mechanics aims to develop new mechanical constructs of physical and functional softness to assure soft and safe interaction between machines and humans. Combining those two technologies, human-centered bio-mimic machines are challenged to create, and paradigm change for the future machine design is also suggested.

2. Research Issues

One of the key performance for human-centered robotics of the future is the softness. The softness of machines has two aspects: functional softness and physical softness. So far, in the field of robotics or mechatronics, functional softness has been improved very much by sensor based intelligent control of skeleton-like rigid structures. However, physical softness of machine has not yet been studied enough, and machines are still hard and rigid even if the functions of machines became soft or flexible. In future society, robots are strongly expected to work with humans to support our everyday life. In such cases, in order to assure safety and affinity, physical softness of machines must be greatly improved. Physical softness is realized by both soft surface of the body and flexible actuation inside. In other words, technical breakthroughs in artificial skin and muscles are strongly required. This means mechatronic machines which are constructed by rigid skeleton and intelligent information processing must evolve to soft machines which are driven by soft actuators and covered by sensor distributed soft surfaces. This requirement may be met by recent rapid progress of micro machine technology, MEMS, and micro-fabrication, which provide novel means for building distributed sensor implanted artificial skin and high precision mechatronic artifacts. I believe that soft-mechanics and micro-mechatronics are the key technologies for creating very advanced mechatronic machines of the future. From that consideration, our project aims to develop those key technologies and to integrate them toward advanced bio-mimetic machines like soft and intelligent robots.

To introduce the research topics for this project very briefly:
(1) Sensor distributed robot skin: Distributed tactile/force/temperature sensors and early signal processors are planted inside the soft material so as to cover whole machine.
(2) Soft actuators: Pneumatic muscle-like actuators; tension control for wire drive mechanism; design, kinematics, and dynamics for multi DOF tendon

driven mechanism.
(3) MEMS robot hand; Precision robot hand, whose size, shape, and structure are mimetic to the human hand will be built by means of MEMS (Micro Electro-Mechanical System) technology. Whole hand will be covered with the sensor distributed soft skin.
(4) Whole-body type mechatronic artifact: Mechatronic artifact of small human/animal shape, which is built by integrating micro-mechatronics and soft-mechanics technology. Whole artifact will be covered by sensor distributed artificial skin. It will be controlled by remote brained robot approach.
(5) Humanoid: Aim for a prototype humanoid, by integrating various technologies such as artificial skin, soft actuators, MEMS robot hand, sensor integrated robot head, and also advanced information processing technology.
(6) Surgery support system based on micro-mechatronics technology and advanced information processing such as visualization, 3D image processing, remote control, and virtual reality technology.

This project attempts to develop artificial skin and muscles for robots. Development of skin-like sensor-distributed soft-cover material, and muscle-like soft actuators promises a technical breakthrough and opens a very fertile approach for future machine design. Not only the hardware development for sensor distributed soft robot skin and tendon drive architecture by a number of muscle-like soft actuators, but also software research for real-time sensory information processing and coordinated control of large number of soft actuators must be studied.

3. Paradigm Change in Machine Design

This project also aims at a paradigm shift in mechanical design. The first paradigm shift is FROM functional softness TO physical softness. In second, the research priority will be changed FROM the research on skeleton plus intelligence TO the research on skin and muscle. Thirdly, the control philosophy will be changed FROM softness via advanced control by means of sensor plus intelligence for rigid actuators TO hardness via advanced control by means of sensor plus intelligence for soft actuators. Consequently, the general design principle of machines will evolve FROM the usual mechanism of the bearing-gears-motor contracts TO a novel bio-mimetic mechanism of the bone-joint-tendons constructs. It will lead to very high degree of freedom mechanisms by using a large number of very simple actuator components. Although the new paradigm relates directly to hardware breakthroughs, the resultant bio-mimetic hardware requires novel approaches for successful control of complex soft structure and distributed sensory information processing. Presumably the hardware breakthroughs also requires another novel software breakthrough.

Concluding Remarks: Towards Evolution of Experimental Robotics [13]

This paper introduces Japanese research programs for intelligent robotics, and attempts to consider about the future evolution of experimental robotics. In part one, the ongoing inter-university research program on intelligent robotics is presented. This program focuses to investigate emerging mechanisms of machine intelligence through real world interactions. Physical understanding of dexterity, active visual recognition in real world behavior, human-robot cooperation, and biologically inspired autonomous systems are the topics to be pursued. A robot is a physical entity that behaves in the real world. Our research interest is focussed onto some sort of intelligence, which is inherently required for robust, efficient, quick interactions between the machine and the real environment. This project aims to find the principles of such mechanical intelligence and demonstrate them by building physical entities.

For robotics research, the development of experimental devices or tools are also very important, because the process of building robots is also very effective approach for understanding robotic principle. If any proper platforms for robotic experiment are available, software libraries will be accumulated and help to accelerate the speed of research. For this purpose, we developed common research tools for distribution. In our project, real time tracking vision system, general purpose DSP based real-time robot vision system, and compact low cost legged robot are developed, and opens the way to the development of general-purpose research platform for the sensing-action coupling approach in experimental robotics. Employing such platform, we can help to examine our own theory, to compare it with others, and to follow up and check various novel attempts presented by others. In such a way, we expect experimental robotics evolves from individual experiments to share common experimental knowledge.

In part two, challenges for new projects are introduced. Humanoid robot, R-cube conception, and network robotics are the topics included in the preliminary investigation for a new robotics project to be initiated by MITI Japan. Right at that time, Honda R&D Co. Ltd. unveiled their humanoid after ten years research efforts. As briefly described in part two, the Honda humanoid forced us to change our general ideas about R&D for human-shaped robot. Honda built a human-shaped walking-working robot as an impressive machine design. Research on humanoid is now ready to step into a new phase. A humanoid is not only a target for grand challenge of development, but also it evolves to a general purpose machine to be used in higher level studies on various robotic behaviors. Based upon a preliminary research for humanoid, a committee organized by MITI Japan is proposing a new national project on humanoid platform for experimental robotics. Honda's humanoid is a very good concrete example for platform design. The committee can estimate the technologies for the near future from the existing model. And a technical specification for the humanoid platform is properly figured out. We expect such a humanoid platform opens new phase of experimental robotics, where robot becomes a general purpose working machine as computers provided us with general means for computing several decades ago.

Acknowledgment: Part one of this work is supported by the Grant-in-Aid for Scientific Research on Priority Areas (Area number 266, Intelligent Robotics) by Ministry of Education, grant number 07245102. The work in part three is supported by The Japan Society for Promotion of Science as a project of the Research for the Future Program, Project number JSPS-RFTF 96P00801.

References

[1] H.Inoue: Research Program on Mechanisms for Emergent Machine Intelligence, G. Giralt and G. Hirzinger Eds., Robotics Research-7th International Symposium, pp.162-172, Springer (1996)

[2] H.Inoue: Inter-University Research Program on Intelligent Robotics, J. Society Instrument and Control Engineers, Vo.35 No.4, pp.237-242, (April 1996)

[3] S.Arimoto: Physical Understanding of Dexterity, ibid, pp.249-255

[4] Y.Shirai: Tightly CoupledSensor and Behavior for Real World Recognition, ibid, pp.256-261

[5] T.Sato: Human Machine Cooperative System, ibid, pp.262-267

[6] S.Yuta: Biologically Inspired Autonomous System, ibid, pp.268-274

[7] Research report of Inter-University Research Program on Intelligent Robotics : Vol.1995, (Feb.1996), Vol.1996,(Feb.1996)

[8] K.Arikawa and S.Hirose: Development of Quadruped Walking Robot TITAN-VIII, Proc. IROS-96 (Nov. 1996)

[9] R.A. Brooks and L.A. Stein: Building Brains for Bodies,Autonomous Robots, Vo.1 No.1, pp.7-25 (1994)

[10] MITI Japan R-Cube Committee Ed.: R-Cube, Nikkan Kogyo Shinbun-sya (1996)

[11] M.Inaba et.al.: Vision-based adaptive and interactive behaviors in mechanical animals using the remote-brained approach, Robotics and Autonomous Systems, Vol.17, pp.35-52 (1996)

[12] H. Inoue: Whither Robotics: Key issues, approaches and applications, Plenary talk at IROS96, Proc. IROS-96, pp.9-14 (Nov. 1996)

Chapter 1

Dexterous Manipulation

Dexterous manipulation is a very active research field due to the increasing requirements of robotics when dealing with complex tasks. The problems to face up to provide a robot with enough ability, concern not only the hand and structure design but also the control strategies that enable the efficient grasping or manipulation.

The paper of Kaneko, Thaiprasert and Tsuji describes an experimental approach for multi-fingered hands grasping. They start with the use of the wedge effect for enveloping grasps of cylindrical objects. Then, an extension to polygonal shaped objects is carried out analyzing the grasping procedure and sensing interaction.

The DLR's multisensory articulated hand constitutes the work presented in the paper by Hirzinger, Butterfaβ, Knock and Liu. The main goal of the hand design is miniaturizing the elements to make possible to integrate in the 4-finger hand with 12 d.o.f, the 112 sensors and the artificial muscles (linear actuators) needed to develop an anthropomorphic hand. After a description of the hand architecture, the control structure of the fingers and the whole hand controller hardware is presented.

The paper by Wellman, Peine, Favarola and Howe deals with the construction of tactile shape alloy wires, SMA. The objective of this research is to convey small scale shape information in teleoperation and virtual environments for robot applications such as remote palpation surgical instruments, or virtual texture experiments. The mechanical description is described based on: first, the configuration requirements and technological possibilities (space between pins, hysteresis, ...); second, thermal characteristics and finally the controller. The experimental results, tested by five subjects, of the effect of bandwidth on performance are shown.

Konno et al. present the humanoid robot Saika and its skillful manipulations. After a brief description of the humanoid modular structure, of the three-fingered hand and of the control system based on vision and tactile sensors, they analize the manipulation abilities of Saika. First, they analyze the behavior of Saika hitting a bounding ball of known size and based on visual analysis and trajectory prediction. Afterwards, the problem of dealing with unknown objects is discussed following two different heuristics, first a grasp that maximizes the fingers-object contact and second, a grasp that maximizes finger's flexion. Finally, they study the problem of catching a thrown ball. These experimentation constitutes a powerful platform to study the humanoid reactive planning behavior and performance.

Experimental Approach on Enveloping Grasp for Column Objects

Makoto Kaneko   Nophawit Thaiprasert   Toshio Tsuji

Industrial and Systems Engineering
Hiroshima University
Higashi-Hiroshima City, JAPAN, 739
kaneko@huis.hiroshima-u.ac.jp
nophawit@huis.hiroshima-u.ac.jp
tsuji@huis.hiroshima-u.ac.jp

Abstract: A grasping strategy for enveloping column objects is presented. In enveloping a cylindrical object placed on a table, a multifingered robot hand can conveniently utilize the wedge effect, in which the object can be automatically lifted up through the slipping motion caused when each finger pushes the bottom part of the object[1],[2]. This paper further extends the strategy to general column objects whose cross sections are polygon. One difficult situation appears when applying the strategy for triangular or rectangular objects, because the wedge effect cannot be expected any more or becomes weak. To cope with this problem, our alternative strategy includes two additional phases before lifting the object, (1)rotating the object and (2)inserting a finger tip into the gap produced by the rotating motion. We precisely discuss how to detect the pushing point for generating the rotating motion. We also show experimental results to verify the idea proposed in the paper.

Key words: Enveloping Grasp, Power Grasp, Active Sensing, Constant Torque Control.

1. Introduction

There have been a number of works concerning multi-fingered robot hands. Most of them address a finger tip grasp, where it is assumed that a part of inner link of finger never makes contact with the object. Enveloping grasp (or power grasp) provides another grasping style, where multiple contacts between one finger and the object are allowed. Such an enveloping grasp can support a large load in nature and is highly stable due to a large number of distributed contact points on the grasped object. While there are still many works discussing enveloping grasps, most of them deal with the grasping phase only, such as contact force analysis, robustness of grasping and contact position sensing. The goal of this work is to provide the sensing and grasping strategy for finally achieving an enveloping grasp for general column objects whose cross sections are polygon, where the object is assumed to be placed on a table.

Lifting up the object is the initial task for finally achieving an enveloping grasp. In order to realize the task, a multifingered robot hand can conveniently utilize the wedge effect, in which the object can be automatically lifted up through the slipping motion caused when each finger pushes the bottom part of the object. In case of either cylindrical objects or the object whose cross section is close to circle, the wedge effect can be easily produced through a simple pushing motion by the finger tip[1, 2]. In case that the object's cross section is triangular, however, no lifting motion is expected by such a pushing motion, because the resultant force will push the object toward the table and does not produce any lifting component. This implies that we are obliged to prepare an alternative grasping strategy for such objects. For preparing the new strategy, a couple of questions coming up are (1)how can the robot detect the failure in lifting up the object by a pushing motion in the horizontal direction, (2)what kind of sensing and grasping strategies are needed for lifting up the object, and (3)how can the robot achieve the target grasp after the object is away from the table.

We first discuss how to detect the failure in lifting up the object by a pushing motion. Once the robot recognizes such a failure, we execute two additional phases before lifting the object, namely, rotating the object around one side of the support polygon, and inserting a finger tip into the gap produced by the rotating motion. In order to obtain a reasonable pushing point for rotation, we introduce an active sensing technique to detect the local shape of object. One emphasis of our work is that we do not assume any tactile sensor except both joint position and joint torque sensors. Generally, when we implement a number of sophisticated tactile sensors over the finger surface, we can easily obtain contact information. At the same time, we have to expect a couple of troubles, such as the damage of sensors due to the direct contact with the object, broken wires due to many signal and power lines, and so forth. On the other hand, an active motion often provides us with new information that can not be obtained without any active motion. We try to make the most advantage of such an active sensing in this work. We utilize the active sensing for obtaining the local shape of object and for finally providing a pushing point for making the object rotate around one side of the support polygon. By rotating the object around an edge, we can produce an enough space to insert one finger tip. The insertion of one finger between the object and the table is the starting point to isolate the object from the table. Once the isolation is completed, a similar grasping strategy as that taken for a cylindrical object is applied for finally achieving an enveloping grasp.

In this paper, both the active sensing and the grasping strategies are precisely discussed. We also verify the proposed grasping strategy experimentally.

2. Related Work

Trinkle and Paul[3] proposed the concept of grasp liftability and derived the liftability regions of a frictionless planar object for use in manipulation planning. Mirrza and Orin[4] applied a linear programming approach to formulate and solve the force distribution problem in power grasps, and showed a significant increase in the maximum weight handling capability for completely enveloping type power grasps. Salisbury[5, 6] has proposed the Whole-Arm Manipulation (WAM) capable of treating a big and heavy object by using one arm which allows multiple contacts with an object. Bicchi[7] showed that internal forces in power grasps which allow inner link contacts can be decomposed into active and passive. Omata and Nagata[8] also analyzed the indeterminate

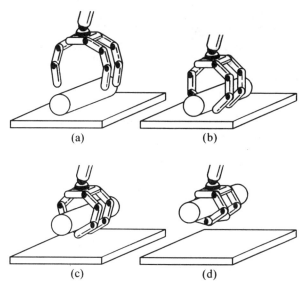

Fig.1 An example of enveloping grasp.

grasp force by fixing their eyes upon that contact sliding directions are constrained in power grasps. Zhang et. al.[9] evaluated the robustness of power grasp by utilizing the virtual work rate for all virtual displacements. Kumar[10] used WAM as an example to explain their minimum principle for the dynamic analysis of systems with frictional contacts. Kleinmann et. al.[11] showed a couple of approaches for finally achieving power grasp from finger tip grasp. In our previous work[1, 2] we have shown the grasping strategy for achieving enveloping grasp for cylindrical objects.

3. Sensing and Grasping Procedures

3.1. Hand and Grasp Model

We assume the three-fingered robot hand as shown in Fig.1. While most of the developed hands has a swing joint at the base of each finger, it is regarded that the swing joint is locked so that each finger can move only in 2D plane. The motion plane of finger is parallel in each other and each joint has a joint position sensor and a joint torque sensor. The joint position sensor is indispensable for determining the finger posture and the joint torque sensor is conveniently utilized for detecting the contact between the finger and the table (or the object) and for realizing either torque control or compliance control. We assume column objects whose cross sections are polygon and they are initially placed on a flat table. We further assume that the palm is already positioned close to the object and, therefore, do not discuss the approach phase of the robot arm itself. Also, it is assumed that the size of the object is roughly given by a visual sensor.

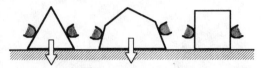

Fig.2 Examples of objects where the upward force is not expected by a simple pushing motion in the horizontal direction.

3.2. Grasping Procedure for Cylindrical Object

Fig.1 explains three phases for grasping a cylindrical object where (a)–(b), (b)–(d), and (d) are the approach phase, the lifting phase, and the grasping phase, respectively. In the approach phase, each finger first takes the designated initial posture (see Fig.1(a)) and then the first link is rotated until the finger tip detects the table. By monitoring a torque sensor output in each joint, we can detect any contact between the finger and the table (table detection). After the table detection, the finger tip is commanded to move along the table until it makes contact with a part of the object (object detection). The approach phase is composed of these two sub-steps. Since the finger tip is commanded to follow the table, it is most probable for the finger tip to make contact with the bottom part of the object. In the lifting phase, the finger tip is further commanded to move along the table to make the most use of the wedge effect. Therefore, there is no real switching point between the approach and the lifting phases. The object height during the lifting phase varies according to the finger tip position, and finally both the first and the second links will make contact with the object (two-points-contact mode). At this moment, the outputs from joint torque sensors abruptly increase, because the degree of freedom of finger along the table surface is no more available under such multiple contacts. By utilizing a large joint torque as a trigger signal, we switch from the lifting phase to the grasping phase. The grasping phase is realized by the natural computation mode, in which a constant torque is commanded in each joint for finally making the object contact with the palm in addition to both the first and the second links. Whether the object really reaches the palm or not and how firmly grasp the object, strongly depend on how much torque command is imparted to each joint.

3.3. Failure Mode in Wedge-Effect Based Grasping

For a cylindrical object, there usually exists an enough space to insert a finger tip between the bottom part of the object and the table, unless the object's diameter is smaller than that of finger tip. As a result, the finger tip can easily produce the upward force. For a general column object, however, depending upon the object's shape, the finger tip forces may balance within the object or they may produce the downward force. Under such situations, the lifting force is not produced, even though we increase the contact force. For example, such situations will be observed for the objects shown in Fig.2. Especially, when the object's cross section has a triangular shape, the vertical force caused by a pushing motion in the horizontal direction always results in the downward, while this is not always the case for the object having quadratic shape. For the objects shown in Fig.3, the wedge effect can not be expected any more. This

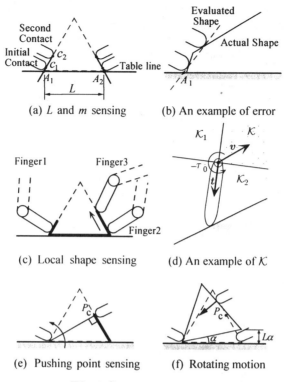

Fig.3 Sensing procedure

situation can be easily detected by the joint torque sensor, because their outputs will sharply increase during a pushing motion in the horizontal direction.

3.4. Human Behavior in Grasping Objects with Triangular Cross Section

Suppose that we achieve an enveloping grasp for a triangular column object placed on the table. Also suppose that the room is dark enough not to be able to utilize our vision. Under such a situation, we try to utilize our tactile information as much as possible by moving our finger over the object. This active motion is for obtaining the information on the object's shape. After we estimate the rough shape of object, we try to grasp the object in an appropriate manner. For a column object having a triangular cross section, however, grasping it from the table may be not always an easy task. Human often tries to produce a space between the table and the object by rotating the object around one side of the support polygon, such that we may insert our fingers into the space. Inserting our fingers into the space produced contribute to easily detaching the object from the table and to lifting it by our fingers. We apply such a procedure to a multi-fingered robot hand.

3.5. Active Sensing by Robot Fingers

When torque sensor outputs sharply increase at the beginning of the lifting phase, the lifting phase is stopped and we start the active sensing phase to detect the local shape of object. One feature in our active sensing is that we do not assume any other sensors except joint position and torque sensors. By imparting an active motion to each finger, we can obtain the tactile information as if there were tactile sensor over each finger tip. We produce such a virtual finger tip tactile sensor by active motion. We believe that this is one of most advantages of active sensing. The information which we intend to detect by active sensing is the geometrical parameters necessary for the robot to rotate the object. Such geometrical parameters are explained step by step based on the sensing procedure as shown in Fig.3.

(a) **Base length L:**
This parameter is indispensable when we determine how much angle should be rotated to ensure the finger tip insertion. In order to evaluate L, we make each finger tip contact with the object with a different height, as shown in Fig.3(a). Since the robot has any tactile sensor at the finger tip, the exact contact point is not provided by each trial. By utilizing two finger postures, however, we can obtain one straight line in contact with both fingers. This line is a candidate for the object surface. We extend this line until it passes through the table line which is already known. (The table line becomes known when each finger tip detects the table.) The intersection, for example, A_1, is a candidate for one side of the polygonal object. Thus, we can evaluate L by simply measuring the distance between both intersections as shown in Fig.3(a). Although this may not provide with the accurate L if two contact points lie on two different lines, respectively as shown in Fig.3(b), we can obtain L accurately enough by achieving the initial contact as close as possible to the table.

(b) **Slope of the bottom part of the object m:**
In order to determine the pushing point, we have to know the partial shape of object to confirm whether the finger tip can really impart an enough moment for rotating the object around one side of the base or not. However, since the robot has no precise knowledge on object's shape, it is difficult to start a tracing motion along the object's surface. To cope with this, we need to know the local slope of the object at the starting point for partial shape sensing. Such a local shape can be obtained simultaneously when we evaluate L. Because the slope information is indispensable for computing the intersection between the table line and the common tangential line. However, as mentioned earlier, the slope obtained is just a candidate. The coincidence between the computed and actual slopes is not guaranteed until we confirm that one finger makes contact with the same tangential line with an arbitrary point between two detected points, C_1 and C_2.

(c) **Pushing point P:**
The goal of the series of sensing is to find the pushing point for rotating the object around one side of the support polygon. For this purpose, we apply an active sensing for detecting the object's shape by utilizing a finger. One difficulty for tactile based active sensing is how to make a motion planning to avoid a large interaction force between the robot and the environment to be sensed. By always assigning at least one compliant (or constant torque) joint for the finger during sensing motion, we can avoid such a large interaction force. Another remark is that pulling a finger tip lying on object's surface can be easily achieved, while pushing it is more difficult due to stick-slip or blocking up by the irregularity existing on the surface. To cope with the difficulty of

pushing motion on frictional surface, the sensing algorithm is based on pulling motion of finger tip as shown in Fig.3(c), where the finger 3 is chosen as a probe finger. For more general discussion, we first define $\mathcal{K}_1$ and $\mathcal{K}_2$.

[**Definition**] *Define $\mathcal{K}_1$ as an assemble of v satisfying $f_1(v) = sgn(v^T t) < 0$, where v and t are vectors expressing the moving direction, and the longitudinal direction of the finger tip, respectively. Also, define $\mathcal{K}_2$ as an assemble of v satisfying $f_2(v) = sgn\{t \otimes v\} > 0$, where $\otimes$ denotes a scalar operator performing $x \otimes y = x_1 y_2 - x_2 y_1$ for two vectors $x = (x_1, x_2)^T$ and $y = (y_1, y_2)^T$.*

A sufficient condition for keeping the pulling condition is given by the following theorem.

[**Theorem**] *Let $\mathcal{K}$ be the area where the last joint can be moved without generating pushing motion. A sufficient condition for achieving a pulling motion based tracing is as follows:*
$\mathcal{K} = \mathcal{K}_1 \cap \mathcal{K}_2$ and $\tau = sgn(v \otimes t)\tau_0$
where τ_0 is the reference torque and the positive direction for torque is chosen in the clockwise direction.

Proof :(Omitted)

The moving direction of the last joint is chosen to satisfy the condition of $\mathcal{K} = \mathcal{K}_1 \cap \mathcal{K}_2$ and the direction of torque applied is determined based on the sign function $sgn(v \otimes t)$. The region of $\mathcal{K}$ and the direction of τ are, for example, shown in Fig.3(d).

Now, let us discuss how to detect the pushing point. Suppose an extreme case, where the friction between the finger tip and the object is zero. In order to produce a rotating moment around one side of the support polygon, we have to impart a pushing force at the upper point than P_c, where P_c is the intersection between the object surface and the normal line from the supporting edge, as shown in Fig.3(e). A sufficient condition capable of imparting a rotating moment, even under a small frictional condition, is to apply the pushing force at the upper point than P_c. However, P_c does not always exist over the object surface and whether P_c exists on the object surface or not strongly depends on the object's geometry. In case of the object with triangle cross section, P_c never exists for the object whose top angle is greater than 90 degrees. When P_c is not detected during the active sensing, the robot anyway tries to impart a rotating moment at the point in which the finger tip can apply the largest moment under a constant pushing force. This approach can work under a significant friction, while it may not under frictionless condition.

(d) Rotation angle α:
The rotation angle α is determined according to an approximate vertical displacement $L\alpha$ produced by the rotation. A sufficient condition for inserting a finger tip into the gap produced by the rotating motion is given by the following inequality.
$$L\alpha > d \qquad (1)$$
where d is the diameter of a finger tip or an equivalent diameter if the cross section is not exactly circle. An example satisfying the sufficient condition is shown in Fig.3(f).

The sensing procedure explained in this subsection can be applied for objects not only with triangular cross section but also with quadratic or other cross sections if they need the sensing phase.

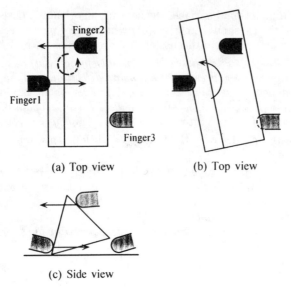

(a) Top view (b) Top view

(c) Side view

Fig.4 Finger Inserting Phase

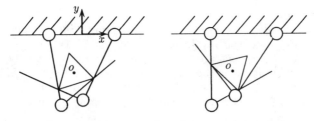

Fig.5 Triangle object enveloped by two jointed fingers

3.6. Finger Inserting Phase

After a sufficient gap is produced, one finger is removed away from the object's surface to be inserted into the gap, as shown in Fig.4(a),(c). At this instant, the contact line between the object and the table has to support the clockwise moment produced by two fingers. If the contact line fails in supporting the moment, the object will rotate around the tip of finger 1, as shown in Fig.4(b). In this work, we assume that the coefficient friction between the object and the table is large enough to avoid such a rotation due to slip.

3.7. Lifting and Grasping Phase

After the finger tip is sufficiently inserted into the gap between the object and the table, we apply the same grasping mode as that taken for a cylindrical object[2]. However, the condition for the object to reach the palm is not as simple as that for a cylindrical object. Because, for a general column object, contact points between the object and finger links change according to the orientation of object, even though the center of gravity of object does not change. This is a big difference between a cylindrical object and a general column one. If the resultant force acted by each contact force is greater than

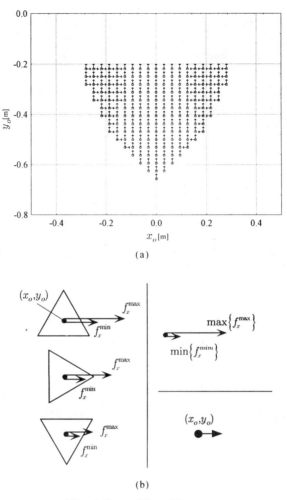

Fig.6 Force Flow Diagram

the gravitation force, irrespective of both orientation and position, then the lifting motion of object is garanteed. For a triangle object as shown in Fig.5, we consider the resultant force acted at each contact point. Under constant torque control, the resultant force is not uniquely determined, because there exist a infinite combinations of contact forces that can balance with a set of given torque. By applying the linear programming technique, we can compute both the maximum and the minimum resultant forces. Fig.6(a) shows the force flow diagram where the positive arrow means that both the maximum and the minimum forces are always positive irrespective of object orientation as shown in Fig.6(b). From Fig.6(a) we can see that the object moves upward until it finally reached the palm under $Mg = -1.0[N]$, $\tau_1 = 2.0$, $\tau_2 = 1.0$, $\tau_3 = 2.0$, $\tau_4 = 1.0$ and the coefficient of friction is 0.18.

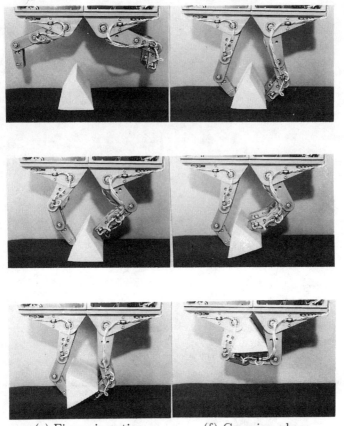

(e) Finger insertion   (f) Grasping phase
Fig.7 An example of grasping for a triangular object.

4. Experiments

Fig.7 shows continuous photos for an experiment, where the object has a triangular cross section. The hand can complete the task quickly, for example, the executing time for finally enveloping the object by the fingers was about 14 [sec]. Fig.8 shows the experimental result during the active sensing, where the object having a triangular cross section is utilized, and the joints shown by a single circle and double circles denote position-controlled and constant torque controlled joints, respectively. The thick lines denote the detected surfaces through the active sensing, where the object shape is shown by the dotted line and P_c is the computed intersection based on the sensing data. Fig.9. shows the experimental result during the rotating motion and the finger tip inserting motion.

5. Conclusions

We discussed sensing and grasping procedure for enveloping an column object placed on a table. For an object having triangular cross section, we pointed

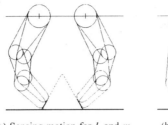
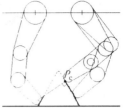

(a) Sensing motion for L and m (b) Sensing motion for P_c

Fig.8 Local object shape obtained by the active sensing

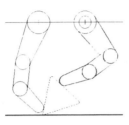
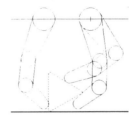

(a) Rotating motion   (b) Finger inserting motion

Fig.9 Rotating and finger tip inserting motion

out that the tactile based active sensing effectively works for obtaining the geometrical information necessary for rotating the object. We described the sensing procedure precisely with a couple of sufficient conditions for achieving the sensing task. We also verified the proposed procedure experimentally by utilizing the Hiroshima Hand. Finally, we would like to express our sincere gratitude to Mr. Y. Hino, Mr. M. Higashimori and Mr. Y. Tanaka for their cooperation for this work.

References

[1] Kaneko, M., Y. Tanaka, and T. Tsuji: Scale-dependent grasp, 1996 *Proc. of the IEEE Int. Conf. on Robotics and Automation*, pp2131-2136

[2] Kaneko, M, Y. Hino, and T.Tsuji: On Three Phases for Achieving Enveloping Grasps, 1997 *Proc. of the IEEE Int. Conf. on Robotics and Automation*, pp385-390

[3] Trinkle, J. C., and R. P. Paul: The initial grasp liftability chart, 1989 *Trans. on Robotics and Automation*, vol.5, no.1, pp47-52

[4] Mirza, K., and D. E. Orin: Control of force distribution for power grasp in the DIGITS system, 1990 *Proc. of the IEEE 29th CDC Conf.*, pp1960-1965

[5] Salisbury, J. K., Whole-Arm manipulation, 1987 *Proc. of the 4th Int. Symp. of Robotics Research*, Santa Cruz, CA, Published by the MIT Press, Cambridge MA

[6] Salisbury, J. K., W. Townsend, B. Eberman, and D. Dipietro, Preliminary design of a Whole-Arm Manipulation System (WAMS), 1988 *Proc. of the IEEE Int. Conf. on Robotics and Automation*, p254

[7] Bicchi, A: Force distribution in multiple whole-limb manipulation, 1993 *Proc. of the IEEE Int. Conf. on Robotics and Automation*, pp196-201

[8] Omata, T., and K. Nagata: Rigid body analysis of the indeterminate grasp force in power grasps, 1996 *Proc. of the IEEE Int. Conf. on Robotics and Automation*, pp1787-1794

[9] Zhang, X-Y., Y. Nakamura, K. Goda, and K. Yoshimoto: Robustness of power grasp, 1994 *Proc. of the IEEE Int. Conf. on Robotics and Automation*, pp2828-2835

[10] Howard, W. S., and V. Kumar: A minimum principle for the dynamic analysis of systems with friction, 1993 *Proc. of the IEEE Int. Conf. on Robotics and Automation*, pp437-441

[11] Kleinmann, K. P., J. Henning, C. Ruhm, and T. Tolle: Object manipulation by a multifingered gripper: On the transition from precision to power grasp, 1996 *Proc. of the IEEE Int. Conf. on Robotics and Automation*, pp2761-2766

DLR's Multisensory Articulated Hand

G. Hirzinger, J. Butterfaß, S. Knoch, H. Liu

DLR

DLR (German Aerospace Research Establishment),
Institute of Robotics and System Dynamics
Oberpfaffenhofen, D-82234 Wessling/Germany
Phone +49 8153 28-2401, Fax: +49 8153 28-1134
email: Gerd.Hirzinger@dlr.de

Abstract: The paper outlines DLR's new 4-finger hand characterized by the fact that all 12 active joints (3 for each finger) are integrated into the fingers or the palm. With 112 sensors, around 1000 mechanical and around 1500 electrical components the new hand is one of the most complex robot hands ever built. Its key element is the „artificial muscle ®" a small but powerful linear actuator, which seems to be one of the first real electromechanical alternatives to hydraulic or pneumatic actuators.

On the overall hand design

For many space operations, i.e. handling drawers, doors and bayonet closures (electric connectors) in an internal lab environment, two-finger grippers seem adequate and sufficient; the appropriate mechanical counterparts in the lab equipment are easily designed and realised even in a very late design stage. For more complex manipulations, however, future space robots (and of course terrestrial servicing robots) should use articulated multifingered hands.

Impressive dexterous robot hands have been built in the past, e.g. the MIT / UTAH hand, the JPL/Stanford hand, the Belgrade hand etc.. However, all of them suffer from one main drawback: if the number of active degrees of freedom exceeds a fairly small number (e.g. 6 dof), there was no chance so far to integrate the drives in the hands wrist or palm: either a number of cables or tubes leads to a separate pneumatic or hydraulic actuator box (e.g. in case of the MIT / UTAH hand) or a mass of bulky motors is somehow mounted at the robot arm, so that the practical use of articulated multifinger hands has often been called into question. Thus, it was our declared goal to build a **multisensory 4 finger hand** with in total twelve degrees of freedom (3 active dof in each finger), **where all actuators (uniformly based on the artificial muscle,** Fig. 1) **are integrated in the hand's palm or in the fingers directly** (Fig. 2 and Fig. 2). Miniaturizing the artificial muscle down to a nut diameter of 10 mm in combination with a specially designed brushless DC-motor with hollow shaft was a first important step. Force transmission in the fingers is realized by special tendons (highly molecular polyethylene), which are optimal in terms of low weight and backlash despite of fairly linear behavior.

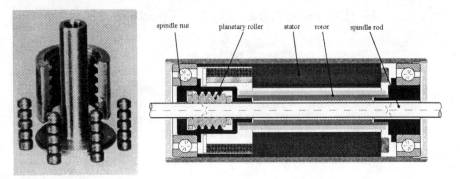

Fig. 1: DLR's planetary roller screw (left) integrated into tiny motors yields the artificial muscle (right)

Each finger shows up a 2 dof base joint realized by artificial muscles (AM) and a third actuator of this type integrated into the bottom finger link (phalanx proximal), thus, actuating the second link (phalanx medial) actively and, by elaborate coupling via a spring, the third link (phalanx distal) passively (Fig. 3). The anthropomorphic fingertips are of crucial importance for grasping and manipulation, thus, they are modular and easily exchangeable with specially adapted versions. Following our mechatronic design principles, literally every millimeter in the fingers is occupied by sensing, actuation and electronic preprocessing technology. **Every finger unit with its 3 active degrees of freedom integrates 28 sensors(!).**

Fig. 1: Our 4 finger hand with its 12 actuators and 112 sensors integrates 1000 mechanical and 1500 electronic components

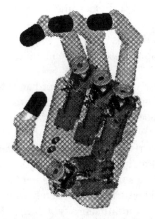

Fig. 2: 3D-CAD-design of the four finger dexterous hand (12 dof)

Sensor type	nr.	data
laser diode	1	wavelength 670 nm
torque sensors	5	range 0-1,8 Nm (9 Bit)
joint position	4	range 110° (9 Bit)
temperature	5	0 - 100°C, res. 0,1°C
rotor position	3	3072 pulses/rotation
light barriers	6	infrared
tactile sensors	4	0,5 - 10 N, res. 35mN

Table 1: A finger integrates 28 sensors

Fig. 3: The 2 degree of freedom base joint

The fingers are position-force-controlled (impedance control), they are gravity compensated and they are prevented from colliding by appropriate collision avoidance algorithms.

Four tactile foils detecting locus (center) and size of external forces cover all links. They are based on FSR (force sensing resistor) technology and arranged as XYZ pads (Fig. 4). The finger tips typically provide a light-projection LED to simplify image processing for the tiny stereo camera system integrated in the hand's palm. Due to the already mentioned modularity the fingertips might be easily exchanged with a version containing e.g. fiber optics. The two-axis torque sensor hereby serves as fast exchange adapter. Signal processing for the tip is integrated in phalanx distal for assuring optimal signal quality.

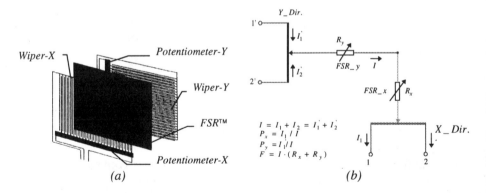

Fig. 4: Construction(a) and simplified electrical model(b) of a XYZ pad sensor

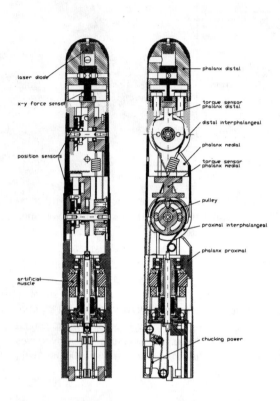

Fig. 6: Angular position sensor

Fig. 5: The finger's construction with integrated artificial muscle (two orthogonal views without 2 dof base joint)

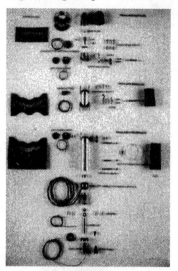

Fig. 7: Finger disassembled

Distal interphalangeal (the upper joint) is equipped with a torque sensor and a specially developed, optically and absolutely measuring angular position sensor. This sensor (Fig. 6) is based on a one-dimensional PSD (Position Sensitive Device), which is illuminated by an infrared LED via an etched spiral-type measurement slot. This measurement principle goes back to the Space-Mouse-development. Using an optimized PCB design exclusively equipped with tiny SMD items and a circuit board with a minimized number *of items used* it was possible to create an optical position sensor with remarkable performance with respect to its size. The sensor measures only 4.8 mm in thickness and 17 mm in diameter. Nevertheless, a voltage regulator and the complete analogous signal conditioning circuit is integrated. The angular resolution is 9 Bit with a linearity error of less than 1 %.

The torque sensor transforms forces at the fingertips into pure torques around the joint axis measured via strain gauges on tiny flexible beams. Miniaturized electronic boards in the finger care for preprocessing of the forces and torques as well.

Proximal interphalangeal (the lower joint) shows up sensor technology nearly identical to that of distal interphalangeal. The base joint (MetaCarpoPhalangeal MCP) as in the human hand shows up two rotary joints with a common center of rotation; other than that, the technology is very similar to that inside the finger (2 artificial muscles, similar position and torque sensors).

Summary the technical characteristics of finger and MCP joint:

max. joint angle finger	105° proximal interphalangeal 110° distal interphalangeal
max. joint angle MCP joint	90° palmar flexion 30° abduction 30° abduction
max. force	10 N at the fingertip
closing time finger	< 0,5 sec
max. frequency	25 Hz
mass of each finger	135 g (without 2 dof base joint)

Table 2: Technical data finger / MCP joint

Finger control structure

The actuation and transmission system of a typical finger joint is shown in Fig. 8. The main components of the single-joint model are: the artificial muscle, the tendon pulley power transmission system, and the joint itself.

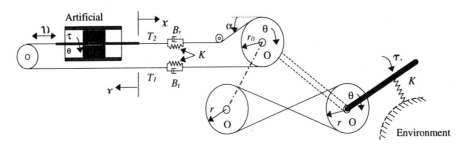

Fig. 8: Schematic diagram of the joint mechanical system

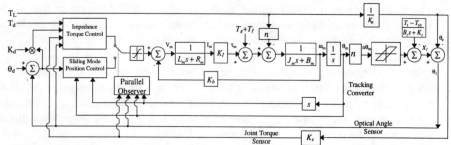

Fig. 9: Block diagram of the parallel torque/position control system

K_I = 9.0x10^{-3} Nm/A	torque constant	t_E = 0.14 ms		electrical time constant
J_m = 1.62x10^{-7} Kg.m^2	total inertia damping	t_M = 4.17 ms		mechanical time constant
B_m = 1.11x10^{-4} Nm/rad/s	coefficient	K_s = 2.73x10^1 Nm/rad		stiffness of sensing joint
K_b = 2.0x10^{-2} V/rad/s	back EMF constant	T_1, T_2		tendon tension
L_m = 1000.0 μH	armature inductance	K_t = 2.86x10^4 N/m		tendon stiffness constant
R_m = 7.0 Ω	ave. Terminal resistance	B_t = 5.6x10^{-4} Nm/rad/s		tendon damping constant
n = 323	reducer ratio	K_r = 1.05		pulley radii reducer ratio

Table 3: Parameters of artificial muscle and transmission system

Parallel torque/position control strategy

The parallel torque/position strategy attempts to combine simplicity and robustness of impedance control and sliding mode control with the ability of controlling both torque and position. The goal is achieved using two controllers parallel, and managing conflicting situations by means of a priority strategy: since the primary aim is to make the resulting system capable to accommodate its motion to environment constraints, the torque controller is designed to prevail over the position controller by a parallel observer. Experimental results based on this strategy are shown in Fig. 10(a),(b), which represent impedance control for torque tracking, and parallel torque/position strategy for the control from free space to constraint environment respectively.

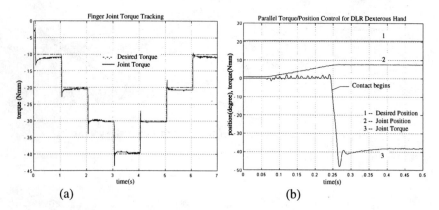

Fig. 10: Experiments on a finger joint (a) joint torque tracking, (b) parallel torque / position control

Controller hardware design for the dexterous hand

Our dexterous robot hand is controlled by a multiprocessor system. The fully modular control architecture is split into two levels, the global hand control level (Fig. 11) and the local finger control level (Fig. 12). The global hand controller is externally located in a workstation running a real time operating system. Thus, high flexibility is provided for implementing various hand control strategies. In contrast to conventional approaches the modular local finger controllers are located next to the hand attached to the manipulator carrying the hand. This design became feasible due to the high degree of integration and miniaturization of the finger controller hardware.

Global hand controller and local finger controllers communicate via SERCOS (*SErial Real time COmmunication System*) by fiber optic links within 1 ms for four fingers. The main goals for this design were high flexibility, expandability and high computational performance. Due to the integration of the local finger controllers as well as the complete drive system including the power converters into the hand-arm system, the amount of cabling carried by the manipulator arm reduces to a fiber optic link ring and four power supply line.

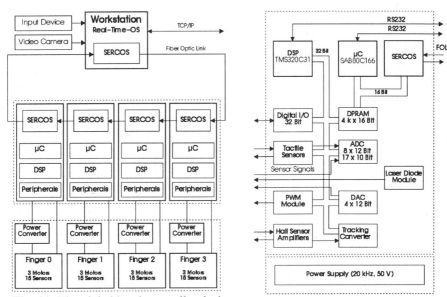

Fig. 11: The global hand-controller design   **Fig. 12:** Finger control structure

For the control of each finger module one finger controller module is necessary. Each finger module incorporates three brushless DC-motors for the three independent dof and various sensors. The controller module is an independent subsystem of the hand system and receives commands from the global control level by SERCOS via a fiber optic link. One µ-controller per finger is responsible for the information management inside the controller module. An additional 60 MFLOPS floating-point DSP provides a control hardware with sufficient computational power for realizing advanced control algorithms for the three actuators in each finger. Identical finger control modules *are put together* by a common power supply to form the complete control system for the local finger level. The modular design allows the use of up to four finger control modules with the current power supply design.

Planning power grasps

A power grasp encloses an object with all fingers and the palm in order to yield optimum contact between object and hand. One first has to find a set of contact points (*candidate grasp*) which result in a stable grasp. Afterwards one may determine the quality of different stable grasps and choose the best one. The general approach towards finding a stable power grasp candidate is to simulate the forces resulting from contacts between the hand and the object to be grasped, and the finger and object movements resulting from these forces.

In a first study dynamic simulation of grasps with our dexterous hand, yielding the resulting contact forces, was performed using SIMPACK (Fig. 13).

Fig. 13: Dynamic simulation of grasps with SIMPACK

Fig. 14: Determining contact forces

In a second approach we try to accelerate the process of generating power grasp candidates by restricting ourselves to the final static situation. So we neglect the influence of both the object mass and the hand mass on the simulation and only regard the contact forces. These are determined by the depth of intersection of the hand and object geometric model. A very efficient method to determine contact forces has been implemented (Fig. 14). Currently we are working on the integration of the contact force determination algorithm in a simulation framework which yields the desired stable power grasp candidate.

Conclusion

We believe that with our new robot hand a real next step in the area of articulated robot hands has been achieved. Presently the hand has about 1,5 times the size of the human hand, but due to the integration of all 12 actuator systems it is modular, i.e. can be mounted on any robot without external energy supplies etc. It would be a challenging goal to build a number of these hands for institutes and refine it later on towards a prosthetic device.

References

[1] J. Pertin-Troccaz. Grasping. A state of the art. In O. et al. Khatib, editor, *Roboics Review,* volume 1, pages 71 - 89. MIT Press, Cambridge, MA 1989

[2] K. B. Shimoga. Robot grasp synthesis: A survey. *Int. Journal of Robotics Research,* 15(3):230 - 266, June 1996

[3] J.K. Salisbury, W.T. Townsend, B.S. Ebermann, D.M. DiPietro, Preliminary Design of a Whole-Arm Manipulation System (WAM), *Proc. 1988 IEEE Int. Conf. on Robotics and Automation*, Philadelphia, PA, April 1988.

[4] S.C. Jacobsen, I.K. Iversen, D. Knutti, R.T. Johnson, K.B. Biggers, Design of the UTAH/MIT Dextrous Hand. *Proceedings IEEE Int. Conf. on Robotics and Automation*, USA, 1986, S. 1520-1532.

[5] J.K. Salisbury, M.T. Mason, Robot Hands and the Mechnics of Manipulation, Cambridge MA, *MIT Press* 1985

[6] T.J. Doll, Entwicklung einer Roboterhand für die Feinmanipulation an Objekten. *Robotersysteme, Springer Verlag*, 3(3):167-174, 1987.

Mechanical Design and Control of a High-Bandwidth Shape Memory Alloy Tactile Display

Parris S. Wellman   William J. Peine
Gregg Favalora   Robert D. Howe
Division of Engineering and Applied Sciences
Harvard University
Cambridge, Massachusetts, USA
parris@hrl.harvard.edu   peine@hrl.harvard.edu
favalora@fas.harvard.edu   howe@deas.harvard.edu

Abstract: We have constructed a tactile shape display which can be used to convey small scale shape in teleoperation and virtual environments. A line of 10 pins spaced 2 mm on center are each actuated with a shape memory alloy wire. A combination of careful mechanical design and liquid cooling allows a simple proportional controller with constant current feed forward to achieve 40 Hz bandwidth. To quantify the value of increased bandwidth, an experiment involving a prototypical search task has been conducted using the display. A digital filter limited the frequency response of the display to three cutoff frequencies: 1, 5 and 30 Hz. Subjects were able to complete the search more than 6 times as quickly with 30 Hz bandwidth than with 1 Hz.

1. Introduction

We are developing systems for conveying small-scale shape information in teleoperation and virtual environments. These tactile display devices consist of an array of pins that are raised and lowered against the finger tip. Our goal is to recreate skin deformations that might be produced by object features such as corners, raised edges, and surface textures. One important application area for these displays is minimally invasive surgery, where the display can reproduce tactile information sensed by instruments within the patient's body. A key example is the localization of tumors, which often appear as hard lumps embedded in soft tissues, such as the lung and the liver [1].

Recent experiments by Peine et al. [2] show that surprisingly high bandwidth is required for effective tactile display. These experiments used an optical tracking system to measure finger speeds while experienced surgeons located simulated tumors in soft rubber models. They found that maximum finger speeds for 90 percent of the population tested were approximately 120 mm/second. This measurement allows us to determine the temporal bandwidth required for effective shape display. Given a shape display that has pins spaced 2 mm apart, the maximum spatial frequency that can be created (in the Nyquist limit, with pins alternating up and down) is 0.25 cycles/mm. If the maximum spatial frequency is scanned across the display at 120 mm/sec, each pin must travel up and down in 33.3 milliseconds. This means that each pin must achieve a temporal bandwidth of at least 30 Hz. If the tactile display device cannot raise and lower the pins in this time interval, then the tactile information will not be correlated with the gross motion of the finger. The lack of correlation between finger motion and tactile information can make it difficult to quickly and accurately locate tactile features. Extremely low bandwidths can lead to a "move and wait" strategy where the user conducts a search using discrete motions followed by pauses to wait for the display to catch up which makes it difficult to quickly and accurately search a space.

In addition to high bandwidth, high spatial resolution is required because Johnson and Phillips [3] have shown that humans can reliably distinguish between two points that are separated by as little as 0.9 millimeters. This implies that pins in a shape display should be spaced as closely as 0.9 millimeters in order to convince users that they have a shape pressed into their fingerpad, rather than an array of pins. In addition to the small spacing requirement, Pawluk et al. [4] have found that human finger pads have a stiffness as high as 3.5 N/mm at 1.2N indentation force when indented with a flat plate. Since we are interested in displaying shape, rather than pressure, we require pins with high stiffness and high force capability. This is especially true when designing displays to be used on force reflecting devices, because the pins will need to support all of the force produced by the user's fingers during manipulation.

A number of researchers have constructed tactile shape displays [5,6,7,8]. These displays have been constructed in many configurations including single pin and multi-pin matrix displays. Unfortunately, these previous displays have all been limited by low bandwidth, low stiffness or lack of static response. Thus, in order to effectively portray small scale tactile shapes a display must have small pin-to-pin spacing, produce large forces, have high pin stiffness and extremely high bandwidth.

2. Mechanical Design

2.1. Configuration Design

Our shape display is intended to be used for virtual texture experiments as well as for use in remote palpation surgical instruments. These two applications require a compact shape display. Using a single row of pins simplifies packaging and mechanical design constraints, yet retains the functionality of a matrix of pins because textures can be easily perceived along the line of the pins. The user can also sweep the display to perceive curvature perpendicular to the line of pins. There is clearly a trade-off between perception of two dimensional curvatures and temporal bandwidth, and we have chosen to concentrate on the latter. Figure 1 shows that the line of pins runs along the finger pad rather than across it. The size of various available components limits the pin spacing to 2.0 mm on center. In order to cover the length of most finger pads, we have chosen to use ten pins.

The small pin to pin spacing, compact size and the extremely high force and stiffness requirements, led us to choose shape memory alloy (SMA) wires to actuate each of the pins. This material is more suited to this task than electromagnetic actuators because of its high force-to-weight ratio and high intrinsic stiffness. It also provides much larger displacements than piezoelectric or magnetostrictive materials. There is a price to be paid for these benefits and SMA has its own set of design challenges which must be addressed.

SMA shortens when it undergoes a phase transition from the Martensitic to the Austenitic phase. This reversible phase transition can be produced by heating the wire above its transition temperature using electric current. It is well known that SMA exhibits hysteresis when cycled through this transition [9]. It is also a relatively slow process because it can take a long time for the wire to cool and lengthen. Previous authors have suggested that it is possible to increase the bandwidth and account for the hysteresis by developing a model of the process and incorporating this in the controller [10,11]. While this undoubtedly will work to some degree, we have chosen to increase the bandwidth through careful thermal design and have minimized the hysteresis through position feedback control of the display.

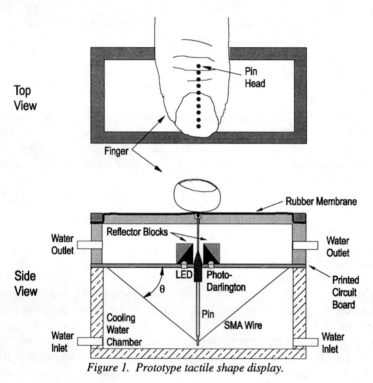

Figure 1. Prototype tactile shape display.

2.2. Mechanical Design

SMA typically undergoes less than 5% strain during the phase transition. In order to compactly package the display we have configured the wires in a V shape [3], as shown in Figure 1. Kinematic analysis reveals that if the original length of the wire is 2L then the displacement of the pin, δ, is

$$\delta = L\sin(\theta) - \sqrt{L^2 \sin^2(\theta) - (1-(1-\gamma)^2 L^2} \qquad (1)$$

where γ is the maximum strain of the wire. The stiffness of the pin, K, is related to the angle, θ, also shown in Figure 1, and the stiffness of the wire, K_w, by the equation

$$K = 2\sin^2(\theta) K_w \qquad (2)$$

K_w is a function of the phase of each wire, its length and its diameter, which should be chosen to maximize heat dissipation. These two equations make it clear that there is a trade-off between compactness, the displacement of each pin and the stiffness of each pin. Thus, they can be solved to determine the required wire length to produce a minimum stiffness, displacement and desired compactness.

In order to facilitate the construction of the display, it uses a layer based modular design that includes all electrical connections on a single printed circuit board. This also minimizes the number of parts that are required and makes for an overall size of 78 mm by 35 mm by 57 mm. Preliminary testing revealed that nickel-titanium shape memory alloy wire can be cycled at 4.8% strain with a 2N load for tens of thousands of cycles with no appreciable loss in performance. Thus, as an extra margin of safety, we chose to limit the SMA to 4.5% strain, which provides the required 3mm displacement when each V in the display is 60 mm wide by 20 mm deep. The more usual number quoted to achieve millions of cycles of is 2% strain we have sacrificed life-span for compactness in this prototype.

During assembly, the SMA wires are first electroplated[*] with copper over a small portion of their length, soldered to the board at one end, and then passed through the pins and the board, where they are tied off. Each knot is electro-plated and then soldered to the board to complete the electrical connection. This electroplating and soldering process produces high strength, high-conductivity connections to the SMA wires, which is a difficult challenge.

The position of each of the pins is measured using an infrared light emitting diode (LED) and photoDarlington transistor pair. As can be seen in Figure 1, the pair is mounted in the printed circuit board, and the light from the LED is reflected past the pin to the photoDarlington by a pair of specially designed reflectors. Dividers separate each pin to prevent crosstalk. Each pin is shaped so that the amount of light it blocks varies with its position, which causes the collector current in the photoDarlington to vary as a monotonic function of pin position.

The circuit board also serves as a bearing surface for one end of each pin, while the top block on the display provides the second bearing surface. Together these constrain the pins to move linearly with minimal side loading. The top of each pin is recessed into the surface of the top block to provide a nearly flush surface for the user's fingertip to press against. Finally, the necessary spring return force for each pin is provided by a latex rubber membrane which also serves as a seal.

2.3. Thermal Design

In this application, the phase transition of SMA is thermally driven, and because it is always possible to apply more electrical current to cause the wire to contract faster through resistive heating, the fundamental limitation is cooling.

---

[*] We use a solution of water and copper sulfate mixed in a 25:1 (H_2O:$CuSO_4$) ratio by weight, and a 5 mA plating current per wire being plated. Each wire is plated for 5 minutes.

Because the rate of heat transfer depends on the ratio of surface area to volume, we use two small (75 micron) wires rather than one larger one to actuate each pin. We also use SMA wire with a transition temperature of 90C because the rate of cooling is linearly related to the temperature difference between the wires and the cooling media.

In previous work, we found that ambient air cooling resulted in a bandwidth of perhaps 1 Hz, while forced air cooling increased this to 5-6 Hz [7]. To obtain the required 30 Hz bandwidth, the present design uses a slowly recirculating bath of water[**]. Water has been chosen as a cooling fluid because of its high thermal conductivity and tremendous heat capacity which allows for a low re-circulation rate within the display. This ensures more uniform cooling of all of the wires in the display, and thus more uniform pin-to-pin performance.

3. Controller Design

Because of the success of the mechanical and thermal design of the display, a simple linear controller has proved adequate to drive each pin. To obtain maximum performance it is desirable to always keep the wires operating between the minimum transition temperature and the maximum transition temperature. We use a proportional controller, with a constant offset current. The current offset is used to heat the wire to near the transition temperature and its magnitude and the gain are determined experimentally. For the experiments reported here they were 0.5 amps and 4 amps/mm, respectively.

In order to achieve maximum performance, the desired position of each of the pins is always kept slightly more than zero to be sure that their temperature remains at approximately the transition temperature. In addition heating the wires to more than their highest transition temperature has no effect on their length, but does require more heat to be dissipated before expansion begins, which slows the response of the display. Therefore we restrict the maximum displacement of the pins to 4.5% strain. Finally, current is limited to 2.5 amps per pin is placed on the display to ensure that wires will not be destroyed by overheating.

4. Display Performance Characterization

The optical sensors were calibrated by placing a potentiometer with a lever arm attached to it on top of each pin. A 50 gram weight was place on the lever arm and a triangular current wave drove each pin. A cubic spline was used to linearize the feedback for each pin. Figure 2 shows the results of the calibration for a representative pin.

Figure 3 shows the response of a representative pin to a 1 Hz triangle wave command with 3 mm of displacement. Figure 3(a) shows the response under two loading conditions: no load, and when the experimenter was pressing on the entire display with finger tip force of at least 10 N. This data is replotted in Figure 3(b) to

---

[**] Actually, we use something of a witch's brew for cooling. A mixture of water and ethylene glycol mixed 8:1 by volume increases the boiling temperature of the water to 110 C. We also mix in less than 0.02% by weight of Surfynol 75, a fumed silica surfactant manufactured by Air Products to promote the migration of bubbles out of the display enclosure.

show the relationship between commanded and actual position; the hysteresis is well under 0.1mm. Figure 3(c) shows the controller current required for each load. In the loaded case, peak current approached 2 amps while in the unloaded case the current reached a more modest level of 1.6 amps. Together, these figures show that the simple proportional controller ensures good performance across the entire range of anticipated load impedance.

Characterizing the bandwidth of this nonlinear system is problematic because the fall time is effectively slew rate limited by the wire cooling rate. One useful performance measure is the output amplitude at each frequency, in response to a triangle wave command at the maximum static displacement. Figure 4 shows the frequency response of the display when driven with a 3mm triangle wave position command for 1 second with a finger pressed firmly (total force approximately 5 N) against the display. The frequency at which the output amplitude is three dB below the commanded input amplitude is approximately 40 Hertz. Although the maximum frequency shown in the figure is 100 Hz, detectable output can still be felt at frequencies approaching 150 Hz, suggesting that the display can be used for vibrotactile feedback [12,13] as well as shape feedback. Figure 5 is a table summarizing the design targets and measured performance of the display.

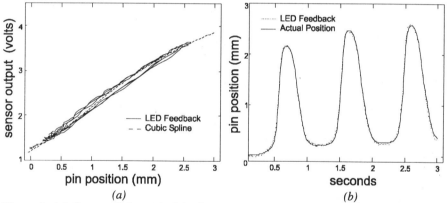

Figure 2 -(a) Output of the optical feedback plotted against the actual position of a pin; (b) output of the optical sensing after fitting cubic spline (dotted line) and actual position (solid line).

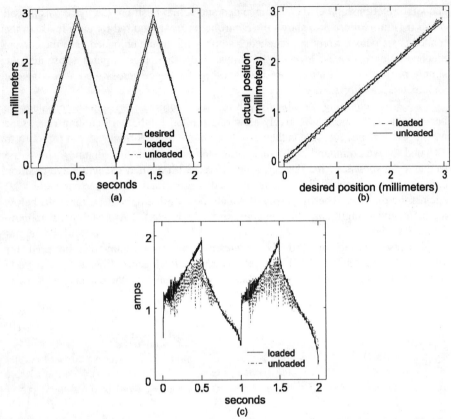

Figure 3 - (a) Desired and actual position of a representative pin under unloaded and loaded (finger pressing at 10N total force) conditions, (b) the same data plotted as desired versus actual position (note lack of hysteresis) and (c) the current required under loaded and unloaded conditions.

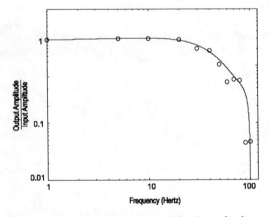

Figure 4 - Frequency response of the shape display.

Parameter	Design Goal	Achieved Performance
Pin Displacement	3.0 mm	3.0 mm
Pin Force	1.2 N	>1.5 N
Pin Stiffness (at maximum displacement)	Rigid	>35 kN/mm
Pin to Pin Spacing	0.9 mm	2.0 mm
Frequency Response	30 Hertz	40 Hertz (-3dB point)

Figure 5 - Design targets and measured performance of the shape display.

5. Effect of Bandwidth on Performance

To quantify the effect of high bandwidth of the shape display on task performance, we asked subjects to perform a prototypical search task while a digital filter with a cutoff frequency of 1, 5 and 30 Hz was applied to the commanded positions for the display. These frequencies were chosen because they correspond to approximately the frequency response reported for shape displays cooled with still air, forced air and water respectively.

The display was mounted to the mouse pointer of a digitizing tablet. Subjects grasped the display with the dominant hand and rested the index fingertip on the pins. The tablet measured finger position as subjects moved within a test area. The computer sampled the display location and raised and lowered the display pins to represent virtual small-scale shapes at fixed locations. Subjects were asked to search a 150 mm diameter circle as quickly as possible to find two small dots (3mm diameter by 2mm height cylinders). They were asked to locate each dot in the search space and center the display on the dot and press a button. Once they found both dots, they were asked to press a second button to end the experiment. The time subjects took to search the space and their average velocity while searching were recorded.

Five subjects (three male, two female, 24 to 29 years of age) were asked to search each of eight virtual environments with two randomly placed dots as described above for the two highest bandwidths, and four virtual textures for the slowest bandwidth. Each subject was presented with the same eight (or four) virtual textures as the other subjects, and these were presented in random order. Subjects were told only that some of the parameters of the display had changed when the frequency bandwidths were altered. Each subject was allowed two, three minute practice periods each time that the frequency response of the display was altered in order to minimize the effect of practice on the results. They were given a maximum of three minutes to complete the search, although this limit was only reached during testing with the 1 Hz filter.

Figures 6 and 7 show the results of the experiment. The average search velocity more than doubled, going from 18 mm/sec to 55 mm/sec when the filter cutoff frequency was increased from 1 Hz to 5 Hz. The effect of going from 5 Hz to 30 Hz was somewhat less pronounced, as the velocity only increased to about 90 mm/second. The search times showed a steady decrease from approximately 150 seconds, to 50 seconds, to 25 seconds when the filter cutoff frequency was changed from 1 to 5 to 30 Hz respectively.

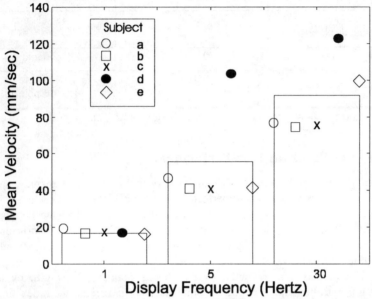

Figure 6 - Mean finger velocity during search for all trials at each test frequency (bars) and mean search velocities for each individual subject for each test frequency (symbols).

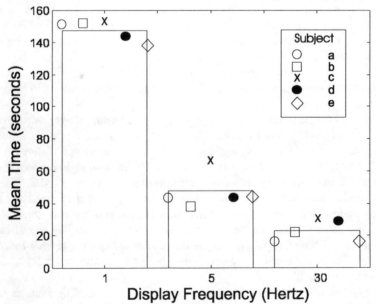

Figure 7 - Mean search time to find both dots (bars) and the mean search time for each of the subjects (symbols).

6. Discussion

This prototypical experiment shows, that at least for some tasks, high bandwidth is essential for good task performance. Subjects described "extreme frustration" when using the display to search the space with the filter set at the lowest cut-off frequency. It is interesting to note that although subject d moved at a much higher average velocity than any of the other 4 subjects during the 5 Hz frequency trials she took nearly the same amount of time to complete the task as the other subjects. This suggests that she did a large amount of wasted searching at high speeds when the display could not move the pins up and down quickly enough. During experimentation she was observed to pass directly over the top of the dots apparently without noticing on more than one occasion. This is a real concern in tasks like tumor localization where finding tumors accurately and quickly is a priority, because it will be very difficult to maintain a regular search pattern and ensure that no tumors will be missed. Thus, it is important that any tumors that may be encountered are felt the first time.

An interesting subjective result of our first studies with this display is that it is relatively easy to perceive curvature along the line of the display, but extremely difficult to perceive curvature perpendicular to it by sweeping the display. This may be because it is necessary to compare two points on the finger pad at the same time in order to have an impression of curvature. A future experiment will ask subjects to evaluate differences in curvature along the line of the display, and perpendicular to it to determine if this has an effect on performance.

7. Acknowledgments

This work was supported by a Whitaker Foundation Biomedical Engineering Research Grant and the Office of Naval Research under grant number N00014-92-J-1887.

8. References

[1] Howe, R.D., Peine W.J., Kontarinis, D.A., and Son, J.S. 1995. "Remote palpation technology," *IEEE Engineering in Medicine and Biology.* 14(3):318-323, May/June.

[2] Peine, W.J., Foucher, K.C. and Howe, R.D. 1997. Finger Speed During Single Digit Palpation. Accepted for publication in *Human Factors*.

[3] Johnson, K.O. and Phillips, J.R. Tactile Spatial Resolution. I. Two-Point discrimination, Gap Detection, Grating Resolution, and Letter Recognition. *Journal of Neurophysiology.* 46.6:1177-1191.

[4] Pawluk, D.T.V. and Howe, R.D. 1996. Dynamic Contact Mechanics of the Human Fingerpad, Part I : Lumped Response. *Submitted to the Journal of Biomechanical Engineering.* December, 1996.

[5] Hasser, C. and Weisenberger J. M. 1993. Preliminary evaluation of a shape memory alloy tactile feedback display. *Symp. Haptic Interfaces Virtual Env. Teleoperator Sys., ASME Winter Annual Meeting*, New Orleans.

[6] Taylor, P.M., Moser, A. and Creed, A. 1997. The Design and Control of a Tactile Display Based on Shape Memory Alloys. *Proceedings of the 1997 IEEE International Conference on Robotics and Automation.* Albuquerque, NM. 1317-1323.
[7] Kontarinis, D.A., Howe R.D. 1993. Tactile Display of Contact Shape in Dexterous Telemanipulation. *DSC-Vol. 49, Advances in Robotics, Mechatronics and Haptic Interfaces*. ASME, 49:81-88.
[8] Fischer, H. Neisus, B. and Trapp, R. 1995. Tactile Feedback for Endoscopic Surgery. *Interactive Technology and the New Paradigm for Healthcare.* K Morgan, R.M. Stava, H.B. Siburg et al. (Eds.) IOS Press and Ohmsha.
[9] Ikuta, K. Micro/miniature shape memory alloy actuator. 1990. *Proceedings of the IEEE International Conference on Robotics and Automation.* Cincinnati. 2156-2161.
[10] Ikuta, K., Takamoto, M. and Hirose, S. 1990. Mathematical Model and Experimental Verification of Shape Memory alloy for Designing Micro Actuator. *Proceedings. IEEE Micro Electro Mechanical Systems.* p.103-8, IEEE.
[11] Grant, D. and Hayward, V. 1997. Controller for a High Strain Shape Memory Alloy Actuator: Quenching of Limit Cycles. *Proceedings of the 1997 IEEE International Conference on Robotics and Automation.* Albuquerque, New Mexico. 254 : 259
[12] Kontarinis, D.A. and Howe, R.D. 1995. Tactile display of vibratory information in teleoperation and virtual environments. *Presence*, 4(4):387-402.
[13] Wellman, P. and Howe, R.D. 1995. Towards realistic vibrotactile display in virtual environments. *Symp. on Haptic Interfaces for Virtual Environment and Teleoperator Systems, Proc. of the ASME Dyn. Sys. and Control Div. of the ASME*, DSC 57.2:713-718.

Toward Dexterous Gaits and Hands

Susanna Leveroni (srl@alphatech.com)
Vinay Shah (vinay@ai.mit.edu)
Kenneth Salisbury (jks@ai.mit.edu)
Artificial Intelligence Laboratory
and
Department of Mechanical Engineering
Massachusetts Institute of Technology
Cambridge MA, USA

Abstract

This paper deals with planning and executing large-scale reorientations of grasped planar objects using a three-fingered robot hand. We begin by presenting a method for planning a gait, a series of finger motions and re-grasps, that can reorient any two-dimensional convex object by any desired amount. The implementation of these gait plans is then discussed, including the control algorithms that were used and how effective they were in achieving the desired orientation. We then discuss the limitations of our approach and present ideas to improve both the algorithms and the hardware. We conclude with a discussion of components of a new modular hand mechanism intended to support future work on spatial reorientation grasp gaits.

1. Introduction

In our previous work, we developed a grasp gait planner which generated a series of finger motions and stable regrasps suitable for reorienting a given object in the plane by an arbitrary amount. Here we discuss the implementation of these grasp gaits using a three-fingered robot system. We present the major components of our system, and then evaluate their effectiveness for achieving the desired object manipulation. The relevant kinematics of the hardware involved, how well the modeling assumptions made in the planning process were met, and the impact of unmodeled effects manipulation process is then reviewed. We discuss the method used to detect contacts by the fingers and how well these methods worked. The of cumulative errors which occur and cause the object to deviate from its intended path are then discussed. We discuss the algorithms developed to estimate actual object position relative to the expected position and how well these algorithms work. We discuss how they could be improved and what impact that would have on the effectiveness of the overall system. Finally, we present the method used to modify the gait to correct for small errors and the limits of this method for error correction. We discuss more complex methods for combining the planning process with the control algorithms so that larger errors can be corrected by on-line replanning.

In implementing the grasp gait algorithm it became clear that accurate contact sensing and force control were essential for executing the gait plans.

Indeed, the object localization and gait adjustment algorithms were added, in part, to correct errors that arose due to poor contact sensing and imprecise force control. This problem is been addressed by our current effort to design an intrinsically force-controllable finger.

The new finger is a three degree-of-freedom design that incorporates intentional mechanical compliance in the joints to mediate and enable measurement of applied forces and achieve better force controllability, rather than using tactile sensors on the fingertips or tension sensors on drive cables. The kinematics are similar to that of a human finger, with intersecting roll and pitch axes at the base and another pitch axis on the distal link.

The compliance is implemented by placing a custom-designed exponentially stiffening spring between each finger joint and its drive motor, producing several beneficial results. First, the exponential spring characteristics produce a large dynamic range of torques that can be sensed and applied, while maintaining a constant percentage resolution over the entire range. In addition, since joint torques are controlled indirectly via spring deflection, the torque control problem has been transformed into one of position control. For the same reason, small, gear-reduced DC motors have been used instead of exotic high performance actuators, without sacrificing sensitivity or force-controllability. In this paper, we present a performance analysis of this finger, including dynamic range and force resolution achieved by introducing compliance to the drive mechanism.

The main drawback of introducing compliance to the mechanism is decreased bandwidth, which limits the acceleration of the fingertip. However, dexterous manipulation tasks typically require slow, gradual movements that are well within the bandwidth of the system. Therefore, it is expected that decreased bandwidth does not affect the design's functionality. In this paper, we will compare the bandwidth of this finger with the compliant transmission to the bandwidth without the compliance. We will then present an empirical analysis of the limitations imposed by decreasing bandwidth.

The finger's modularity will allow hands to be constructed with varying placement and number of fingers. It is a goal of this research to implement the planar grasp gait algorithm described above with a 3-fingered hand, and then to extend the algorithm to include 4-fingered gaits as well as out-of-plane (spatial) rotations.

2. Planning Grasp Gaits

To continually reorient an object using a robot hand with finite finger workspaces, a series of finger grasps and object repositionings are required. An object *repositioning* is achieved by moving the robot fingers in a coordinated way while they are in contact with the object, such that the object moves in a desired way. A new finger *grasp* is achieved when one of the fingers on the object is lifted off and then replaced on the surface in a new location. One or more of these regrasps may be necessary before the object can be reoriented again. We call this sequence of object repositionings and regrasps a finger *gait*.

Given a particular object shape, an initial stable grasp, and a set of finger workspace limits, there is no guarantee that a new grasp will exist which will allow the object to be rotated further than it could be with just the initial grasp. With large coefficients of friction, large finger workspaces, and object of high symmetry, finding new grasps may be easy. However, without these simplifying assumptions, finding new grasps may be difficult or impossible. The existence of a new grasp depends on the given object and finger geometry as well as on the preceding series of motions leading to the current state. Therefore, finding gaits to reorient objects requires solving a planning problem. In [2] we develop a framework and a solution to the problem of finding two-fingered gaits to continually reorient two-dimensional objects in the plane. In [3] we discuss the implementation of many of these gaits to reorient a variety of objects using a three-fingered robot hand. In Section 3 we summarize this approach. In Section 4 we then discuss the results of our approach, considering for which cases it was sufficient and for which cases modeling simplifications or errors led to grasp failures. Finally, we consider what improvements could be made which would lead to more robust gait implementation.

3. Implementing Gaits

To implement these gaits, we use the control architecture shown in Figure 1, summarized below. This method is discussed more fully in [3] and in [4]. The inputs to the gait planner are the object geometry, finger workspace limits, the coefficient of friction between fingers and object, and a desired reorientation amount. The output is a series of finger motions, finger grasps and grip forces which comprise the gait. Position (PD) control with a feed-forward grip force is used to carry out these motions and forces. Contact detection is estimated by setting a threshold for the commanded servo force, above the amount required to move the fingers through free space. When this threshold force is exceeded, contact is assumed to have occured. Above this basic architecture, we add the following features, in order to close the loop at the action level. We use a minimum error estimator to estimate the actual object position, using the three contact points given by the fingers as three known points on the object. We then search a local space near the expected position of the object to find the position which gives minimum error assuming all three fingers are in contact. We then correct finger motions to account for the change in object position. We also correct for grip force errors, by increasing the commanded forces to remain above the grip force threshold and also to lie within the friction cones of the fingers. Therefore, if the object is displaced from its expected position by an external force (which occurs if one tries to pull the object out of the robots grip), the grip force is increased, resisting slip and also maintaining finger contact at all points.

4. Gait Results

The control algorithms described above were successful in implementing a variety of gaits generated for a number of different convex objects. Among the objects tested were a circle, five and six sided polygons, and ellipses of varying eccentricity. These objects were reoriented 360 degrees using several dif-

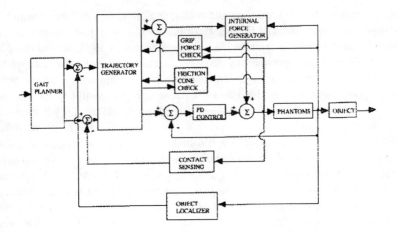

Figure 1. Gait Controller Architecture.

ferent gait types. During these reorientations, the objects were pushed and pulled slightly (introducing unmodeled disturbance forces, and the grip force was appropriately adjusted to avoid slipping. As the disturbance forces were increased, causing the objects to slip out of position, the object detection algorithm was successful in detecting and quantifying the position errors, and the gait adjustment algorithm was successful in correcting them.

However, there were also a subset of cases where errors were introduced, either intentionally or through unmodeled kinematics and dynamics, and the control algorithms were insufficient in either detecting or correcting them for them. Below we discuss some of the known unmodeled kinematics and when their effects they may lead to gait failure. We also speculate about the cause of other possible errors which may lead to gait failure. In both cases we suggest ways in which these errors may be corrected or mitigated.

4.1. Known Modeling Errors

4.1.1. Contact location of fingers on object

The point of contact between the object and a finger can be specified by an angle and a radius for both the object and the finger. Given the position and orientation of both the object and the finger, we approximate the point of contact by conceptually growing the object by the finger's radius and shrinking the finger to a point. This is an approximation, whose error depends on the geometry of the object at the point of contact. If a given force is then applied by the finger to the object, the direction of the force is correct, but the point of application is not. For a given grasp, this may introduce a small unmodeled moment applied to the object, which may lead to the object settling in at a new equilibrium, or it may cause the grasp to become unstable. In addition, the deviation of the actual contact locations from the expected locations may cause the grasp to be one that is not force closure, where no set of forces exist such that they can be applied to keep the object in equilibrium. This could occur for gaits which were not conservative, in that they required grasps which were close to the boundary of the stable regions in the grasp map. For gaits which were more conservative, new grasps would still remain force closure, even though the correct forces were not applied. These errors could be corrected by solving more accurately for the correct contact location on the object, whereby the direction of the applied force could be adjusted to maintain equilibrium.

4.1.2. Rolling

Another known modeling error results from assuming that the contact points are stationary relative to the object during reorientations. This would occur if the fingertips were points, or if the fingertips had some shape but rotated by the same amount about there centroid as their center rotated about the center of the object rotation. In our case the fingertips were spheres and rotated as they moved in a direction that exacerbated the rolling problem. Therefore a grasp that began as force closure could roll outside of the stable region leading to grasp failure.

4.1.3. Adjustment to plan

Errors were also known to persist within the gait adjustment algorithm. These errors were, more specifically, a failure to fully correct for known object position errors. The gait adjustment algorithm detects the deviation between the expected and actual object position. The commanded finger position for the next finger to be placed on the object is then adjusted so that when it makes contact, it will touch the object in the desired location on the object, rather than at a spurious one, which would occur if the object position error was not taken into account. Fingers that are already in contact are assumed to be at the correct contact location. Given this assumption, even with a perfect object localization algorithm, if previous finger placement errors or slipping occured, the new grasp could possibly not be force closure, since previous fingers placed might not be in their expected location. This could be corrected if actual contact location for all fingers are solved for once object location is determined, and then the gait is readjusted for all the fingers, either bringing the present grasp back to the expected one, or modifying the gait such that it converges back to the original plan after several grasps.

4.1.4. Object Localization

Errors in the object localization algorithm are also known to persist. We assume that there is some noise in finger position measurements relative to the object. These may be due to encoder errors or unmodeled compression of the fingers or object. Therefore, when the object localization algorithm finds the local minimum in the error metric, the estimated object position is updated using a variable filter, which is dependent on the sensitivity of the algorithm to displacements in orthogonal directions. Therefore, even if the true object position is found by the minimization scheme, the estimated object position will only approach the true object position as the sensitivity of the algorithm approaches infinity.

4.2. Possible Additional Errors

In addition to the errors discussed above, which result from known modeling approximations, other errors may occur as well. Calibration errors may occur for each of the fingers, which will cause the transformations between finger reference frames and the common object frame to be inaccurate. This would lead to both spurious finger placements on the object as well as errors in the direction of the forces applied.

5. A Compliant Robot Hand

An effort is currently underway in our lab to design an intrinsically compliant and force-controllable robot hand capable of performing dexterous manipulation tasks, such as the grasp gaits described above. Studies of human hand function and performance, as well as an implementation of the grasp gait algorithm with the PHANToM [6] haptic interface device, have revealed several essential characteristics for a dexterous manipulator. Key among these are the ability to sense and exert a large dynamic range of forces, an intrinsically variable joint stiffness, precise fingertip force control, and accurate contact sensing.

Human fingers appear to achieve their great dexterity through an extremely large dynamic range and an intrinsically variable compliance. The dynamic range, defined to be the ratio of the largest to smallest exertable or sensible force, of a human finger has been determined to be on the order of 10000:1, [12]. In contrast, a very good multi-DOF robot like the PHANToM is limited to a dynamic range of only about 100:1 [7], a difference of 2 orders of magnitude. Furthermore, humans have a nearly constant percentage resolution of force perception and exertion throughout this entire range [8]. That is, humans can perceive (or exert) changes in force of around 10% [1], regardless of whether the forces are very small or very large. This gives human fingers extremely fine resolution when forces are very small, but also adequate resolution when forces are large. In addition, humans are able to modulate the intrinsic compliance of their fingers by co-activation of muscles across joints or between fingers, enabling them to achieve a widely-varying grasp stiffness. It is believed that these key characteristics help to make human hands far superior to any existing mechanical hand.

In implementing the grasp gait algorithm with the PHANToM haptic interface device, it became clear that accurate force control and contact sensing are essential for performing dexterous manipulation tasks [4]. Inadequacies in these areas caused the object's motion to deviate from the intended path during the gait. Imprecise application of contact forces caused the object to be pushed slightly away from the desired position. The same type of displacement occurred due to inadequate contact sensing. Object localization and gait adjustment algorithms were added to the grasp planner to increase robustness to small errors of these types; however, grasp failures were not eliminated.

The robot hand we are currently designing will be better suited for performing grasp gaits and other dexterous manipulation tasks. It will be composed of identical, modular, intrinsically compliant, 3-DOF fingers which each possess the essential traits described above. The modularity of the fingers decreases their per unit cost, while increasing flexibility. Each finger can be used singly for palpation and perception experiments, as well as in combinations of variable number and arrangement of fingers for grasping and grasp gait experiments.

At the heart of each finger are custom-designed, nonlinear compliant elements that provide the intrinsic mechanical elasticity and help the finger achieve the desired performance characteristics. Our work compliments and extends the results of other researchers ([9], [13] [11]) who have achieved performance enhancements using linear compliance.

6. A Compliant Element

The compliant element is a compact, exponentially-stiffening spring that is placed between each drive motor and its associated finger joint, producing a non-linearly compliant transmission. The deflection of this transmission determines the torque applied to each joint; therefore, precise force control can be achieved by accurately controlling the deflection of the compliant element. In this way, the force control problem has been transformed into the well-understood problem of position control.

The compliant transmission also imparts the essential performance characteristics of dynamic range, variable stiffness, force control, and contact sensing. The large dynamic range is achieved through the exponentially-stiffening nature of the spring, since a huge range of output torque is possible for a given displacement range without compromising torque resolution. In addition, a stiffening spring allows modulation of grasp stiffness by varying the contact forces, since the spring's stiffness changes with torque. Finally, excellent contact sensing can be achieved because a very small change in force at low force levels produces a relatively large, and easily detectable, change in spring deflection.

6.1. Design of Compliant Element

The compliant element is an elastic mechanism designed to have the desired exponential stiffening characteristic, and to be easily instrumentable, compact, and durable. It consists of two parallel aluminum disks with short, radial walls extending from each disk, and 1/16"D buna rubber balls that fill the resulting pie-shaped pockets between the disks. Relative motion between the disks causes the rubber balls to be compressed between the walls, providing the non-linear elasticity. As the balls are initially compressed, the contact area between each ball and the surrounding walls is extremely small producing an very large compressive stress, and therefore a quite small effective stiffness. As the ball deforms further, the contact area increases, thereby decreasing the stress, increasing the effective stiffness, and producing the nearly exponential torque/displacement relationship shown in Figure 2. The spring's displacement, and therefore the applied torque, is directly measured by an integrated potentiometer. The instrumented mechanism is .75"D and .50" thick, contains only one moving part, and is composed of durable and inexpensive aluminum and rubber components.

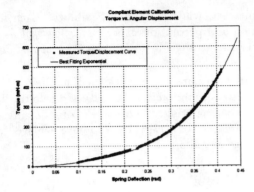

Figure 2. Torque vs. angular deflection of the compliant element: actual data points and the best fitting exponential curve.

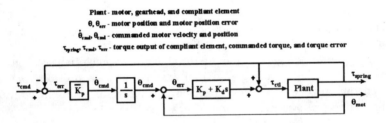

Figure 3. Torque controller consisting of an inner velocity loop and an outer loop that commands velocities proportional to the torque error.

6.2. Performance of Compliant Element

To measure the performance improvements that the compliant element provides, a simple single-axis actuator, consisting of a motor and a compliant element, was constructed to simulate the finger in contact with a rigid object. In this system, the output side of the compliant element was held fixed, while the input side was connected to a Maxon RE016 DC motor with an 84.3:1 gear-reduction and a rear-mounted Hewlett Packard optical encoder. A torque controller, whose block diagram is shown in Figure 3, was implemented. This control structure, inspired by [10], closes a high-gain velocity control loop around the motor's motion, taking advantage of the clean and collocated nature of its encoder. The response of this system to a step and sinusoidal input are shown in Figure 4. Experiments were then performed to determine the dynamic range, torque resolution, and bandwidth of the system.

The torque dynamic range of this actuator was increased from 35 to 1000, a factor of almost 30, when the compliant element was used. This increase could

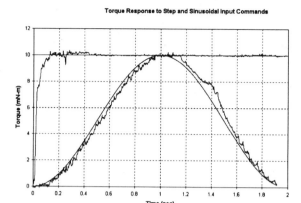

Figure 4. Torque output in response to step and sinusoidal input commands.

only be accomplished by decreasing the minimum torque that can be applied, since the maximum torque is fixed by the maximum allowable motor current. The minimum torque is normally equal to the friction torque in the gear head (15mN-m). However, when the compliant element is used, the minimum torque is limited only by the resolution of the sensor and backlash in the gear head, since motor position control can eliminate the effects of friction. The minimum torque has been reduced to .5mN-m by using the compliant element.

The torque resolution is governed by the resolution of the potentiometer that measures the deflection of the compliant element. This potentiometer is accurate to within .25deg, which corresponds to a 6.5% torque resolution for low torques and 4.8% for high torques. This percentage resolution is not exactly constant because the spring curve is not a pure exponential; the torque is zero at zero displacement, so the curve is approximately an exponential minus a constant.

The bandwidth of this actuator, which is directly related to the stiffness of the transmission, is nearly one decade lower, as shown in Figure 5, due to the added flexibility of the compliant element. However, the stiffness of the compliant element increases with the applied torque, and so the bandwidth can actually increase as the applied torque rises. This allows modulation of grasp stiffness, by changing the grasp forces. It is expected that the decrease in bandwidth will not have an adverse effect on grasping performance, since grasping and grasp gaits generally do not require rapid, high acceleration movements.

7. Finger Design

The finger, shown in Figure 6, was designed to be kinematically similar to a human finger, but with one fewer degree of freedom and a larger overall range of motion. As with human fingers, this mechanical finger has (nearly) intersecting roll and pitch axes at the base joint, and another pitch axis on the distal link. Unlike human fingers, however, the third pitch axis and associated

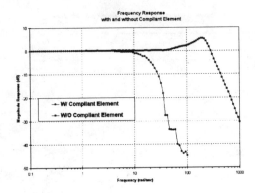

Figure 5. Frequency response of the actuator with and without the compliant element.

Figure 6. Preliminary design of a modular, 3-DOF robot finger.

link has been omitted from the design because we believe that this freedom is not vital for fingertip manipulation. Furthermore, the finger was designed to have 180 degrees of motion on the two base axes and 270 degrees on the third axis, compared to only 30 degrees on the base roll axis and 90 degrees on the two pitch axes for a human finger. The larger range of motion increases the workspace of the finger, which can simplify object reorientations by allowing larger rotations before a regrasp is required.

The link lengths were selected so that a locus of isotropic points (points where the Jacobian condition number is 1) would exist in a desirable location within the workspace (as was done in [5]). The three joints are powered by three of the motor/compliant element pairs that were used in the single axis experiments. These actuators are removed from the finger to reduce the mass and size of the links; instead, they are located directly below the base link, and power is transmitted to the joints via steel cables running over pulleys. This cable drive causes cross-axis coupling between the motor and joint motions and adds a small amount of additional flexibility and friction to the transmission.

The finger will be controlled by combining three of the single-axis controllers developed in Section 6.2 and inverting the cross-axis coupling introduced by the cable drive. It is expected that performance of this finger, as measured by dynamic range, force resolution, force control, and contact detection, will far exceed the performance using the customary linearly stiff transmissions.

8. Conclusions

Our development and implementation of grasp gait planning and control techniques has shown the viability of our approach. It also pointed up the need for intrinsically higher performance finger mechanisms. To address this an exponentially stiffening compliant element has been designed and constructed to create a nonlinear elastic transmission for use in our new fingers. Use of this compliant element has been shown to drastically enhance the actuation system's performance as measured by torque dynamic range, torque resolution, torque control, and contact sensing. This actuation system forms the heart of a modular, 3-DOF robot finger currently being designed for complex finger-based manipulation tasks, such as grasp gaits. These tasks require excellent performance in the metrics mentioned above. In the future we expect apply these new fingers to the implementation of 3 and 4 finger gaits for reorienting objects.

Acknowledgments

The authors would like to gratefully acknowledge the financial support of the Office of Naval Research, University Research Initiative Program, Grant N00014-92-J-1814.

References

[1] Clark, F.J., Horch, K.W. "Kinesthesia, vol. 1 of Handbook of perception and human performance - sensory processes and perception." Wiley and Sons, New York, NY, 1986.

[2] Leveroni, Susanna and K. Salisbury, "Reorienting Objects with a Robot Hand Using Grasp Gaits," *Proceedings of the 7th International Symposium on Robotics Research*, Munich, Germany, Oct 1995. Springer-Verlag.

[3] Leveroni, S.R. and J.K. Salisbury, "Cooperative Control of Multiple Robots to Manipulate Objects," Proceedings of SPIE Photonics East '96 Symposium on Sensor Fusion and Distributed Robotic Agents, Boston, Nov 1996.

[4] Leveroni, Susanna R., "Grasp Gaits for Planar Object Manipulation," PhD thesis, Department of Mechanical Engineering, MIT. Sep 1996.

[5] Mason, Matthew T., Salisbury, J. Kenneth. "Robot hands and the mechanics of manipulation." MIT Press, Cambridge, MA, 1985.

[6] Massie, Thomas. "Design of a 3-DOF force reflecting haptic interface". B.S. Thesis, MIT Department of Electrical Engineering and Computer Science, 1993.

[7] Morrell, John B., Salisbury, J. Kenneth. "In pursuit of dynamic range: using parallel coupled actuators to overcome hardware limitations". Proc. 4th International Symposium on Experimental Robotics, pp. 263-273, 1995.

[8] Pang, X.D., Tan, H.Z., Durlach, N.I. "Manual discrimination of force using active finger motion". Perception and Psychophysics, vol 49, no. 6, pp. 531-540, 1991.

[9] Pratt, Gill, Williamson, Matt, ea. "Stiffness isn't everything". Proc. 4th International Symposium on Experimental Robotics, pp. 253-262, 1995.

[10] Salisbury, J. Kenneth. "Design and control of an articulated hand." Proc. International Symposium on Design and Synthesis, pp. 459-466, 1984.

[11] Spong, M.W. "Modeling and control of elastic joint robots." Journal of Dynamic Systems, Measurement, and Control, v. 109, no. 4, pp. 310-319, 1987.

[12] Srinivasan, Mandayam A., Chen, Jyh-shing. "Human performance in controlling normal forces of contact with rigid objects". ASME Dynamic Systems and Control: Advances in Robotics, Mechatronics, and Haptic Interfaces, DSC-49, pp. 119-125, 1993.

[13] Sugano, S., Tsuto, S., Kato, I. "Force control of the robot finger equipped with a mechanical compliance adjuster." Proc IEEE/RSJ Conference on Intelligent Robots and Systems, pp. 2005-2013, 1992.

Dexterous Manipulations of Humanoid Robot *Saika*

Atsushi Konno, Koichi Nishiwaki, Ryo Furukawa, Mitsunori Tada,
Koichi Nagashima, Masayuki Inaba and Hirochika Inoue
Department of Mechano-Informatics, University of Tokyo
7-3-1 Hongo, Bunkyo-ku, Tokyo 113, JAPAN
E-mail: konno@jsk.t.u-tokyo.ac.jp

Abstract: This article addresses the development of a humanoid robot named *Saika* and skillful manipulations performed by *Saika*. The developed humanoid robot *Saika* has a two-DOF neck, dual five-DOF upper arms, a torso and a head which consists of two eyes and two ears. *Saika* has human-size dimension and weighs eight kilograms. This article also presents three kinds of manipulations performed by *Saika*: (1) hitting a bounding ball, (2) grasping unknown objects by groping and (3) catching a thrown ball. Those tasks are chosen to study behavior-based movement control and intelligence.

1. Introduction

Humanoid robots of the same size and figure as human beings' are considered one of the forms of the robots of the coming generation. Such humanoid robots have an advantage in handling tools in the same way as human beings handle them. Furthermore, due to their human-like bodily form, there is a possibility of learning skills by imitating human behavior. Also in the field of brain science, humanoid robots would bring inspiration to the researchers. Therefore, some researchers have noticed the importance of humanoid robots and expected them to be good research platforms.

The first humanoid robot in the world, *WABOT-1* (WAseda roBOT 1), was developed in 1973 by Kato and others [1]. *WABOT-1* had a human-like form with two legs and two arms. In 1985, the robot musician *WABOT-2* was developed [2] and was displayed in a world exhibition held in Tsukuba JAPAN. Brooks hypothesized that humanoid intelligence required humanoid interactions with the world, and developed the humanoid *Cog* to verify the hypothesis [3, 4]. Hollerbach and Jacobsen have developed a lot of interesting humanoid robots such as Navy Teleoperated System, the Disney robots and the Ford robot over the past few decades [5].

Although a lot of humanoid robots have been developed so far, humanoid robots that are lighter, less expensive and easier to handle are expected to be developed. This article addresses the development of a low-cost, light-weight humanoid robot named *Saika* ("outstanding intelligence" in Japanese), and presents skillful manipulations done by *Saika*: (1) hitting a bounding ball [6], (2) grasping unknown objects by groping [7] and (3) catching a thrown ball [8].

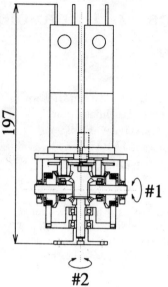

Table 1. Specification of the developed joint module.

DOF	2
Weight	1.01 [Kg]
Gear ratio	Bevel gear 1:2
	Motor gear head 1:99.7
Rated torque	100 [Kgcm] for each DOF
Sensor	Multi-turn potentiometer
Motor driver	Titech Robot Driver [9]
Reachable region	#1: ±100° #2: ±240°

Figure 1. Mechanism of the developed joint module.

Those manipulations are chosen as examples that require a reactive planning and the ability to behave in the real world.

2. Humanoid Robot *Saika*
2.1. A two-DOF joint module

To reduce the developing cost and to make maintenance easy, the humanoid robot *Saika* is modularized. A compact and light-weight two-DOF joint module was developed in advance.

Figure 1 shows the mechanical design of the developed joint module. A differential gear mechanism is used to produce the rotations #1 and #2 as shown in Figure 1. A differential gear mechanism brings efficient motor driving, which makes it possible to decrease the size of actuator. Rotations of motor axes are sensed by five-turn type potentiometers, and are used to compute the joint angles #1 and #2. Specification of the developed servo module is presented in Table 1.

2.2. Mechanical Design of the humanoid robot *Saika*

To weaken pressure and fear that human beings feel against the humanoid robot, the humanoid robot is designed to be as light as possible given the dimension similar to a human's. Figure 2 shows the specified joints-location and the size of the humanoid robot *Saika*. Two-DOF joint modules presented in Section 2.1 are used in the places circled by dotted lines shown in Figure 2.

Due to the design of the two-DOF joint module, it becomes possible to install the DC motors inside the arms and the torso of *Saika*. Existence of motors does not hinder the robot from performing tasks.

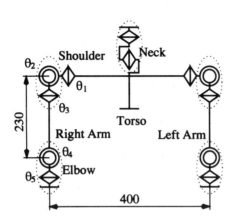

Figure 2. Location of degrees-of-freedom of *Saika*.

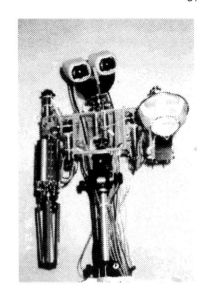

Figure 3. Humanoid robot *Saika*.

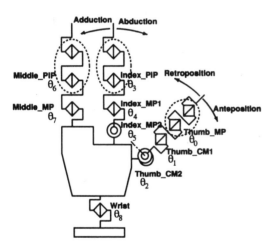

Figure 4. Distribution of degrees-of-freedom of the robot hand.

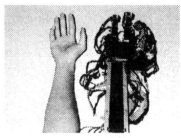

Figure 5. An appearance of the developed hand.

An appearance of *Saika* is shown in Figure 3. In Figure 3, a bowl is attached to the *Saika*'s left forearm to perform the catch a ball task.

2.3. A Three-Fingered Hand

Considering the movement of human fingers, the distribution of DOFs of the robot hand is designed as shown in Figure 4. The joints circled by dotted lines are coupled. Figure 5 shows an appearance of the developed three-fingered robot hand. Eighty-seven touch sensors are distributed over the surface of the

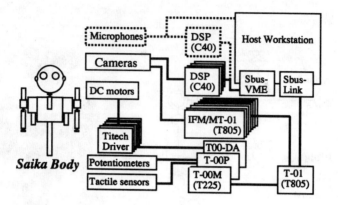

Figure 6. The transputer-based control system.

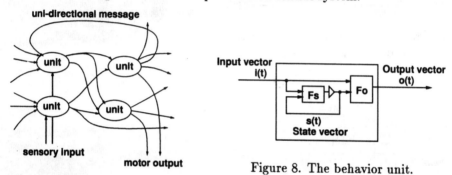

Figure 7. BeNet model (the figure is from [11]).

Figure 8. The behavior unit.

palm and the fingers.

2.4. Control system

The control system consists of transputer(T805)-based vision system (JSK-IFM/MT-01 [10]), transputer(T225)-based motor control system and transputer networks. A host workstation (SUN Ultra Sparc 1) is used for booting transputers, human interface programming and upper level programming. Communication between the host workstation and the transputers is handled via a Sbus-Link card (Planetron).

The control system is outlined in Figure 6. The parts of the system indicated by dotted lines are still under development.

2.5. Asynchronous Parallel Process Networks

Oka has proposed the network model of asynchronous parallel processes named "BeNet" for robots' brain [11] (Figure 7).

Saika's control programs are developed on the basis of the BeNet. This section briefly describes BeNet. Please see the reference [11] for more details.

The BeNet is a network of BUs (Behavior Units). A BU consists of an input message vector $i(t)$, an output message vector $o(t)$ and a state variable

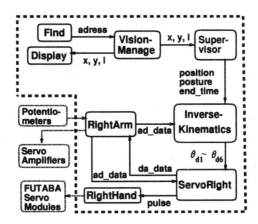

Figure 9. The BeNet for hitting a bounding ball.

vector $s(t)$ as shown in Figure 8. The output message vector $o(t)$ and the state variable vector $s(t)$ are computed by following equations;

$$s(t + \Delta t) = F_s(i(t), s(t)) \quad \text{and} \tag{1}$$

$$o(t + \Delta t) = F_o(i(t), s(t)), \tag{2}$$

where sampling period Δt and functions F_s and F_o of the BU are given by the programmer.

In Saika's control programs BUs are placed both on the host-workstation and on the transputers. Communication between BUs takes place through router programs that is running both on the host-workstation and on each transputer [12].

3. The Behavior of *Saika*'s Hitting a Bounding Ball

This section presents the behavior of *Saika*'s hitting a bounding ball. Figure 9 shows the BeNet for the behavior.

For simplicity, a blue-colored ball is used while the background is white. The diameter of the ball is given in advance (the diameter of the ball used in the experiment was 7 [cm]). Only the left eye is used to find the ball. For 3D localization of the ball in the world coordinate system, the BU does not use stereo visual information but use the diameter of the ball.

Three BUs *Find*, *Vision-Manage* and *Supervisor* are programmed to find and localize a ball. The BU *Find* will find the ball and output the left and right edges of the ball to the BU *Vision-Manage*. Sampling points are located at each of the 16 pixels in the view image (Figure 10). The BU *Find* investigates the brightness of each sampling point. If the brightness of a sampling point is less than a given threshold, the BU *Find* judges that the sampling point is on the ball's image, and investigates the surrounding area of the sampling point to detect the left and right edges of the ball. In the case where the BU cannot detect the left or right edge of the ball or the sampling points whose brightness

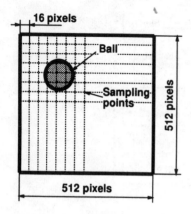
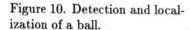
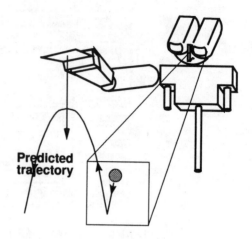

Figure 10. Detection and localization of a ball.

Figure 11. Generation of the hand's trajectory.

is less than the given threshold, the BU *Find* waits until the next view image is obtained.

Once the BU *Vision-Manage* receives the left and right edges of the ball from the BU *Find*, the BU *Vision-Manage* will compute the center x, y and diameter l of the ball in the camera image. The BU *Supervisor* will localize the ball in the world coordinate system using x, y and l received from the BU *Vision-Manage*. The BU *Supervisor* predicts the ball's path in the world coordinate system from the current and previous positions of the ball. If the top of the predicted path is in the reachable area of the humanoid robot *Saika*, the BU *Supervisor* produces the hand's trajectory which intersects the top of the ball's predicted path (Figure 11).

The behavior of *Saika*'s hitting a bounding ball is shown in Figure 12. For clarity, the ball is circled by solid lines in the pictures.

4. Grasping Unknown Objects by Groping

A strategy for grasping unknown objects by groping is discussed in this section. The aim of the strategy is to achieve an artificial power grip without having any models of the object. The strategy is composed of two behavioral processes: (I) active abduction or adduction of the thumb and the index finger to find the optimal grasping configuration that maximizes a performance index $J(\theta_i)$, and (II) mild flexion round the object. Both behavioral processes of (I) and (II) utilize the touch sensory information.

Two heuristics are considered here: (a) grasp so as to maximize the contact between the fingers and the object (Figure 13 (a)), and (b) grasp so as to maximize the fingers' flexion (Figure 13 (b)). Considering these heuristics, a performance index $J(\theta_i)$ is given by:

$$J(\theta_i(t)) = N(\theta_i(t)) + \alpha(\theta_{i-1}^{contact}(t) - \theta_{i-1}^{div}) \quad (i = 2, 5). \quad (3)$$

The subscript i takes either 2 or 5, where $\theta_2(t)$ and $\theta_5(t)$ correspond to the

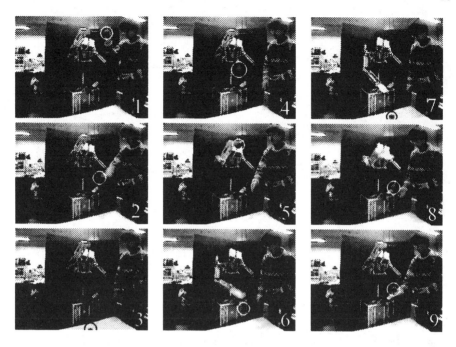

Figure 12. The behavior of *Saika*'s hitting a ball.

abduction/adduction of the thumb and the index finger, respectively, while $\theta_1(t)$ and $\theta_4(t)$ correspond to the flexion. $N(\theta_i(t))$ $(6 \geq N(\theta_i(t)) \geq 0)$ is the number of contact points obtained by the set of touch sensors when $\theta_i = \theta_i(t)$. $\theta_1^{contact}(t)$ and $\theta_4^{contact}(t)$ are the phalangeal joint angles of Thumb_CM1 and Index_MP1 (see Figure 4) when the control system detects the contacts between the object and fingers. A contact is detected by seeing the touch sensory information. However, if the touch sensors do not detect the contact with the object, the system judges whether there are contacts or not from the difference between the reference position command θ_{i-1}^{ref} and current position θ_{i-1} detected by the potentiometer. The threshold $\theta_{i-1}^{threshold}$ is set to be 10 [°] in the experiment. θ_{i-1}^{div} is the pre-shaping divergence configuration before grasping and is given here by

$$[\theta_1^{div} \quad \theta_4^{div}] = [-40 \quad 0] \; [°]. \tag{4}$$

α in eq. (3) is a weight to give a priority to estimation. α is set to be 1/90 that gives the number of contact points high priority (because the second term of the right hand side of eq. (3) becomes less than 1.0).

In this strategy, after finding the configurations of θ_2 and θ_5 that maximize the performance index $J(\theta_i(t))$ given by eq. (3), mild flexion round the object is performed. To clarify the discussion, the strategy is illustrated in the flowchart (Figure 14).

A can, a ball, a cone, a plate, and a cube are chosen as unknown objects for the hand to grasp in the experiments. Once the robot hand detects contact

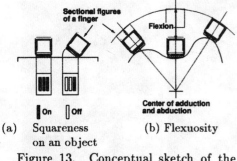

(a) Squareness on an object   (b) Flexuosity

Figure 13. Conceptual sketch of the grasping strategy.

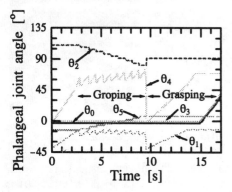

Figure 15. Joint trajectory of fingers.

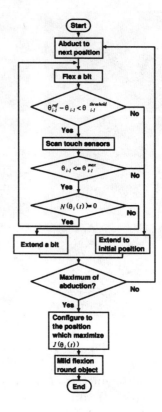

Figure 14. Strategy flowchart.

with an object, it starts grasping after groping. Figure 15 plots the phalangeal joint angle trajectories while the hand is groping and grasping a can. Saw-edged trajectories of θ_1 and θ_4 indicate groping.

5. The Behavior of *Saika*'s Catching a Thrown Ball

This section presents the behavior of *Saika*'s catching a thrown ball. The ball is thrown two meters away from the humanoid. A yellow-colored ball is used. The same algorithm described in Section 3 is used to find and localize the ball.

5.1. The Ball's Path Prediction

The thrown ball's path is projected on the x-z plane and y-z plane (see Figure 16). The ball's path in the x-z plane is approximated by a first-order-equation, while the path in the y-z plane is approximated by a second-order-equation.

The ith sampled ball's position in the world coordinate system is denoted by (x_i, y_i, z_i). In the prediction of the ball's path, higher priority is given to the more fresh data introducing the weighting coefficient p_i for (x_i, y_i, z_i).

The ball's path in the x-z plane is approximated by

$$x = a_0 + a_1 z. \tag{5}$$

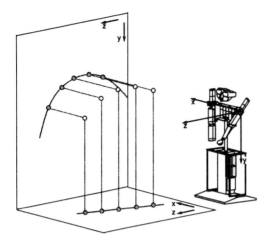

Figure 16. Projection of the ball's path on the x-z plane and y-z plane.

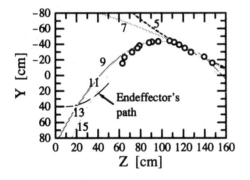

Figure 17. Predicted path from the data.

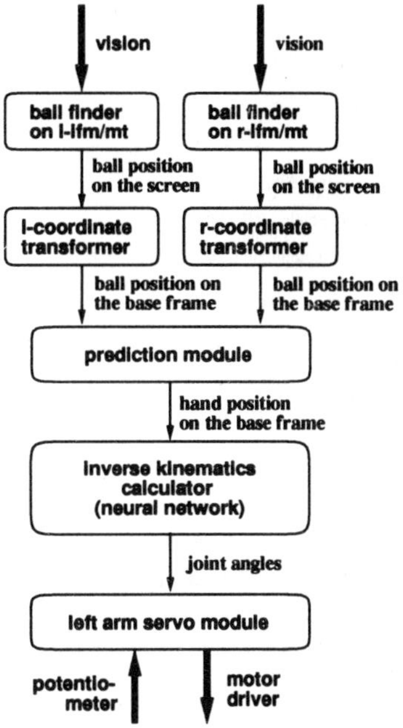

Figure 18. The BeNet for catching a thrown ball.

Given the n sampled ball's position data, the coefficients a_0 and a_1 are determined so as to minimize the sum of the square error E_{xz} defined by the following equation:

$$E_{xz} = \sum_{i=1}^{n} p_i (a_0 + a_1 z_i - x_i)^2. \tag{6}$$

By solving $\partial E_{xz}/\partial a_0 = \partial E_{xz}/\partial a_1 = 0$, a_0 and a_1 are given by:

$$a_1 = \frac{S_x^n S_z^n - S_0^n S_{xz}^n}{(S_z^n)^2 - S_0^n S_{z^2}^n}, \quad a_0 = \frac{S_{xz}^n - a_1 S_{z^2}^n}{S_z^n}, \tag{7}$$

where

$$S_0^n = \sum_{i=1}^{n} p_i, \quad S_x^n = \sum_{i=1}^{n} p_i x_i, \quad S_z^n = \sum_{i=1}^{n} p_i z_i,$$

$$S_{z^2}^n = \sum_{i=1}^n p_i z_i^2, \quad S_{xz}^n = \sum_{i=1}^n p_i x_i z_i.$$

If the weighting coefficient p_i is given here by:

$$p_i = r^{n-1} \quad (0 < r \leq 1), \tag{8}$$

then S_0^n, S_x^n, $\cdots$ can be expressed by the following recurrent formulas:

$$S_0^n = 1 + rS_0^{n-1}, \quad S_x^n = x_n + rS_x^{n-1}, \quad \cdots.$$

The ball's path in the y-z plane is predicted in the same way. The path is approximated by:

$$y = b_0 + b_1 z + b_2 z^2. \tag{9}$$

The coefficients b_0, b_1 and b_2 are determined so as to minimize the sum of the square error E_{yz} defined by the following equation:

$$E_{yz} = \sum_{i=1}^n p_i (b_0 + b_1 z_i + b_2 z_i^2 - y_i)^2. \tag{10}$$

The correction of the ball's path prediction is done when the ball's fresh position data is sampled. Figure 17 shows the correction process in the ball's path prediction in the y-z plane. The dotted line by the number '5' in Figure 17 shows the ball's path predicted from the first five sampled data of ball's position. In the same way, the curve by the number '15' shows the path predicted from the fifteen sampled data.

5.2. Positioning the End-effector

Considering the workspace of the humanoid robot *Saika*, the end-effector's path is determined in a heuristic way and given by

$$y = 40 - 0.01875(z - 10)^2, \tag{11}$$

which is plotted in Figure 17.

Two intersections are obtained by solving eqs. (9) and (11). However, the smaller solution is chosen as the reference position for the end-effector. The solution and eq. (5) give the end-effector's reference x position.

5.3. BeNet for the Task of Catching a Thrown Ball

Figure 18 shows the BeNet for the behavior of catching a thrown ball. Both of the cameras are fixed. To make the total field of view wide, the left and right cameras are set to see the different fields of view. The thrown ball is detected in the BUs *ball_finder_on_l-ifm/mt* or *ball_finder_on_r-ifm/mt*, and localized in the corresponding BUs *l-coordinate_transformer* or *r-coordinate_transformer*. The BU *prediction_module* predicts the ball's path and outputs the reference end-effector's position for catching the ball to the BU *inverse_kinematics_calculator*. An inverse kinematics neural network model was acquired in advance in the BU *inverse_kinematics_calculator*. The arm of the humanoid *Saika* is controlled by the conventional PD feedback controller.

The behavior of *Saika*'s catching a thrown ball is shown in Figure 19. For clarity, the ball is circled by solid lines in the pictures.

Figure 19. The behavior of *Saika*'s catching a thrown ball.

6. Conclusion

A light-weight low-cost humanoid robot *Saika* was developed. The developed humanoid could be a good research tool for robotics, artificial intelligence and brain science.

Three tasks were implemented to study the reactive planning and behaviors and to demonstrate the performance of the developed humanoid. The tasks were: (1) hitting a bounding ball, (2) grasping unknown objects by groping and (3) catching a thrown ball. Programs are developed on the basis of the BeNet, a network model of asynchronous parallel processes.

Next research topic using *Saika* would be interactions with human such as learning by imitation, playing games and non-verbal communication.

Acknowledgment

This work is supported by the Japan Society for the Promotion of Science Grant for Research For The Future JSPS-RFTF96P00801, and The Ministry of Education, Science, Sports and Culture Grant-in-Aid for Developmental

Scientific Research 06555069.

References

[1] Kato I 1973 Development of WABOT-1. *Biomechanism 2*. The University of Tokyo Press, Tokyo pp 173–214 (In Japanese)

[2] Kato I, Ohteru S, Shirai K, Narita S 1985 A Robot Musician "WABOT-2" (WAseda roBOT-2). *Journal of the Robotic Society of Japan*. 3(4):337–338 (In Japanese).

[3] Brooks R A, Stein L A 1994 Building Brains for Bodies. *Autonomous Robots*. 1:7–25

[4] Brooks R A 1996 Prospects for Human Level Intelligence for Humanoid Robots. *Proc. of the First Int. Symp. on HUmanoid RObots (HURO '96)*. pp 17–24

[5] Hollerbach J M, Jacobsen S C Anthropomorphic Robots and Human Interactions. *Proc. of the First Int. Symp. on HUmanoid RObots (HURO '96)*. pp 83–91

[6] Furukawa R 1997 Study on learning of a robot arm's movement based on visual information. Graduation Thesis, University of Tokyo

[7] Tada M 1997 Study on the grasping unknown objects using multi-fingered robot hand. Graduation Thesis, University of Tokyo

[8] Nishiwaki K 1997 Study on the catch a ball behavior using a humanoid. Graduation Thesis, University of Tokyo

[9] http://mozu.mes.titech.ac.jp/research/TitechRobotDriver/TitechRobotDriver.html

[10] Inoue H, Inaba M, Mori T, Tachikawa T 1993 Real-time robot vision system based on correlation technology. *Proc. of Int. Symp. on Industrial Robots (ISIR '93)*. pp 675–680

[11] Oka T, Takeda K, Inaba M, Inoue H 1996 Disigning Asynchronous Parallel Process Networks for Desirable Autonomous Robot Behaviors. *Proc. of the 1996 IEEE/RSJ Int. Conf. on Intelligent Robots and Systems (IROS '96)*. pp 178–185

[12] Nagashima K, Konno A, Inaba M, Inoue H 1997 Development of BeNet System for Multi-thread EusLisp and Transputer Network. *Proc. of JSME Conf. on Robotics·Mechatronics (ROBOMECH '97)*

Chapter 2

Dynamics and Control

Robot environment interaction implies the use of control models that consider not only the position and orientation but also the contact conditions. Different kinds of contacts and tasks give place to very diverse problems that require new control techniques and robot-environment contact models.

The paper of Caccavale, Natale, Siciliano and Villane focuses on the study of different impedance control schemes for 6 d.o.f. robots interacting with their environment. To determine the robot end-effector pose, the authors use two different representations of Euler angles and unitary quaternions. The study considers not only the position and orientation but also its displacement. Besides the classical control schemes and reference frames, the spatial impedance concept and the control implementation results are presented.

Nakawaki, Joo, Yoshikuwa and Miyazaki study the kip movement. Based on the analysis of a gymnast performing these movements on various apparatus, a three link underactuated mechanism is developed. From the study of the pendulum control techniques two different optimization techniques are used. The first uses global optimization of kinematics redundancy, the second uses parameter optimization and evaluates joint torques, dynamic energy and the change in kinetics energy of the shoulder and hip joints, learning from the movements. It is shown how both techniques may be put together to improve the kip.

Also in the trend of analyzing human movements, Morizono and Kawamura develop a virtual environment for sports training. The system uses a head mounted display to visualize the world, a parallel wire mechanism to display forces and an auditory display device. A combination of the problem of studying forces and high speed movements is experimented in sports such as tennis, in which a high speed reaction and force control is required.

In the work of Featherstone, Sonck and Khatib a study is done on how to generalize the Raibert-Craig contact model, to develop tasks where force and motion control is not sufficient. Besides the configuration space that considers the contact constraints, the N and T spaces (normal and tangential), two new spaces N' and T' are defined. The new representation combines the contact model with dynamics enabling the description of any kind of contact. Then, it is possible to control tasks including multiple points of contact.

Foulon, Fourquet and Renaud face up the problem of objects manipulation by a 6 d.o.f. robot manipulator mounted on a mobile platform. The need to control simultaneously two elements, the mobile base and the arm to follow a given trajectory has motivated the development of a control algorithm. The combination of either direct, inverse and pseudo-inverse methods has produced a robust control algorithm, first simulated and then tested over a non-holonomic mobile platform of the Hilare type

Experiments of Spatial Impedance Control

F. Caccavale   C. Natale   B. Siciliano   L. Villani
PRISMA Lab
Dipartimento di Informatica e Sistemistica
Università degli Studi di Napoli Federico II
Via Claudio 21, 80125 Napoli, Italy
{caccavale,natale,siciliano,villani}@disna.dis.unina.it

Abstract: The goal of this work is to present the results of an experimental study of impedance control schemes for a robot manipulator in contact with the environment. Six-degree-of-freedom interaction tasks are considered which require the implementation of a spatial impedance described in terms of both its translational and its rotational part. Two representations of end-effector orientation are adopted; namely, Euler angles or quaternions, and the implications on the choice of different orientation displacements are discussed. The controllers are tested on an industrial robot with open control architecture in a number of case studies.

1. Introduction

One of the most reliable strategies to manage the interaction of the end effector of a robot manipulator with a compliant environment is impedance control [1]. The majority of interaction control schemes refer to three-degree-of-freedom (dof) tasks in that they handle end-effector position and contact linear force. On the other hand, in order to perform six-dof tasks, not only is a representation of end-effector orientation required, but also a suitable definition of end-effector orientation displacement to be related to the contact moment should be sought.

The usual minimal representation of orientation is given by a set of Euler angles. These three coordinates, together with the three position coordinates, allow the description of end-effector tasks in the so-called operational space [2]. A drawback of this description is the occurrence of representation singularities of the analytical Jacobian [3]. These can be avoided by resorting to a four-parameter singularity-free description of end-effector orientation, e.g. in terms of a unitary quaternion. Such a description has already been successfully used for the attitude motion control problem of spacecrafts [4, 5] and manipulators [6].

The goal of this work is to present six-dof impedance controllers, where the impedance equation is characterized for both its translational part and its rotational part. In the framework of an operational space formulation, an analytical approach is pursued first where the end-effector orientation displacement is merely given by the difference between the actual and the desired set of Euler angles.

Then, in order to relate the rotational parameters of the impedance to the task geometry, a different approach is followed where the end-effector orientation displacement is extracted from the mutual rotation matrix between the actual and the desired end-effector frame. This can be obtained in terms of either three Euler angles or a unitary quaternion [7], where the latter allows the mechanical impedance to be derived from suitable energy contributions of clear physical interpretation [8].

In order to obtain a configuration-independent desired impedance, an inverse dynamics strategy with contact force and moment measurement is adopted leading to a resolved acceleration scheme. A modification of the basic scheme by the inclusion of an inner loop acting on the end-effector position and orientation error is devised to ensure good disturbance rejection [9], e.g. unmodeled friction.

The proposed controllers are tested and critically compared in a number of experiments on a setup comprising a 6-joint industrial robot Comau SMART-3 S with open control architecture and a 6-axis force/torque wrist sensor ATI FT 130/10.

2. Representations of Orientation

The location of a rigid body in space is typically described in terms of the (3×1) position vector p and the (3×3) rotation matrix R describing the origin and the orientation of a frame attached to the body with respect to a fixed reference frame.

A minimal representation of orientation can be obtained by using a set of three Euler angles $\phi = [\varphi \ \vartheta \ \psi]^T$. Among the 12 possible definitions of Euler angles, the XYZ representation is considered leading to the rotation matrix

$$R(\phi) = R_x(\varphi) R_y(\vartheta) R_z(\psi) \qquad (1)$$

where R_x, R_y, R_z are the matrices of the elementary rotations about three independent axes of successive frames.

The relationship between the time derivative of the Euler angles $\dot\phi$ and the body angular velocity ω is given by

$$\omega = T(\phi)\dot\phi \qquad (2)$$

where the transformation matrix T corresponding to the above XYZ representation is

$$T(\phi) = \begin{bmatrix} 1 & 0 & \sin\vartheta \\ 0 & \cos\varphi & -\sin\varphi\cos\vartheta \\ 0 & \sin\varphi & \cos\varphi\cos\vartheta \end{bmatrix}. \qquad (3)$$

Notice that a representation singularity occurs whenever $\vartheta = \pm\pi/2$. Also, in view of the choice of Euler angles XYZ, it is

$$T(0) = I \qquad (4)$$

which will be useful in the following.

A singularity-free description of orientation can be obtained by resorting to a four-parameter representation in terms of a unitary quaternion

$$\eta = \cos\frac{\theta}{2} \tag{5}$$

$$\epsilon = \sin\frac{\theta}{2}r, \tag{6}$$

where θ and r are respectively the rotation and the unit vector of an equivalent angle/axis description.

The relationship between the time derivative of the quaternion and the body angular velocity is established by the so-called quaternion propagation:

$$\dot{\eta} = -\frac{1}{2}\epsilon^T\omega \tag{7}$$

$$\dot{\epsilon} = \frac{1}{2}E(\eta,\epsilon)\omega \tag{8}$$

with

$$E = \eta I + S(\epsilon), \tag{9}$$

being $S(\cdot)$ the skew-symmetric matrix operator performing vector product.

3. Spatial Impedance

When the manipulator moves in free space, the end-effector is required to match a desired frame specified by p_d and R_d. Instead, when the end effector interacts with the environment, it is worth considering another frame specified by p_c and R_c; then, a mechanical impedance can be introduced which is aimed at imposing a dynamic behaviour for the position and orientation displacements between the above two frames.

The mutual position between the compliant and the desired frame can be characterized by the position displacement

$$\Delta p = p_c - p_d. \tag{10}$$

Then, the translational part of the mechanical impedance at the end effector can be defined as

$$M_p\Delta\ddot{p} + D_p\Delta\dot{p} + K_p\Delta p = f \tag{11}$$

where M_p, D_p, K_p are symmetric positive definite matrices describing the generalized mass, translational damping, translational stiffness, respectively, and f is the contact force at the end effector; all the above quantities are referred to a common base frame.

On the other hand, the mutual orientation between the compliant and the desired frame can be characterized in different ways. With reference to an operational space formulation, the end-effector orientation displacement can be computed as

$$\Delta\phi = \phi_c - \phi_d \tag{12}$$

where ϕ_c and ϕ_d denote the set of Euler angles that can be extracted from $\boldsymbol{R}_c$ and $\boldsymbol{R}_d$, respectively. Then, the rotational part of the mechanical impedance at the end effector can be defined as

$$\boldsymbol{M}_{o,\Delta}\Delta\ddot{\boldsymbol{\phi}} + \boldsymbol{D}_{o,\Delta}\Delta\dot{\boldsymbol{\phi}} + \boldsymbol{K}_{o,\Delta}\Delta\boldsymbol{\phi} = \boldsymbol{T}^{\mathrm{T}}(\boldsymbol{\phi}_c)\boldsymbol{\mu} \qquad (13)$$

where $\boldsymbol{M}_{o,\Delta}$, $\boldsymbol{D}_{o,\Delta}$, $\boldsymbol{K}_{o,\Delta}$ are symmetric positive definite matrices describing the generalized inertia, rotational damping, rotational stiffness, respectively, $\boldsymbol{\mu}$ is the contact moment at the end effector; all the above quantities are referred to a common base frame, and the matrix $\boldsymbol{T}^{\mathrm{T}}(\boldsymbol{\phi}_c)$ is needed to transform the moment into an equivalent operational space quantity.

Notice that, differently from (11), the impedance behaviour for the rotational part depends on the actual orientation of the end effector through the matrix $\boldsymbol{T}^{\mathrm{T}}(\boldsymbol{\phi}_c)$. Equation (13) becomes ill-defined in the neighbourhood of a representation singularity for $\boldsymbol{\phi}_c$; in particular, at such a singularity, moment components in the null space of $\boldsymbol{T}^{\mathrm{T}}$ do not generate any contribution to the dynamics of the orientation displacement, leading to a possible build-up of high values of generalized forces at the contact.

The effect of the matrix $\boldsymbol{T}^{\mathrm{T}}(\boldsymbol{\phi}_c)$ is best understood by considering the elastic contribution of the moment, i.e.

$$\boldsymbol{\mu}_E = \boldsymbol{T}^{-\mathrm{T}}(\boldsymbol{\phi}_c)\boldsymbol{K}_{o,\Delta}\Delta\boldsymbol{\phi}. \qquad (14)$$

In the case of small orientation displacement about a constant $\boldsymbol{\phi}_d$, at first approximation it is

$$\Delta\boldsymbol{\phi} \simeq \boldsymbol{T}^{-1}(\boldsymbol{\phi}_d)\boldsymbol{\omega}_c\mathrm{d}t \qquad (15)$$

where $\boldsymbol{\omega}_c\mathrm{d}t$ is the infinitesimal angular displacement of the compliant frame. Plugging (15) into (14) gives

$$\boldsymbol{\mu}_E \simeq \boldsymbol{T}^{-\mathrm{T}}(\boldsymbol{\phi}_d)\boldsymbol{K}_{o,\Delta}\boldsymbol{T}^{-1}(\boldsymbol{\phi}_d)\boldsymbol{\omega}_c\mathrm{d}t, \qquad (16)$$

revealing that the equivalent rotational stiffness between the angular displacement and the physical moment depends on the desired end-effector orientation.

A geometrically consistent expression for the end-effector orientation displacement can be derived by characterizing the mutual orientation between the compliant and the desired frame directly in terms of the rotation matrix

$$\tilde{\boldsymbol{R}}^d = \boldsymbol{R}_d^{\mathrm{T}}\boldsymbol{R}_c \qquad (17)$$

where the superscript evidences that the matrix is referred to the desired frame. If a minimal representation of the end-effector orientation displacement is sought, a set of Euler angles $\tilde{\boldsymbol{\phi}}$ can be extracted from $\tilde{\boldsymbol{R}}^d$. Then, the rotational part of the mechanical impedance at the end effector can be defined as

$$\boldsymbol{M}_{o,\phi}\ddot{\tilde{\boldsymbol{\phi}}} + \boldsymbol{D}_{o,\phi}\dot{\tilde{\boldsymbol{\phi}}} + \boldsymbol{K}_{o,\phi}\tilde{\boldsymbol{\phi}} = \boldsymbol{T}^{\mathrm{T}}(\tilde{\boldsymbol{\phi}})\boldsymbol{\mu}^d \qquad (18)$$

where $\boldsymbol{M}_{o,\phi}$, $\boldsymbol{D}_{o,\phi}$, $\boldsymbol{K}_{o,\phi}$ are defined as in (13), $\boldsymbol{\mu}^d$ is expressed in the desired frame, and $\boldsymbol{T}^{\mathrm{T}}(\tilde{\boldsymbol{\phi}})$ is needed to transform the moment into a quantity consistent with $\dot{\tilde{\phi}}$ via a kineto-static duality concept. An advantage with respect to (13) is that the impedance behaviour for the rotational part does not depend on the actual end-effector orientation but only on the orientation displacement through the matrix $\boldsymbol{T}^{\mathrm{T}}(\tilde{\boldsymbol{\phi}})$.

If the XYZ representation of Euler angles in (1) is adopted, the transformation matrix $\boldsymbol{T}$ satisfies (4) for $\tilde{\boldsymbol{\phi}} = \boldsymbol{0}$, i.e. when the compliant frame is aligned with the desired frame. Also, representation singularities have a mitigated effect since they occur for large end-effector orientation displacements.

By developing an infinitesimal analysis similar to (14)–(16), the orientation displacement

$$\mathrm{d}\tilde{\boldsymbol{\phi}} = \tilde{\boldsymbol{\omega}}^d \mathrm{d}t \tag{19}$$

can be considered where (4) has been exploited. This corresponds to the angular velocity

$$\tilde{\boldsymbol{\omega}}^d = \boldsymbol{\omega}_c^d - \boldsymbol{\omega}_d^d \tag{20}$$

with

$$\tilde{\boldsymbol{\omega}}^d = \boldsymbol{T}(\tilde{\boldsymbol{\phi}})\dot{\tilde{\boldsymbol{\phi}}}. \tag{21}$$

Then, the elastic contribution of the moment is given by

$$\boldsymbol{\mu}_E^d = \boldsymbol{T}^{-\mathrm{T}}(\mathrm{d}\tilde{\boldsymbol{\phi}})\boldsymbol{K}_{o,\phi}\mathrm{d}\tilde{\boldsymbol{\phi}} \simeq \boldsymbol{K}_{o,\phi}\mathrm{d}\tilde{\boldsymbol{\phi}} \tag{22}$$

where (4) has been exploited again. Equation (22) with (19) shows how the equivalent rotational stiffness has a clear geometric meaning and is constant on condition that both the angular displacement and the contact moment are expressed in the desired frame.

An alternative representation of the end-effector orientation displacement can be extracted from (17) by resorting to the unitary quaternion defined in (5),(6), i.e.

$$\tilde{\eta} = \cos\frac{\tilde{\theta}}{2} \tag{23}$$

$$\tilde{\boldsymbol{\epsilon}}^d = \sin\frac{\tilde{\theta}}{2}\tilde{\boldsymbol{r}}^d, \tag{24}$$

where the vector part of the quaternion has been referred to the desired frame for consistency. It can be shown that the rotational part of the mechanical impedance at the end effector can be defined as [7]

$$\boldsymbol{M}_{o,\epsilon}\dot{\tilde{\boldsymbol{\omega}}}^d + \boldsymbol{D}'_{o,\epsilon}\tilde{\boldsymbol{\omega}}^d + \boldsymbol{K}'_{o,\epsilon}\tilde{\boldsymbol{\epsilon}}^d = \boldsymbol{\mu}^d \tag{25}$$

where

$$\boldsymbol{D}'_{o,\epsilon} = \boldsymbol{D}_{o,\epsilon} - \boldsymbol{M}_{o,\epsilon}\boldsymbol{S}(\boldsymbol{\omega}_d^d) \tag{26}$$

$$\boldsymbol{K}'_{o,\epsilon} = 2\boldsymbol{E}^{\mathrm{T}}(\tilde{\eta}, \tilde{\boldsymbol{\epsilon}}^d)\boldsymbol{K}_{o,\epsilon} \tag{27}$$

are the resulting time-varying rotational damping and stiffness matrices derived from an energy-based formulation, with $\boldsymbol{E}$ as in (9).

By developing an infinitesimal analysis similar to the above, the elastic contribution of the moment is given by

$$\mu_E^d \simeq K_{o,\epsilon}\tilde{\omega}^d dt \tag{28}$$

which allows the rotational stiffness to be clearly related to the task geometry, as for (22),(19). Compared to the previous Euler angles descriptions, though, a breakthrough is represented by the avoidance of representation singularities thanks to the use of a unitary quaternion.

4. Control Implementation

For a 6-dof rigid robot manipulator, the dynamic model can be written in the well-known form

$$B(q)\ddot{q} + C(q,\dot{q})\dot{q} + d(q,\dot{q}) + g(q) = u - J^T(q)h_e, \tag{29}$$

where q is the (6×1) vector of joint variables, B is the (6×6) symmetric positive definite inertia matrix, $C\dot{q}$ is the (6×1) vector of Coriolis and centrifugal torques, d is the (6×1) vector of friction torques, g is the (6×1) vector of gravitational torques, u is the (6×1) vector of driving torques, $h_e = [f_e^T \ \mu_e^T]^T$ is the (6×1) vector of contact forces exerted by the end effector on the environment, and J is the (6×6) Jacobian matrix relating joint velocities $\dot{q}$ to the (6×1) vector of end-effector velocities $v_e = [\dot{p}_e^T \ \omega_e^T]^T$, i.e.

$$v_e = J(q)\dot{q}, \tag{30}$$

which is assumed to be nonsingular.

According to the well-known concept of inverse dynamics, the driving torques are chosen as

$$u = B(q)J^{-1}(q)(a - \dot{J}(q,\dot{q})\dot{q}) + C(q,\dot{q})\dot{q} + \hat{d}(q,\dot{q}) + g(q) + J^T(q)h_e, \tag{31}$$

where $\hat{d}$ denotes the available estimate of the friction torques, and h_e is the measured contact force. Notice that it is reasonable to assume accurate compensation of the terms in the dynamic model (29), e.g. as obtained by a parameter identification technique [10], except for the friction torques.

Substituting the control law (31) in (29) and accounting for the time derivative of (30) gives

$$\dot{v} = a - \delta \tag{32}$$

that is a resolved end-effector acceleration for which the term

$$\delta = JB^{-1}(d - \hat{d}) \tag{33}$$

can be regarded as a disturbance. In the case of mismatching on other terms in the dynamic model (29), such a disturbance would include additional contributions.

The new control input can be chosen as $\boldsymbol{a} = [\,\boldsymbol{a}_p^{\mathrm{T}}\ \ \boldsymbol{a}_o^{\mathrm{T}}\,]^{\mathrm{T}}$ where $\boldsymbol{a}_p$ and $\boldsymbol{a}_o$ are designed to match the desired impedance for the translational and the rotational part, respectively. In view of (11), $\boldsymbol{a}_p$ is taken as

$$\boldsymbol{a}_p = \ddot{\boldsymbol{p}}_c + k_{Vp}(\dot{\boldsymbol{p}}_c - \dot{\boldsymbol{p}}_e) + k_{Pp}(\boldsymbol{p}_c - \boldsymbol{p}_e) \tag{34}$$

where k_{Vp}, k_{Pp} are suitable positive gains of the position loop, while $\boldsymbol{p}_c$ and its associated derivatives can be computed by forward integration of the differential equation (11).

As regards the orientation loop, instead, $\boldsymbol{a}_o$ can be taken according to the three different representations of orientation displacement illustrated above. Then, with reference to (13), it is

$$\boldsymbol{a}_{o,\Delta} = \boldsymbol{T}(\boldsymbol{\phi}_e)\big(\ddot{\boldsymbol{\phi}}_c + k_{Vo,\Delta}(\dot{\boldsymbol{\phi}}_c - \dot{\boldsymbol{\phi}}_e) + k_{Po,\Delta}(\boldsymbol{\phi}_c - \boldsymbol{\phi}_e)\big) + \dot{\boldsymbol{T}}(\boldsymbol{\phi}_e)\dot{\boldsymbol{\phi}}_e \tag{35}$$

where $k_{Vo,\Delta}$, $k_{Po,\Delta}$ are suitable positive gains, $\boldsymbol{\phi}_e$ is the set of Euler angles that can be extracted from the rotation matrix $\boldsymbol{R}_e$ expressing the orientation of the end-effector frame with respect to the base frame; note that the presence of $\boldsymbol{T}$ and its time derivative is originated from the time derivative of (2). Further, $\boldsymbol{\phi}_c$ and its associated derivatives can be computed by forward integration of the differential equation (13).

Next, with reference to (18), it is

$$\boldsymbol{a}_{o,\phi} = \dot{\boldsymbol{\omega}}_d + \boldsymbol{T}_d(\tilde{\boldsymbol{\phi}}_e)\big(\ddot{\tilde{\boldsymbol{\phi}}} + k_{Vo,\phi}(\dot{\tilde{\boldsymbol{\phi}}} - \dot{\tilde{\boldsymbol{\phi}}}_e) + k_{Po,\phi}(\tilde{\boldsymbol{\phi}} - \tilde{\boldsymbol{\phi}}_e)\big) + \dot{\boldsymbol{T}}_d(\tilde{\boldsymbol{\phi}}_e)\dot{\tilde{\boldsymbol{\phi}}}_e \tag{36}$$

where $k_{Vo,\phi}$, $k_{Po,\phi}$ are suitable positive gains, $\tilde{\boldsymbol{\phi}}_e$ is the set of Euler angles that can be extracted from $\boldsymbol{R}_d^{\mathrm{T}}\boldsymbol{R}_e$, and the matrix

$$\boldsymbol{T}_d(\tilde{\boldsymbol{\phi}}_e) = \boldsymbol{R}_d \boldsymbol{T}(\tilde{\boldsymbol{\phi}}_e) \tag{37}$$

is needed to refer the angular velocity in (21) to the base frame, which then generates the presence of $\dot{\boldsymbol{T}}_d$ in (36) too. Further, $\tilde{\boldsymbol{\phi}}$ and its associated derivatives can be computed by forward integration of the differential equation (18).

Finally, with reference to (25), it is

$$\boldsymbol{a}_{o,\epsilon} = \dot{\boldsymbol{\omega}}_c + k_{Vo,\epsilon}(\boldsymbol{\omega}_c - \boldsymbol{\omega}_e) + k_{Po,\epsilon}\boldsymbol{\epsilon}_{ce} \tag{38}$$

where $\boldsymbol{\epsilon}_{ce}$ is the vector part of the quaternion that can be extracted from $\boldsymbol{R}_e^{\mathrm{T}}\boldsymbol{R}_c$ when referred to the base frame. Further, $\boldsymbol{R}_c$, $\boldsymbol{\omega}_c$, $\dot{\boldsymbol{\omega}}_c$ can be computed by forward integration of the differential equation (25).

Notice that the inner position and orientation loops in the previous equations are used in order to provide robustness to the disturbance in (33), which otherwise could not be effectively counteracted through the impedance parameters [9].

5. Experiments

The laboratory setup consists of an industrial robot Comau SMART-3 S. The robot manipulator has a six-revolute-joint anthropomorphic geometry with nonnull shoulder and elbow offsets and non-spherical wrist. The joints are actuated by brushless motors via gear trains; shaft absolute resolvers provide motor position measurements. The robot is controlled by an open version of the C3G 9000 control unit which has a VME-based architecture with a bus-to-bus communication link to a PC Pentium 133. This is in charge of computing the control algorithm and passing the references to the current servos through the communication link at 1 ms sampling rate. Joint velocities are reconstructed through numerical differentiation of joint position readings.

A 6-axis force/torque sensor ATI FT 130/10 with force range of ± 130 N and torque range of ± 10 Nm is mounted at the wrist of the robot manipulator. The sensor is connected to the PC by a parallel interface board which provides readings of six components of generalized force at 1 ms.

An end effector has been built as a steel stick with a wooden disk of 5.5 cm radius at the tip. The end-effector frame has its origin at the center of the disk and its approach axis normal to the disk surface and pointing outwards. The environment is constituted by a flat plexiglas surface. The resulting translational stiffness at the contact between the end effector and the surface is of the order of 10^4 N/m, while the rotational stiffness for small angles is of the order of 20 Nm/rad.

The dynamic model of the robot manipulator has been identified in terms of a minimum number of parameters, where the dynamics of the outer three joints has been simply chosen as purely inertial and decoupled. Only joint viscous friction has been included, since other types of friction (e.g. Coulomb and dry friction) are difficult to model.

The three proposed impedance controllers have been tested in two case studies which consist in taking the disk in contact with the surface at an angle of unknown magnitude. The parameters of the translational part of the mechanical impedance in (11) —which is common to all three controllers— have been set to $\boldsymbol{M}_p = 10\boldsymbol{I}$, $\boldsymbol{D}_p = 600\boldsymbol{I}$, $\boldsymbol{K}_p = 1000\boldsymbol{I}$. The parameters of the rotational part of the mechanical impedance in (13),(18),(25) have been set to: $\boldsymbol{M}_{o,\Delta} = \boldsymbol{M}_{o,\phi} = \boldsymbol{M}_{o,\epsilon} = 0.25\boldsymbol{I}$, $\boldsymbol{D}_{o,\Delta} = \boldsymbol{D}_{o,\phi} = \boldsymbol{D}_{o,\epsilon} = 3.5\boldsymbol{I}$, $\boldsymbol{K}_{o,\Delta} = \boldsymbol{K}_{o,\phi} = \boldsymbol{K}_{o,\epsilon} = 2.5\boldsymbol{I}$. Notice that the stiffness matrices have been chosen so as to ensure a compliant behavior at the end effector (limited values of contact force and moment) during the constrained motion, while the damping matrices have been chosen so as to guarantee a well-damped behaviour.

The gains of the various control actions in (34),(35),(36),(38) have been set to: $k_{Pp} = k_{Po,\Delta} = k_{Po,\phi} = k_{Po,\epsilon} = 2025$, $k_{Vp} = k_{Vo,\Delta} = k_{Vo,\phi} = k_{Vo,\epsilon} = 60$.

An analysis of the computational burden for the three controllers has been carried out for the available hardware, leading to a total time of 0.354 ms, 0.345 ms, 0.286 ms depending on which expression of the rotational part of the mechanical impedance among (13),(18),(25) is utilized.

In the first case study, the end-effector desired task consists of a straight line motion with a vertical displacement of -0.24 m along the z-axis of the base frame. The trajectory along the path is generated according to a 5th-order interpolating polynomial with null initial and final velocities and accelerations, and a duration of 7 s. The end-effector desired orientation is required to remain

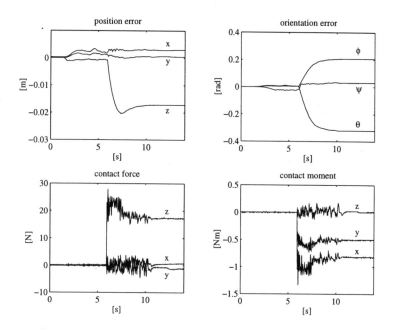

Figure 1. Components of $p_d - p_e$, $\phi_d - \phi_e$, f_e, μ_e with impedance controller based on (13) in the first case study.

constant during the task. The surface is placed (horizontally) in the xy-plane in such a way as to obstruct the desired end-effector motion.

The results in Figures 1 to 3 show the effectiveness of all the three controllers. After the contact, the component of the position errors along the z-axis significantly deviate from zero, as expected, while small errors can be seen also for the components along the x- and the y-axis due to contact friction. As for the orientation errors, all the components significantly deviate from zero since the end-effector frame has to rotate with respect to the base frame after the contact in order to comply with the surface. Also, in view of the imposed task, a prevailing component of the contact forces can be observed along the z-axis after the contact, while the small components along the x- and the y-axis arise as a consequence of the above end-effector deviation. As for the contact moments, the component about the z-axis is small accordingly. It can be recognized that all the above quantities reach constant steady-state values after the desired motion is stopped. The oscillatory behavior during the transient can be mainly ascribed to slipping of the disk on the surface after the contact. In sum, it can be asserted that a compliant behaviour is successfully achieved for all the three impedance controllers. This can be explained because the actual end-effector orientation keeps far from a representation singularity.

The second case study is aimed at testing the performance of the three impedance controllers when the actual end-effector orientation is close to a representation singularity of T. The end-effector desired task consists of a straight line motion with a horizontal displacement of 0.085 m along the x-axis of the

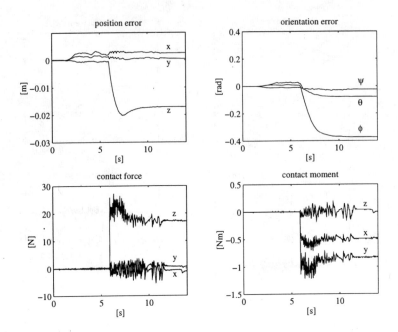

Figure 2. Components of $p_d - p_e$, $\tilde{\phi}_e$, f_e, μ_e with impedance controller based on (18) in the first case study.

base frame. The trajectory along the path is generated according to a 5th-order interpolating polynomial with null initial and final velocities and accelerations, and a duration of 5 s. The end-effector desired orientation is required to remain constant during the task. The surface is now placed vertically in such a way as to obstruct the desired end-effector motion. Differently from above, no impedance control is accomplished for the positional part.

The results in Figures 4 to 6 show that significant differences occur in the performance of the scheme based on (13) with respect to the other two schemes. Large values of contact force and moment are generated since the equivalent rotational stiffness in (16) suffers from ill-conditioning of the matrix $T(\phi_d)$.

6. Conclusion

Three spatial impedance controllers have been presented in this work, where different definitions of the end-effector orientation displacement have been adopted. The results of an experimental investigation on an industrial robot have shown the superiority of the scheme based on the unitary quaternion representation, both to clearly relate the rotational parameters of the impedance to the task geometry and to avoid the possible occurrence of end-effector representation singularities.

Acknowledgements

This work was supported in part by *MURST* and in part by *ASI*.

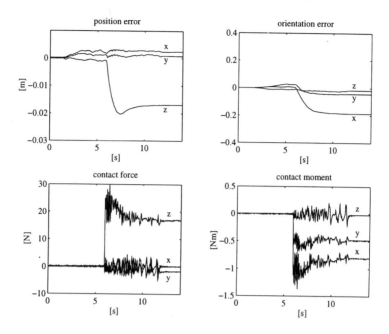

Figure 3. Components of $p_d - p_e$, ϵ_{ce}, f_e, μ_e with impedance controller based on (25) in the first case study.

References

[1] Hogan N 1985 Impedance control: An approach to manipulation, Parts I–III. *ASME J Dyn Syst, Meas, Contr.* 107:1–24

[2] Khatib O 1987 A unified approach for motion and force control of robot manipulators: The operational space formulation. *IEEE J Robot Automat.* 3:43–53

[3] Sciavicco L, Siciliano B 1996 *Modeling and Control of Robot Manipulators.* McGraw-Hill, New York

[4] Wen J T-Y, Kreutz-Delgado K 1991 The attitude control problem. *IEEE Trans Automat Contr.* 36:1148–1162

[5] Egeland O, Godhavn J-M 1994 Passivity-based adaptive attitude control of a rigid spacecraft. *IEEE Trans Automat Contr.* 39:842–846

[6] Yuan J S-C 1988 Closed-loop manipulator control using quaternion feedback. *IEEE J Robot Automat.* 4:434–440

[7] Siciliano B, Villani L 1997 Six-degree-of-freedom impedance robot control. In: *Proc 8th Int Conf Advanced Robot.* Monterey, CA

[8] Fasse E D 1995 Simplification of compliance selection using spatial compliance control. In: *Proc ASME Dyn Syst Contr Div.* vol 57-1, pp 193–198

[9] Bruni F, Caccavale F, Natale C, Villani L 1996 Experiments of impedance control on an industrial robot manipulator with friction. In: *Proc 5th IEEE Int Conf Contr Appl.* Dearborn, MI, pp 205–210

[10] Caccavale F, Chiacchio P 1994 Identification of dynamic parameters and feedforward control for a conventional industrial manipulator. *Contr Engnr Pract.* 2:1039–1050

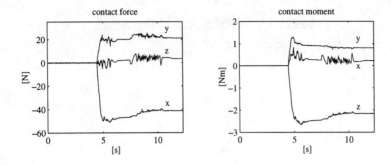

Figure 4. Components of f_e, μ_e with impedance controller based on (13) in the second case study.

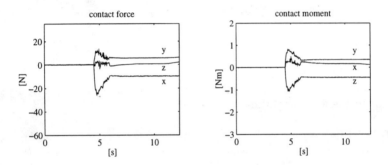

Figure 5. Components of f_e, μ_e with impedance controller based on (18) in the second case study.

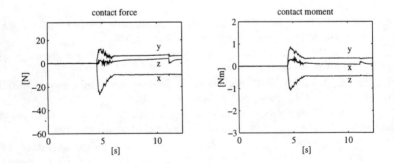

Figure 6. Components of f_e, μ_e with impedance controller based on (25) in the second case study.

Optimal Control based Skill Development System for the Kip

Darrell Nakawaki, Sangwan Joo, and Fumio Miyazaki
Faculty of Engineering Science, Osaka University
1-3 Machikaneyama-cho, Toyonaka city, Osaka 560, JAPAN
e-mail: nakawaki@robotics.me.es.osaka-u.ac.jp

Abstract: A skill development system which uses a pendulum model and kinematic redundancy to determine the torques of a 3 link system is discussed. This 3 link model performs an acrobatic motion called the kip. A kip pattern of a gymnast's center of mass is optimized by using a variable length pendulum. We input the results of this model into the 3 link system and by using kinematic redundancy considering system dynamics, we can find the optimal joint trajectories. The gymnast uses primarily torques generated in the hip and shoulder joints to perform the kip. We show that we can reduce the torque in the wrist joint such that it is less than the torques generated in the hip and shoulder joints of the 3 link robot and successfully perform the kip. Additionally, analyzing the trajectory of the center of mass may provide useful coaching information to the beginner.

1. Introduction

The kip is an elementary movement which is common in gymnastic performance on various apparatus at all levels of proficiency. The gymnast with the body extended, initiates the kip with a forward swing on the horizontal bar. When the gymnast reaches a point well beyond the bottom of the swing, the hips are flexed and the ankles are lifted close to the bar by the hands. Although the kip is an elementary movement in gymnastics, it has been shown to be very difficult for the beginner [4].

The person performing the kip must use torques generated primarily in the shoulder and hip joints which must be timed and controlled in such a way to generate enough energy to move his body over horizontal bar. A system as shown in Figure 1 that could take data from a beginner attempting the kip and output an analysis or evaluation on how to improve his performance would be very useful in improving current coaching techniques for the kip.

1.1. Pendulum based control technique

In this paper, we apply an approach to study the kip movement via a 3 link model of a human being as shown in Figure 2.

We propose that the human body's center of mass moves in the same way as a pendulum with a varying length. Our rather simple approach uses the path of a varying length pendulum based on the kip performed by an expert gymnast as an input to a set of dynamic equations for the 3 link model.

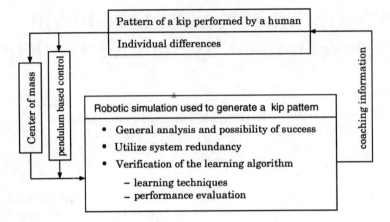

Figure 1. Skill development system for the kip

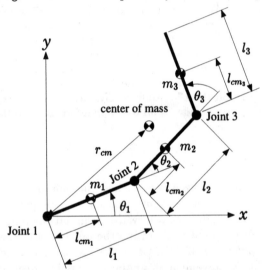

Figure 2. 3 link manipulator.

In short, what we have proposed can be summarized in terms of data taken from an expert gymnast, the pendulum phenomena, and kinematic redundancy.

1. From [4], we can get information about the trajectory of the center of mass from an expert gymnast performing the kip. We can also obtain the desired joint angle ranges throughout the kip movement.

2. In [6], it was shown that a pendulum varies in amplitude with each swing if its length is varied simultaneously.

3. An input vector can be used to plan the joint trajectories of a kinematic redundant system [7]. A potential function can be used as an input vector for applications such as obstacle, joint limit and singularity avoidance.

By using the information in 1. as a method to vary the pendulum's length of 2., we can model the kip's trajectory, and obtain information about the position, velocity, and acceleration of the center of mass. We then rewrite the Jacobian of 3. in terms of the mass center of the 3 link system and develop an input vector which can reduce the torque of joint 1 while maintaining controlled movement of the system. A potential function is also used for joint limit avoidance.

This pendulum technique gives us a simple way to control the 3 link system and also has the advantage of reducing the differences of individuals performing the kip to the total mass differences of individual human bodies and the overall location of the center of mass. Additionally, our analysis of the kip can be simpified by using this pendulum technique. We can summarize the advantages of our approach as follows.

1. The variable length pendulum model gives us a simple way to control the 3 link system.

2. Using the pendulum model allows us to reduce the differences of individuals performing the kip to the total mass differences of individual human bodies and the overall location of the center of mass, which may simplify the kip analysis.

3. We can generate various patterns in which a 3 link system can successfully perform the kip by using the kinematic redundancy of the system based on the system center of mass.

1.2. Kip analysis and coaching information

In the near future, we can consider the next section of our system which can evaluate, and analyze parameters such as the joint torques, the dynamic energy of the system center of mass, and the change in kinetic energy of the shoulder and hip joints. We can also obtain other useful coaching information. For example, we can take the mass, length and location of the center of mass for the arms, torso, and legs of the beginner and set them to the parameters of our 3 link system. Using this information and taking advantage of the kinematic redundancy, we can generate a kip pattern which suits that particular person moving along the gymnast's trajectory.

Most likely, the beginner will not successfully complete the kip on the first try. We've thought about taking the beginner's trajectory from the end of the trajectory which would be located somewhere below the horizontal bar and connecting it to the end. We would then move our three link robot to see what kind of torque is necessary to complete the kip trajectory. If we can show that the kip can be performed with a slight addition of torque, we may be able to convince the beginner that performing the kip is possible.

Also, we may be able to provide a goal for the beginner to aim for when attempting the kip. In this paper, we have shown that it may be possible to determine the region around the location of the maximum forward swing point in which the kip can be successfully completed. In this way, the beginner may improve his probability of a successful completion.

1.3. Previous research

There have been a number of previous studies of related to gymnastics and underactuated mechanical systems; only a few of which are presented here. The Acrobot [1] uses only a 2 link system and was used to perform a swing up algorithm to a balancing position over the bar using partial linear feedback. In [2,3], a similar mechanism was developed using sinusoidal torque input signals while varying the phase to raise the system above the bar. This method uses a 3 link model as shown in Figure 2. It has been shown in [4] that the human body can be approximated as a 3 link model when performing the kip.

2. A kip trajectory based pendulum model

By analyzing the energy of the overall motion of the kip's center of mass, we can model the system as a pendulum with a varying length (Figure 3). This system has the following equation

$$\ddot{\theta} + \left(\frac{2}{l}\dot{l} - \frac{c}{ml^2}\right)\dot{\theta} + \frac{g}{l}\sin\theta = 0 \qquad (1)$$

where θ is the angle, l is the length, g is the acceleration of gravity, and c is the frictional coefficient. We use the trajectory data from an expert gymnast performing the kip to vary l. This coefficient is varied in order to approximate the trajectory of the kip. Considering a viscosity term ($c > 0$), we can obtain a more reasonable pendulum model which gradually decreases in angular velocity as its mass swings above the fulcrum.

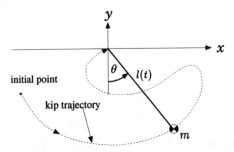

Figure 3. A simple pendulum with a varying length following a kip trajectory.

2.1. Optimizing the kip trajectory using a variable length pendulum

In this section, we optimize a variable length pendulum dynamic equation which is based on the concept developed in Section 2. Choosing three points on the kip trajectory and using the constraints of the kip trajectory, we can determine an optimized kip trajectory. We rewrite the pendulum dynamic equation as a set of state equations.

$$\dot{\theta} = \beta \qquad (2)$$

$$\dot{\beta} = -\frac{1}{l}\left(2\beta\dot{l} - g\sin\theta + \frac{c}{ml}\dot{\theta}\right) \qquad (3)$$

We define the state vector $\Omega = (\alpha, \beta, \gamma)^T$, where $\alpha = \theta$, $\gamma = \frac{1}{l}$. We then rewrite the pendulum dynamics equation using the above state vector.

$$\dot{\Omega} = \begin{pmatrix} \beta \\ 2\beta\frac{u}{\gamma} - g\gamma\sin\alpha - \beta c\gamma^2/m \\ u \end{pmatrix} = f(\Omega, u) \quad (4)$$

where $u = \dot{\gamma} = -\gamma^2 \dot{l}$. We choose three points along the kip trajectory which are the initial, final point and the maximum forward swing point to define our boundary conditions. We then optimize the trajectory using the following boundary conditions and criterion function.

$$\Omega(t_o) = \Omega_o, \Omega(t_f) = \Omega_f \quad (5)$$

$$C_T = \frac{1}{2}\int_{t_o}^{t_f} u^2 dt \quad (6)$$

We minimize u in order to determine a center of mass trajectory which can minimize the overall torques generated in joints 2 and 3. l^2 is proportional to the pulling energy of the pendulum string and $\frac{1}{l^4}$ is a weighting term which increases the effect of u as l decreases. By minimizing u over the kip trajectory, we can determine a trajectory that optimizes the pulling energy of the string and thus minimize the pulling energy of the overall torques generated in joints 2 and 3. By applying Pontryagin's maximum principle, the Hamiltonian becomes

$$H = -\frac{1}{2}u^2 + f^T\psi \quad (7)$$

Our system is reduced to a set of two-point boundary value equations.

$$\dot{\Omega} = \frac{\partial H}{\partial \psi} = f(\Omega, u) \quad (8)$$

$$\dot{\psi} = \frac{\partial H}{\partial \Omega} = -\frac{\partial f^T}{\partial \Omega}\psi \quad (9)$$

We can easily simulate this set of equations using a Newton-like method as shown in [5] and our results are shown in Figure 4 and Figure 5. Figure 4 shows that a variable length pendulum based optimized trajectory possesses similar phase plot and system energy characteristics to a pendulum trajectory based on the gymnast's center of mass. From the optimized trajectory, we can numerically determine $\ddot{r}_{cm}(t)$ which is similar to that of a person performing the kip and input this value into Eq.(11).

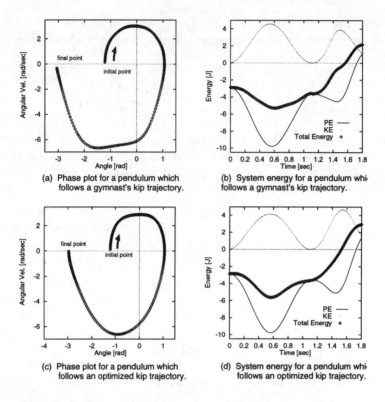

Figure 4. Phase plots and system energy for a pendulum.

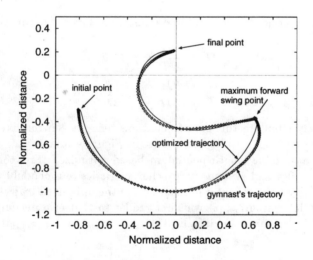

Figure 5. Plot of system center of mass for both optimized and gymnast kip trajectory.

Figure 4 shows that the trajectory becomes increasingly large indicating an increase in energy as the center of mass moves from the beginning position to the end. We can also see that the pendulum which follows the path of a gymnast demonstrates an increase in potential and total energy during the kip movement. Comparing the trajectories in Figure 5, we can see that by choosing three points on the kip trajectory and by optimizing the trajectory in between these points, we can approximate the kip trajectory with our optimization approach.

3. Kinematic redundancy considering system dynamics

A direct method can be used to determine the torques of this 3 link system and at the same time reduce the torque generated in joint 1, however, the torques of joint 2 and 3 are relatively large and the system is difficult to control. It has been shown that, as a kinematically redundant manipulator moves along a given trajectory, the posture of the system can be determined by using a potential function as an input vector to avoid obstacles, singularities and joint limits [7]. In Section 3.1, we introduce an input vector which is able to reduce the torque generated in joint one while maintaining controlled movement of the three link system. The dynamics of robot manipulators is generally represented by the following equation:

$$\boldsymbol{\tau} = \boldsymbol{A}(\boldsymbol{\theta})\ddot{\boldsymbol{\theta}} + \boldsymbol{B}(\boldsymbol{\theta}, \dot{\boldsymbol{\theta}}) + \boldsymbol{C}(\boldsymbol{\theta}) \tag{10}$$

where $\boldsymbol{A}(\boldsymbol{\theta}) \in R^{3\times 3}$ is an inertia matrix, $\boldsymbol{B} \in R^{3\times 1}$ is a torque vector caused by centrifugal and Coriolis forces, $\boldsymbol{C} \in R^{3\times 1}$ is a gravitational torque vector, $\boldsymbol{\tau} \in R^{3\times 1}$ is a joint torque vector, and $\boldsymbol{\theta} \in R^{3\times 1}$ is the joint angle vector. When the dynamics is described by a second-order differential equation, the kinematic redundancy can be explicitly represented as follows:

$$\ddot{\boldsymbol{\theta}} = \boldsymbol{J}_{cm}^{\#}(\ddot{\boldsymbol{r}}_{cm}(t) - \dot{\boldsymbol{J}}_{cm}\dot{\boldsymbol{\theta}}) + (\boldsymbol{E} - \boldsymbol{J}_{cm}^{\#}\boldsymbol{J}_{cm})\boldsymbol{Y} \tag{11}$$

where $\boldsymbol{r}_{cm}(t) \in R^{3\times 1}$ is the position of the system's center of mass, $\boldsymbol{J}_{cm} \in R^{2\times 3}$, is the Jacobian matrix written with respect to the system's center of mass, $\boldsymbol{J}_{cm}^{\#}$ is its pseudoinverse, $\boldsymbol{E} \in R^{3\times 3}$ is an identity matrix, and $\boldsymbol{Y} \in R^{3\times 1}$ is an input vector. We can determine $\ddot{\boldsymbol{r}}_{cm}(t)$ from our optimized pendulum trajectory.

By introducing a vector $\boldsymbol{X} = (\boldsymbol{\theta}^T, \dot{\boldsymbol{\theta}}^T)^T \in R^{6\times 1}$, we rewrite $\boldsymbol{\tau}$ with respect to the state vector $\boldsymbol{X}$ to obtain the following equation.

$$\boldsymbol{T}(\boldsymbol{X}, \boldsymbol{Y}, t) = \boldsymbol{U}(\boldsymbol{X}, t) + \boldsymbol{V}(\boldsymbol{X})\boldsymbol{Y} \tag{12}$$

where $\boldsymbol{T}(\boldsymbol{X}, \boldsymbol{Y}, t) \in R^{3\times 1}$ is the joint torque vector written with respect to state vector $\boldsymbol{X}$ and

$$\begin{aligned}
\boldsymbol{U}(\boldsymbol{X}, t) &= \boldsymbol{A}\boldsymbol{J}_{cm}^{\#}(\ddot{\boldsymbol{r}}_{cm}(t) - \dot{\boldsymbol{J}}_{cm}\dot{\boldsymbol{\theta}}) + \boldsymbol{B} + \boldsymbol{C} \\
&\in R^{3\times 1}
\end{aligned} \tag{13}$$

$$\begin{aligned}
\boldsymbol{V}(\boldsymbol{X}) &= \boldsymbol{A}(\boldsymbol{E} - \boldsymbol{J}_{cm}^{\#}\boldsymbol{J}_{cm}) \\
&\in R^{3\times 3}
\end{aligned} \tag{14}$$

3.1. Torque reducing input vector

One of the characteristics of the kip movement is that the required torques is generated primarily in the joint 2 and joint 3. In order to reduce the torque generated in joint 1, we set T_1 of Eq.(12) to $T_1 = T_{reduce}$, and we write our input vector as follows.

$$\boldsymbol{Y}_a = \begin{pmatrix} Y_1 \\ 0 \\ 0 \end{pmatrix} \quad (15)$$

$$Y_1 = \kappa \left(-\frac{U_1(\boldsymbol{X},t)}{V_{11}(\boldsymbol{X})} + \frac{T_{reduce}}{V_{11}(\boldsymbol{X})} \right) \quad (16)$$

We substitute this input vector $\boldsymbol{Y}_a$ into Eq.(20).

3.2. Joint limit avoidance input vector

We also add a joint limit avoidance input vector which was developed in [8] in order to obtain controlled movement of the 3 link system such that the joint limits are not exceeded. We input the following $\boldsymbol{Y}_b$ as an input to Eq.(20)

$$\boldsymbol{Y}_b = \eta \left(\frac{\partial p}{\partial \boldsymbol{\theta}} \right) \quad (17)$$

where,

$$p = \frac{1}{3} \sum_{i=1}^{3} \left(\frac{\theta_i - a_i}{a_i - \theta_{i_{max}}} \right)^2 \quad (18)$$

$$a_i = \frac{\theta_{i_{max}} - \theta_{i_{min}}}{2} \quad (19)$$

and η is an adjusting coefficient. We set $\eta = 20.0$ and the joint angle limits of

Table 1. Joint angle limits.

joint	θ_{min} [rad]	θ_{max} [rad]
1	-4.600	-0.700
2	-0.100	3.000
3	-0.250	2.100

this 3 link system are shown in **Table 1**.

3.3. Overall input vector

Our overall input vector which reduces the torque of joint 1 and provides a means for joint limit avoidance is defined as follows.

$$\boldsymbol{Y} = \boldsymbol{Y}_a + \boldsymbol{Y}_b \quad (20)$$

We substitute this input vector into Eq.(11) which allows us to guide the 3 link system to successfully perform the kip.

4. Simulation results

We divide the overall kip movement into two phases which are the forward and backward swings and analyze their torque responses. We define our initial and final values for both the pendulum and 3 link system in **Table 2**.

Table 2. Initial and final condition parameters.

	Parameters	t_o	t_f
Pendulum	θ_{cm_0} [deg]	-160.0	-270.0
	$\dot{\theta}_{cm_0}$ [deg]	0.000	-3.210
	l [normalized]	0.851	0.203
3 link system	θ_1 [deg]	-183.5	94.00
	θ_2 [deg]	47.38	146.1
	θ_3 [deg]	50.87	72.76
	$\dot{\theta}_1$ [deg]	0.000	-0.512
	$\dot{\theta}_2$ [deg]	0.000	-0.414
	$\dot{\theta}_3$ [deg]	0.000	-0.224

Using our input vector, we are able to realize the kip movement as shown in Figure 6. The torque data for the optimized pendulum trajectory is shown

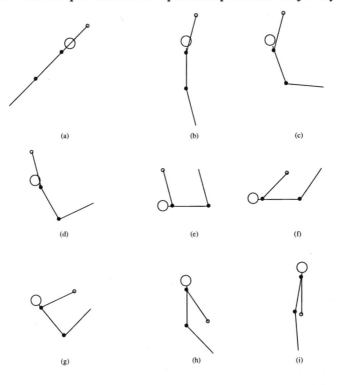

Figure 6. A successful kip movement.

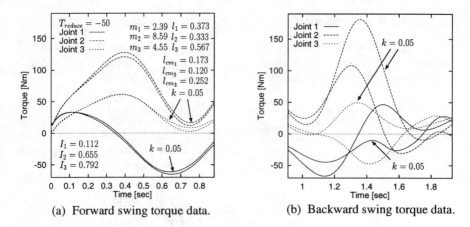

(a) Forward swing torque data.

(b) Backward swing torque data.

Figure 7. Torque data for the optimized kip trajectory.

in Figure 7. We can see only a slight change in the torque joint 1 as we increase the weighting factor κ in the first phase of the kip, but in the second phase we can see a much greater effect. There is a much greater reduction in the torque of joint 1 which causes significant changes in the torques of joints 2 and 3.

We also varied the maximum forward swing point values to determine the effective optimization range as shown in Figure 8. For example, if maximum

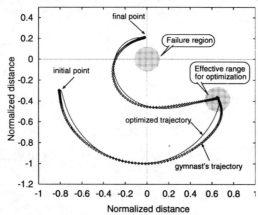

Figure 8. Plot of system center of mass trajectory indicating the range for effective optimization and failure region.

forward swing point is set to a point before entering this region, the trajectory is optimizable to the maximum forward swing point, but the final point cannot be reached. In the same way, if it is set to a point after entering this region, the maximum forward swing point cannot be reached and thus the trajectory cannot be optimized. This information could be valuable as coaching information to the beginner as a goal for the forward swing. If he is able move his center of mass into this region, the beginner may have a high probability of completing the kip.

The center of mass for this 3 link model is located such that its minimum distance to the horizontal bar is 0.152. We have seen that if the optimized trajectory results in a path entering this region, then the kip cannot be completed. From Eq.(11) in Section 3, we see that the term $\ddot{r}_{cm}(t) - \dot{J}_{cm}\dot{\theta}$ will show an increase in $\ddot{\theta}$ as l drops below 0.152. In other words, as the body nears the horizontal bar the more difficult it is to maneuver its joints and may require more torque. Thus, we can also indicate a failure region near the horizontal bar to avoid when attempting the kip.

5. Conclusion

In this paper, we presented a pendulum based control for the kip movement. We were able to show that we could reduce the torque in joint 1, while successfully performing the kip. By varying the boundary conditions of our optimized kip trajectory and adjusting our torque reducing parameter we can obtain various torque patterns for the kip. In addition, by making use of the effective range for optimization, we may be able to help beginners determine a target to aim for when attempting the kip. In the future, we plan to develop the other sections of our kip coaching system which may include learning control and feedback to the beginner to improve his performance.

References

[1] M.W. Spong, The Swing Up Control Problem For the Acrobot, *IEEE Control Systems*, pp. 49-55, 1995.

[2] Takashima, S., Control of Gymnast on a High Bar, *IEEE Int'l Workshop on Intel. Robots and Systems*, Proc. of IROS '91, pp. 1424–1429, Osaka, Japan, 1991.

[3] Takashima, S., Dynamic Modeling of a Gymnast on a High Bar, *IEEE Int'l Workshop on Intel. Robots and Systems*, Proc. of IROS '90, pp. 955–962, Osaka, Japan, 1990.

[4] Komatsu, The Kip and the Changes in Position in Motion and the Grip Force, *Osaka Physical Education Research*, Vol 30, pp. 20–27, 1991. (in Japanese)

[5] Uno, Kawahito, Repetitive Learning Control for an Optimal Trajectory of a Robotic Manipulator, *Measurement and Automation*, Vol. 24, No 8, 1988.(in Japanese)

[6] Toda, "Vibration Theory", New Physics Series, Baifukan, 1968. (in Japanese)

[7] Nakamura,Y., *Advanced Robotics–Redundancy and Optimization*, Addison-Wesley Publishing Company, Inc., 1995.

[8] Liegeois, A., Automatic Supervisory Control of the Configuration and Behavior of Multibody Mechanisms, *IEEE Transactions on Systems, Man and Cybernetics*, Vol. SMC-7, No 12, 1977.

Toward Virtual Sports with High Speed Motion

Tetsuya Morizono and Sadao Kawamura
Department of Robotics, Ritsumeikan University
Kusatsu, Shiga, Japan
{gr010952, kawamura}@bkc.ritsumei.ac.jp

Abstract: In this paper, we consider virtual sports systems as a useful application of virtual reality. A force display device used in such an application has to possess abilities of high speed motions and a large motion area. To satisfy those abilities, parallel wire mechanism is adopted for this device. Through some basic experimental results, we demonstrate that this device is capable of high speed motions enough to play virtual sports. Also we describe implementation of some virtual sports systems.

1. Introduction

Recently, many applications of virtual reality are studied on several fields. Sports training is one of useful applications of virtual reality. In this paper we focus on virtual sports, especially virtual ball games such as virtual playing catch or virtual tennis. In this application, force feedback from a virtual environment is necessary to achieve high quality of the virtual sports system. Therefore, at first we consider requirements for the force display device used in this application.

By observing many sports in the real world, we note that arms of the player move in a large space with high speed. From this fact, the force display device used in the virtual sports system especially requires following two abilities: capacity for high speed motions and a large motion area. In addition, the force display device also has to possess some general abilities such as player's safety and structural transparency[1].

Although several types of force display devices have been proposed so far, it is difficult for them to satisfy those four abilities simultaneously. For example, one way to achieve a large motion area is to adopt a serial link mechanism. However, such a mechanism inherently has large inertia. Though this large inertia can be compensated with higher-powered actuators, for such a device it is difficult to maintain safety in case that the device gets out of control. To overcome this difficulty, a new type of force display device have been proposed so far, which is utilizing *parallel wire mechanism*[2]. Wires have very small masses against their lengths. This useful characteristic gives some advantages on this mechanism.

Basically a parallel mechanism can be equipped with a moving part (we call a grip handle in this paper) whose mass is small. Moreover, because of

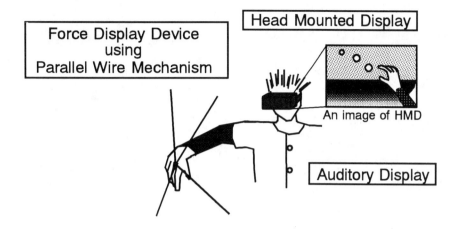

Figure 1. The Sketch of the Virtual Playing Catch System

small masses of wires supporting a grip handle, just lower-powered actuators can be utilized to drive wires. As a result, the whole system also has low inertia. This is a fundamental characteristic to achieve player's safety and structural transparency under high speed motions. Secondly, this characteristic is maintained even in the case of the device which has a large motion area, because increase of inertia in the device is negligibly small even if we employ long wires to realize a large motion area.

There are some studies dealing with this mechanism[3][4]. However in those studies, this mechanism has been treated as manipulators. On the other hand, in the force display device category, although some studies have been reported[5], capability of high speed motions on this kind of device has not been clarified.

In this paper, considering virtual sports systems as a suitable application for the force display device using parallel wire mechanism, at first we describe implementation of this system and the detail of our equipment. Secondly, we investigate high speed motion capacity of the force display device using parallel wire mechanism through some basic experiments. From a viewpoint of velocity and acceleration, it is demonstrated that this mechanism is capable of achieving high speed motions. Finally, some issues to improve this device are discussed.

2. Virtual Playing Catch System
2.1. Summary

At first, we show the sketch of the *virtual playing catch system* in Figure 1 and the realized system in Figure 2. This system has three display devices: a HMD(Head Mounted Display) as a visual display device, the force display device using parallel wire mechanism and an auditory display device. The specification for this system is described as follows:

Figure 2. The Realized Virtual Playing Catch System

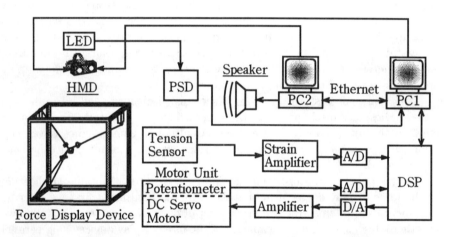

Figure 3. The Control Diagram of the Virtual Playing Catch System

- Through a HMD, a player can see a virtual environment, a ball, an opponent player and an arm of the player.
- Through the force display device, a player can feel mass of the ball and reaction forces from the ball when he/she pitches and catches it.
- Through an auditory display device, a player can hear sounds generated in the virtual environment.

The control diagram is shown in Figure 3.

2.2. Updating images in HMD

To generate images which should be displayed for a player, position and orientation of a player's head are needed. As shown in Figure 3, they are obtained

Figure 4. An Example Image of the Virtual Environment

through signals of the PSD(Position Sensitive Device) from LEDs attached onto the HMD. On the other hand, trajectories of moving virtual objects such as a ball and an opponent player are calculated using the DSP(Digital Signal Processor). These signals and trajectories are sent to one of personal computers(PC-9821Xa13), and also to another computer via ethernet. Based on them, each computer generates the image of the virtual environment to be displayed for each eye of a player. These images are separately sent to each display in the HMD. By generating both images with some disparity, a player can see the 3-dimensional view.

One of images is shown in Figure 4. Walls, a ball, an opponent player and an arm of the player are displayed.

2.3. Force Display Device using Parallel Wire Mechanism

2.3.1. Structure

We show the sketch view of the force display device using parallel wire mechanism in Figure 5. This device has four components: a frame, four motor units, four wires including four tension sensors and a handle. According to [2] and [3], we can realize a 3-DOF device with four wires.

To obtain a large motion area of the device, we need a large frame. Then we prepare a frame whose each length is approximate 2,400[mm]. Naturally, by using a larger frame, a larger motion area can be realized.

Four motor units are attached onto this frame. As shown in Figure 6, the structure of each unit is simple. Each unit has a 60[W] DC actuator, a reduction mechanism, a wire driving pulley and a potentiometer to measure a length of the wire.

One end of each wire is connected to the wire driving pulley and another end is attached to the handle. The handle is a ring style one. A player inserts his/her finger into this. Mass of the handle is just 15[g]. In the middle of each wire, a tension sensor is set. To maintain an advantage of low inertia of the device, this sensor is made as small as possible, whose size is 20×10[mm] and

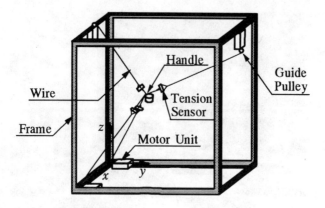

Figure 5. The Force Display Device using Parallel Wire Mechanism

Figure 6. The Motor Unit of the Device

its thick is 2[mm].

2.3.2. Control

The Control diagram of the force display device is already shown in Figure 3. All signals such as lengths and tensions are sent to the DSP. Based on these signals, the DSP calculates desired wire tensions and input currents for actuators by implementing a control law. Desired wire tensions are determined according to the desired force to be displayed for a player[6]. The DSP also calculates the handle position using direct kinematics of the device. Then this position is sent to personal computers to draw an arm of a player in images of the virtual environment.

3. Basic Experiments on Force Display Device

In this section, we especially focus on the force display device using parallel wire mechanism. To investigate high speed motion ability of this device, we carried out some basic experiments.

3.1. Control Laws

Simplicity of a control law gives some advantages to a sampling time and stability of the device. Based on this idea, in this paper we adopt following two control laws which are easy to implement: a tension feedback law and a combined control law of both tension feedback and dynamic compensation of motor units. We describe details of implementing those control laws below.

3.1.1. Tension Feedback Law

This control law is written as

$$u_i = K_f(f_{di} - f_i) + f_{di} \quad (i = 1, \cdots, 4), \tag{1}$$

where u_i, $K_f > 0$, f_{di} and f_i denote the input current for the ith motor unit, a feedback gain, a desired tension for the ith wire and an actual tension obtained from the ith tension sensor, respectively.

Naturally, to obtain better performance of the device with this control law, a higher feedback gain K_f is needed.

3.1.2. Combined Control Law

This is the control law in which dynamic compensation for motor units are added to the tension feedback law mentioned before.

This control law is described as

$$u_i = K_f(f_{di} - f_i) + f_{di} + \hat{D}_i \dot{\theta}_i + \hat{c}_{0i} \text{sgn}(\dot{\theta}_i), \tag{2}$$

where $\hat{D}_i$ and $\hat{c}_{0i}$ denote estimated coefficients of viscous and coulomb frictions when we assume the model of the ith motor unit as

$$J_i \ddot{\theta}_i + D_i \dot{\theta}_i + c_{0i} \text{sgn}(\dot{\theta}_i) = \tau_i - T_i, \tag{3}$$

where J_i denotes the inertia coefficient of the ith motor unit, θ_i denotes the angle of the actuator in the ith motor unit, and τ_i means the exerted torque on the actuator which is given by $\tau_i = k_t u_i$ (k_t: the torque constant). T_i is the equivalent torque caused by the ith wire tension f_i ($T_i = nr_p f_i$, n: the reduction ratio and r_p: the radius of the wire driving pulley). We assume that J_i, D_i and c_{0i} are constants during our experiments. For implementation of the combined control law eq.(2), we estimate each coefficient. Table 1 shows estimated coefficients for each motor unit.

It is noted that in eq.(2), while viscous and coulomb frictions are compensated, the inertia term is not compensated. One reason is that in general we need exact acceleration signals to compensate the inertial term. However, in the experiment using this control law, we are rather interested in investigating how effective the mechanical advantage of low inertia is in case that parallel wire mechanism is adopted for the force display device. Therefore, though we assume the model of each motor unit as eq.(3), we implement compensation of dynamics of motor units without inertia terms as shown in eq.(2).

Table 1. Estimated Coefficients of Motor Units

Unit Number	J_i ($\times 10^{-6}$)[Nms^2]	D_i ($\times 10^{-5}$)[Nms]	c_{0i} ($\times 10^{-3}$)[Nm]
0	8.60	6.18	5.17
1	9.51	8.35	6.70
2	8.62	3.27	11.6
3	9.49	2.96	20.8

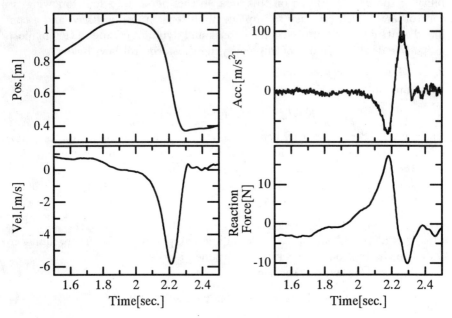

Figure 7. A Result Obtained using the Tension Feedback Law

3.2. Procedure

The procedure of experiments is quite simple. Firstly, we set the desired force for the player as *zero* in all experiments. And the player moves the handle as fast as possible. The direction of handle motions is negative one along the y axis in Figure 5. Then all results shown below are also plotted along the y axis.

During the handle movement, handle positions and reaction forces were measured. Velocities and accelerations are calculated using numerical differentiation. To demonstrate obtained results intelligibly, we handle all of them with a digital filter whose cut off frequency is 10[Hz].

3.3. Desirable Results

As mentioned in introduction, one of our purpose for these experiments is to investigate high speed motion ability of the force display device using parallel wire mechanism. Therefore, values themselves of achieved velocities and accelerations are very important in this paper. Naturally, results of high speed motions (large velocities and accelerations) are better ones.

On the other hand, different from experiments for manipulators[4], in ex-

Table 2. The Maximum Values Obtained through Experiments

Type of Control Law	Gain K_f	Velocity [m/s]		Acceleration [m/s^2]		Reaction force [N]	
Tension Feedback	4.0	5.8		70		17.2	
	7.5	6.2	(1.07)	85	(1.21)	12.2	(0.709)
Combined Control	4.0	7.5	(1.29)	105	(1.40)	16.9	(0.983)

* Each record is presented as an absolute value. Values inside parentheses denote each ratio of records compared to those in the case of the tension feedback law with $K_f = 4.0$.

periments for the force display device we have to pay our attention also to reaction forces from the device. Unnecessary forces generated due to the structure or characteristics of the device taint quality of virtual reality. Ideally speaking, since the desired force for a player is set as zero, the force display device must not give any forces to a player against any handle motions. Practically, it is better that a higher velocity and a higher acceleration are realized with a lower reaction force.

3.4. Experimental Results

At first we show an example of experimental results in Figure 7. This result was obtained using the tension feedback law eq.(1) with $K_f = 4.0$. From Figure 7, we can observe following values: the maximum velocity of 5.8[m/s] of the handle (we describe them as absolute values here), the maximum acceleration of approximate 70[m/s^2] and the maximum reaction force of 17.2[N] while the handle is accelerated. In this way, we pick up maximum velocities, accelerations and reaction forces while the handle is in motion from three results, and summarize them into Table 2. In this table, types of used control laws and set feedback gains also appear.

3.4.1. Capability for High Speed Motions

By observing velocities and accelerations, we can see that this device is capable of motions with velocities of more than 5[m/s] and accelerations of more than 70[m/s^2]. Especially in case of the combined control law, a player could realize the maximum velocity of 7.5[m/s] and the maximum acceleration of approximate 105[m/s^2] of his arm with this device. Probably, it seems that this handle motion is the fastest records at present in the force display device category.

3.4.2. Performance Improvement on Tension Feedback Law

Next, from Table 2 we can also see improvement of performance by an increased feedback gain. We focus on the case of using the tension feedback law. Comparing the result with $K_f = 7.5$ to the one with $K_f = 4.0$, it is understood that higher velocity and acceleration are realized with a lower reaction force. However, unfortunately, many oscillations with high frequency were observed after stopping the handle (Figure 8). There are some reasons to explain those oscillations. For example, the gain of $K_f = 7.5$ may be beyond the margin in which the device maintains stability, or it is possible that slight nonlinear

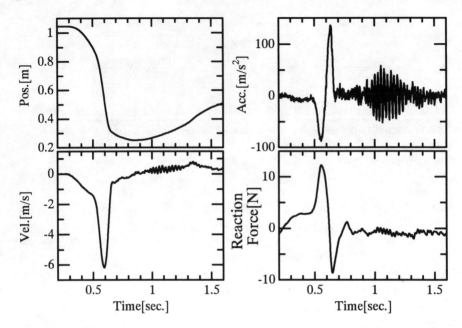

Figure 8. A Result in case of the Tension Feedback Law with a Higher Gain $K_f = 7.5$

elasticity of the force display device causes such oscillations under a higher feedback gain.

3.4.3. Performance Improvement on Combined Control Law

Here, we compare the result in the case using the combined control law with the one obtained by using the tension feedback law. The feedback gain of $K_f = 4.0$ is the same in both laws. From this comparison, we can see improvement on performance of the device. That is, higher velocity and acceleration are realized by the combined control law and with almost same reaction force. Moreover in this case, the problem on the oscillation has not occurred. In this comparison, possibility of improving performance of the device has been demonstrated. Further discussion will be in the final section.

4. Virtual Tennis System

4.1. Summary

Finally, we briefly explain another application of virtual sports, the *virtual tennis system*. The realized system is shown in Figure 9. The specification and the control diagram of the system are similar to those of the virtual playing catch system except the force display device.

The sketch of the force display device used in the virtual tennis system is shown in Figure 10. This device employs seven wires and realizes 6-DOF with a different wire configuration from the device in the virtual playing catch system. To avoid interferences between a player and wires, we use a rod. Wires are connected to both ends of the rod. Inherently, feasible rotating ranges of

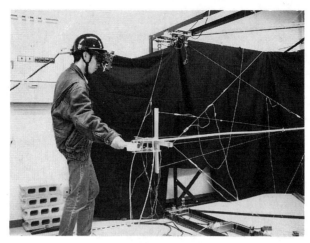

Figure 9. The Virtual Tennis System

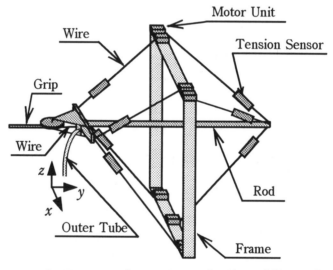

Figure 10. The Force Display Device in the Virtual Tennis System

parallel mechanisms are small compared with serial mechanisms. Therefore, by attaching an additional 1-DOF rotating type display to the rod, we obtain a large rotation area enough for forehand and backhand swings of the racket. Since this is a 7-DOF device, we can display not only forces but also moments to the grip.

4.2. Experimental Results

We carried out a basic experiment for this device to investigate high speed motion ability. Also in this experiment, the desired force for a player is set as zero. Then the player oscillates the grip along the x axis in Figure 10. While the grip is in motion, we measure the grip position.

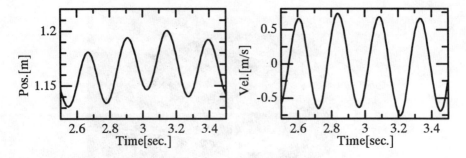

Figure 11. A Result for oscillating the grip of the 6-DOF Device

The position and the velocity of the experimental result are shown in Figure 11. This experiment resulted in only the maximum velocity of 0.7[m/s] and the maximum acceleration of approximate 17[m/s^2]. It is presumed that this poor performance is caused due to viscous and coulomb frictions of the device. In fact, we employ not the combined control law but the tension feedback law in this experiment. Therefore, it is expected that better performance will be obtained if we utilize compensation against viscous and coulomb frictions.

5. Conclusion and Future Works

In this paper, we have considered the force display device which especially requires abilities of high speed motions and a large motion area. To realize such a device, parallel wire mechanism was adopted for the force display device. To investigate high speed motion ability for this kind of device, some basic experimental results based on two control laws have been demonstrated in this paper. Moreover, we described implementation of virtual sports systems such as the virtual playing catch system and the virtual tennis system as suitable applications of this force display device.

Through basic experiments, we have demonstrated that this force display device is surely capable of high speed motions which sometimes appear in many sports. However, for example in case of playing catch, it is said that the initial velocity of the ball tends to 30[m/s] when an adult male pitches it. Therefore, improvement on high speed motion ability is one of our future works.

Our more important future work is how to reduce reaction forces under such high speed motions. In this paper, we examined two control laws. Using the tension feedback law, we checked improvement on performance of the device. However, many oscillations were observed at the same time. Although theoretical explanation will be needed in future, it is natural consideration that there is some margin of the feedback gain and then improvement on performance has a limit quantitatively.

On the other hand, in this paper we have also shown that utilizing the compensation on viscous and coulomb frictions of motor units is effective. It means that viscous and coulomb frictions have significant values in comparison with the inertia term. To reduce the reaction force and obtain better performance of the device, we have to consider some improvement of our method. In this paper, we assumed that coefficients of motor units were constants and estimated them in off-line. However, in general it is natural that those coefficients vary in every experiment and sometimes with time in one experiment. This consideration suggests that an adaptive compensation is needed and useful for this kind of force display device. Furthermore, since inertia terms of motor units must affect performance of the device with higher acceleration, we may also need a good way to compensate those terms exactly.

References

[1] Hui R. and Gregorio P., The Virtual Handle, *Proc. of the 1995 IEEE/RSJ Int. Conf. on Intelligent Robots and Systems*, pp.127-132, 1995.

[2] Kawamura S. and Itoh K., A New Type of Master Robot for Teleoperation Using A Radial Wire Drive System, *Proc. of the 1993 IEEE/RSJ Int. Conf. on Intelligent Robots and Systems*, pp.55-60, 1993.

[3] Higuchi T. and Ming A., Study on Multiple Degree-of-Freedom Positioning Mechanism Using Wires, *Proc. of the Asian Conf. on Robotics and Its Application*, pp.101-106, 1991.

[4] Kawamura S., Choe W., Tanaka S. and Pandian S.R., Development of an Ultra-high Speed Robot FALCON using Wire Drive System, *Proc. of IEEE Int. Conf. on Robotics and Automation*, pp.215-220, 1995.

[5] Ishii M., Nakata M. and Sato M., Networked SPIDAR: A Networked Virtual Environment with Visual, Auditory and Haptic Interactions, *Presence*, Vol.3, No.4, pp.351-359, 1994.

[6] Kawamura S., Ida M., Wada T. and Wu J.L., Development of a Virtual Sports Machine using a Wire Drive System — A Trial of Virtual Tennis —, *Proc. of the 1995 IEEE/RSJ Int. Conf. on Intelligent Robots and Systems*, pp.111-116, 1995.

[7] Morizono T., Kurahashi K. and Kawamura S., Realization of a Virtual Sports Training System with Parallel Wire Mechanism, *Proc. of the 1997 IEEE Int. Conf. on Robotics and Automation*, pp.3025-3030, 1997.

[8] Morizono T., Ida M., Wada T., Wu J.L. and Kawamura S., A Trial of Virtual Tennis using a Parallel Wire Drive System, *J. of the Robotics Society of Japan*, Vol.15, No.1, pp.153-161, 1997(in Japanese).

A General Contact Model for Dynamically-Decoupled Force/Motion Control

Roy Featherstone*
Department of Engineering Science
Oxford University
Parks Road, Oxford OX1 3PJ, England
roy@robots.oxford.ac.uk

Stef Sonck
Aerospace Robotics Laboratory
Department of Aeronautics and Astronautics
Stanford University
Stanford, CA 94305, USA
ssonck@sun-valley.Stanford.EDU

Oussama Khatib
Robotics Laboratory
Department of Computer Science
Stanford University
Stanford, CA 94305, USA
khatib@cs.Stanford.EDU

Abstract: This paper integrates a general first-order kinematic model of rigid-body contact with the equations of motion of the manipulated objects and robot arms. The more general kinematics allows us to model tasks that cannot be described using the Raibert-Craig model; a single Cartesian frame in which directions are either force- or motion-controlled is not sufficient. The integration with the object and manipulator dynamics allows us to generalize the concept of projection matrices in force/motion control and related applications. The model is developed using an invariant formulation based on the duality between motion and force vectors. Experimental results are presented showing a manipulation that involves controlling the force in two separate face-vertex contacts while performing motion. These multi-contact compliant motions often occur as part of an assembly and cannot be described using the Raibert-Craig model.

1. Introduction

The modern concept of hybrid control, as described in the work of Raibert and Craig [15], has attracted a great deal of interest over the years. In addition to the various improvements, extensions and practical implementations that have

---

*Supported by EPSRC Advanced Research Fellowship number B92/AF/1466.

been proposed, two theoretical errors in the original formulation have received attention: the non-invariant formulation of the original contact model [5, 12] and the phenomenon of kinematic instability caused by an incorrect filtering of error signals into force and motion components [1, 8, 18].

Another problem with the Raibert-Craig model, which appears to have been neglected, is that it lacks sufficient generality to describe an arbitrary state of contact between two rigid bodies. Other published models vary on this point: some suffer the same problem (*e.g.* [4, 6, 10]) while others are completely general (*e.g.* [2, 3, 9, 16, 17]).

This paper presents a dynamic contact model that solves all three problems: it can describe any state of frictionless contact between two rigid polyhedral bodies. Its projection matrices are invariant with respect to choice of units and coordinate systems. In addition, if the manipulator's dynamics is compensated by a dynamic control structure then these projections exhibit a dynamic-decoupling property that has not been recognized before: the actual contact force depends only on the input to the force projection, and the relative acceleration only on the input to the motion projection.

The rest of this paper is organized as follows. First, we show that the Raibert-Craig model is not general and give an example of a contact that it can not handle. Then, after a brief review of dual vector systems, we describe the new model and derive the equations of the force and motion projection matrices. The next section presents an analysis of the equations of motion of a manipulator in contact and defines the *dynamically consistent projection matrices*. To validate the usefulness of these projection matrices, we present experimental results showing a robot performing a compliant motion task that can not be described by the Raibert-Craig model. The task involves controlling the contact force between the manipulated object and the environment in two face-vertex contacts, while executing a motion.

2. Generality of Contact Models for Force/Motion Control

A state of contact between a robot's end effector and its environment defines a constraint surface in the robot's operational space such that all points on the surface satisfy the contact constraints. At any given instant, this surface defines two vector spaces: a space of tangent vectors containing all permissible motions (velocities, infinitesimal displacements, or accelerations after compensation for velocity-product effects), and a space of normal vectors containing all permissible contact forces. A mathematical model of contact must include a means of describing these two spaces.

The contact model used by Raibert and Craig was based on the theoretical work of Mason [13], and consists of a Cartesian coordinate frame, called the constraint frame, and a compliance selection matrix. The constraint frame defines three lines of pure force aligned with the coordinate axes, three couples parallel to the axes, three translations parallel to the axes and three rotations about the axes. Unfortunately, any contact model that uses only six geometric parameters is lacking in generality.

Figures 1 and 2 show examples of contact states that can not be described

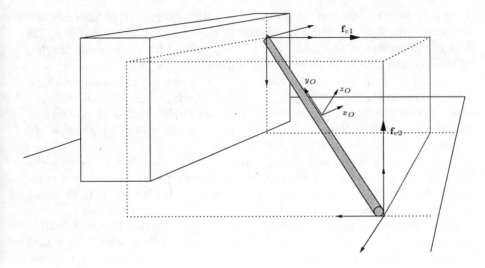

Figure 1. A general two-point contact with skew, non-intersecting contact normals.

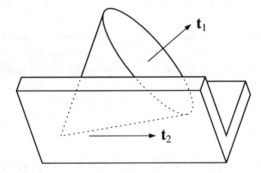

Figure 2. The cone has two motion freedoms: rotation about its axis (t_1) and translation along the groove's axis (t_2). The two axes are neither parallel nor perpendicular.

by the Raibert-Craig model. For example, in figure 1 it is not possible to find a location for the constraint frame such that the space spanned by the two contact normals along forces f_{c1} and f_{c2} equals a space spanned by two constraint-frame directions.

3. Dual Vector Spaces

In this section, we briefly review the mathematics of dual vector spaces. The basic idea is that we have two separate vector spaces, one space contains force-type vectors, the other contains motion-type vectors. We shall call the two spaces F^6 and M^6. The scalar product defined between them is called the reciprocal product, it is the work done by a force-type vector acting on a motion-type vector.

A Cartesian coordinate frame defines two separate bases: $\{f_1 \ldots f_6\}$ for F^6 and $\{m_1 \ldots m_6\}$ for M^6. The vectors $f \in \mathsf{F}^6$ and $v \in \mathsf{M}^6$ are represented by the column matrices $\mathbf{f}$ and $\mathbf{v}$ in these bases. They must satisfy the reciprocity condition

$$f_i \cdot m_j = \begin{cases} 1\,W & \text{if } i = j \\ 0 & \text{otherwise} \end{cases},$$

where W is the appropriate unit of work. This condition enforces consistency of units and ensures that the scalar product $f \cdot v$ can be expressed as $\mathbf{f}^T \mathbf{v}$. By convention, the two bases comprise three unit forces, three unit couples, three unit linear motions and three unit angular motions, all arranged along/about the coordinate frame's axes.

If $\mathbf{f}_P, \mathbf{f}_Q, \mathbf{v}_P$ and $\mathbf{v}_Q$ are representations of f and v in Cartesian coordinate frames P and Q, and $\mathbf{X}_f$ and $\mathbf{X}_m$ are coordinate transformation matrices from P to Q coordinates for force-type and motion-type vectors, respectively, then

$$\mathbf{v}_Q = \mathbf{X}_m \mathbf{v}_P, \quad \mathbf{v}_P = \mathbf{X}_m^{-1} \mathbf{v}_Q, \tag{1}$$

$$\mathbf{f}_Q = \mathbf{X}_f \mathbf{f}_P, \quad \mathbf{f}_P = \mathbf{X}_f^{-1} \mathbf{f}_Q. \tag{2}$$

The transformational properties of the two spaces are constrained by the need to preserve invariance of $\mathbf{f}^T \mathbf{v}$:

$$\mathbf{X}_f = (\mathbf{X}_m^{-1})^T. \tag{3}$$

Given the two vector spaces, we can define four types of linear mapping between them. Physical examples are: inertia, inverse inertia, force projection and motion projection, respectively. To preserve invariance, these mappings must transform as follows.

$$\mathsf{M}^6 \mapsto \mathsf{F}^6 : \quad \mathbf{A}_Q = \mathbf{X}_f \mathbf{A}_P \mathbf{X}_m^{-1}, \tag{4}$$

$$\mathsf{F}^6 \mapsto \mathsf{M}^6 : \quad \mathbf{B}_Q = \mathbf{X}_m \mathbf{B}_P \mathbf{X}_f^{-1}, \tag{5}$$

$$\mathsf{F}^6 \mapsto \mathsf{F}^6 : \quad \mathbf{C}_Q = \mathbf{X}_f \mathbf{C}_P \mathbf{X}_f^{-1}, \tag{6}$$

$$\mathsf{M}^6 \mapsto \mathsf{M}^6 : \quad \mathbf{D}_Q = \mathbf{X}_m \mathbf{D}_P \mathbf{X}_m^{-1}. \tag{7}$$

4. A General First-order Model of Contact

A general state of contact between two rigid bodies defines a constraint surface in their relative configuration space. At any given instant, this surface defines two vector spaces: an r-dimensional space of contact normal vectors $N \subseteq \mathsf{F}^6$ and a $(6-r)$-dimensional space of tangent vectors $T \subseteq \mathsf{M}^6$, where r is the degree of motion constraint. These spaces can be modeled by means of a $6 \times r$ matrix $\mathbf{N}$ and a $6 \times (6-r)$ matrix $\mathbf{T}$ such that

$$N = \text{range}(\mathbf{N}), \quad T = \text{range}(\mathbf{T}).$$

The columns of $\mathbf{N}$ are any r linearly independent force vectors in N, and the columns of $\mathbf{T}$ are any $6-r$ linearly independent motion vectors in T. There is no need for any of these vectors to be normalized or orthogonal in any sense.

One of the basic properties of a contact constraint is that a constraint force does no work against an infinitesimal displacement that is consistent with the constraint. In other words, the scalar product of any member of N with any member of T is zero. This property can be expressed as

$$N \perp T,$$

or, in terms of matrices,

$$\mathbf{N}^T \mathbf{T} = \mathbf{0}.$$

In many situations, a suitable value for $\mathbf{N}$, $\mathbf{T}$ or both can be obtained by inspection. For the contact shown in Figure 1, a suitable value for $\mathbf{N}$ is $\mathbf{N} = [\mathbf{f}_{c1}\ \mathbf{f}_{c2}]$; and for the contact shown in Figure 2, a suitable value for $\mathbf{T}$ is $\mathbf{T} = [\mathbf{t}_1\ \mathbf{t}_2]$.

5. The Projection Matrices

Let us introduce two more spaces, N' and T', and their matrix representations $\mathbf{N}'$ and $\mathbf{T}'$, satisfying

$$N \oplus N' = \mathsf{F}^6, \quad T \oplus T' = \mathsf{M}^6,$$

where $\oplus$ means direct sum. N' can be any $(6-r)$-dimensional force subspace with no non-zero element in common with N, and T' can be any r-dimensional motion subspace with no non-zero element in common with T. It is important to recognize that N' and T' are not defined by the contact.

Each possible value of N' defines a unique decomposition of a general force vector into components in N and N'. Similarly, each possible value of T' defines a unique decomposition of a general motion vector into components in T and T'. Thus, given $f \in \mathsf{F}^6$ and $v \in \mathsf{M}^6$, we have

$$\mathbf{f} = \mathbf{f}_1 + \mathbf{f}_2 = \mathbf{N}\,\boldsymbol{\alpha}_1 + \mathbf{N}'\,\boldsymbol{\alpha}_2, \tag{8}$$

$$\mathbf{v} = \mathbf{v}_1 + \mathbf{v}_2 = \mathbf{T}\,\boldsymbol{\beta}_1 + \mathbf{T}'\,\boldsymbol{\beta}_2, \tag{9}$$

where $\boldsymbol{\alpha}_1$, $\boldsymbol{\alpha}_2$, $\boldsymbol{\beta}_1$ and $\boldsymbol{\beta}_2$ are all uniquely determined.

We are now in a position to define the projection matrices, $\boldsymbol{\Omega}_f$ and $\bar{\boldsymbol{\Omega}}_f$, which split a force vector $\mathbf{f}$ into its components $\mathbf{f}_1 = \boldsymbol{\Omega}_f \mathbf{f} \in N$ and $\mathbf{f}_2 = \bar{\boldsymbol{\Omega}}_f \mathbf{f} \in N'$, and the motion projection matrices $\boldsymbol{\Omega}_m$ and $\bar{\boldsymbol{\Omega}}_m$. These projections are not uniquely defined by N and T, but also depend on N' and T'. Premultiplying both sides of Eq. 8 by $\mathbf{T}^T$ eliminates the $\boldsymbol{\alpha}_1$ term. As $\mathbf{T}^T \mathbf{N}'$ is always invertible, we may solve for $\boldsymbol{\alpha}_2$ giving

$$\boldsymbol{\alpha}_2 = (\mathbf{T}^T \mathbf{N}')^{-1} \mathbf{T}^T \mathbf{f}.$$

Rearranging Eq. 8 into $\mathbf{f}_1 = \mathbf{f} - \mathbf{N}'\boldsymbol{\alpha}_2$ and substituting for $\boldsymbol{\alpha}_2$ gives

$$\mathbf{f}_1 = (\mathbf{1} - \mathbf{N}'\,(\mathbf{T}^T \mathbf{N}')^{-1}\,\mathbf{T}^T)\,\mathbf{f},$$

hence

$$\boldsymbol{\Omega}_f = \mathbf{1} - \mathbf{N}'\,(\mathbf{T}^T \mathbf{N}')^{-1}\,\mathbf{T}^T. \tag{10}$$

Similar processing of Eq. 9 yields

$$\Omega_m = 1 - \mathbf{T}'(\mathbf{N}^T \mathbf{T}')^{-1} \mathbf{N}^T. \tag{11}$$

Ω_f and its complementary projection $\bar{\Omega}_f = 1 - \Omega_f$ are both force-to-force mappings, and therefore transform according to Eq. 6. Similarly, Ω_m and $\bar{\Omega}_m$ transform according to Eq. 7.

Equations 10 and 11 represent every possible invariant projection matrix that is consistent with the given contact constraint. Without loss of generality, it is possible to represent N' and T' in the form

$$N' = \mathbf{A}\,T = \text{range}(\mathbf{A}\,T), \quad T' = \mathbf{B}\,N = \text{range}(\mathbf{B}\,N),$$

where $\mathbf{A}$ and $\mathbf{B}$ are positive-definite 6×6 matrices representing linear mappings from M^6 to F^6 and F^6 to M^6 respectively. To preserve invariance, $\mathbf{A}$ must transform like inertia (Eq. 4) and $\mathbf{B}$ like inverse inertia (Eq. 5). Using this representation, all possible projection matrices for a given contact constraint can be expressed as

$$\Omega_f(\mathbf{A}) = N\,(N^T \mathbf{A}^{-1} N)^{-1} N^T \mathbf{A}^{-1}, \tag{12}$$
$$\bar{\Omega}_f(\mathbf{A}) = \mathbf{A}\,T\,(T^T \mathbf{A}\,T)^{-1} T^T, \tag{13}$$
$$\Omega_m(\mathbf{B}) = T\,(T^T \mathbf{B}^{-1} T)^{-1} T^T \mathbf{B}^{-1}, \tag{14}$$
$$\bar{\Omega}_m(\mathbf{B}) = \mathbf{B}\,N\,(N^T \mathbf{B}\,N)^{-1} N^T. \tag{15}$$

Eqs. 13 and 15 are obtained from Eqs. 10 and 11 by substituting $\mathbf{A}\,T$ for N' and $\mathbf{B}\,N$ for T'. Eqs. 12 and 14 are obtained from Eqs. 8 and 9 by premultiplying both sides by $N^T \mathbf{A}^{-1}$ (or $T^T \mathbf{B}^{-1}$) to eliminate the α_2 (or β_2) term and solving for α_1 (or β_1).

Some useful relationships are independent of the choice of $\mathbf{A}$ or $\mathbf{B}$

$$\Omega_f(\mathbf{A}_1)\,\Omega_f(\mathbf{A}_2) = \Omega_f(\mathbf{A}_2), \tag{16}$$
$$\Omega_m(\mathbf{B}_1)\,\Omega_m(\mathbf{B}_2) = \Omega_m(\mathbf{B}_2), \tag{17}$$

$$\bar{\Omega}_m(\mathbf{A}_1)\,\Omega_m(\mathbf{A}_2) = \Omega_m(\mathbf{A}_3)\,\bar{\Omega}_m(\mathbf{A}_4) = 0, \tag{18}$$
$$\bar{\Omega}_f(\mathbf{B}_1)\,\Omega_f(\mathbf{B}_2) = \Omega_f(\mathbf{B}_3)\,\bar{\Omega}_f(\mathbf{B}_4) = 0. \tag{19}$$

6. Combining the Contact Model with Dynamics

The basic operational-space equation of motion for the object and the manipulator(s) (i.e. the augmented object model) [11] is

$$\Lambda_0(x)\,\dot{\vartheta} + \mu_0(x,\vartheta) + p_0(x) + \mathbf{f}_c = \mathbf{f} \tag{20}$$

where x is a vector of operational-space coordinates, $\vartheta, \dot{\vartheta} \in M^6$ are end-effector velocity and acceleration vectors, Λ_0 is the operational-space inertia matrix, $\mu_0, p_0 \in F^6$ are vectors of velocity-product and gravitational terms, $\mathbf{f}_c \in F^6$ is the contact force applied by the robot to the environment, and $\mathbf{f} \in F^6$ is the force command to the robot.

We model the environment by the equation of motion

$$\mathbf{a}_e = \mathbf{\Phi}_e \, \mathbf{f}_c + \mathbf{b}_e. \tag{21}$$

$\mathbf{a}_e$ is the environment's acceleration, $\mathbf{\Phi}_e$ is its inverse inertia (a positive semidefinite matrix), and $\mathbf{b}_e$ is its bias acceleration (the acceleration it would have in the absence of a contact force) [7]. Equation 21 models any environment that accepts an arbitrary applied force, and is functionally equivalent to the model used in [3]. A fixed, stationary environment can be modeled with $\mathbf{\Phi}_e = 0$, $\mathbf{b}_e = 0$.

To simplify subsequent equations, we introduce the symbol

$$\dot{\vartheta}' = \dot{\vartheta} - \mathbf{a}_e - \dot{\mathbf{T}}\beta,$$

which is the robot's relative acceleration with the velocity-product terms removed.

The contact imposes constraints on the relative acceleration between the end effector and environment, and on the contact force. These constraints are

$$\dot{\vartheta}' = \mathbf{T}\,\dot{\beta}. \tag{22}$$

$$\mathbf{f}_c = \mathbf{N}\,\alpha, \tag{23}$$

where $\mathbf{T}$ and $\mathbf{N}$ model the contact's tangent and normal subspaces, $\dot{\beta}$ and α are unknown acceleration and force vectors, β is a known velocity vector, and $\dot{\mathbf{T}}\beta$ is a velocity-product term.

Combining Eqs. 20–22 and rearranging gives

$$\dot{\beta} = (\mathbf{T}^T \mathbf{\Lambda}_{rel}\, \mathbf{T})^{-1}\, \mathbf{T}^T\, \mathbf{\Lambda}_{rel}\, \mathbf{\Lambda}_0^{-1}\, \mathbf{f}', \tag{24}$$

$$\alpha = (\mathbf{N}^T \mathbf{\Lambda}_{rel}^{-1} \mathbf{N})^{-1}\, \mathbf{N}^T \mathbf{\Lambda}_0^{-1}\, \mathbf{f}', \tag{25}$$

where

$$\mathbf{f}' \triangleq \mathbf{f} - \mu_0 - \mathbf{p}_0 - \mathbf{\Lambda}_0\,(\dot{\mathbf{T}}\beta + \mathbf{b}_e),$$

and

$$\mathbf{\Lambda}_{rel} \triangleq (\mathbf{\Lambda}_0^{-1} + \mathbf{\Phi}_e)^{-1}.$$

$\mathbf{\Lambda}_{rel}$ is the relative inertia between the manipulator and the environment—the tensor that relates contact force to relative acceleration. Note that if $\mathbf{\Phi}_e = 0$ then $\mathbf{\Lambda}_{rel} = \mathbf{\Lambda}_0$. Substituting Eqs. 24 and 25 back into Eqs. 22 and 23 gives (the factor $\mathbf{\Lambda}_{rel}\,\mathbf{\Lambda}_0^{-1}$ disappears if $\mathbf{\Phi}_e = 0$)

$$\dot{\vartheta}' = \mathbf{\Lambda}_{rel}^{-1}\,\bar{\mathbf{\Omega}}_f^*\,\mathbf{\Lambda}_{rel}\,\mathbf{\Lambda}_0^{-1}\,\mathbf{f}' \;=\; \mathbf{\Omega}_m^*\,\mathbf{\Lambda}_0^{-1}\,\mathbf{f}', \tag{26}$$

$$\mathbf{f}_c = \mathbf{\Omega}_f^*\,\mathbf{\Lambda}_{rel}\,\mathbf{\Lambda}_0^{-1}\,\mathbf{f}' \;=\; \mathbf{\Lambda}_{rel}\,\bar{\mathbf{\Omega}}_m^*\,\mathbf{\Lambda}_0^{-1}\,\mathbf{f}', \tag{27}$$

where the *dynamically consistent projection matrices* are (see Eqs. 12 and 14)

$$\mathbf{\Omega}_f^* = \mathbf{\Omega}_f(\mathbf{\Lambda}_{rel}), \tag{28}$$

$$\mathbf{\Omega}_m^* = \mathbf{\Omega}_m(\mathbf{\Lambda}_{rel}^{-1}). \tag{29}$$

The applied force has been split into two components by means of the dynamically consistent projections Ω_f^* and $\bar{\Omega}_f^*$, with one component responsible for the manipulator's relative acceleration and the other responsible for the contact force. Although there are an infinite number of force decompositions that are compatible with a contact constraint treated in isolation (given by Eqs. 12–15), there is only one that models its dynamic behavior correctly in any given circumstance.

Now let us consider the effect of controlling the manipulator via the following dynamic control structure (any commanded $\mathbf{f}'$ can be written in this form)

$$\mathbf{f}' = \Lambda_0 \dot{\vartheta}'_{comm} + \Lambda_0 \Lambda_{rel}^{-1} \mathbf{f}_{ccomm}, \qquad (30)$$

with $\dot{\vartheta}'_{comm}$ the commanded relative acceleration and $\mathbf{f}_{ccomm}$ the commanded contact forces. This results in the equations of motion

$$\dot{\vartheta}' = \Omega_m^* \dot{\vartheta}'_{comm} + \Lambda_{rel}^{-1} \bar{\Omega}_f^* \mathbf{f}_{ccomm}, \qquad (31)$$

$$\mathbf{f}_c = \Omega_f^* \mathbf{f}_{ccomm} + \Lambda_{rel} \bar{\Omega}_m^* \dot{\vartheta}'_{comm}. \qquad (32)$$

The actual relative acceleration is therefore the commanded relative acceleration filtered through Ω_m^* plus a disturbance from the $\bar{\Omega}_f^*$ component of $\mathbf{f}_{ccomm}$; and the actual contact force is the commanded contact force filtered through Ω_f^* plus a disturbance from the $\bar{\Omega}_m^*$ component of $\dot{\vartheta}'_{comm}$. If the commanded force and acceleration are such that the two disturbance terms are always zero then the system is dynamically decoupled.

7. Experimental Application of the model

The model can be used in several ways to build force/motion controllers that can deal with contact tasks that go beyond the Raibert-Craig model.

7.1. The experimental testbed

Fig. 3 shows the experimental platform: a dual-arm robotic workcell, consisting of two SCARA-type manipulators and an overhead vision system which tracks the position of the arms and the objects that the arms are in contact with. In this experiment, both arms cooperatively grasp a metal rod (approximately 60 cm long). Hinges in the grippers allow the bar to rotate with respect to the arm end effector. Each individual arm has four actuated degrees of freedom; both arms together can move the rod in five degrees of freedom. The two end-points of the rod are in contact with the environment, following the configuration of Fig. 1.

A joint-level controller, running at 350 Hz, uses joint torque sensors to compensate for (exaggerated) joint-flexibility and the fact that the actuators are not ideal torque sources [14]. The dynamics of the two arms and the manipulated object are combined in the operational space and are of the form of Eq. 20, where $\Lambda_0 = \Lambda_{0l} + \Lambda_{0r} + \Lambda_{0o}$ is the total inertia of the two arms and the manipulated object, $\mathbf{f}$ is the total torque from both arms, and $\mathbf{f}_c = \mathbf{f}_{c1} + \mathbf{f}_{c2}$ is the total effect of the two contacts. Using the transformations of Eqs. 4

Figure 3. The dual-arm robotic workcell and the two-contact task

to 7, all vectors and matrices are expressed in an operational frame which is (arbitrarily) located in the middle of the rod.

7.2. Decoupled Architecture for Force/Motion Control

Figure 4 shows the global architecture of the force/motion controller. In this implementation, the contact model is represented by a 6×2 matrix $\mathbf{N}$. It is continuously updated as the arms and/or objects move.

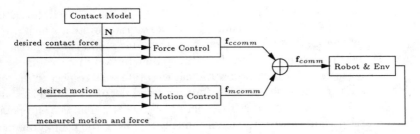

Figure 4. Decoupled Architecture for Force/Motion Control

7.3. Projection of the measured forces

The 6-dof force-sensors used in this experiment sit between the arm end-points and grippers holding the manipulated object described by the matrix Λ_{0o}. The measured forces are filtered through the projection matrix $\Omega_f(\Lambda_{0o})$ which is dynamically consistent with the inertia Λ_{0o} of the manipulated object.

The experiment of Fig. 7.3 shows how the reciprocity condition holds when the contact constraints are respected. The top part of the plot is the total instantaneous power $\boldsymbol{\vartheta}^T \mathbf{f}'_{meas}$, which includes the power resulting from object motion. The bottom plot is the power $\boldsymbol{\vartheta}^T \left(\boldsymbol{\Omega}_f(\boldsymbol{\Lambda}_{0o}) \mathbf{f}'_{meas} \right)$ delivered by the forces projected in the contact space. Because of reciprocity, this power should be zero (small errors are mainly due to friction in the contacts).

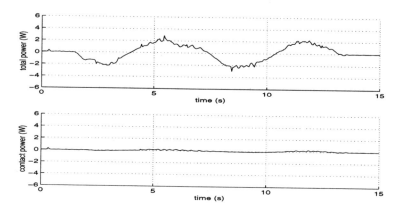

Figure 5. Power with and without projection while contacts are maintained

Figure 7.3 shows how the reciprocity condition is violated when the constraints are not respected. Every peak corresponds to a collision between the rod and the environment.

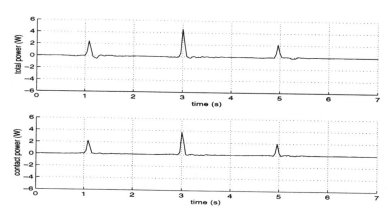

Figure 6. Power with and without projection when contacts are *not* maintained

7.4. Force Control

Fig. 7 shows the basic force control structure. A generalized inverse of $\mathbf{N}$ transforms the projected measured force in the two-dimensional contact space. The controller of the contact forces compares the measured contact forces $\boldsymbol{\alpha}_{meas}$ with the desired contact forces $\boldsymbol{\alpha}_d$ for $\mathbf{f}_{c1}$ and $\mathbf{f}_{c2}$. Finally, the commanded forces are multiplied by $\mathbf{N}$ to return to the operational space. From Eq. 31, it is clear that this controller will not disturb the control of motions.

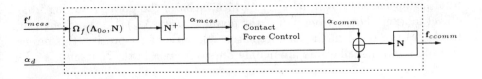

Figure 7. Force Control in the **N** space

Figure 8 shows the desired and measured contact forces $\mathbf{f}_{c1}$ and $\mathbf{f}_{c2}$. These forces are controlled in the two-dimensional space N (described by matrix **N**), which can simply not be described by the Raibert-Craig model.

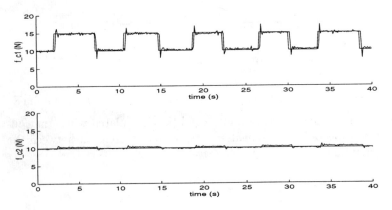

Figure 8. Multi Contact Force Control

In practice, additional velocity damping (using the projection matrix $\bar{\mathbf{\Omega}}_m$) is added so that the controller gracefully re-acquires a broken contact.

8. Conclusion

This paper extends the Raibert-Craig model to a more general framework in which any contact state between the manipulated object and the environment can be described (to first order). This kinematic model is combined with the dynamics of the arm and the manipulated objects. We have also discussed all possible motion and force projection matrices and defined what it means for a projection to be *dynamically consistent*.

This approach makes it possible to model and control force/motion tasks that include multiple points of contact which cannot be tackled with the Raibert-Craig model. The experimental results show how the reciprocity condition holds for the contact forces that are obtained by a dynamically consistent projection of the measured forces. The results also show how the force control of the two contact points is decoupled from the object motion.

References

[1] An, C. H., and Hollerbach, J. M., Kinematic Stability Issues in Force Control of Manipulators, IEEE Int. Conf. Robotics and Automation, 1987.

[2] Bruyninckx, H., Demey, S., Dutré, S., De Schutter, J., Kinematic Models for Model-Based Compliant Motion in the Presence of Uncertainty, *Int. Jnl. Robotics Research*, Vol. 14, No. 5, pp. 465–482, 1995.

[3] De Luca, A., and Manes, C., Modelling of Robots in Contact with a Dynamic Environment, *IEEE Trans. Robotics and Automation*, Vol. 10, No. 4, pp. 542–548, 1994.

[4] De Schutter, J., and Van Brussel, H., Compliant Robot Motion I. A Formalism for Specifying Compliant Motion Tasks, *Int. Jnl. Robotics Research*, Vol. 7, No. 4, pp. 3–17, 1988.

[5] Duffy, J., The Fallacy of Modern Hybrid Control Theory that is Based on "Orthogonal Complements" of Twist and Wrench Spaces, *Jnl. Robotic Systems*, Vol. 7, No. 2, pp. 139–144, 1990.

[6] Faessler, H., Manipulators Constrained by Stiff Contact—Dynamics, Control and Experiments, *Int. Jnl. Robotics Research*, Vol. 9, No. 4, pp. 40–58, 1990.

[7] Featherstone, R., *Robot Dynamics Algorithms*, Kluwer Academic Publishers, Boston/Dordrecht/Lancaster, 1987.

[8] Fisher, W. D., and Mujtaba, M. S., Hybrid Position/Force Control: A Correct Formulation, *Int. Jnl. Robotics Research*, Vol. 11, No. 4, pp. 299–311, 1992.

[9] Jankowski, K. P., and ElMaraghy, H. A., Dynamic Decoupling for Hybrid Control of Rigid-/Flexible-Joint Robots Interacting with the Environment, *IEEE Trans. Robotics & Automation*, Vol. 8, No. 5, pp. 519–534, Oct. 1992.

[10] Khatib, O., A Unified Approach for Motion and Force Control of Robot Manipulators: The Operational Space Formulation, *IEEE Jnl. Robotics & Automation*, Vol. 3, No. 1, pp. 43–53, 1987.

[11] Khatib, O., Inertial Properties in Robotic Manipulation: An Object-Level Framework, *Int. Jnl. Robotics Research*, Vol. 14, No. 1, pp. 19–36, 1995.

[12] Lipkin, H., and Duffy, J., Hybrid Twist and Wrench Control for a Robotic Manipulator, *ASME Jnl. Mechanisms, Transmissions & Automation in Design*, vol. 110, No. 2, pp. 138–144, June 1988.

[13] Mason, M. T., Compliance and Force Control for Computer Controlled Manipulators, *IEEE Trans. Systems, Man & Cybernetics*, Vol. SMC-11, No. 6, pp. 418–432, June 1981.

[14] Pfeffer, L. E., and Cannon, R. H., Experiments with a Dual-Armed, Cooperative, Felxible-Drivetrain Robot System, IEEE Int. Conf. Robotics and Automation, pp. 601–608, Atlanta, GA, 1993.

[15] Raibert, M. H., and Craig, J. J., Hybrid Position/Force Control of Manipulators, *ASME Jnl. Dynamic Systems, Measurement & Control*, Vol. 103, No. 2, pp. 126–133, June 1981.

[16] West, H., and Asada, H., A Method for the Design of Hybrid Position/Force Controllers for Manipulators Constrained by Contact with the Environment, IEEE Int. Conf. Robotics and Automation, pp. 251–259, St. Louis, MO, 1985.

[17] Yoshikawa, T., Sugie, T., and Tanaka, M., Dynamic Hybrid Position Force Control of Robot Manipulators — Controller Design and Experiment, *IEEE Jnl. Robotics & Automation*, Vol. 4, No. 6, pp. 699–705, 1988.

[18] Zhang, H., Kinematic Stability of Robot Manipulators under Force Control, IEEE Int. Conf. Robotics and Automation, pp. 80–85, Scottsdale, AZ, 1989.

Control of a Rover-Mounted Manipulator

Gilles Foulon, Jean-Yves Fourquet and Marc Renaud
LAAS-CNRS
7, Avenue du Colonel Roche, 31077 Toulouse Cédex 4
FRANCE
foulon@laas.fr

May 20, 1997

Abstract

In this paper, we present a control algorithm for a mechanical system built from a six revolute joint robot manipulator mounted on a nonholonomic mobile platform. We describe a global control of this mechanical system which calculates the joint values for both the robot manipulator and the mobile platform. We want the end-effector *location* (position+orientation) to evolve between starting and final required locations and to follow a required trajectory. First, we present simulated results to illustrate the efficiency of the control algorithm. Then, the controls are implanted and checked on the real mechanical system.

1 Introduction

When we write across or erase a blackboard, we locate our body so that our arm is in the most comfortable configuration to be able to write. We place our body, in an instinctive manner, in the admissible workspace of our arm by moving our legs. So we consider our legs like a mobile platform which is able to locate and move our torso in a parallel to the board. We can describe two different ways to move and realize our goal : first, by moving the platform towards the board, keeping the robot manipulator stopped, and then by moving the robot manipulator ; secondly, by moving both the robot manipulator and the mobile platform as a human does.
Now, we consider our mechanical system built from a six revolute joints robot manipulator mounted on a non-holonomic mobile platform, of the HILARE [1] type. For our system, the first way to move needs to synchronize two controls and, as far as we know, this solution is rather slow for a complex system. For example, we studied a *plane scraping robot* [2] which took as much time to locate the platform as the robot manipulator. However, many algorithms were developed on this scheme and we can use robust and well-known algorithms.

The second way, faster and more difficult, is to consider the entire mechanical system and to solve the combinated evolution of the platform and the robot manipulator so that the end-effector follows the trajectory in a required location. In the litterature, the study of a global control is very limited.

In the case of a holonomic platform, the approach of O. Khatib [3] to control redundant systems is based on two models : an end-effector dynamic model obtained by projecting the mechanism dynamics into the operational space and a dynamically consistent force/torque relationship. It allows a decoupled control of the joint motions.

In the case of a non-holonomic platform, K.Nagatami [4] applies the concept of action primitives to the mobile manipulator control system. Action primitives are defined as unit elements of a complex behavior (such as door-opening behavior) which control a robot according to a sequence of planned motion primitives. Y.Yamamoto [5][6] defines a control algorithm for the platform so that the robot manipulator is always located at the preferred configurations measured by its manipulability [7]. Nevertheless, for the experiment, he only used the first three joints of a six revolute joints robot manipulator. And to complete this list, H. Seraji [8] defines a kinematic model of the system which incorporates the non-holonomic constraint with the end-effector task. The user has to assign weighting factors to the rover movement and manipulator motion to solve each task specification. The non-holonomic constraint, the desired end-effector motion and the user-specified redundancy tasks are combined to form a set of differential kinematic equations which is solved by calculating a pseudo-inverse. This method is robust but it is rather difficult to define the right weighting factors or the user-specified redundancy tasks.

In this paper, we present a simpler control algorithm which only needs to define, for our system, two additional tasks (or equations) to form a set of differential kinematic equations. First, we introduce the notations of the system. Secondly, we define our methodology and then, we present the results.

2 Modelling of the mechanical system

2.1 Notations

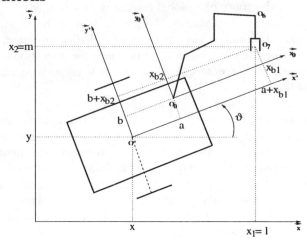

Figure 1 : Notations

$\mathcal{R}=(O,\vec{x},\vec{y},\vec{z})$ fixed frame,
$\mathcal{R}'=(O',\vec{x'},\vec{y'},\vec{z'})$ moving frame attached to the platform,
$\mathcal{R}_0=(O_0,\vec{x_0},\vec{y_0},\vec{z_0})$ moving frame attached to the robot manipulator,
$\mathcal{R}_6=(O_6,\vec{x_6},\vec{y_6},\vec{z_6})$ moving frame attached to the end-effector, with the origin point O_6.
The point O_7, center of the end-effector, is fixed relative to the moving frame $\mathcal{R}_6$.
The moving frames $\mathcal{R}_0$ et $\mathcal{R}'$ are parallel.
The three generalized coordinates of the non-holonomic mobile platform are defined by : $\mathbf{q}_p = (q_{p1}\ q_{p2}\ q_{p3})^t = (x\ y\ \vartheta)^t$, and the 6 generalized coordinates of the robot manipulator by : $\mathbf{q}_b = (q_{b1}\ q_{b2}\ q_{b3}\ q_{b4}\ q_{b5}\ q_{b6})^t$.
The nine generalized coordinates of the mechanical system are defined by :
$\mathbf{q} = (q_1\ q_2\ q_3\ q_4\ q_5\ q_6\ q_7\ q_8\ q_9)^t = (x\ y\ \vartheta\ q_{b1}\ q_{b2}\ q_{b3}\ q_{b4}\ q_{b5}\ q_{b6})^t$.

2.2 Kinematics equations of the mechanical system

2.2.1 Mobile platform subsystem

The non-holonomic constraint can be written in the form :
$$sin\vartheta.\dot{x} - cos\vartheta.\dot{y} = 0$$

Then :
$$\underbrace{\begin{bmatrix} sin\vartheta & | & -cos\vartheta & | & 0 \end{bmatrix}}_{\mathcal{G}(\mathbf{q}_p)} \begin{bmatrix} \dot{x} \\ \dot{y} \\ \dot{\vartheta} \end{bmatrix} = 0$$

Which can be written easier $\mathcal{G}(\mathbf{q}_p)\dot{\mathbf{q}}_p = 0$.
It follows that :

$$\begin{bmatrix} \mathcal{G}_{1\times 3} & | & \mathbb{O}_{1\times 6} \end{bmatrix} \begin{bmatrix} \dot{\mathbf{q}}_p \\ \dot{\mathbf{q}}_b \end{bmatrix} = 0 \qquad (1)$$

where $\mathbb{O}_{n\times m}$ is a matrix block which values are nil.
It follows : $[\mathcal{G}_{1\times 3}|\mathbb{O}_{1\times 6}]\dot{\mathbf{q}} = 0$

2.2.2 Manipulator subsystem

The end-effector location relative to $\mathcal{R}_0$ can be defined by :
$\mathbf{x}_b = (x_{b1}\ x_{b2}\ x_{b3}\ x_{b4}\ x_{b5}\ x_{b6})^t$
where the first three components are (x_{b1}, x_{b2}, x_{b3}) which are the position coordinates of the point O_7, and $(x_{b4}, x_{b5}, x_{b6}) = (\psi, \theta, \varphi)$ are the Euler's angles [9].
Then, the end-effector location relative to $\mathcal{R}'$ is $\mathbf{x}'_b = (a+x_{b1}\ b+x_{b2}\ x_{b3}\ \psi\ \theta\ \varphi)$.

The Direct Differential Model [10] of the robot manipulator relative to $\mathcal{R}_0$ can be written :

$$\dot{\mathbf{x}}_b = J_b(\mathbf{q}_b)\dot{\mathbf{q}}_b \qquad (2)$$

As $\mathcal{R}'$ is obtained from $\mathcal{R}_0$ by a translation characterized by the two constant parameters a et b, we can write $\dot{\mathbf{x}}_b = \dot{\mathbf{x}}'_b = J_b(\mathbf{q}_b)\dot{\mathbf{q}}_b$.

2.2.3 Mechanical system

We can characterize the end-effector location relative to $\mathcal{R}$ by the following vector : $\mathbf{x} = (x_1\ x_2\ x_3\ x_4\ x_5\ x_6)^t = (l\ m\ n\ \beta\ \theta\ \varphi)^t$ where the three components l, m and n are the cartesian coordinates of the point O_7 relative to $\mathcal{R}$ and : $\beta = \vartheta + \psi$. Then :
$l = x + (a + x_{b1})\cos\vartheta - (b + x_{b2})\sin\vartheta$
$m = y + (a + x_{b1})\sin\vartheta + (b + x_{b2})\cos\vartheta$
$n = x_{b3}$
From equations (1) and (2), we obtain in matrix form the following set of equations :

$$\begin{bmatrix} \mathcal{G}_{1\times 3} & \mathbb{O}_{1\times 6} \\ A_{6\times 3} & B_{6\times 6} \end{bmatrix} \underbrace{\begin{bmatrix} \dot{\mathbf{q}}_p \\ \dot{\mathbf{q}}_b \end{bmatrix}}_{\substack{\dot{\mathbf{q}} \\ 9\times 1}} = \underbrace{\begin{bmatrix} 0 \\ \dot{\mathbf{x}} \end{bmatrix}}_{7\times 1} \qquad (3)$$

where $A = [\frac{\partial \mathbf{x}}{\partial \mathbf{q}_p}]$ and $B = [\frac{\partial \mathbf{x}}{\partial \mathbf{q}_b}]$. Then, the previous system is kinematically redundant [10] of degree 2.

3 Methodology

The methodology presented in this paper consist in characterizing two additional equations (or tasks), to rise the former set of equations (3) in a square form and inverse it. We mark these two additional equations :

$$J_a \dot{\mathbf{q}} = \mathbf{w_a} \qquad (4)$$

where J_a is a 2 × 9 matrix such as :

$$J_{a_{2\times 9}} = \left[\begin{array}{c|c|c|c} J_{a11} & J_{a12} & J_{a13} & J_{ap_{1\times 6}} \\ J_{a21} & J_{a22} & J_{a23} & J_{as_{1\times 6}} \end{array} \right]$$

and $\mathbf{w_a}$ a 2 × 1 vector. Then, we obtain the following set :

$$\left[\begin{array}{cc} \mathcal{G}_{1\times 3} & \mathbb{0}_{1\times 6} \\ A_{6\times 3} & B_{6\times 6} \\ & J_{a_{2\times 9}} \end{array} \right] \dot{\mathbf{q}} = \left[\begin{array}{c} 0 \\ \dot{\mathbf{x}} \\ \mathbf{w_a} \end{array} \right] \qquad (5)$$

Let K be the 9x9 matrix which describes the mechanical system :

$$K = \left[\begin{array}{cc} \mathcal{G}_{1\times 3} & \mathbb{0}_{1\times 6} \\ A_{6\times 3} & B_{6\times 6} \\ & J_{a_{2\times 9}} \end{array} \right]$$

3.1 Analysis of the matrix K

First, we propose to define exactly the matrix K, then, we try to simplify this matrix and to define feasible tasks. Let us calculate $\dot{\mathbf{x}}$.

$$\dot{\mathbf{x}} = \left[\begin{array}{c} \dot{x}_1 \\ \dot{x}_2 \\ \dot{x}_3 \\ \dot{x}_4 \\ \dot{x}_5 \\ \dot{x}_6 \end{array} \right] = \left[\begin{array}{c} \dot{x} - \{(a+x_{b1})sin\vartheta + (b+x_{b2})cos\vartheta\}\dot{\vartheta} + \dot{x}_{b1}cos\vartheta - \dot{x}_{b2}sin\vartheta \\ \dot{y} + \{(a+x_{b1})cos\vartheta - (b+x_{b2})sin\vartheta\}\dot{\vartheta} + \dot{x}_{b1}sin\vartheta + \dot{x}_{b2}cos\vartheta \\ \dot{x}_{b3} \\ \dot{\vartheta} + \dot{\beta} \\ \dot{\theta} \\ \dot{\varphi} \end{array} \right]$$

Then :

$$\dot{\mathbf{x}} = \left[\begin{array}{c} \dot{x} - \{(a+x_{b1})sin\vartheta + (b+x_{b2})cos\vartheta\}\dot{\vartheta} + (J_{b1}cos\vartheta - J_{b2}sin\vartheta)\dot{q}_b \\ \dot{y} + \{(a+x_{b1})cos\vartheta - (b+x_{b2})sin\vartheta\}\dot{\vartheta} + (J_{b1}sin\vartheta + J_{b2}cos\vartheta)\dot{q}_b \\ \dot{x}_{b3} \\ \dot{\vartheta} + \dot{\beta} \\ \dot{\theta} \\ \dot{\varphi} \end{array} \right]$$

where J_{bk} is the line number k of J_b. We obtain the expression of K :

$$K = \left[\begin{array}{cc|cc|c} sin\vartheta & -cos\vartheta & 0 & & 0 \\ 1 & 0 & -\{(a+x_{b1})sin\vartheta + (b+x_{b2})cos\vartheta\} & & \{J_{b1}cos\vartheta - J_{b2}sin\vartheta\} \\ 0 & 1 & \{(a+x_{b1})cos\vartheta - (b+x_{b2})sin\vartheta\} & & \{J_{b1}sin\vartheta + J_{b2}cos\vartheta\} \\ 0 & 0 & 0 & & J_{b3} \\ 0 & 0 & 1 & & J_{b4} \\ 0 & 0 & 0 & & J_{b5} \\ 0 & 0 & 0 & & J_{b6} \\ J_{a11} & J_{a12} & J_{a13} & & J_{ap} \\ J_{a21} & J_{a22} & J_{a23} & & J_{as} \end{array} \right]$$

We must notice that the non-holonomic constraint gives a relation between the motion of the mobile platform on the $(O,\vec{x})$ axis and the $(O,\vec{y})$ axis. In fact, the rank of the matrix K is at the most 8. Then, we can do the following change of parameter :

$$\begin{bmatrix} v \\ 0 \end{bmatrix} = \begin{bmatrix} cos\vartheta & sin\vartheta \\ -sin\vartheta & cos\vartheta \end{bmatrix} \begin{bmatrix} \dot{x} \\ \dot{y} \end{bmatrix}$$

where v defines the velocity of the point O' relative to the fixed frame $\mathcal{R}$. The non-holonomic constraint is necessarily verified if we take the following parameters $(v, \dot{\vartheta}, \dot{q}_{b1}, \cdots, \dot{q}_{b6})$ instead of $(\dot{x}, \dot{y}, \dot{\vartheta}, \dot{q}_{b1}, \cdots, \dot{q}_{b6})$. Then, we can eliminate the first line of the set of equations and we obtain a set of 8 equations with 8 parameters :

$$\bar{K}\eta = \begin{bmatrix} \dot{\mathbf{x}} \\ \mathbf{w_a} \end{bmatrix} \quad (6)$$

with $\eta = [v, \dot{\vartheta}, \dot{q}_{b1}, \dot{q}_{b2}, \dot{q}_{b3}, \dot{q}_{b4}, \dot{q}_{b5}, \dot{q}_{b6}]^t$ and :

$$\begin{bmatrix} \dot{x}_1 \\ \dot{x}_2 \\ \dot{x}_3 \\ \dot{x}_4 \\ \dot{x}_5 \\ \dot{x}_6 \\ w_{a1} \\ w_{a2} \end{bmatrix} = \begin{bmatrix} cos\vartheta & -\{(a+x_{b1})sin\vartheta + (b+x_{b2})cos\vartheta\} & \{J_{b1}cos\vartheta - J_{b2}sin\vartheta\} & v \\ sin\vartheta & \{(a+x_{b1})cos\vartheta - (b+x_{b2})sin\vartheta\} & \{J_{b1}sin\vartheta + J_{b2}cos\vartheta\} & \dot{\vartheta} \\ 0 & 0 & J_{b3} & \dot{q}_{b1} \\ 0 & 1 & J_{b4} & \dot{q}_{b2} \\ 0 & 0 & J_{b5} & \dot{q}_{b3} \\ 0 & 0 & J_{b6} & \dot{q}_{b4} \\ \{J_{a11}cos\vartheta + J_{a12}sin\vartheta\} & J_{a13} & J_{ap} & \dot{q}_{b5} \\ \{J_{a21}cos\vartheta + J_{a22}sin\vartheta\} & J_{a23} & J_{as} & \dot{q}_{b6} \end{bmatrix}$$

3.2 Definition of the additional tasks

Let us remember that our aim is to obtain the motion of the joints of the mechanical system defined by $[v, \dot{\vartheta}, \dot{q}_{b1}, \dot{q}_{b2}, \dot{q}_{b3}, \dot{q}_{b4}, \dot{q}_{b5}, \dot{q}_{b6}]^t$.

Thus, we must calculate $\bar{K}^{-1}$ and solve : $\eta = \bar{K}^{-1}\begin{bmatrix} \dot{\mathbf{x}} \\ \mathbf{w_a} \end{bmatrix}$.

The methodology consists in defining two additional tasks which allow us to specificate J_a et $\mathbf{w_a}$. However, we cannot define new geometric relations between the mobile platform and the robot manipulator, such as $\dot{q}_{b4} = \dot{q}_{b5}$, without overconstraining them. Then, we must define more specifical tasks, if it is possible, to follow the required trajectory [10]. Of course, we must take into account the type of trajectory, the obstacle avoidance, the dynamics parameters of the mechanical system $\cdots$ to define w_{a1} and w_{a2}. For a parabolic trajectory, one solution consists in giving the linear motion of the mobile platform, noted v_d, and the angular motion, noted $\dot{\vartheta}_d$. Then, the choice of the values is :

$J_{a11} = cos\vartheta, J_{a12} = sin\vartheta, J_{a13} = J_{ap} = 0$
$J_{a21} = 0, J_{a22} = 0, J_{a23} = 1, J_{as} = 0$

The set of equations to be solved can be written :

$$\begin{bmatrix} \dot{x}_1 \\ \dot{x}_2 \\ \dot{x}_3 \\ \dot{x}_4 \\ \dot{x}_5 \\ \dot{x}_6 \\ v_d \\ \dot{\vartheta}_d \end{bmatrix} = \begin{bmatrix} cos\vartheta & -\{(a+x_{b1})sin\vartheta + (b+x_{b2})cos\vartheta\} & \{J_{b1}cos\vartheta - J_{b2}sin\vartheta\} \\ sin\vartheta & \{(a+x_{b1})cos\vartheta - (b+x_{b2})sin\vartheta\} & \{J_{b1}sin\vartheta + J_{b2}cos\vartheta\} \\ 0 & 0 & J_{b3} \\ 0 & 1 & J_{b4} \\ 0 & 0 & J_{b5} \\ 0 & 0 & J_{b6} \\ 1 & 0 & 0 \\ 0 & 1 & 0 \end{bmatrix} \begin{bmatrix} v \\ \dot{\vartheta} \\ \dot{q}_{b1} \\ \dot{q}_{b2} \\ \dot{q}_{b3} \\ \dot{q}_{b4} \\ \dot{q}_{b5} \\ \dot{q}_{b6} \end{bmatrix}$$

Thus, by indicating the conditions of the linear motion and the angular motion of the mobile platform, we try to locate the robot manipulator so that it is able to follow the required location. Furthermore, the rank of the matrix $\bar{K}$ is 8 and we can calculate $\bar{K}^{-1}$.

4 Solution for singularity

If we define a complex trajectory, the mechanical system could be brought in a singular configuration that means the rank of the matrix $\bar{K}$ is lower than 8. So, it is impossible to inverse directly $\bar{K}$. In such a case, we have to define an other solution which brings the mechanical system in a non-singular configuration.

Feasible solutions

We have previously defined our methodology where our aim consists in giving two additional equations to obtain a square matrix. As we did it, we can define others equations and others matrix $\bar{K}$ that we will use according to the singularity to treat. If we cannot find other square matrix, we use an pseudo-inverse methodology. In this case, the methodology that H. Seraji developed [8] consists in solving the values of the generalized coordinates even in case of singularity. The following figure [Fig.2] illustrates the algorithm we use on the mechanical system.

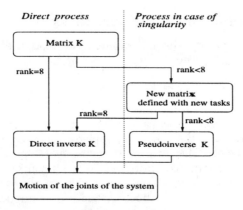

Figure 2 : Control algorithm

5 Computer simulation

We realized the simulation of our mechanical system (six revolute joint robot manipulator mounted on a non-holonomic mobile platform of the HILARE type) in different configurations : complex trajectories, singular configurations or different steps for the points of the trajectories. We report the results of a parabolic trajectory to compare next with the real results.

The mobile platform is first located at the origin of the fixed frame and turned towards the (O, $\vec{x}$) axis. The robot manipulator is in the configuration illustrated on the following figure [Fig.3].

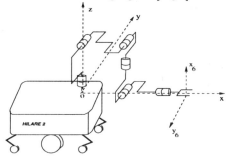

Figure 3 : Starting configuration

Parabolic trajectory, fixed orientation

1st case : Direct inverse

The average step between two points is 5 cms. The orientation of the end-effector is fixed and parallel to the fixed frame $\mathcal{R}$.

The value of the position error depends on the step. For example, for a 20 cms step we obtain a 18 cms maximal error of the end-effector position. In the case of a 5 cms step we obtain a 4.5 cms maximal error.

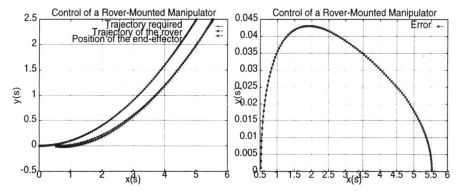

Figure 4 : Direct inverse method

2nd case : Pseudo-inverse

Let us remember the formula of the pseudo-inverse that H.Seraji developped :

$$\dot{q} = [J^t W J + W_v]^{-1}[\dot{x}_d + C(x_d - x)]$$

We defined for this pseudo-inverse based method, the two task weighting factor matrix (W et W_v) which take into account the maximal velocities of the mechanical system [8]. The corrective weighting factor C is equal to 1 and x_d is the required location. We notice that the results in this case are worse than in the case of the direct inverse. In fact, it is really difficult to exactly define the weighting factors and it is why we do not have good results. We notice that we must define weighting factors for each type of trajectory.

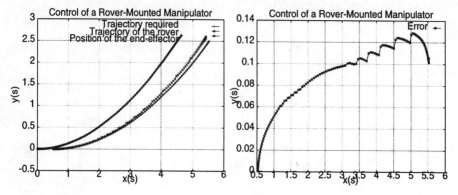

Figure 5 : Pseudo-inverse method

6 Experimental results

6.1 Technical data

The following figure [Fig.6] illustrate the non-holonomic mobile platform of the HILARE [1] type, named H2bis and the robot manipulator, named GT6A.

Figure 6 : Mobile platform and robot manipulator

6.1.1 Mobile subsystem

Technical data[1] of H2bis :
Dimensions : length 1,3 m * wide 0,8 m * height 0,97 m
Weight : 400 kgs
Maximum acceleration : linear 1 m/s^2, angular $3\pi/2$ rd/s^2
Maximum velocity : linear 0,9 m/s, angular 1 rd/s

---
[1]m : meter , s : second , rd : radian

6.1.2 Manipulator subsystem

The robot manipulator is a six revolute joint robot manipulator built by GT Robotique and the basic control named Albatros developped by Robosoft [11]. Technical data of the robot manipulator :
Workspace : 0,78 m radius sphere
Weight : 35 kgs
Maximum velocity : 1,4 rd/s (axis 1,2,3) and 2,0 rd/s (axis 4,5,6).

6.2 Programming

The programming of the control algorithm is implemented using the C Programming Language. We also use the VxWorks Control and Programming Language [12] to control in real time the mechanical system. The interface between the SPARC computers and the ALBATROS computer uses an interface board RSVME.

6.3 Results

We previously test our control algorithm on the robot manipulator GT6A not mounted on the platform and the mobile platform H2bis alone. The aim was to verify our results in a free space and to avoid collisions between the robot manipulator and the mobile platform. However, the good results shows that we can now put the two subsystems together.
The following figure [Fig.7] illustrates the robot manipulator response time in the case of a parabolic trajectory.

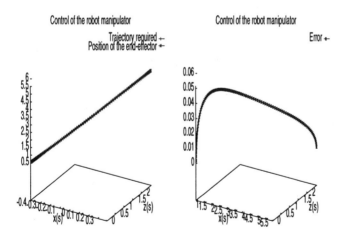

Figure 7 : Time position response of the robot manipulator

The following figure [Fig.8] illustrates the mobile platform response time in the case of a parabolic trajectory.

The values xRefG, yRefG define the position and the value ThetaRefG defines the orientation values for the mobile platform (the profil of the value controls has the form of stairs). The values xCPos and yCPos are the position of the center of the mobile platform and we obtain a smooth time response.

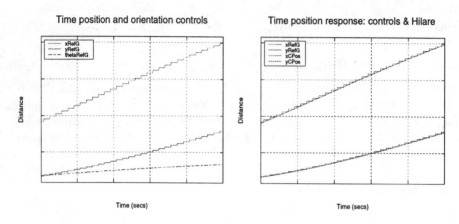

Figure 8 : Time position response of the mobile platform

6.4 Development to come

The time processing of the control algorithm of the robot manipulator is fixed by the ALBATROS computer and is about 30 ms. The location controls cannot be interrupted except with the emergency stop.

Because of these problems, we decided to define a new control and to take off the ALBATROS controls. This new control only takes 2 ms to compute the process and it can be interrupted. We are implementing this new control on the mechanical system and defining basic controls such as : moving point to point with a specific velocity in the cartesian or configuration space . Thus, we should have the same time computer process for both the mobile platform and the robot manipulator.

7 Concluding remarks

In this paper, a global control method of a six revolute joint robot manipulator mounted on a non-holonomic mobile platform is discussed. The major difficulty of this method consits in giving two additional tasks relative to the kinematic of the mechanical system and to obtain the square matrix form of the set of equations.

We proved that the end-effector of the mechanical system follows the required trajectory for different type of trajectories. By combination of both direct inverse and pseudo-inverse methods, we obtain a robust control algorithm. The tests of the two subsystems validate the control but we have to improve the dynamic behavior of the mechanical system. The work to come consists in defining new tasks and also, to fix a video camera on the end-effector. Thus, the control of the camera will give to our mechanical system control other specific trajectories to follow.

References

[1] G.Giralt et al., 1984, *An integrated navigation and motion control system for autonomous multisensory mobile robots* In proceedings of the first International Symposium of Robotics Research, Cambridge, MA, USA.

[2] G.Foulon, 1996, *Commande d'une plateforme mobile non holonome associée à un bras manipulateur redondant*, Rapport interne LAAS 96005.

[3] O.Khatib, K. Yokoi, K. Chang, D. Ruspini, R. Holmberg and A. Casal, 1996, *Coordination and Decentralized Cooperation of Multiple Mobile Manipulators*, Journal of Robotic Systems.

[4] K.Nagatami, S. Yuta, 1996, *Door-Opening Behavior of an Autonomous Mobile Manipulator by Sequence of Action Primitives*, Journal of Robotic Systems.

[5] Y.Yamamoto, X. Yun, 1993 *Coordinating Locomotion and Manipulation of a Mobile Manipulator* ICRA.

[6] Y.Yamamoto, 1994, *Control and Coordination of Locomotion and Manipulation of a Wheeled Mobile Manipulator*, Ph. D. Dissertation, University of Pennsylvania.

[7] T. Yoshikawa, 1990, *Foundations of Robotics: Analysis and Control*, The MIT Press, Cambridge, Massachusetts.

[8] H.Seraji, 1993, *An On-Line Approach to Coordinated Mobility and Manipulation*, IEEE.

[9] R.A. Wehage, 1994, *Quaternions and Euler parameters-a brief exposition NATO ASI series*, Vol. F9, Springer Verlag, Berlin, Heidelberg.

[10] G. Foulon, J-Y. Fourquet and M. Renaud, 1997, *On coordinated tasks for nonholonomic mobile manipulators*, SYROCO'97, Nantes, FRANCE.

[11] Albatros
Six dof Manipulator User's Manual V1.2.

[12] VxWorks, 1995, *Programmer's Guide V 5.3*, Wind River System.

Chapter 3

Haptic Devices

Human robot cooperation is becoming a common practice in advanced robotics. For such systems haptic devices and virtual reality constitute significant and necessary tools to enable an efficient man-machine interaction.

In the paper of Yoshikava and Ueda a systematic method of constructing a virtual world is proposed. A modular architecture is designed combining dynamic modules that represent elements of the virtual world, and interaction modules. The structure of the virtual world is analyzed, as well as the kind of information required for world modeling and the adequate human interaction. The modular structure allows to represent worlds with multiple virtual objects, being possible to add or to eliminate elements when required.

The problem of teaching robots to develop complex tasks is faced up by Bordegoni, Cugini and Rizzi. Their goal is to provide robots with sufficient skill to be able to manipulate flexible objects, coping with different situations. Skill acquisition from human behavior is not trivial. The availability of haptic devices for sensing human actions and its integration with a phantom haptic device to perform a simulated task enables to record the task evaluation. Different simulations based on the task model have been carried out for tasks of wire manipulation. This experimentation constitutes the first step towards robot learning from human skill.

The paper of Ellis, Zion and Tso addresses the problem of understanding the force/displacement relationship of elements of biological joints. To advance in this study they develop and test a method for interacting with a dynamic model of the knee. The analysis of biological joints, quite different from robotic ones, leads to a kinematics and dynamic study of the human knee. The aim of the system are learning and pre-operative exploring in total knee replacement surgery or ligament reconstruction.

Module-Based Architecture of World Model for Haptic Virtual Reality*

Tsuneo Yoshikawa and Hitoshi Ueda**

Department of Mechanical Engineering
Kyoto University, Kyoto 606, Japan
yoshi@mech.kyoto-u.ac.jp

Abstract

Recently the force-feedback has been recognized to be important for virtual reality systems, and many studies have been done in this field. Although we have proposed several methods for displaying the operating feel of a virtual object considering its dynamics in previous works, we have not paid much attention to cases of multiple objects or operators. In more advanced applications of haptic virtual reality, however, the case of multiple operators and multiple dynamic objects will become more important. In this paper, we propose a systematic method of constructing a dynamic virtual world by combining dynamics modules and interaction modules, the former representing elements of the virtual world — virtual objects and real operators — and the latter representing interactions between them. Since this method makes it possible to regard these elements as equal, we can construct a flexible and extendible virtual world to which we can easily add new elements or remove some of its elements. An experimental result using a prototype force display system shows the validity of the proposed method.

1. Introduction

In the early studies on virtual reality, the visual and audio senses were mainly displayed. Recently, however, the importance of displaying force or sense of touch has been recognized and many studies have been made in the field of haptic virtual reality.

Previously studies were made on displaying shapes[1], forces for deforming objects[2], weights[3] and so on, but there were no studies which tried to display the dynamics of virtual objects. Thinking of such applications, however, as examining operating feel of a virtual product in stead of a real product, or as exercising in the virtual world, the dynamics of the virtual object cannot be neglected. In order to take dynamics into consideration, it is necessary to pay attention to the relation between forces and moments in the virtual world. We have taken this factor into

---
*This paper is a revised version of [8].
**Currently with NTT, Japan

consideration, and have proposed several algorithms for displaying operating feel of a virtual object considering its dynamics[4][5][6].

Although a method to express a virtual object with joints is described in [5], the virtual object is assumed to be single in these studies. Concerning the operator, a study has been made on collaborative operation of two operators for one virtual object[7], but no attempt is made to formulate a general method to treat more complex cases. Generally speaking, the case of multiple dynamic objects and multiple operators has not been paid much attention so far. In more advanced applications of haptic virtual reality, however, the case of multiple operators and multiple dynamic objects will become more important. Hence it is essential to establish a universal method to construct the virtual world for this case.

In this paper, we propose a systematic method of constructing a virtual world by combining dynamics modules and interaction modules. The dynamics modules include not only the virtual objects but also real objects such as operators and tools which operators manipulate. The interaction modules represent interaction among dynamics modules such as interaction force due to elasticity and surface friction characteristics. Since all real and virtual objects are regarded as equal, the method makes it possible to construct a virtual world which is quite flexible and extendible in the sense that we can easily add new elements to it, or we can easily remove some of its elements from it. Some experiments using a prototype force display system have been done to show the validity of the proposed method.

2. Structure of the Virtual World
2.1. Image of the Virtual World

In the case of one dynamic virtual object and one operator, a conventional image of the haptic virtual reality system would be something like that shown in Fig.1. At least during the work of [4][5][6] the authors had this kind of image. Our straightforward extension of this image to the case of two objects and two operators was Fig.2. In trying to find a systematic way of dealing with this case and of constructing a virtual world, however, we became to recognize that Fig.3 is a better and more natural image of the system. Thus we consider not that the virtual world exists in the real world, but that the virtual objects and the operators (who are part of the real world) exist in the virtual world.

This way the operators and the virtual objects can be regarded as the same kind of dynamic subsystems receiving some input from other subsystems and sending out some outputs to them.

2.2. Modules Constituting the Virtual World

It is the information of force and motion that plays the role of input or output of each of the dynamic subsystems mentioned in the preceding section. We assume that the configurations of the elements are expressed in generalized coordinates, and that the information of motion and force is expressed in generalized coordinates and generalized forces.

We can think of two types of subsystems representing the elements. One type receives force as the input and sends out motion as the output, and the other type receives motion as the input and sends out force as the output[4]. We adopt the former one in consideration of convenience in connecting these subsystems in a systematic way. We call this kind of subsystems the *dynamics module*, which is shown in Fig.4. The dynamics module is considered representing an entity in the virtual world. A dynamics module for a movable virtual object represents its dynamics. A dynamics module for an operator, which includes the real world, can be regarded as representing its dynamics in the virtual world, because this module specifies how it moves when certain forces are exerted on it.

We introduce another kind of module which represents the interaction between two dynamics modules. We call this kind of modules the *interaction module*. Since the dynamics module receives force as the input and sends out motion as the output, it is derived that the interaction module should receive motions from two dynamics modules as the input, and should send out force to them as the output (Fig.5). The interaction module is considered representing not an entity, but, as it were, a phenomenon.

The image in Fig.3 can be represented by Fig.6 using the dynamics modules and interaction modules.

Thus the virtual world can be constructed quite naturally, independently of the number or the kind of the dynamics modules. This is because the dynamics module has been decided to receive forces as the input and to send out motions as the output. That is to say, because the forces can be added and the motions can be distributed, we can combine dynamics modules and interaction modules without any logical problem.

Since each module can be specified independently of any other ones, the modules can be easily added or removed, and are easily re-utilized. It is also expected that this modularization would be of help in dividing the computational load among multiple computers connected in a network.

2.3. Dynamics Modules

The dynamics modules are not just limited to virtual objects and operators in the future applications of haptic virtual reality technology. They can be classified into three types: virtual-object type, real-object type, and mixed virtual/real-object type.

Virtual-object type dynamics modules are those which represent purely virtual objects. The behavior of the object should be simulated completely. In the case of movable virtual objects, their dynamics should be specified. The object can have joints which can be either active or passive. Note that there are some virtual objects whose motion can be assumed to be independent of the exerted forces such as non-movable walls and belt conveyers.

Real-object type modules are those which exchange the input and the output with the real world via display devices. We should note that the elements represented by the real-object type dynamics modules are not always operators. For example, a deforming real object provided by a suitable force display device and a motion measuring instrument can be represented by this kind of dynamics module instead of a deforming virtual object which may require a lot of complex calculations.

Mixed virtual/real-object type modules represent an object a part of which exists in the real world and the rest is virtual. For example, as shown in Fig.7 a hammer whose gripper is provided in the real world and whose head should be simulated in the virtual world can be represented by a mixed virtual/real-object type module.

2.4. Interaction Modules

Interaction modules calculate the force working on each of two dynamics modules connected to it, using the motion information from these two dynamics modules.

The contact forces, caused by elasticity or viscosity, or the potential forces, exerted in the flow field or the magnetic field, can be expressed by static interaction modules.

It is often the case with the operation in the virtual world that the operators manipulate the virtual objects with their hands. The interaction force in this case, or the frictional contact force, cannot be determined by their current relative motion only. The information on the past relative motion is also needed. This difficulty, however, can be solved by a dynamic interaction module with a state value called contact state[6]. An example of this situation will be presented in section 4.

3. Examples of Dynamics Modules

In this section, an example of dynamics modules of real-object type and virtual-object type will be described to show the details of the dynamics modules.

3.1. Real-Object Type

3.1.1. Display Device

Consider the display device consisting of robot arms fixed in the environment. Each arm is assumed to have 6-degrees-of-freedom.

An encoder is attached to the axis of each joint, and the displacements of the joints can be measured. A force sensor is placed between the finger ring and the robot arm, and the force exerted by the operator's fingertip can be measured. A motor is attached to each joint axis, which drives the display device.

3.1.2. Display of Force

Let q_h denote a 6-dimensional generalized position and orientation vector representing the configuration of the fingertip in the virtual world. q_h is equivalent to the configuration of the endpoint of the robot arm. The input of this module is the 6-dimensional generalized force F_h corresponding to q_h.

Consider a generalized force f_h exerted on the endpoint of the robot arm by the operator's fingertip, which corresponds to q_h. The equation of motion of the robot arm is given by

$$\tau_l = M_l(q_l)\ddot{q}_l + \hat{h}_l(q_l, \dot{q}_l) - J_{hl}^T(q_l)f_h \tag{1}$$

where τ_l is the joint driving torque, q_l the joint vector, $M_l(q_l)$ the inertia matrix of the robot arm, and $\hat{h}_l(q_l, \dot{q}_l)$ the centrifugal, Coriolis, and gravity force term. $J_{hl}(q_l)$ is the Jacobian matrix of q_h with respect to q_l.

In order to compensate for the dynamics of the robot arm, we define the joint driving torque as follows:

$$\tau_l = M_l(q_l)\ddot{q}_l + \hat{h}_l(q_l, \dot{q}_l) - J_{hl}^T(q_l)u_f \tag{2}$$

By making suitable compensations, such as PI control law, to F_h for its errors, and substituting it into u_f in equation (2), we can display the input force F_h to the operator.

3.1.3. Measurement of Motion

The configuration vector q_h of the fingertip, or of the endpoint of the robot arm, is given by

$$q_h = f(q_l) \tag{3}$$

where the function f is generally nonlinear. Differentiating equation (3), we obtain

$$\dot{q}_h = J_{hl}(q_l)\dot{q}_l \tag{4}$$

The output from this dynamics module is the 6-dimensional configuration vector q_h and the 6-dimensional generalized velocity $\dot{q}_h$.

3.2. Virtual-Object Type

We assume that, although the virtual object has an elastic surface, the deformation of the object due to elasticity is so small that we can use the equation of motion of a rigid body representing the object.

The equation of motion of the virtual object is given by

$$M_o(q_o)\ddot{q}_o + \hat{h}_o(q_o, \dot{q}_o) = F_o \tag{5}$$

where q_o is the generalized vector denoting the configuration of the virtual object, which is 6-dimensional in the case of a single rigid body, or $(6 + n)$-dimensional in the case of a virtual object having n joints. $M_o(q_o)$ is the inertia matrix and $\hat{h}_o(q_o, \dot{q}_o)$ is the centrifugal, Coriolis, and gravity force term. F_o is the generalized force working on the virtual object.

The input to this module is the generalized force F_o. Substituting it into equation (5), $\ddot{q}_o$ is obtained, and then $\dot{q}_o$ and q_o, by integration. These are the output from this module.

4. Example of Interaction Modules

In this section, an interaction module for representing a frictional contact between two polyhedra will be outlined as an example of interaction modules.

In [1], an algorithm for displaying contact forces corresponding to elasticity is proposed, by dividing the contact between two polyhedra into those between a vertex and a face, and those between a pair of edges of the polyhedra. Here we will present an interaction module which can represent not only the elastic contact but also the frictional one between two polyhedra. This module consists of sections of contact detection, frictional contact model, and conversion into the generalized force.

In the section of contact detection, it is checked if the two objects are in contact or not, based on the input information on the motions of two objects, which are outputted from the two dynamics modules. When they are in contact, the contact pairs of features — a vertex and a face, or a pair of edges — are determined.

In the section of frictional contact model, the interaction forces and moments are determined according to the contact models shown in Fig.8 and 9 (the reference points in static frictional contact state, or the points to be connected with two ends of the springs and dampers, are those at the moment of the beginning of static frictional contact). Here the concept of contact state variable is introduced which can take one of the three values — non-contact state, static frictional contact state, and dynamic frictional contact state. Fig.10 shows the transition of the contact state for each pair of features of the polyhedra. Whether "slipping starts" or not is determined by whether the static frictional force (or moment) is larger than the maximum static frictional force (or moment) which is proportional to the normal force.

In the section of conversion into the generalized force, the force and the moment working at the contact point are converted into the generalized forces in the object coordinates of the polyhedra. They are added together and outputted from this interaction module.

5. Experiments

Based on the proposed method, a virtual world was constructed for a prototype haptic virtual reality experimental system we developed[5] to display the operating feel while manipulating a virtual sphere with two fingers. Details of the system and some experimental results are given below.

5.1. Display System

The structure of the experimental display system is shown in Fig.11, and the display devices — the force display and the graphic display — are shown in Fig.12.

A pair of 3 d.o.f. miniature robot arms driven by geared DC motors are used for the force display devices, which are located 300 mm apart. Force sensors and finger rings are attached to the tips of the arms.

Personal computers are used as a host computer (i486DX2, 66MHz) and a graphic computer (i486DX, 33MHz).

A stereoscopic display of time-sharing polarizing type is used for the graphic display.

5.2. Dynamics and Interaction Modules

The virtual object is a sphere with radius 30 mm and the mass 2000 g, and is represented by a virtual-object type dynamics module described in section 3.2. Since the gravitational acceleration is set $9800 \times 0.05 = 490$ mm/s^2 (one twentieth of the normal gravity on the earth), the weight of the virtual object is 0.98 N. The operator manipulates the virtual object with two fingers. Each fingertip is represented by a real-object type dynamics module described in section 3.1.

We approximate the contacts between the two fingertips and the sphere by vertex-face contacts and use two interaction modules described in section 4. We assume that the rotational static friction is negligible for simplicity, and that no interaction works between the fingertips.

Hence the experimental virtual world consists of three dynamics modules and two interaction modules.

5.3. Results

The task we performed was first picking up the sphere from the floor, then waving it horizontally, and finally throwing it up and catching it when it falls.

The results are shown in Fig.13. Fig.13 (a) is the position of the sphere (the output of the dynamics module for the sphere), (b) is the sum of the calculated forces working on the sphere (the outputs of the interaction modules), and (c) is the sum of the forces measured by the force sensors (approximately equal to the forces felt by the operator at the fingertips).

In the beginning the operator was not touching the sphere ($\sim$A). Next he picked it up (A$\sim$B), and waved it from side to side (B$\sim$C). Then he threw it up and caught it twice (D$\sim$E, F$\sim$G). The x, y, and z axes are in right, up, and out-from-the-screen direction as the operator faces to the display device. The graphs in the z axis are omitted, because we tried to operate the object in x-y plane.

From a good coincidence of (b) and (c), we can see a good performance of the force display device. Furthermore, from (a) and (b), we can see the following intuitively natural features of the operation.

- In the interval B$\sim$C, the phases of the position and force in the x axis are in reverse. In the y axis, the force around 0.98 N is exerted, which is the weight of the virtual object.

- In the intervals D∼E and F∼G, impulsive forces are observed at the moments of throwing up and catching.

Although we do not show the results concerning friction, we succeeded in displaying a natural operating feel for the case where slipping starts in grasping weakly and slipping stops in grasping strongly.

6. Conclusion

In this paper, we have described a method of constructing the virtual world by combining the dynamics modules, representing the operators or the virtual objects, and the interaction modules, representing the interactions between them. This method makes it possible to construct the virtual world naturally even in cases where multiple operators manipulate multiple dynamic virtual objects. Since each module is defined to be fairly independent of the other ones, it is easy to alter or to re-construct the virtual world.

Examples of the real-object type and the virtual-object type dynamics modules have been presented. There are many varieties of the real-object type and the mixed real/virtual-object type dynamics modules, because they are dependent on display devices. All of the virtual-object type dynamics modules may be based on the description in this paper. In the case of virtual objects with joints, however, when frictions work in these joints, or when both sides of joints interfere with each other, the dynamics module would be more complex.

The interaction module representing frictional contact between polyhedra has been presented as an example. It realizes the operations or the work in the virtual world utilizing friction. Because these are quite general operations, we believe that this frictional contact model will be useful in many applications.

We have done some experiments of displaying the operating feel of a virtual object utilizing these example modules. We have confirmed the validity of the proposed method.

The proposed method can be applied to various cases. For example, trainings or sports in the virtual world are conceived. As we have mentioned, a dynamics module including the real world could replace a virtual object, using not an operator but a real object as a real entity. Therefore, for example, making all tools for teleoperation out of "virtual" objects at first, the operator can go into training with no danger. Then letting them including "real" tools step by step, the operator can be used to the operation step by step. The proposed method is also useful when we divide the calculation among multiple computers connected in a network, in order to, for example, construct a virtual mall system in which we can do shopping, taking goods in our hands.

References

[1] K. Tanie and T. Kotoku : Force Display Algorithms, *Lecture Notes for IEEE Workshop on Force Display in Virtual Environments and its Application to Robotic Teleoperation*, pp. 60–78, 1993.

[2] K. Yamamoto, A. Ishiguro, and Y. Uchikawa : A Development of Dynamic Deforming Algorithms for 3D Shape Modeling with Generating Interactive Force Sensation, *VRAIS '93*, pp. 504–511, 1993.

[3] H. Iwata : Artificial Reality with Force-feedback: Development of Desktop Virtual Space with Compact Master Manipulator, *Computer Graphics*, Vol. 24, No. 4, pp. 165–170, 1990.

[4] T. Yoshikawa, Y. Yokokohji, T. Matsumoto, and X.-Z. Zheng : Display of Feel for the Manipulation of Dynamic Virtual Objects, *Trans. ASME, J. DSMC*, Vol. 117, No. 4, pp.554–558, 1995.

[5] T. Yoshikawa, X.-Z. Zheng, and T. Moriguchi : Display of Operating Feel of Dynamic Virtual Objects with Frictional Surface, *Proceedings of the IEEE/RSJ/GI International Conference on Intelligent Robots and Systems*, pp. 731–738, 1994.

[6] T. Yoshikawa and H. Ueda : Display of 3-Dimensional Operating Feel of Dynamic Virtual Object with Frictional Surface, *Proceedings of International Conference on Virtual Systems and Multimedia '95 in Gifu*, pp. 183–188, (1995).

[7] M. Ishii, M. Nakata, and M. Sato : Networked SPIDAR : A Networked Virtual Environment with Visual, Auditory, and Haptic Interactions, *PRESENCE*, Vol. 3, No. 4, pp. 351–359, 1994.

[8] T. Yoshikawa and H. Ueda : Construction of Virtual World Using Dynamics Modules and Interaction Modules, *Proceedings of the IEEE International Conference on Robotics and Automation*, pp. 2358–2364, 1996.

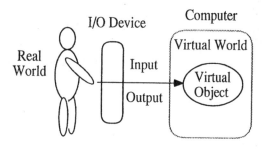

Figure 1. Conventional Image of the Haptic Virtual Reality System

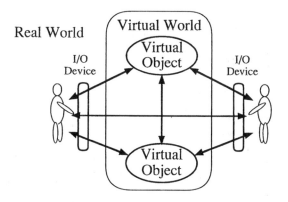

Figure 2. An Image of the Haptic Virtual Reality System for Two Virtual Objects and Two Operators

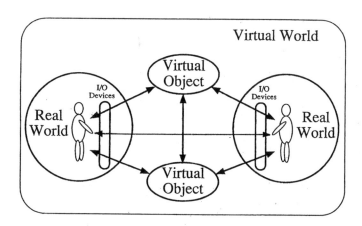

Figure 3. A New Image of the Haptic Virtual Reality System for Two Virtual Objects and Two Operators

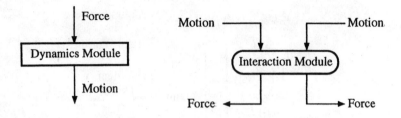

Figure 4. Dynamics Module   Figure 5. Interaction Module

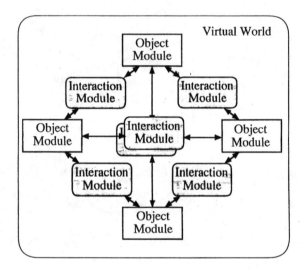

Figure 6. Representation of the System in Fig.3 Using Dynamics Modules and Interaction Modules

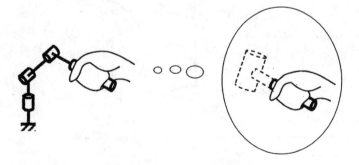

Figure 7. An Example of Mixed Virtual/Real-Object Modules

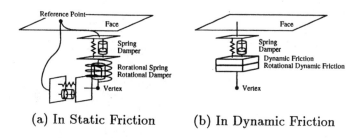

Figure 8. Vertex-Face Frictional Contact Model

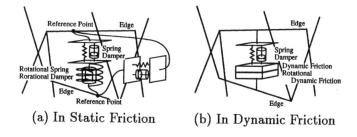

Figure 9. Edge-Edge Frictional Contact Model

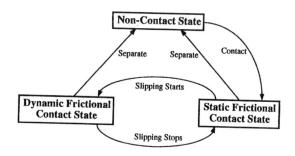

Figure 10. Transition of Contact State

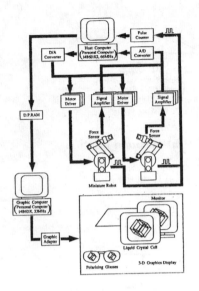

Figure 12. Overview of Display Device

Figure 11. Structure of Display System

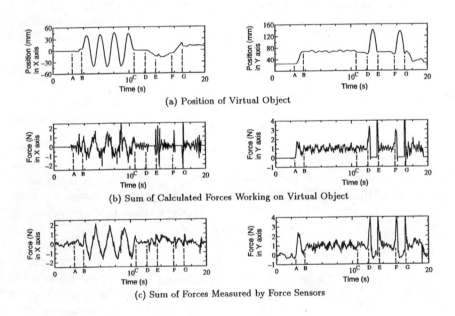

(a) Position of Virtual Object

(b) Sum of Calculated Forces Working on Virtual Object

(c) Sum of Forces Measured by Force Sensors

Figure 13. Experimental Results

Haptic Augmented Simulation supporting teaching Skill to Robots

Monica Bordegoni, Umberto Cugini and Caterina Rizzi
Dipartimento di Ingegneria Industriale
Università di Parma
Viale delle Scienze, 43100 Parma - Italy
E-mail: [mb, cugini, rizzi]@ied.eng.unipr.it

Abstract: This paper presents the preliminary results of the experimentation done at the University of Parma, and partially carried out as part of the Brite/Euram III Project Skill-MART, aiming at studying the use of digital simulation and haptic interaction for the purpose of analysing human skill and mapping it into a skilled robotics system.

1. Introduction

The objective of the research work done within the Brite/Euram Project Skill-MART [1] is to provide robots with enough skill in order to be able to manipulate flexible objects. Typically, grasping and manipulation tasks require a certain level of skill. If we think of skill only at a physical or effectual level, robots can be taught a sequence of actions to be performed for achieving the task. Skill is in the capability of the end-effectors, in its ability of coordinating the movements, etc. Therefore, we can talk of automation of the task, without implying any intelligence or learning from the robots. In these systems, the robot cannot deal with situations that have not been previously planned in the task sequences.

When dealing with flexible objects, it is necessary to think of strategies for coping with unexpected situations. For example, if the robot is manipulating wires and cables, it might happen that the wire starts sliding away. Which force is required in order to keep the wire tightly? When grasping a bag full of liquid, which pressure has to be exerted on the bag in order to firmly holding it, and avoiding its breakage?

It is not efficient and quite complex to teach robots the sequence of actions to perform for coping with several situations. Alternatively, we can think of a robotics system provided with general instructions on how to manipulate objects, and knowledge for operating also in front of unexpected situations, and also for learning from them. Therefore, the system is required to be skilled at a higher level: the conceptual level. In order to be able to teach conceptual level skill to the robot, it is basic to know which is the skill required to accomplish a certain task.

This paper presents the approach followed for acquiring and analysing human skill, and how it has been validated through the use of haptic interaction and graphic simulation of the handling process. This has required the use of proper software environment to model and simulate the behaviour of nonrigid objects, as wires and infusion bags.

2. Skill acquisition

The identification of the skill required to accomplish a certain task is a non trivial issue. The idea pursued in our research work has been to begin by analyzing the human behavior when manipulating flexible objects [2]. By watching and recording human actions while performing a task, some spatio-temporal features that characterize the skill can be extracted. Several technologies can be used for recording hand movements and gestures. For example, it can be made through some video recording, or through the use of some hand-input device [3]. In any case, the acquisition is done in terms of tracking of the fingers and hand movement in space. By analysing the acquired data is possible to extract information on the spatio-temporal features of the hand movements while performing a task.

In order to have the possibility to experiment several situations, a simulation of the environment and the physical behavior of objects in the environment has been considered. For example, we can think of simulating a set of wires, each having a specific physical characteristic. An operator can be asked to grasp the wire visualised in the simulated environment, and insert one of its ends into a fixed hole. An analysis of the operator's gestures leads to a classification of grasping skill, according to the type of wire. For defining the simulation environment it has been used the system for modelling and simulating flexible materials behaviour developed at the University of Parma, named SoftWorld [4]. The system provides an interactive and graphical simulation environment for supporting the automation of industrial processes dealing with handling of flexible products.

The use of simulation is more flexible in respect to the use of a real environment, where changes in the environment set-up might require long setting up time. Moreover, it allows us to easily perform a "what if" analysis. For example, we are able to detect "what if" the wire starts sliding away from the operator's fingers.

Some information on the forces taking part in the task can be deduced by the values recorded during the gesture acquisition. A more precise analysis is required for providing an accurate description of the skill required for accomplishing the tasks. The following chapter describes the approach adopted for the analysis of the forces involved.

3. Augmented simulation approach

It was realised that a more accurate analysis of the forces exerted by the human while manipulating objects was necessary. That in order to be able to better identify the skill required for that specific task.

The idea has been to use some haptic device for detecting and recording the forces exerted by the operator during the performance of the manipulation tasks. To this purpose, the nonrigid material simulator has been integrated with the Phantom haptic device by SensAble [5]. The device acts in a twofold way. On the one side it detects the intensity of the forces exerted by the user on a simulated object. On the other side, it provides the user a feedback of the resistance of that object material to that force.

Differently from using a simulation only based on vision, through the use of a haptic device the operator has a feeling of what he is touching, and can react consequently and properly [6]. Experiments with an operator wearing the haptic device have been performed. Also in this case, it has been possible to derive the human skill from the analysis of the behaviour of the operator facing a set of different simulated situations.

4. Skill formalization

The approach followed for skill identification and formalization is depicted in figure 1. The analysis of the hand gestures provides information on the human skill. In particular, the first acquisition thread permits to gather information on trajectory and sequence of operations to be performed, in order to fulfill the goal. Through the second thread we acquire force values detected by the haptic device. For example, information on the forces to be exerted in order to firmly grasp a wire or a bundle of wires are derived.

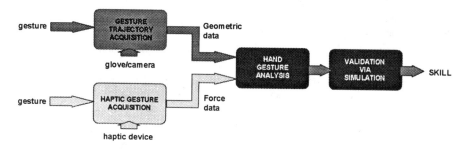

Figure 1. Phases of the research activity on human skill analysis

On the basis of the study of several situations, it can be derived an hypothesis on the effective skill required for a task. For example, how the hand fingers are locally adapted according to the object shape for a firm grasping. The analysis can also provide information on some strategies used for performing a task. That refers to the cognitive skill used for facing differing situations. For example, which is the best way for approaching an object and grasping it, how to recover the catching if the object is falling down, etc.

The knowledge providing conceptual skill to the robot has been build upon the information gathered from the analysis of hand gestures and on the basis of some assumptions. This means that the model on which the knowledge is built is mostly theoretical, and requires to be tested. Again, we have thought of using the simulation for validating the skill encoded. The behavior of an operator wearing a haptic device and asked to operate with different digital representations of nonrigid objects has been recorded. The actual values can be compared with the strategy intended to be taught to the robot, providing an opportunity to validate the theoretical assumptions through a real validation.

5. Experimentation

To demonstrate our ideas we have considered as study-case an application simulating a physical layout for wire looming, where the operator wears the Phantom haptic devices for grasping and manipulating the wires. Experiments can be made by changing the mechanical wire properties, the location of obstacles within the layout, etc.

We are dealing with this type of operation since automated wire laying is becoming more and more important in several industrial sectors, such as automotive and aerospace.

First, we have concentrated on the grasping operation, and we have started a preliminary test to study the manipulation process. The simulation tasks concern the behavior of the wires with different physical characterization when grasped by the haptic device, deriving information on how to perform correctly the operation (e.g. avoiding wire sliding) and forces to be applied. We have considered the possibility to interact with the digital model through the use of two haptic devices (two Phantom devices) the user wears on two fingers.

Wire behavior simulation can be carried out at different levels of accuracy (and computational cost) depending on the problem requirements. The software simulator used, SoftWorld, is based on the particle base model [7], by which a deformable object is described as a set of mechanical elements (particles) and forces (acting among particles), representing the macro-behaviour of the material (e.g., elasticity, plasticity). Therefore the simulation has required, first, the study of the right model able to describe the physical characteristics of the wire and find out which is the most suitable with reference to the goal of the simulation, in this case, the gripper action and related wire deformation. Different experiments have been done applying different force values. Figure 2 portrays the end-user wearing the haptic devices and grasping the wire. Figure 3 shows a detail of the wire behavior when grasped and the deformation of the wire due to the gripper action.

Forces exerted on the cable can now be detected, and inserted into the skill analysis.

Preliminary tests concerning automatic wire laying have been carried out. We have supposed to grasp a wire with a gripper and lay it on a flat panel in presence of some obstacles. The wire is solidary with the gripper and is constrained to follow the trajectory imposed by the gripper. Figure 4 shows a step of the simulation process.

Figure 2. End-user grasping wires wearing the haptic devices

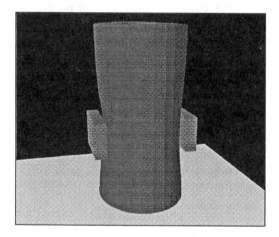

Figure 3. Simulation of wire handling

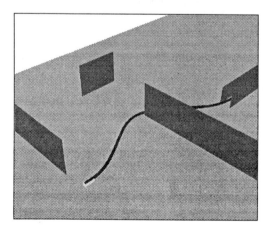

Figure 4. Simulation of automatic wire laying

Conclusions

This paper has presented the approach adopted for acquisition and validation of manipulation skill. The approach is based on skill acquisition through human demonstration and validation through visual and haptic simulation. The results of preliminary experiments can be considered positive and encourage us to proceed following this approach, even if some issues are still open, such as computational time related to the particle based model.

The results provide input for the definition of knowledge for effectual and conceptual skill for the robotics system, and eventually for the definition of the gripper that best achieve that catching.

Acknowledgments

The research work presented in the paper was partially carried out as part of the Brite/Euram III project Skill-MART n. 1564, funded by the Commission of the European Union. The authors wish to thank all the Skill-MART project partners, for contributing to this work.

References

[1] Skill-MART (Skilled Multi-Arm RoboTs), Brite/Euram III Project No. BE-1564, Technical Annex, 1996.
[2] Kaiser M. and Dillmann R., Building elementary robot skills from human demonstration, In *IEEE International Conference on Robotics and Automation*, Minneapolis, Minnesota, USA, 1996.
[3] *Progress in Gestural Interaction*, Proceedings of Gesture Workshop '96, York, Springer Verlag.
[4] Denti P., Dragoni P., Frugoli G., Rizzi C., SoftWorld: A system to simulate flexible products behaviour in industrial processes, in *Proceeding ESS 1996*, Vol. II, Genova, 24-26 October 1996.
[5] Massie T.H., Initial Haptic Explorations with the Phantom: Virtual Touch Through Point Interaction, Master thesis, Massachusetts Institute of Technology, February 1996.
[6] Bordegoni M., Enhancing human computer communication with multimodal interaction techniques, in *Proceedings 1st International Conference on Applied Ergonomics (ICAE'96)*, May 21-24, 1996, Istanbul, Turkey.
[7] Witkin A. et al. 1995 "Physically based modeling." In *SIGGRAPH 95 Course Notes* (Los Angeles, California, USA, Aug. 8-10), no. 34.

Interactive Visual and Force Rendering of Human-Knee Dynamics

R. E. Ellis*[†]   P. Zion*   C. Y. Tso*

* Department of Computing and Information Science
† Department of Mechanical Engineering
Queen's University, Kingston, Canada K7L 3N6

Abstract

The kinematics and force/displacement relationships of elements of biological joints are notoriously difficult to understand. In particular, the human knee has bearing surfaces of complex geometry that are connected by interacting sets of compliant tissues. It is proposed that knee motion and reaction forces can be usefully presented, during virtual manipulation of the joint, by simultaneous visual and haptic rendering. A prototype system for visuo-haptic rendering of the human knee included an extension of a common model of knee motion, a solution of the extended inverse kinematics, and an approach to the computation of bearing-surface reaction forces which balances the two needs of speed and accuracy.

Knee kinematics were rendered with a graphical animation, and constraint forces were rendered with a haptic interface. A two-degree-of-freedom planar haptic interface detected the position of a handle and applied forces via the handle to the human operator. Multimodal presentation of motion and forces is a promising technique for studying ligament balancing and reconstruction, and may also be useful in studying implantation of artifical joint components.

1 Introduction

The force/displacement relationships of elements of biological joints is notoriously difficult to understand. In particular, the human knee has articulating surfaces of complex geometry that are connected by tissues of highly nonlinear mechanical behavior. The kinematics of the knee are generally understood, but presenting the dynamics (motion plus forces) is nontrivial.

The goal of this work is to develop and test a method for interacting with a dynamics model of the knee, especially of the knee following total joint replacement. The target user group is composed of practicing surgeons, as well as physiotherapists and students. These people need to have a direct and intuitive interface that permits them to explore the effects of differing geometries, component placement, and ligament states. For example, ligament laxity is typically examined in a patient by manipulation but tactile feedback is not available from ordinary computed simulations.

To explore the motion of the knee and the corresponding forces, a multimodal interface is required. The kinematics can be rendered with a graphical animation, and the forces can be rendered with a haptic interface. We had available in the laboratory a two-degree-of-freedom planar haptic interface, which was about the size of a typical computer mouse pad. It detected the position of a handle and applied forces via the handle to the human operator (torques were not possible to render).

In the present work we develop a technique for interacting directly with a dynamics simulation of a mechanism by means of inverse kinematics, and extend a classical model of human-knee kinematics to permit a general planar solution to its motion. The forces rendered to the operator of the haptic interface arise from ligament constraints and bearing-surface constraints; the latter are calculated using a fast new hierarchical method that maintains high update rates without introducing excessive artificial forces as the hierarchy is traversed. The techniques have been implemented and preliminarily tested in our laboratory.

2 Computed System Dynamics and Haptic Interfaces

The dynamics of a serial mechanism with n degrees of freedom can be expressed as a linear system of the form

$$M\ddot{q} = c(\dot{q}, q) + g(q) + \tau$$

in which the generalized variables are components of the vector q, the Coriolis and centripetal forces $c(\cdot)$ are functions of position and velocity, the potential terms $g(\cdot)$ are functions of position, and the generalized forces at each joint are τ. This system can be numerically simulated in linear time [6] even with arbitrary forces and torques applied to each mechanism link [11], and can be accurately simulated even when subject to a system of Pfaffian or holonomic constraints [3].

Interacting with such a dynamics simulation by means of a haptic interface requires correct transference of effort and flow between the two systems. A natural formalization of a haptic interface, and of a numerical simulation of dynamics, are as black boxes that have forces and torques (effort) as input and which produce positions and orientations (integrated flow) as output. Typically, transference is accomplished by introducing a simple (linear spring-damper) system between the haptic interface and an element of the dynamics simulation, so that operator motions produce forces and torques on the element which are then injected into the forward-dynamics solution. This method, although straightforward, introduces artificial compliance and damping that can interfere with the haptic presentation of the simulation and can also adversely affect the numerical computations of the dynamics.

An alternative is to use the output of the haptic interface to specify the configuration of one or more links of the mechanism. In general this requires computation of the inverse kinematics and dynamics for part of the mechanism, computation of the forward dynamics for the remainder of the mechanism, and a way of managing the forces of constraint on the elements in question.

The present work considers the specific problem of human-knee motion, for which there is a simple yet plausible model as a two-element multiple-degree-of-freedom mechanism. One element can be regarded as fixed, and the kinematic configuration of the independent element can be computed as a function of the instantaneous configuration of the haptic interface. The forces at any instant are the sum of external forces and the forces of constraint. Here, the external forces are only those from ligaments that constrain the knee (gravity is disregarded) and the constraint forces are only from interpenetration of the bearing surfaces.

3 Human-Knee Kinematics

As is frequent in the literature [9, 5], passive joint kinematics is defined as a succession of quasi-static positions. Biological joints are notably unlike robotic joints in that they are not usually composed of conforming surfaces, i.e. they are not lower-pair joints. Instead they are composed of non-conforming smooth surfaces that are constrained by soft tissue. There is a large body of literature on mathematical models of the human knee (cf. Hirokawa [10] for an excellent review). The present work will follow and extend the model proposed by Goodfellow and O'Connor [8], which is a widely known and well established planar model.

The dominant motion of the human knee is flexion-extension. Taking full extension as $0°$, flexion is measured positively and can range up to $170°$ in deep kneeling. Modeling the knee as a planar mechanism in the sagittal plane, the joint can be described as composed of two bearing surfaces that are connected by a complicated set of passive nonlinear viscoelastic elements (ligaments) and nonlinear actuators and connective elements (muscles and tendons). It is traditional to first model the passive kinematics, when there are no internal muscular loads or external loads, and to then consider external loads as deviating the nominal passive kinematics by straining the connective tissue.

We model the knee as described in Figure 1. The femur and tibia are modeled as rigid bodies, and the principal connective tissues are taken to be the cruciate ligaments. Previous work by one of the authors (REE [12]) indicated that using these ligaments alone produces a good overall model of knee motion.

Goodfellow and O'Connor proposed that the knee be modeled as a four-bar linkage. One bar is the line segment connecting the ACL and PCL attachments on the femur, a second bar is the line segment connecting the ACL and PCL attachments on the tibia, and the other two bars are the ligaments. Their model requires the assumption that the ligaments are inextensible elements that simply rotate about their insertion points, and is diagrammed in Figure 2. There is a reference frame on the tibia, a reference frame on the femur, and the relative motions are described by four-bar linkage equations.

To express all points in the tibial frame, a point $\mathbf{x}$ with femoral coordinates $\mathbf{x}^f$ has tibial coordinates $\mathbf{x}^t$ given as

$$\mathbf{x}^t = R(\theta)\mathbf{x}^f + \mathbf{p} \qquad R = \begin{bmatrix} \cos(\theta) & -\sin(\theta) \\ \sin(\theta) & \cos(\theta) \end{bmatrix} \qquad (1)$$

where R the rotation matrix is $\mathbf{p}$ is the displacement between the origins.

The traditional four-bar model has fixed lengths of the cruciate ligaments. Let L_a denote the ACL length, and L_p denote the PCL length; these can be specified in the tibial frame as

$$L_p = \|\mathbf{D}^t - \mathbf{A}^t\| \qquad L_a = \|\mathbf{C}^t - \mathbf{B}^t\| \qquad (2)$$

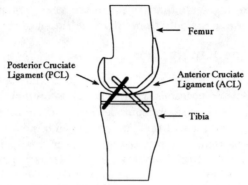

Fig. 1. Elements of the knee model

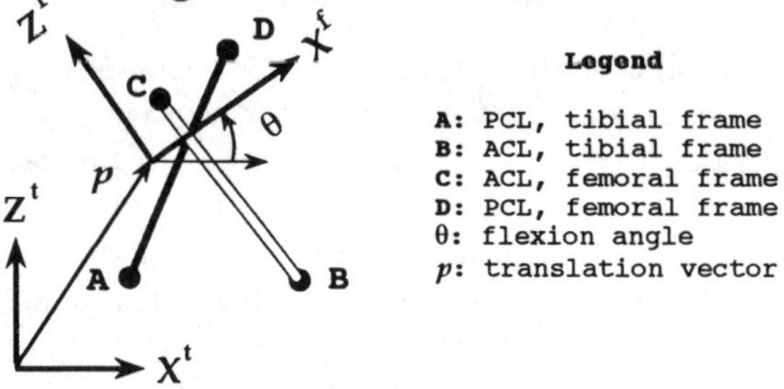

Legend
- A: PCL, tibial frame
- B: ACL, tibial frame
- C: ACL, femoral frame
- D: PCL, femoral frame
- θ: flexion angle
- p: translation vector

Fig. 2. A four-bar linkage model of the knee

The forward kinematics of the traditional model can therefore be specified as the determination of the frame displacement vector **p** for a given flexion angle θ. This is a simple geometric transformation – described more fully in Section 3.1 – for known values of $\mathbf{A}^t$, $\mathbf{B}^t$, $\mathbf{C}^f$, $\mathbf{D}^f$, L_a, and L_p.

A four-bar linkage has a single degree of freedom, but interaction with a haptic interface requires a less constrained kinematic model. Our extension to the traditional model is to permit the ligaments to be of varying lengths, which permits calculation of the forward and inverse kinematics in the sagittal plane. We will present the solution for the forward and inverse cases, applying constraints that are appropriate for the present problem of interest.

3.1 Forward Kinematics

The key observation for developing the general cruciate-based kinematics of the human knee is to permit the ligaments to have variable length. This is equivalent to introducing prismatic joints into the four-bar linkage, so the linkage will have three degrees of freedom in nonsingular configurations. Let these degrees of freedom be specified by the relative angle θ and the coordinate displacement **p**.

Interaction of a knee model with a haptic interface requires finding geometric and force relations in the joint, given a pair of inputs that represent the coordinates of a particular point in one reference frame. It is most convenient here to state the forward problem as:

$$\text{Given } \theta, L_a, \text{ and } L_p \qquad \text{Find } \mathbf{p}$$

We do this because there is no other independent variable that is of clinical interest in this system.

For any fixed set of anatomical landmarks $\mathbf{A}^t$, $\mathbf{B}^t$, $\mathbf{C}^f$, $\mathbf{D}^f$, the kinematic equations can be determined by requiring $\mathbf{D}^t$ to lie on a circle of radius L_p centered on $\mathbf{A}^t$ and requiring $\mathbf{C}^t$ to lie on a circle of radius L_a centered on $\mathbf{B}^t$.

For a given flexion angle θ, converting Equations 2 to tibial coordinates using Equation 1 gives

$$(R(\theta)\mathbf{D}^f + \mathbf{p} - \mathbf{A}^t)^2 = L_p^2 \qquad (R(\theta)\mathbf{C}^f + \mathbf{p} - \mathbf{B}^t)^2 = L_a^2 \qquad (3)$$

which can be used to solve for the two components of $\mathbf{p}$.

Let $\mathbf{G} = R(\theta)\mathbf{D}^f - \mathbf{A}^t$ and let $\mathbf{H} = R(\theta)\mathbf{C}^f - \mathbf{B}^t$. Equations 3 can be expanded to find that

$$G_x + p_x = \alpha_f + \beta_f p_z \qquad (4)$$

where

$$\alpha_f = G_x + \frac{L_p^2 - L_a^2 - G_x^2 + H_x^2 - G_z^2 + H_z^2}{G_x - H_x} \qquad \beta_f = \frac{H_z - G_z}{G_x - H_x}$$

Substituting the term from Equation 4 into the expansion of Equation 3 gives a quadratic equation in p_z, and resubstituting the numerical values of p_z into Equation 4 gives two values of p_x.

The two solutions of the quadratic correspond to distinct configurations of the extended linkage, but the anatomically consistent solution can always be identified by requiring that the $\mathbf{Z}$ component of $\mathbf{p}$ places the femur above the tibia.

3.2 Inverse Kinematics

The inverse kinematics are of more interest in the present work than are the forward kinematics, because these determine the joint geometry for a given position of the haptic interface. The interface provides two independent variables, which can be mapped to the $(\mathbf{X},\mathbf{Z})$ coordinates of a point of interest. Since there are only two input variables but three potential output variables (θ, L_p, and L_a) an additional constraint is needed.

We propose using a simple linear relationship between the ligament lengths. Suppose that the lengths of the ligaments are given as

$$L_p = \rho_p S \qquad L_a = \rho_a S \qquad (5)$$

where the independent variable S is a dimensionless scale from which the lengths are calculated, and ρ_p and ρ_a are proportionality constants for the PCL and ACL respectively. The inverse problem can then be posed as:

Given p,   Find θ and S

for fixed values of the anatomical landmarks and fixed moduli of the cruciate ligaments.

The governing equations in Section 3.1 can be used to solve this problem. Let $\mathbf{J} = \mathbf{p} - \mathbf{A}^t$ and let $\mathbf{K} = \mathbf{p} - \mathbf{B}^t$. The governing relations of Equations 2 plus the constraints of Equations 5 can be expanded and simplified; eliminating S from then gives a trigonometric relation which can be abbreviated as

$$\alpha_i \cos(\theta) + \beta_i \sin(\theta) = \gamma_i \tag{6}$$

where

$$\alpha_i = \rho_a^2(2D_x J_x + 2D_z J_z) - \rho_p^2(2C_x K_x + 2C_z K_z)$$
$$\beta_i = \rho_a^2(2D_x J_z - 2D_z J_x) - \rho_p^2(2C_x K_z - 2C_z K_x)$$
$$\gamma_i = \rho_p^2(C_x^2 + C_z^2 + K_x^2 + K_z^2) - \rho_a^2(D_x^2 + D_z^2 + J_x^2 + J_z^2)$$

Equation 6 has the two solutions

$$\theta = \mathrm{atan2}(\beta_i, \alpha_i) \pm \mathrm{atan2}(\sqrt{\alpha_i^2 + \beta_i - \gamma_i^2}, \gamma_i) \tag{7}$$

The valid solution to this problem has $0 \leq \theta < \pi$, which can be substituted into the expanded form of Equation 2 to find S.

4 Testing Intersections of Prosthetic Surfaces

Using our method to interact with a dynamics simulation by means of a haptic interface requires calculation of reaction forces at joints. The bearing surfaces of prosthetic knee joints differ from natural knee surfaces in shape, material characteristics, and especially by the lack of a compliant meniscus interposed between the femoral and tibial components. For the purposes of kinematics and quasi-static studies, the surfaces can be modeled as rigid bodies or as bodies with straightforward mechanical characteristics.

Suppose that a simple model of material compliance is used, e.g. the reactive force due to interpenetration is calculated as that of a linear spring or of Gaussian contact. For such simple models, the force is a function of the distance of interpenetration and of assumed material properties. The computational problem is rapidly finding an acceptable approximation to the maximum geometrical intersection of two given regions.

The component geometries can be modeled in many ways; here, we assume that the contours are specified as ordered point sets and that analytic models are fit to the point sets. The simplest contour model is a piecewise polynomial. This model has good local error characteristics, but intersection calculations between models of M and N points respectively is of $O(MN)$ if a brute-force algorithm is used. For models with thousands of points, current affordable computers (UNIX workstations or 34MFLOP DSP computers) cannot perform the calculations at the frequencies required for vivid haptic presentation.

One efficient way to calculate geometric interpenetration is to use a hierarchical (coarse-to-fine) strategy, also called a multiresolution strategy [13]. Requirements of speed and accuracy are conflicting, but can be resolved in the following manner. The system begins at the coarsest level, and finds the deepest interpenetration at that level of resolution. If, at the next time step, the point sets have not undergone much relative motion then a local interpenetration test at the next finest level of resolution can find the interpenetration more accurately. An example of a multi-resolution model is given in Figure 3(a).

A problem with calculating forces from a linear model is that the penetration depth and direction to the nearest model segment vary irregularly between different model resolutions and also between model elements at the same level of resolution. Consider moving a single point towards and through the model, beginning above the model at coordinate $x = 0.85$ and moving vertically downward at 2 cm/s. Figure 3(c) shows the direction to the nearest model element and Figure 3(e) shows the penetration depth; use of such a model can produce a poor haptic rendering because of the rapid vector changes.

An alternative is to model the bearing contours as multi-resolution overlapping circular arcs, illustrated in Figure 3(b) for the same points as in Figure 3(a). The same point trajectory produces the directions plotted in Figure 3(d) and penetration depths plotted in Figure 3(f). Except for an abrupt initiation this contour model can produce a much smoother and more realistic haptic rendering of compliance.

5 Implementation and Results

The prototype system for visuo-haptic rendering of human-knee kinematics used the extended inverse kinematics to resolve knee motion, plus a combination of interpenetration calculations and simple-spring modeling of ligaments to determine forces that were reflected to the operator. The system was implemented on a network of four computers: a PC for the haptic interface, two DSP processors for interpenetration calculations, and a SUN workstation for graphical display.

Initially, the operator was presented with a visual display of the knee joint. After selecting an operational point on a bone (which became the origin of the local frame) the system was activated. Thereafter, the handle position of the planar haptic interface determined the displacement **p** of the bone's frame, and the inverse solution was used to find the rotation θ and thus the lengths of the ligaments. The rotation and displacement controlled the visual presentation directly. The inverse solution of Section 3.2 was used to determine the lengths of the ligaments, and also the depth and direction of interpenetration using the hierarchical arc method described in Section 4. For the purposes of force reflection the ligaments were modeled as springs that could be slack: if the neutral length of a ligament was l_0, then at length l the force $F(l)$ was calculated using its stiffness k as:

$$F(l) = \begin{matrix} 0 & \text{if } l \leq l_0 \\ k(l - l_0) & \text{if } l > l_0 \end{matrix} \qquad (8)$$

and the interpenetration distance d was reflected as $F(d) = kd$.

The stiffness for the bearing surfaces was typically more than 20 times the stiffness of the ligaments, so that hard and soft constraints were well presented. Computations could be performed at up to 500 Hz in both the visual and haptic domains.

6 Discussion

The prototype system was tested using contours derived from proprietary information provided by a knee-prosthesis manufacturer, with two operators who were familiar with the kinematics of the human knee during total-joint replacement surgery. The tests were of a preliminary nature, ensuring mainly that the rendering was sufficiently realistic for potential use in training and simulation of subsequent surgical procedures.

The posterior cruciate ligament insertions were determined from published anatomical data ([12], case 2), and anterior cruciate ligament insertion were determined from atlas predictions. Two tests were done, one in which the placement of the components were varied and one in which the mechanical properties of the ligaments were varied.

The kinematics were in concordance with expectations. The linear-segment model of the bearing surfaces provided a poor haptic rendering – a "scratchy" feeling and severe oscillation were often present. The circular-arc model provided a much smoother feeling. Because the bearing-surface stiffness was nearly at the rendering limit of the haptic device, we used the sample-estimate-hold method of Ellis et al. [4] to minimize energy-transfer errors between the operator and the haptic device. The interaction was not perfectly smooth but was vivid.

Preliminary testing was performed by varying the locations of components and the mechanical properties of the ligaments. A full psychophysical study has not been done, but pilot results are promising. It was not easy to distinguish kinematic or dynamic changes due to component displacements of a few millimeters (such changes are implicated in wear of artificial joint components) but changes in mechanical properties of the ligaments were dramatic with force feedback. Especially noticeable to the surgeon who tested the system was the balancing of ligament forces: as the knee moved into deep flexion the relative contributions of the cruciate ligaments were readily felt. The actions of these ligaments are difficult to visualize but were rendered well by the system.

Realistic rendering is important because the results of total knee replacement surgery depend on restoration of knee kinematics nearly to those of a normal knee. This process is sensitive to the type of components used, to the placement of the components, and to ligamentous balance [1, 2]. Although several studies have been conducted to quantify the effect of specific variables on alignment, range of motion or stability of the joint [7, 14], the optimization of the implantation requires an overall and detailed analysis of the resulting kinematics. An improved evaluation of the restored kinematics can help a surgeon to choose the optimal position and size of the components, and can also assist in balancing the ligaments for each individual patient.

The system we developed may be useful for education and for pre-operatively exploring possible consequences of total knee replacement surgery or ligament reconstruction. Future work should be conducted on optimizing the calculations, and on a better psychophysical evaluation of the usefulness of visuo-haptic rendering in education and training.

Acknowledgments

This research was supported in part by the Natural Sciences and Engineering Research Council of Canada, by the federal government under the Institute for Robotics and Intelligent Systems (which is a National Centres of Excellence programme), and by the Information Technology Research Centre of Ontario.

References

1. T. A. Andriacchi, T. S. Stanwyck, and G. J. O. "Knee biomechanics and total knee replacement". *Journal of Arthroplasty*, 1(3):211–219, 1986.
2. E. Y. S. Chao, E. V. D. Neluheni, H. R. W. W., and D. Paley. "Biomechanics of malalignment". *Orthopedic Clinics of North America*, 25(3):379–386, 1994.
3. R. E. Ellis and S. L. Ricker. "Two numerical issues in simulating constrained dynamics". *IEEE Transactions on Systems, Man, and Cybernetics*, 24(1):19–27, 1994.
4. R. E. Ellis, N. Sarkar, and A. Jenkins, M. "Numerical methods for the force reflection of contact". *ASME Journal of Dynamic Systems, Measurement, and Control*, 1997. In press.
5. J. R. Essinger, P. F. Leyvraz, J. H. Heegard, and D. D. Robertson. "A mathematical model for the evaluation of the behaviour during flexion of condylar-type prostheses". *Journal of Biomechanics*, 22(11):1229–1241, 1989.
6. R. Featherstone. *Robot Dynamics Algorithms*. Kluwer Academic Publishers, Norwell MA, 1985.
7. A. Garg and P. S. Walker. "Prediction of total knee motion using a three-dimensional computer-graphics model". *Journal of Biomechanics*, 23(1):45–58, 1990.
8. J. W. Goodfellow and J. J. O'Connor. "The mechanics of the knee and prosthesis design". *Journal of Bone and Joint Surgery*, 60B:358–369, 1978.
9. M. S. Hefzy and E. S. Grood. "Review of knee models". *Applied Mechanics Review*, 41(1):1–23, 1988.
10. S. Hirokawa. "Biomechanics of the knee joint: A critical review". *Critical Reviews in Biomedical Engineering*, 21(2):79–135, 1993.
11. O. M. Ismaeil and R. E. Ellis. "Effects of non-tip external forces and impulses on robot dynamics". In *Proceedings of the IEEE International Conference on Systems, Man, and Cybernetics*, pages 479–485, 1992.
12. S. Martelli, R. E. Ellis, M. Marcacci, and S. Zaffagnini. "Total knee replacement kinematics: Computer simulation and intraoperative evaluation". *Journal of Arthroplasty*, 1997. (Conditionally accepted February 3, 1997).
13. D. K. Pai and L.-M. Reissel. "Touching multiresolution curves". In *Proceedings of the ASME Dynamic Systems and Control Division, DSC-Vol. 58*, pages 427–432, 1996.
14. D. D. Rhoads, P. C. Noble, J. D. Reuben, and H. S. Tullos. "The effect of femoral component position on the kinematics of total knee arthroplasty". *Clinical Orthopaedics and Related Research*, 286:122–129, 1993.

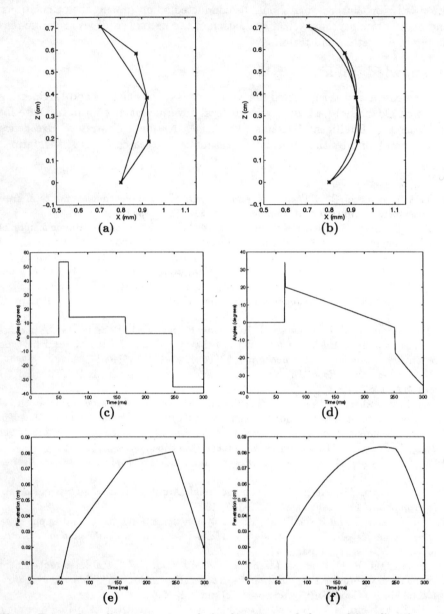

Fig. 3. Points fit to line segments and circular arcs, and the resulting direction and magnitude of interpenetration

Chapter 4

Mobile Robots Navigation

The four papers of this chapter deal with indoor and outdoor mobile robots navigation.

The paper by Maeyama, Ohya and Yuta presents the basic concept behind the navigational system for the small Yamabiko robot. It is the hallmark of Professor Shinichi Yuta's laboratory at the University of Tsukuba. The robot was human guided, learning long path segments in outdoor, flat terrain environments. The subsequent autonomous repetition of these paths by playback allows to recognize and learn navigation cues. A multi-agent architecture was used to recognize and match vision and ultrasonic cues (extracted during training) with natural landmarks during the autonomous repetition stage. Path segments totaling approximately 2 km are the target of this functional system which carries all sensors, computational support, motion control electronics and power sources onboard a differentially steered vehicle weighing approximately 12 kg.

The paper by Ray Jarvis reported work on extending a previously constructed indoor mobile robot navigation system by replacing an umbilical cable by a radio ethernet link to a wired LAN of Silicon Graphics Work Stations. The navigation system is capable of relatively brisk navigation from a nominated start point to a specified goal position in an initially unknown and relatively unstructured environment and can accommodate environmental variability. A Denning bar-code beacon-based localizer and a Erwin Sick time-of-flight scanning range finder were the only two sensors used. A distance transform based global path planner is used in continuous mode to update the best remaining path plan as the unknown and perhaps changing environment is being mapped and updated. There are no computing resources on-board except those embedded in the radio ethernet unit, the serial line server and the micro-controller used for wheel control. Standard wheelchair motor/gear sets were used in differentially steered mode. The system is very robust and can carry a payload of up to 150 kg for up to 6 hours between rechargings of the on-board batteries. The system shows considerable commercial viability and can be delivered complete for approximately US$25K.

Cozman and Krotkov present a mobile robot operating in space. The main interest is the interface that aids human operators to teleoperate rovers by localizing the position of the mobile. The position estimation algorithm using a constructed figure of merit that reflects the effect of disturbances in the image and can include information about a priori knowledge about position.

The paper by Hait, Simeon and Taix deal with the problem of combining terrain modeling with 3D motion planning using specified landmarks. Both the validity constraints concerning the climbing viability and stability of the six wheeled, articulated body, Russian built vehicle and the visibility constraints concerning the view ability of the landmarks, were taken into account. A global planner defines sub-goals, transfers between which are then handled by a local planner which calculates and chooses between feasible trajectories. Edge costs evaluated during optimal path
selection include both path length and terrain difficulty. This project was undertaken within the framework of the EDEN project at LAAS-CNRR in Toulouse, France. The system is capable of fully autonomous navigation in very rugged and obstacle cluttered environments. A number of simulated results were shown to illustrate the method. The motion planner presented is currently being integrated into the EDEN experimental test-bed. A Marsokhod vehicle, called LAMA, equipped with a range finder and stereo vision system, will be used.

Long distance outdoor navigation of an autonomous mobile robot by playback of Perceived Route Map

Shoichi MAEYAMA   Akihisa OHYA   and   Shin'ichi YUTA
Intelligent Robot Laboratory
Institute of Information Science and Electronics
University of Tsukuba
Tsukuba, Ibaraki, 305 JAPAN
phone:+81-298-53-5155 fax:+81-298-53-6471
Email: {maeyama, ohya, yuta}@roboken.esys.tsukuba.ac.jp

Abstract: In this paper, we report a development of an autonomous mobile robot for long distance outdoor navigation in our university campus. We propose how to generate a long distance Perceived Route Map (PRM), a position-based navigation algorithm using PRM, and incremental integration of the robot system by multiple processors and multiple agents. Furthermore, we developed an experimental robot system and conducted experiments of autonomous navigation using PRM in our university campus. Finally, from experimental results, we discuss open problems and future work in outdoor navigation.

1. Introduction

We present a research on long distance outdoor navigation of an autonomous and self-contained mobile robot. The objective of this research is the development of a robot to achieve outdoor navigation over a distance of about 2km in our university campus (Figure 1). The target environment is the paved or tiled walkway shown in Figure 2. The walkway can be assumed to be a two dimensional plane along with a wall, a hedge or a tree etc., which can be utilized as landmarks. A walker or a bicycle can be assumed to be a moving obstacle. In response to these moving obstacles, the robot waits for them to go away, because the robot is much slower than the obstacles.

A walkway is a much more unstructured environment than highway, in which the vehicle can be controlled by following white lane. In conventional works, the boundary edge of the road area or the shape of the detected road area is used for navigation control[1][2]. These approaches focus on the road following technique like an autonomous highway vehicle. But, it is difficult to treat a shadow on the road, detect intersections and change the way. On the other hand, positioning is essential for long distance walkway navigation. Position information is also useful for locomotion control, since it is easy to decide a turning point and the goal. However, in conventional works, it has not been actively used.

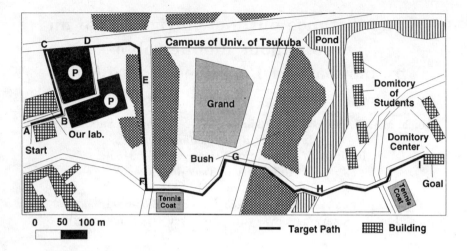

Figure 1. Schematic map of the target environment. (A,B,...,H,I are passing points.)

In this paper, we propose the position-based outdoor navigation using the Perceived Route Map, which includes path from the start to the goal taught by an operator and landmarks acquired automatically by robot itself. Then, we discuss some problems on walkway navigation from experimental results.

2. Navigation using the Perceived Route Map (PRM)

In our basic strategy, while the operator controls the robot from the start to the goal, the robot generates the route map which is perceived by own internal and external sensors (Figure 3). We call such a map the Perceived Route Map (PRM). After that, the position-based autonomous navigation is done by playback of the PRM. On a walkway, the question "what kind of properties in this environment can be utilized for robust navigation" itself is a important problem. If the robot is navigated a long distance using the PRM, it will become obvious what kind of information is needed for walkway navigation.

2.1. Acquisition of the PRM

For position-based navigation, a precise route map must be given to the robot in advance. However, it is difficult for a human to make a route map over a distance of about 2km in outdoor environment. So, we want that the robot makes the route map by itself. But, exploration by the robot itself in outdoor consume much time. Our interest is not map building by exploration. Therefore, We propose route map generation by natural landmark acquisition through human route teaching. In this method, a human operator controls the robot to the goal at first. Then, the robot remembers its own trajectory as a path and the location of landmarks to correct its position on the way to the goal. The operator teaches only the path from the start to the goal with a manual controller. The robot generates the route map which is perceived by own internal and external sensors. As the result, the route map is generated

Figure 2. Target environment which has a paved road along with trees and hedges. (B, D, E, G and H are some passing points shown in Figure 1.)

with the expression suitable for the robot. The relative location between path and landmarks is recorded in the route map.

A landmark is the mark to confirm the position. Therefore, the objects for landmarks must satisfy the requirements, that they can be detected at the same location with the same characteristics even if the robot has displacement. The robot must detect the objects which satisfy with such conditions. But, of course the robot does not know where such objects are in advance. So, we propose the PRM generation by parallel execution of multiple agents (Figure 4). At first, we suppose that the robot has the functions of the sensors to measure the distance or/and direction of the objects. Furthermore, the robot should have a locomotion controller and position estimator. Then, we prepare Landmark Agent, Path Agent and Radio controller Agent. The Landmark Agent is the agent to detect the landmark candidate from the measurements by sensors and store the sensing point and landmark location and so on when it is repeatedly detected at the same landmark position even if the robot travels over a distance. The Path Agent is the agent to store the passing point of the robot. The Radio controller Agent is the one to interpret the human operation and to send the control command to the locomotion controller. In this system,

Figure 3. Outdoor navigation using Perceived Route Map (PRM) : PRM generation by natural landmarks acquisition and autonomous navigation using the PRM.

agents work in parallel to cope with the process which has a different time constant. Then, the PRM can be generated without missing the landmarks in the target environment.

2.2. Position based navigation using the PRM

If the robot has a path from start to goal and can estimate the precise position, the robot can arrive at the goal by feedback control of the estimated position. So, the position estimation is the dominant subject for navigation. From recent work of position estimation of mobile robots, the Kalman Filter (KF) is well known for the good performance. KF has a reasonable redundant sensor data

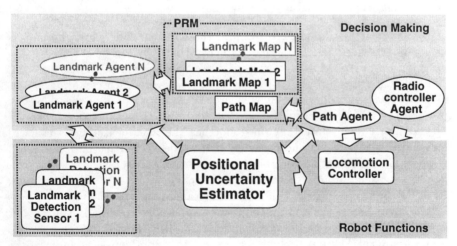

Figure 4. Parallel execution of multiple agents for autonomous navigation using PRM.

fusion with low calculation and small amount of saving data, step by step, and utilizes the low dimensional observation with the correlation of the position parameters [3][4][5][6][7][8]. We also confirmed the performance of this position estimation technique in outdoor application[9][10].

So, we use this technique for navigation using PRM. The PRM data includes the robot inherent errors such as wheel diameter, tread and the other off set. Therefore, if the robot can compensate the un-repeatable error such as the interaction from the road surface by observing landmarks, navigation using the PRM will succeed.

Position based navigation using PRM is also done by multiple agents, shown in Figure 4. The Path Agent gets the path to be followed from the Path Map and sends the command to the locomotion controller. Each Landmark Agent gets the set of sensing points and landmark locations, from each Landmark Map. If the robot crosses the sensing point of a landmark, the Landmark Agent searches the landmark. If the Landmark Agent finds the landmark, the information of observation is sent to the positional uncertainty estimator. Then, the estimated robot position and the error covariances are corrected by Kalman Filter[10]. The threshold for identification of the observed landmark and the landmark in Landmark Map is determined based on the value of the covariance of the estimated position. If the difference of the location of the observed landmark and the map's landmark is less than the threshold, the Landmark Agent decides that they are the same one. Thresholding is very useful to avoid the misunderstanding of landmarks. Each Landmark Agent works independently. But, through the resulting covariance of the estimated position, each Landmark Agent communicates with each other implicitly. More reliable navigation system can be developed by adding more Landmark Agent incre-

Front   Back

Figure 5. Photographs of the mobile robot YAMABICO NAVI. Dimension (WxHxD) is about 450x600x500 mm. Weight is about 12 kg. Wheel diameter is about 150 mm. Tread is about 400 mm.

mentally. The Radio controller Agent is not used for autonomous navigation.

3. Experimental autonomous mobile robot YAMABICO NAVI

We implemented the proposed navigation system in the experimental mobile robot *YAMABICO NAVI* (Figure 5). Figure 6 shows the system configuration of this robot. The robot has a dead reckoning system by fusion of odometry and gyro[11], SONAVIS [1] to detect landmarks, sonar to detect obstacles, a valve regulated lead acid battery (12V 7Ah) and two DC motors to drive the wheels.

The controller is distributed on multiple CPUs. The *Master* and the other functional modules have a connection like a star with Dual Ported Memory. The *MASTER* is a CPU module to control a total behavior of the robot. Information for decision making is gathered into the *MASTER*. The *MASTER*

---

[1] SONAVIS is a landmark detection sensor with ultrasonic range sensor and vision mounted on a turn table.

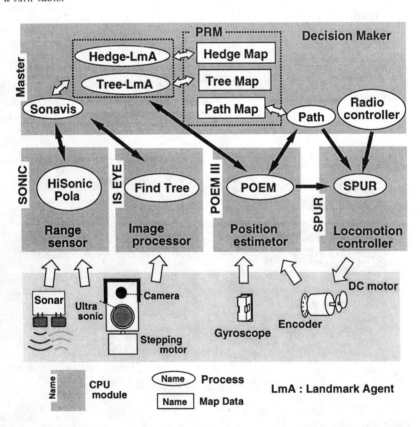

Figure 6. System configuration of YAMABICO NAVI.

decides next motion from these information. Then, the *MASTER* gives the commands to the other functional modules. *YAMABICO NAVI* has functions of Locomotion control (SPUR[12]), Position estimation (POEM III[10]), Image processing (ISEYE) and Ultrasonic range sensing (SONIC).

The multi-agent system for navigation is implemented on the *MASTER*. *Hedge-LmA* is the Landmark Agent to detect hedges as landmarks when the distance measured at the same direction by ultrasonic sensor mounted on SONAVIS is almost same distance while traveling over 90 cm. *Tree-LmA* is one to detect trees as landmarks when the tree is detected at the almost same location by SONAVIS while traveling over 60 cm. *Sonavis Agent* is one to arbitrate *Hedge-* and *Tree-LmA*, since these two Landmark Agents use the same sensor property SONAVIS.

4. Experiments

We conducted some experiments with our robot *YAMABICO NAVI* mentioned above.

At first, the experiment for PRM generation was done. The generated PRM is shown in Figure 7. This is the PRM from A to G in Figure 1 around 5:00 pm in the beginning of May. The weather of these days was fine. The maximum speed of the robot in this experiment was 30 cm/s. Total distance was about 810 m. *Hedge-LmA* detected 95 landmarks which include hedges,

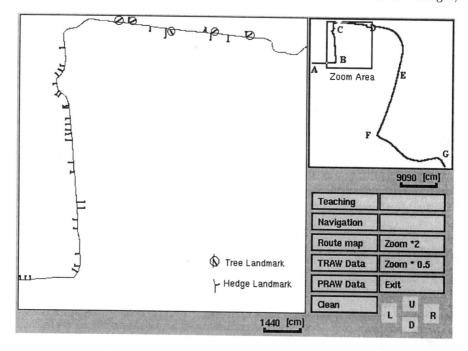

Figure 7. An example of the generated PRM from A to G in Figure 1. (This figure is a window of the debug monitor system.)

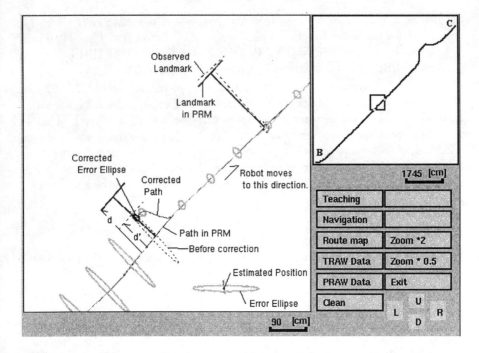

Figure 8. Position correction by hedge landmark.

walls, fences and bushes. 26 trees were detected as landmark by *Tree-LmA*. The maximum distance between landmarks was about 45 m. The average distance between landmarks was about 5 m. From the experiment, we found there are three ramps, which can not be traveled over for the reason of small wheel diameters.

Next, we tried to do some experiments of autonomous navigation using the PRM. Figure 8 shows the position correction by *Hedge-LmA*. In this figure, d is the distance from the robot to the hedge expected from *Hedge Map*, which is included in the PRM. d' is the measured distance of the hedge. So, $d - d'$ is the difference of the map's and the measured distance. If $d - d'$ is smaller than the threshold for the identification, *Hedge-LmA* sends this value to the POEM III function module. The threshold is the deviation 1σ about the estimated position of the sensing direction. The corrected path is automatically generated by the control algorithm of the SPUR function module and the robot returned to the desired path in *Path Map*. The position correction by *Tree-LmA* is similer to *Hedge-LmA*. We reported it in [9].

5. Discussion

The robot sometimes failed in the following cases : (1) Insufficent number of landmarks, (2) Change of sunlight during a day and (3) Dynamic change of environment.

In case (1), when there are insufficient number of landmarks or the robot does not acquire enough landmarks, the robot loses its position. The reason is

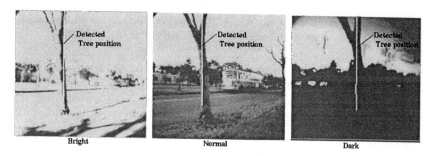

Figure 9. The detected tree position in the case of varying sunlight.

misunderstanding of landmarks or a large error of the estimated direction. As the distance the robot moves without landmark increases, the rate of misunderstandings and missings of following landmarks will be higher, because the positional uncertainty is growing up.

In the next case (2), the sunlight during a day is changing gradually (Figure 9). The tree in the captured image is detected at different positions in varying sunlight. When the measurement error generated from the change of sunlight can not be ignored, the robot must have different a PRM for morning, afternoon and evening.

In the last case (3), unfortunately, the hedge and the bush along the walkway are cut down for maintenance. Some other objects in the environment also move or remove for maintenance or some other reasons. In such cases, the robot must generate this part of the PRM again.

6. Conclusions

We presented a long distance outdoor navigation system for an autonomous mobile robot in this paper. We realized an experimental robot system for generating a long distance Perceived Route Map (PRM) and navigation using the PRM, realized as a multi-agent system. Furthermore, we conducted experiments of autonomous navigation using the PRM in our university campus. Up to now, we have tried to do autonomous navigation using the partial PRMs, which are the PRMs from one passing point to the next. We found some problems such as too few landmarks, the change of environment and the connection of partial PRMs.

In future work, we want to overcome the above stated problems. We will make more various Landmark Agents, robust against the change of environment, and we will connect these partial PRMs.

Acknowledgement

We thank Dr. M. Rude for many helpful suggestion to draw up the paper.

References

[1] Nishikawa K and Mori H, "Rotation Control and Motion Estimation of Camera

for Road Following" Proc. of IEEE/RSJ International conference on Intelligent Robots and Systems, Vol.2, pp.1313-1318 (1993)

[2] Thorpe C (Ed.) *Vision and Navigation : The CMU Navlab*, Kluwer Academic Publishers, (1990)

[3] Kam M, Zhu X and Kalata P, "Sensor Fusion for Mobile Robot Navigation", Proceedings of the IEEE, Vol.85 No.1, pp.108-119, (1997)

[4] Crowly J L, "World Modeling and Position Estimation for Mobile Robot Using Ultrasonic Ranging", Proc. of IEEE International Conference on Robotics and Automation Vol 2, pp.674-680, (1989)

[5] Kriegman D J, Triendl E and Binford T O, "Stereo Vision and Navigation in Buildings for Mobile Robots", IEEE Transaction on Robotics and Automoation, Vol.5, No.6, pp.792-680, (1989)

[6] Leonard J J and Durrant-Whyte H F, "Mobile Robot Localization by Tracking Geometric Beacons", IEEE Transaction on Robotics and Automoation, Vol.7, No.3, pp.376-382, (1991)

[7] Komoriya K, Oyama E and Tani K, "Planning of Landmark Measurement for the Navigation of a Mobile Robot", Proc. of IEEE/RSJ International Conference on Intelligent Robots and Systems, Vol.2, pp.1476-1481, (1992)

[8] Kosaka A and Kak A C, "Fast Vision-Guided Mobile Robot Navigation Using Model-Based Reasoning and Prediction of Uncertainties", CVGIP: Image Understanding Vol.56, No.3, pp.271-329, (1992)

[9] Maeyama S, Ohya A and Yuta S, "Positioning by tree detection sensor and dead reckoning for outdoor navigation of a mobile robot", Proc. of IEEE International conference on Multisensor Fusion and Integration for Intelligent systems, pp.653-660 (1994)

[10] Maeyama S, Ohya A and Yuta S, "Non-stop outdoor navigation of a mobile robot - Retroactive positioning data fusion with a time consuming sensor system -", Proc. of IEEE/RSJ International conference on Intelligent Robots and Systems, Vol.1, pp.130-135 (1995)

[11] Maeyama S, Ishikawa N and Yuta S, "Rule based filtering and fusion of odometry and gyroscope for a fail safe dead reckoning system of a mobile robot", Proc. of IEEE International Conference on Multisensor Fusion and Integration for Intelligence Systems, pp.541-548 (1996)

[12] Iida S and Yuta S, "Vehicle command system and trajectory control for autonomous mobile robots", Proc. of IEEE/RSJ International Workshop of Intelligent Robots and Systems, pp.212-217 (1991)

Etherbot - An Autonomous Mobile Robot on a Local Area Network Radio Tether

Ray Jarvis
Intelligent Robotics Research Centre
Monash University
Wellington Road
Clayton   Victoria   3168
Australia
E-mail: Ray.Jarvis@eng.monash.edu.au

Abstract

It is well understood in the robotics research community that an autonomous mobile robot which is physically tethered to extend battery power limits, harness more powerful computational resources, allow convenient monitoring of sensor and process components or to increase the pay load is, at best, just a laboratory curio since the tether renders the vehicle more or less useless in many real working situations because of potential entanglement and actual range limitations.

This paper demonstrates that it is feasible and advantageous to build a mobile robot which can navigate competently in an initially unknown and time-varying obstacle cluttered environment with most of the above advantages of tethered operation but without a physical tether; instead, a virtual tether based on a radio ethernet bridge is used.

Introduction

The actualisation of a complete mobile robot system, if it is to exhibit autonomous and useful behaviour in initially unknown environments, requires that the capabilities of localisation, environmental mapping, path planning and motion control be properly integrated [see Figure 1.]. Each of these sub-systems can be built using a variety of different approaches and an even greater variety of instrumentation. The reliability of such a system often depends upon the quality of its functional components and the simplicity with which they are forged together.

When the question of physically packaging a mobile robot system is confronted, a considerable number of alternative configurations must be chosen between. The choice made will largely determine aspects of reliability, operational periods, ease of maintenance, flexibility of control, payload, computer upgrade ability, fragility and application scope and thus cannot be treated lightly. Unfortunately, in emphasising sophistication and advanced capabilities of such systems in research laboratory environments, many designers neglect to pay proper attention to the packaging question and their 'inventions' are unable to survive the rigours of practical application in a real work place. Transforming a 'hot house' robot to a working machine is not often possible; the system must be designed for its intended working environment from the beginning.

There has been, in recent times, much debate on the 'behaviour' versus 'planning' based approach to autonomous robot navigation (alternatively referred to as the 'reaction' versus 'reason' approaches). It has been argued that the fragility of many existing robot systems, when confronted with real life practical situations, relates to the inadequacy of complexity management as it relates to the integration of sensor, locomotion and planning systems associated with the 'reason' approach [1] and that this can be partially avoided by the 'reaction' approach [2,3] with its intrinsic bias towards relatively independent multiple sensor to action subsystems. However, it is often difficult to direct a 'reaction' based system to carry out specific and complex tasks. The combination of the two approaches is likely to be the way forward so that both reaction and reasoning might interrelate as the appropriate levels.

In this paper the emphasis is on the 'reason' approach but this is not to say that the addition of some levels of the 'reaction' approach might lead to even more robust performance. For both approaches simplicity of intent and actualisation should be the guiding light if performance and reliability are to be simultaneously realised.

The following section addresses the question of physical packaging, citing examples from our own past projects (in the Intelligent Robotics Research Centre at Monash University) and indicates the general characteristics of various packaging modes. Then follows a section describing the radio ethernet link used in this current project. The instrumentation used for localisation and environmental mapping is next described. Then follows a section on Distance Transform path planning. The performance of the Etherbot system is presented next. The paper closes with discussion and conclusions.

Physical Packaging of a Mobile Robot System

Energy, size/weight and communications requirements dominate the packaging question. Regarding the energy factor there are three regular electrical sources, battery, fossil fuel generator or mains supply and three energy sinks, locomotion, computational power and instrumentation (including communications units). If a robot must be capable of carrying a large payload and also be smart and fast a great deal of energy is required. Locomotion energy follows the normal laws of physics in considering payload and speed questions but relating computer electrical power consumption to smartness is not so easy, since information is a type of by-product when computers turn electrical energy into heat. However, if a robot need not be smart, less electrical energy is generally needed to power its computational requirements.

Considering which things to have on-board and which off-board is the main game. There are real engineering compromises which have to be executed when making such decisions.

Figure 2 illustrates some test cases we have had direct experience with. The robot photographs of Figure 3 correspond to examples of these cases, one to one. Case

2(a) represents a tetherless robot that does not need to be very smart. Our experience in this class is with an Automated Guided Vehicle (AGV) [4] which is taught paths by a human pushing it around and then follows a selected path until stopped. It is battery powered and has only two sensors, a scanning beacon localiser and a bumper switch. An IBM clone personal computer is sufficient to meet its computational needs which are to calculate its position, to control the motion along a pre-taught path and to react appropriately to bumper signals (e.g. stop for a few seconds and then proceed). The AGV can operate for approximately six hours continuously between recharging and can carry a 200 kg payload. The localisation relies upon passive beacons at known positions around the walls of the work environment. No environmental mapping or path planning is involved. Fleet control is not possible without additional communications instrumentation.

Case 2(b) represents an autonomous mobile robot with full navigational capabilities in initially unknown environments [5] with considerable instrumentation on-board and a low bandwidth (9600 baud) serial radio link to a stationary workstation. An on-board PC collects range data, pre-processes it and sends edge line component data to the home station on command. Also on command it transmits localisation data and receives and executes motion instructions. Batteries on-board can drive the system for between 1 to 2 hours, adequate for research experimentation but not as a working machine. This machine was first developed to have a beacon localisation mode of operation but was extended to allow for pure natural landmark based navigation. This machine has essentially no payload capabilities, moves at approximately 20 cm/sec and weighs 200 kg. A video transmitter is used to convey PC screen status data to a TV monitor near the home station.

Case 2(c) represents a wire tethered autonomous mobile robot [6] which has been on active exhibition at a participatory science museum since August of 1996. It carries batteries which power the wheels and instruments (range finder and localiser) but these are trickle charged by an on-board battery charger which is powered by a mains lead on the tether cable. Thus the charging process is continuous and requires no attention until the batteries finally need replacing.

The sensing instruments are connected to the home computer along the tether. The control system receives analogue signals on the tether. Also on-board are a stereo video camera pair for allowing a vision based teleoperation mode to be used instead of autonomous mode when chosen. The camera's signals are also carried on the tether.

This packaging configuration was partly chosen to allow the mapping, planning and path following process to be fully monitored on the screen of the workstation so that spectators/operators might easily observe the computational support of the demonstration and so that mode switching could also occur at the workstation. A second reason was to allow long hours of continuous use with little operational hassle for museum staff. A third reason was to render the robot itself physically robust as many 'participators' deliberately drive it into obstacles in teleoperation mode.

However, it is interesting to note that, although we have had little or no trouble with the computer software/hardware or instruments, the tether cable has had to be repaired on a number of occasions despite its being carried overhead on a swinging boom.

Case 2(d) is the physical tether-free situation which is the main focus of this paper and will be discussed in detail in what follows.

Note that, in general, fleet control can be more easily formulated using a home base which communicates with multiple mobile robots rather than devising distributed protocols and inter-communications amongst the robots.

The Etherbot Project

This project is in direct line of experimental succession to the wire tethered robot (GO-2) described in Case 2(c), above, and depicted in Figure 3(c). The way the project developed was as follows:

In the process of testing GO-2 in the laboratory before commissioning it at the Spotswood ScienceWorks Museum in Melbourne, the anticipated inconvenience caused by tethers in general was confirmed in particular. Setting up an overhead boom was a nuisance but a necessary construction to remove any possibility of it catching on obstacles or being run over by the robot. This limited the range of the robot and also meant that the robot had to be programmed to 'unwind' when rotations in any one direction accumulated beyond an acceptable limit. But the worst objection was the ungainliness of the whole thing and the obvious 'low-tech' stigma attached to this mode of operation despite the sophistication of the localiser and range sensor used, the elegance of the path planning algorithm applied and the robustness of the programming for the whole system. The total effect was not convincing enough as a quality demonstration of autonomous navigation; it just looked too much like a trick.

In practical terms, the range of operation limitation of the physical tether was the most objectionable short-coming. The application of a radio ethernet 'tether' turned out to have several advantages in addition to expanding the operational range considerably.

Consider the system schematic shown in Figure 4. A Silicon Graphics workstation, one of ten or so linked by an ethernet local area network in the Intelligent Robotics Research Centre, is the only traditional computational support for Etherbot. There are no computational resources on-board the robot at all except for embedded microprocessors in the mobile radio ethernet node, the multiple serial line server and the wheel motor controller. On-board battery power is almost entirely devoted to driving the wheels; the sensors, server and the ethernet node take relatively little current. The motor/gear sets are standard units used for motorised wheel chairs and can thus carry a load of up to 150 kg for up to six hours between battery recharges. This period could be easily doubled by adding more batteries.

A pair of radio ethernet nodes act as the junction points for a radio bridge between two local area networks, one being made up of the set of Silicon Graphics (and other) workstations on wired ethernet and the other made up only of a multiple serial ethernet server with the capacity of providing eight serial line connections, one of which is used only to set up the configuration (speeds etc.) of the server. Three of the additional sever lines are used in the Etherbot project. One of these is connected to a Denning scanning localiser [7], another to an Erwin Sick active infra-red sheet scanning range finder and the third to an M68HC11A microcontroller which drives the proportional, integrative and differential (P.I.D.) controllers for wheel rotation. These controllers also include dynamic braking which allows the robot to stop smartly when not being powered.

Differential steering about the mid point between the wheel hubs allows the robot to turn on the spot about its centre of gravity. A set of casters stop the robot from tipping over.

The radio ethernet transfers data at a 2 Mb rate and can operate over 200 metres with standard stub antennas and up to 6 km with special directional antennas.

Localisation and Environmental Mapping

Localisation is achieved using a Denning scanning remote bar code reading system which can identify and angularly locate passive bar code reflectors (made up of vertical retro-reflective strips on a board about 20 cms wide and 20 cms high) up to 100 feet away. The angular positions of a minimum of three identified bar codes are sufficient to determine the location and orientation of the robot. The scanning system inside the unit rotates ten times per second and this defines the maximum localisation update rate. Typically 6 to 8 bar codes are detected and the redundancy exploited to improve accuracy and reliability. Various combinations of bar code obscurances can thus be tolerated. The Denning system can identify up to 32 different codes. However, since the rotational direction of the scan is known, codes can be reused in different parts of an extensive working area by locating them to produce unique order-of-reading sequences, provided no two identically coded cards can be read from the same location. Thus, many interconnected largish rooms could be easily set up with many more than 32 bar code cards; overall, the robot could operate in building sized environments. Typically, localisation accuracies of ± 3 cm location and $\pm 1^0$ orientation can be achieved.

Environmental mapping is achieved using a Erwin Sick sheet scanning active range finder set-up to make 180^0 horizontal sweeps approximately 10 cm above the floor. Readings of up to 50 metres range are returned at $1/2^0$ intervals over the entire 180^0 in approximately 1 second and are transferred to the Silicon Graphics workstation via the RS232 serial line multiplexor radio ethernet bridge at 9600 band.

It is assumed that no overhanging parts of obstacles that cannot be detected at this height off the floor exist. Figure 5 shows a scan of some boxes in a laboratory where the scan is set at a height so as to intercept the two boxes on top of the pile in the foreground; the gap between them is clearly detected. An accuracy of ± 3 cm is typically achieved with this sensor.

A rectangularly tessellated grid map of the working environment is filled in as obstacles are detected, cells being emptied when they are no longer being occupied by obstacles are detected. An accumulation calculation is used to stabilise additions and deletions [6]. Thus, mapping proceeds incrementally as the robot moves through its work environment.

Distance Transform Based Path Planning

The floor space within the work environment of the robot is initially regarded as a blank sheet of tessellated cells. The robot's position and orientation on this sheet is provided by the Denning localiser. One or more goal cells are selected by the operator. A tour of the goals, nearest first at each stage, will be carried out. A goal which happens to be placed where part of an obstacle exists cannot be reached unless the obstacle moves away.

For simplicity, two restrictions concerning the sampling of environmental data and the moving behaviour of the robot were imposed. Range data is only collected when the robot is stationary; this avoids data skew which could occur because of the scanning nature of the range data collection and also reduces the communication and processing demands on the system. Further, the robot only moves on straight line trajectories and turns only at junctions of straight lines after a short stop when it collects range data. The robot is also stopped after a specified length of straight line movement. The localiser data is used to correct trajectories to keep them on the planned straight line paths. These restrictions do not cause any real problems in the overall operation of the robot even though there are occasions whilst the robot is moving that a moving obstacle could intersect its path without being detected. In practice, if the intervals between stops (at which range data is collected) are small and the moving obstacle speed is slow, collisions are avoided.

As obstacle face location is detected by the range scanner the obstacle cells in the map are inflated by on cell in all directions where the cell size is approximately the radius of the robot; this allows the robot to be considered a point in a grown obstacle field and simplifies path planning.

Path planning is carried out dynamically at a global level using Distance Transform methodology developed and expanded by the author and colleagues over a number of years [8]. For any one time snap-shot of robot location, goal locations and so far detected obstacles (grown by the radius of the robot) distances are propagated out from goals and fill all of free space; the shortest path to the nearest goal is given by walking as steeply as possible down the distance contours. Details can be found in [9]. The computation is fast and can be repeated continuously so that new path plans result as new obstacles appear or previous ones are erased as the robot moves. Unknown space at any instance is simply considered empty - this we have called the 'optimistic' strategy. The robot simply follows these planned paths as they evolve and change to accommodate new data. At every stage, subject to the 'optimistic' strategy, a globally optimal path (shortest path) is always maintained.

As each goal is reached, the nearest remaining goal to it will be targeted and so on until the tour is complete. This scheme is capable of dealing with very complex, slowly time-varying environments in a relatively straightforward way.

The workstation screen shows the on-going status of the tour. As new obstacles are detected or old ones detected as no longer being where they were, the map is altered and the current robot position/orientation as well as the current planned path are shown on screen. The screen view looks like a simulation but is actually monitoring a real physical event in real-time.

Etherbot Performance

This is hard to describe anywhere as well as can be demonstrated using a video clip (which will be presented). Suffice it to say that the system has proven extremely effective, safe and reliable and has been exhaustively trialled in many complex arrangements of obstacles in our laboratory. The robot moves at about 20 cms/second, stops only briefly whilst collecting range data ($\approx$ 1 sec) and avoids slowly moving obstacles crossing its currently planned path quite nicely. Since it has no standard computer on-board and the radio ethernet and multiple serial line server are quite small and light, the robot can carry a considerable pay load (up to 150kg) and has the space to carry it on. It is also very rugged and can take considerable knocks. In Figure 3(d), as something of a joke, two tea cups/saucers, a teapot and a jug are shown as a token pay load as if the robot were taking tea to laboratory staff. The robot can operate for long periods at a time (6 hours - even more with additional batteries). Video cameras can be mounted on top of the dome of the Denning localiser and radio video transmitters could pass the images to a workstation if this were useful. Also, the workstation can be upgraded at any time with no concern for changes of power requirements, weight or size.

Figure 6 depicts several stages of map building, path planning and motion for Etherbot. The way in which new data is accommodated into a developing map and the appropriate path plan adjustments made are easily noted. These figures are not for a simulation but monitor a real experiment in the laboratory.

In all, the Etherbot system has proven sound and effective and potentially commercialisable for a variety of applications. It can be build for approximately US$25,000 including the Silicon Graphics workstation.

Discussion and Potential Applications

A bonus which comes with the Etherbot concept which was not initially appreciated is that any Silicon Graphics workstation on the stationary side of the radio ethernet bridge can, if loaded with the appropriate program, take over the control of the robot. Thus the Etherbot could be time-shared amongst a group of people occupying the same building, perhaps for carrying mail, refreshments, electronic components or whatever. It can also be controlled from a more remote site also on the net for surveillance of a building. If video cameras are used the images can be transmitted to a close by workstation and then via the ethernet to the remote station. Since it can carry a person, its mobility being provided by standard wheelchair motor/gear/wheel sets, a chair construction could be added for the transport of elderly or disabled people throughout a hospital or senior citizen home. An individual could start out near a workstation and program a tour of goals and then be driven on that tour. Manual overrides can be easily provided if thought necessary.

The combination of multi-point remote control, physical tether-free operation over a large complex of rooms and high pay load/operating period/ruggedness properties make the Etherbot an almost ideal autonomous mobile robot system for a very large range of practical applications.

It is also fairly obvious that multiple robots can be controlled from a single workstation or amongst several, each radio ethernet linked serial line server having its own unique address on the network.

Lastly, the workstations which are capable of controlling the robot are not dedicated to this project and can be used most of the time on other tasks. In effect, the robot is a sharable peripheral which can be used when needed, if it is free.

Conclusions

This paper has presented the details of an autonomous mobile robot system which operates without an on-board computer using a radio ethernet connected to a local area network of workstations, any one of which can take over the control of the robot. This system has proven effective and reliable, rugged and capable of carrying high pay loads, is relatively inexpensive to make and easily customised for a large range of marketable applications.

Bibliography

1. Chatila, R. and Lacroix, S. Adaptive Navigation for Autonomous Mobile Robots, Proc. The Seventh International Symposium on Robotics Research (edited by G. Guralt and G. Hirzinger), 1996, pp450-458.

2. Brooks, R.A. A Robust Layered Control System for a Mobile Robot, IEEE Journal of Robotics and Automation, Vol. RA-2, No.1, March 1986, pp14.

3. Brooks, R.A. Elephants Don't Play Chess, Journal of Robotics and Autonomous Systems, Vol. 6, 1990, pp3-15.

4. Hendry, B., Jarvis, R.A. and Bridger, I. Floor Cleaning Using an Automated Guided Vehicle with Beacon Localisation, Proc. IARP First International Workshop - Robots for the Service Industries, Sydney, 18-19 May, 1995, pp.13-20.

5. Jarvis, R.A. and Byrne, J.C., An Automated Guided Vehicle with Map Building and Path Finding Capabilities, invited paper, Fourth International Symposium of Robotics Research, University of California at Santa Cruz, 9-14 Aug., 1987, pp. 155-162.

6. Jarvis, R.A. and Lipton, A.J., GO-2 :- An Autonomous Mobile Robot for a Science Museum, Proc. 4th International Conference on Control, Automation, Robotics and Vision, Westin Stamford, Singapore, 3-6 December, 1996, pp.260-266.

7. Jarvis, R.A. Environmental Modelling and Localisation for Mobile Robot Navigation, invited paper, Sixth International Symposium on Robotics Research, Oct. 2-5, 1993, Hidden Valley, Pennsylvania, U.S.A. Post conference proceedings published by IFRR in 1994, pp.595-624.

8. Jarvis, R.A., Collision-Free Trajectory Planning Using Distance Transforms, Proc. National Conference and Exhibition on Robotics - 1984, Melbourne 20-24 August 1984. Also in Mechanical Engineering Transactions, Journal of the Institution of Engineers, Vol.ME10, No.3, Sept. 1995, pp.187-191.

9. Jarvis, R.A. On Distance Transform Based Collision-Free Path Planning for Robot Navigation in Known, Unknown and Time-Varying Environments, invited chapter for a book entitled 'Advanced Mobile Robots' edited by Professor Yuan F. Zang World Scientific Publishing Co. Pty. Ltd. 1994, pp. 3-31.

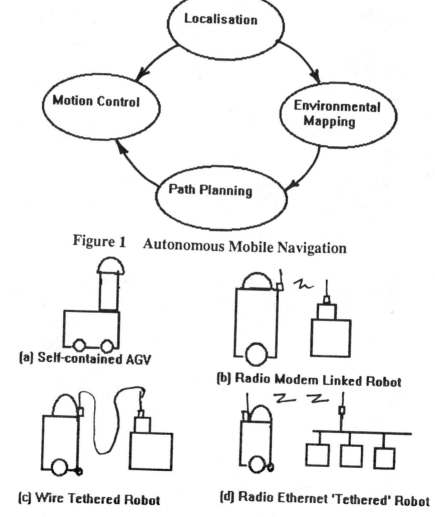

Figure 1   Autonomous Mobile Navigation

(a) Self-contained AGV

(b) Radio Modem Linked Robot

(c) Wire Tethered Robot

(d) Radio Ethernet 'Tethered' Robot

Figure 2   Packaging Alternatives

Figure 3 A Suite of Mobile Robots

(a) Tetherless Automated Guided Vehicle

(b) Serial Line Radio Link Autonomous Mobile Robot

(c) GO-2 : Tethered Mobile Robot

(d) Etherbot: Ethernet 'Tethered' Mobile Robot

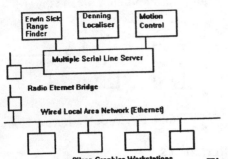

Figure 4 Etherbot Schematic

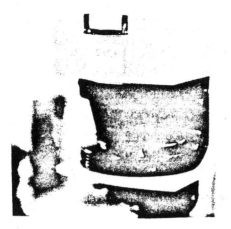

(a) The Instrument

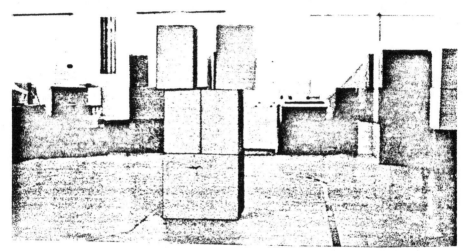

(b) Typical Time-of-Flight Horizontal Range Scan of Indoor Scene

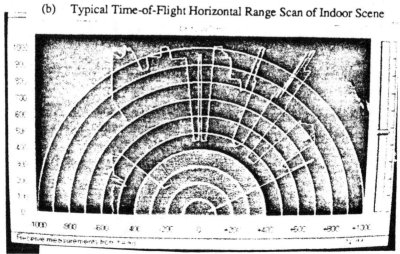

Figure 5   Erwin Sick Range Finder

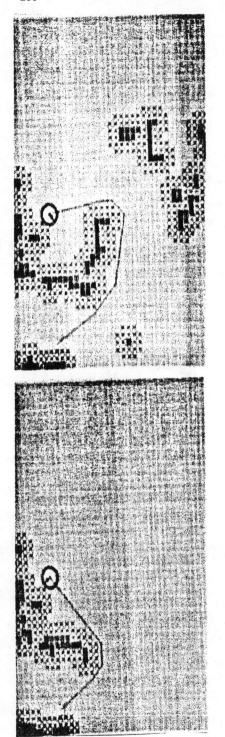
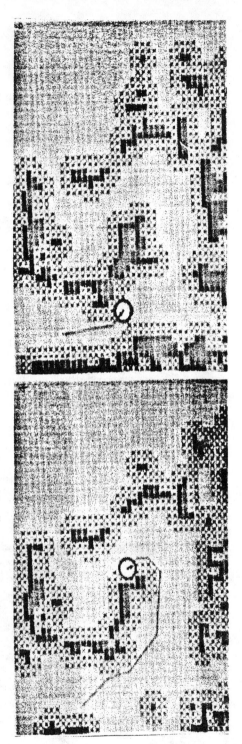

Figure 6   Etherbot Experiment

Automatic Mountain Detection and Pose Estimation for Teleoperation of Lunar Rovers[*]

Fabio Cozman   Eric Krotkov
Robotics Institute, Carnegie Mellon University
Pittsburgh, PA, 15213
{ fgcozman, epk }@cs.cmu.edu

Abstract: This paper presents a system aimed at mobile robot operations in space: we discuss an interface which receives and analyzes images sent by a rover operating in a distant environment. We are particularly interested in long-duration space missions, where rovers interact with human operators on Earth. The position estimates are presented to the operator so as to increase situational awareness and prevent loss of orientation. The system detects mountains in images and automatically searches for mountain peaks in a given topographic map. We introduce our mountain detector algorithm and present a large number of illustrative results from images collected on Earth and on the Moon (by the Apollo 17 mission). We present an algorithm for position estimation which uses statistical descriptions of measurements to produce estimates, and discuss results for scenery from Pennsylvania, California, Utah, Atacama desert and the Apollo 17 site. The implemented system achieves better estimation performance than any competing method due to our quantitative approach and better time performance due to our pre-compilation of relevant data.

1. Introduction

Only with great difficulty can human operators teleoperate rovers in an unfamiliar environment based solely on imagery sent by the rover, even with maps of the rover's environment [1, 6, 10]. Teleoperating a rover presents further challenges when the rover is on a lunar mission [7], because of the 5-second round-trip communication delay [12], coupled with the unfamiliarity of the environment, less gravity, and variable surface properties. For example, astronauts in the Apollo missions had great difficulty determining distances from mountains and craters [5].

This paper presents a system that assists operators driving remote vehicles. The basic idea is to offload navigation functions, permitting the remote driver to concentrate on pilot functions without getting disoriented or lost. Figure 1 summarizes the idea. The operator observes images from the rover and looks at a topographic map of the imaged area. Position of the robot is unknown but constrained to lie in a region the size of the map. The images are analyzed; structures found in the map are marked. In this paper we report on an automatic detector of mountains and a position estimator that operates from the detected peaks. The ultimate goal is to overlay position information on the maps and on rover-acquired images, just as "augmented reality" systems for training and medical applications do [3].

---
[*]This article is an updated version of a paper (with identical name) presented at the 1997 IEEE International Conference of Robotics and Automation.

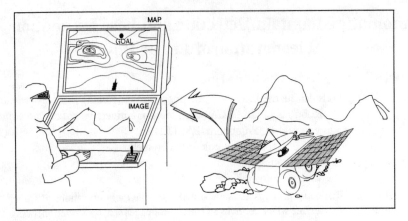

Figure 1. An interface for teleoperation of mobile robots

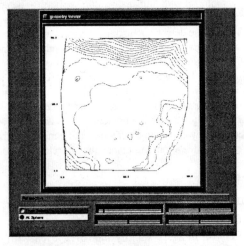

Figure 2. The map window, with a top view of the Apollo 17 topographic map (North and South Massifs appear on top and bottom respectively)

The interface presents three windows to the operator. The first window carries video; we currently use a standard video display in a Silicon Graphics workstation to look at our footage. The second window displays panoramas formed from selected images, and indicates the results of the mountain detector. The third window, depicted in Figure 2, carries the map information. The map can be seen from above, as displayed, or rendered as seen from the ground. The map in Figure 2 shows the topography of the Apollo 17 site on the Moon, generated from the Apollo 17 Landing Area topophotomap [11].

The next section discusses the basic requirements of our system; we then present our vision-based mountain detector and results collected on terrestrial and lunar data. The position estimator is then described, together with results obtained for data in the Pittsburgh East and Dromedary Peak USGS quadrangles. We show the most complete set of test images to date and we report improvements in speed and accuracy relative to previous approaches. The implemented system achieves better estimation performance than any competing method due to our quantitative approach and better time performance due to our pre-compilation of relevant data.

2. Localization from Outdoor Imagery

A variety of methods can be used to determine position, some using internal measurements (dead reckoning) or external measurements of motion. Since dead reckoning degrades over time, external measurements are necessary. A mobile robot performing a long traverse on the Moon or other celestial bodies poses a great challenge for position estimation technologies, due to the absence of the GPS infrastructure. On celestial bodies, attitude measurements can be generated accurately from a star sensor, since the visibility of stars from bodies without atmosphere is excellent, but position accuracy is low, on the order of kilometers [19]. In short, relative position and absolute orientation are available, but absolute position is not.

Our system uses mountains to fix position, as usually done by human sailors in marine navigation. Other researchers have studied the possibility of vision-based localization in outdoor applications [4, 9, 14, 16, 17], but real data analysis has been scarce (the only available data sets have been produced by Thompson and his group [17]), probably due to the complexity of collecting outdoor imagery coupled with reliable ground-truth data. In this paper we present a rich collection of images obtained by a customized platform, which allows us to tag every image with six dimensional ground-truth (position and orientation).

Each image feature can be converted to an angle in the robot's coordinate system; the angle is called a *bearing* [2]. The image contains a set R of m image bearings. The map contains a number n of landmarks $l_j = [X_j, Y_j]$. The pose of the robot in the global coordinate system is $\Gamma = [x, y, \phi]$ (ϕ is orientation). An interpretation I of the image is a set of correspondences between image features and map features.

To be able to construct an interpretation, the system must have information about the position of mountains in a topographic map of the rover's environment. To test our system with Earth images, we use 7.5 minute Digital Elevation Maps (DEMs) provided by the United States Geographical Survey (USGS), covering approximately 140km^2. Topographic maps are pre-processed and mountains in such maps are marked in an off-line stage, which automatically detects local maxima in the 7.5 minute DEM [18]. For example, the map in Figure 3 covers approximately 37 km^2 over the city of Pittsburgh, Pennsylvania. The black dots indicate the points where the system found mountains. The user has the ability to change or correct mountains detected at this stage, but we have never had to do that to obtain our results.

As the images are received by the system, mountains must be detected and position estimates have to be generated. We do not operate directly on a stream of video; individual images are selected from the video window and combined into a mosaic. Figure 4 shows panoramas produced by the system. In the system described in this paper, the images are first projected into a cylindrical surface, and then the translational displacement between images is calculated using the Kuglin/Hines method [8, 15]. The method uses the 2D Fourier transform of two images, and calculates the translational displacement by subtracting phases and transforming back to the image domain. A more accurate and robust system has replaced our first implementation;

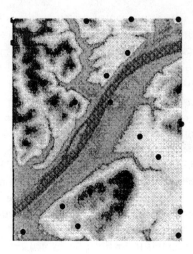

Figure 3. Topographic map of the Pittsburgh area with detected mountains

the last panorama in Figure 4 is produced by the new system[1]. In the current implementation, images are projected onto a sphere; their displacement is first calculated by the Kuglin/Hines method and then refined through an optimization, iterative process. The system picks each image and glues it into the existing mosaic by smoothing the overlapping regions; note the seamless character of the last panorama.

The mountain detector works on the mosaics. The mosaic is first segmented in two regions, one corresponding to the sky, another to the ground, and then the boundary between the two regions is searched for mountains. The theory and implementation of this mountain detector was described in a previous publication [2], together with preliminary results. Here we concentrate on a description of results obtained with a more comprehensive set of tests. The mountain detection system takes at most 2 seconds per full panorama in the current implementation on a Silicon Graphics workstation, depending on the complexity of the scenes.

Figure 4 shows a gallery of panoramas processed by our system. Images come from Pennsylvania, Utah, California, the Atacama desert in Chile and the Apollo 17 site on the Moon. Detected mountains are enclosed in rectangles; the detected position of a mountain peak is marked through a vertical line inside the rectangle.

The panorama (1) composes a sequence of images taken by the Apollo 17 Lander Module. The lunar cart can be seen at the extreme right of the panorama. The first large structure is the South Massif, followed by two smaller mountains and then the North Massif. The system detects all of those mountains plus a smaller formation which follows the North Massif. For lunar images, we have found that a very simple thresholding operation is enough to segment the sky from the ground reliably; that was used in panorama (1).

Panoramas (2) and (3) bring images from California; the first is by the Don Pedro Resort and the second is by the East entrance of the Niles Canyon. Mountains are very distinct in the Don Pedro area, but not so in the Niles Canyon. The large structures

---

[1] The current version of the mosaicing algorithm was implemented by Carlos Guestrin.

in this latter panorama are captured, but the peak in the extreme right has a skewed position because of trees on the top of the mountain. These images were acquired under bright sunlight, and the resulting panoramas suffer some effects of the camera's auto-iris operation: the sky shows "waves" of high brightness. This would make it impossible to use a simple thresholding operation as in the lunar images, but our sky segmentation algorithm [2] works without problems.

Panoramas (4) and (5) were acquired by Prof. William Thompson and his group and are publicly available. The detector has no problems, except in the detection of very small sequences of mountains that are quite distant.

Panoramas (6), (7) and (8) were obtained in Pittsburgh, by the Allegheny river. All distinct mountains are detected, except the big mountain in panorama (6), where the system gets confused by the pole in the extreme left. The system has proven to be quite reliable and flexible to adjustments, but some occasional misses or false detections occur as a result of unforeseen events like the pole in panorama (6). We plan to give some latitude to the operator, so that the system's conclusions can be modified when mistakes occur.

Panorama (9) was acquired in the Atacama desert in Chile; the location displayed in the panorama will be the scenario for the incoming Atacama Desert Trek mission [20]. The purpose of this mission is to have the *Nomad* rover conduct a long duration, semi-autonomous traverse of a planetary analog terrain, during which several key technologies will be tested. The system described here will be tested during the mission in the Atacama desert.

3. The Estimation Algorithm

Given the bearings, we must generate estimates of position. Two decisions must be made. First, what is the space in which we conduct the search. Second, how to establish a convenient figure of merit to evaluate the possible estimates of position.

The search for best estimates can be conducted in the space of interpretations [17], but only when using few peaks and image features; otherwise too many interpretations have to be visited. Use of few features compromises accuracy, which requires the use of large *and* small peaks in the horizon. Another search strategy is to look at the space of possible renderings of the map [14]. In order to conduct such a search, speed-up techniques like quantization must be used, again causing a loss of accuracy.

A more promising approach is to search the space of positions. To handle a USGS topographic quadrangle, we need to go through 4×10^4 positions and establish a figure of merit for each. This seemingly daunting task has been pursued by Talluri and Aggarwal [16]; to speed up the search, they have used a single point in the image as a measure of "goodness of fit" between image and map. In real images such a simplified figure of merit is unrealistic, since it does not take into account the noise and artifacts of real images.

We search in the position space, but we construct a figure of merit that reflects the various disturbances in the image and can also include prior knowledge about position. The key idea to speed-up our algorithm is to simplify the search procedure by pre-computing virtually all the calculations that must be performed during search. We automatically create (off-line) a table containing all the peaks that can be found in the map. We also create (off-line) a table containing all the peaks that are visible from ev-

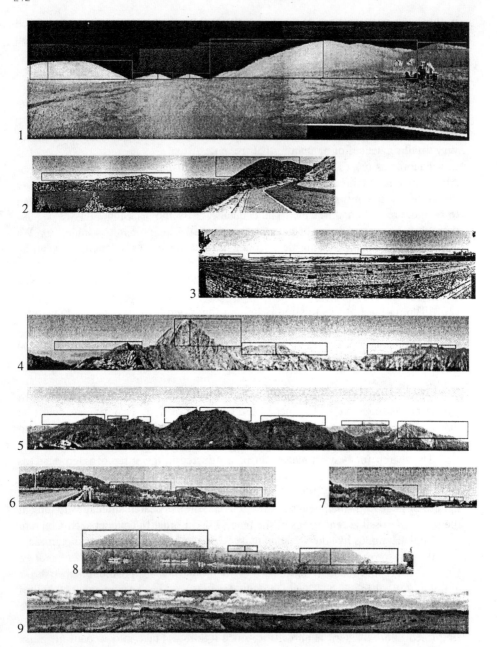

Figure 4. Peak detection results from a wide variety of scenes. From top to bottom: (1) Apollo 17 site (Moon), (2) Don Pedro Resort (California), (3) Niles Canyon (California), (4,5) Dromedary Peak (Utah), (6,7,8) Allegheny river at Pittsburgh (Pennsylvania), (9) Atacama desert (Chile)

ery possible position. During search, we need only access the latter table for retrieving the index of visible peaks, access the former table for retrieving characteristics of the peaks, and compute the posterior probability for position.

Our estimation objective is to maximize the posterior probability of position given the bearings, $p(x, y|R)$. Position is discretized in intervals of 30 meters, agreeing with the discretization of Digital Elevation Maps. At first, we must specify a prior density for position, $p(x, y)$. Currently we use a uniform distribution to signify absence of prior knowledge. Secondly, we must specify likelihood of bearings (the measurements) given position, $p(R|x, y)$. This distribution is constructed by assuming independence of bearings given a particular interpretation, and Gaussian distributions for the distribution of bearings [2]. Error that exceed 18 degrees are discarded as mistakes and assumed distributed uniformly on the interval [0, 360] degrees. We construct the posterior distribution by using Bayes rule on these models. Since bearings do not need to be associated to all possible mountains, the posterior distribution can be quickly calculated. As an example, computation of the posterior probability for the more than 3×10^4 positions in the Pittsburgh map of Figure 2 takes 3 seconds running in an Impact SGI workstation.

4. Experiments

In the current version of the system, images are obtained from a customized platform featuring a camera, an electronic compass, and a differential GPS system, all mounted on a tripod. The compass provides absolute orientation measurements; once calibrated, accuracy is ±0.5 degree. A differential GPS device obtains ground-truth measurements of position, with accuracy of 3 meters. The camera is calibrated [13], and errors of ±2 degrees are introduced by calibration inaccuracies. The images are stored on tape and played back when testing the system; every image is tagged with six dimensional ground-truth.

We have run the estimation procedure with data obtained in Pittsburgh using sequences of images as in the last two panoramas of Figure 4. Two mountain peaks are detected in the first panorama, three in the second. The bearings, in degrees, are $\{177.6, 131.1, 92.4, 102.3, 65.3\}$. The system estimates position with an error of 87 meters. We have also run the system on data from Dromedary Peak, Utah; the panoramas are shown in Figure 4. The five bearings detected automatically are (in degrees): $\{224.5, 198.8, 163.7, 107.2, 100.8, 96.3\}$. A square area of 6km by 6km was used the Dromedary Peak quadrangle, to make the test similar to the Pittsburgh tests. The system estimated position with accuracy of 95 meters, which can be compared to the 71,700m^2 obtained by Thompson [17]. We benefit greatly from our reliance on all available features, large and small, present in the image. We also have results for other panoramas presented in Figure 4. The Apollo 17 panorama generated an estimate with 180 meters of error, starting from the same 6km by 6km area. Data from the Don Pedro Resort, California, generated an estimate with 130 meters of error.

5. Conclusion

We have presented a quantitative, feature-based approach for pose estimation from outdoor visual information. Our objective is the construction of an interface for rover teleoperation that can intelligently process rover imagery and help human operators. We presented results with real data which demonstrate an improvement over the state of the art on outdoor position estimation.

The superiority of our implementation, as compared to others, stems from two factors. First, we allow several mountain features, large and small, nearby and far away, to be used by the estimator. To manage the complexity created by this rich set of measurements, we have to impose some quantitative structure; our posterior distributions give this structure. Second, we have a fast, efficient implementation of the pre-compilation stage, where all possible visibility relationships are calculated and stored. Both factors are responsible for the accuracy and time performance of our method.

We are currently extending our matching algorithm to incorporate a match for the full skyline, not just the few bearings that are associated to mountains. The extended system will be tested during the Atacama Desert Trek [20], a long duration mission for the *Nomad* rover. Other developments should be incorporated in the future. A stream of images must be presented to the user and processed by the system, and results of position estimation must be overlayed on the images so they can be assimilated by the operator. Such achievements will make it easier to remotely drive a rover in a wide variety of environments, with a particular impact on lunar missions.

Acknowledgements

This research is supported in part by NASA under Grant NAGW-1175. Fabio Cozman was supported under a scholarship from CNPq, Brazil.

We thank Prof. W. Thompson for his assistance in the processing of the Dromedary Peak quadrangle data, Yoichi Sato and Mark Maimone for his help with the mosaicing algorithm, Carlos Guestrin for his invaluable help with the imaging system and mosaicing algorithm, and Andrew Johnston and Jim Zimbelman from the Museum of Air and Space in Washington DC (USA), for their generous assistance in providing lunar data, images and maps.

References

[1] W. Aviles, T. Hughes, H. Everett, A. Umeda, S. Martin, A. Koyamatsu, M. Solorzano, R. Laird, and S. McArthur. Issues in mobile robotics: The unmanned ground vehicle program teleoperated vehicle (TOV). *Proc. of Mobile Robots SPIE*, V:587–597, November 1990.

[2] F. G. Cozman and E. Krotkov. Position estimation from outdoor visual landmarks for teleoperation of lunar rovers. *Proc. Third IEEE Workshop on Applications of Computer Vision*, pages 156–161, December 1996.

[3] S. Feiner, B. Macintyre, and D. Seligmann. Knowledge-based augmented reality. *Communications of the ACM*, 36(7):53–62, 1995.

[4] M. Fishler and R. Bolles. Random sample consensus: a paradigm for model fitting with applications to image analysis and automated cartography. *Communications of the ACM*, 24(6):381–395, June 1981.

[5] G. Heiken, D. Vaniman, and B. French, editors. *Lunar Sourcebook*. Cambridge University Press, 1991.

[6] G. Kress and H. Almaula. Sensorimotor requirements for teleoperation. Technical Report R-6279, Corporate Technology Center, FMC Corporation, Santa Clara, CA, December 1988.

[7] E. Krotkov, J. Bares, L. Katragadda, R. Simmons, and R. Whittaker. Lunar rover technology demonstrations with Dante and Ratler. In *Proc. Intl. Symp. Artificial Intelligence*,

Robotics, and Automation for Space, Jet Propulsion Laboratory, Pasadena, California, October 1994.

[8] C. D. Kuglin and D. C. Hines. The phase correlation image alignment method. *Proc. of the IEEE Int. Conf. on Cybernetics and Society*, pages 163–165, September 1975.

[9] T. Levitt, D. Lawton, D. Chelberg, and P. Nelson. Qualitative navigation. In *Proc. Darpa Image Understanding Workshop*, pages 319–465, 1987.

[10] D. McGovern. Experiences and results in teleoperation of land vehicles. Technical Report SAND90-0299, Sandia National Laboratories, Albuquerque, NM, April 1990.

[11] NASA. Apollo 17 landing area topophotomap. sheet 43D1S1(50), 1972.

[12] R. Newman. Time lag considerations in operator control of lunar vehicles from Earth. In C. Cummings and H. Lawrence, editors, *Technology of Lunar Exploration*. Academic Press, 1962.

[13] L. Robert. Camera calibration without feature extraction. *Proc. Intl. Conf. Pattern Recognition*, 1994.

[14] F. Stein and G. Medioni. Map-based localization using the panoramic horizon. In *Proc. IEEE Intl. Conf. Robotics and Automation*, pages 2631–2637, Nice, France, May 1992.

[15] R. Szeliski. Image mosaicing for tele-reality applications. Technical Report CRL94/2, DEC Cambridge Research Lab, May 1994.

[16] R. Talluri and J. Aggarwal. Position estimation for an autonomous mobile robot in an outdoor environment. *IEEE Trans. Robotics and Automation*, 8(5):573–584, October 1992.

[17] W. Thompson, T. Henderson, T. Colvin, L. Dick, and C. Valiquette. Vision-based localization. In *DARPA Image Understanding Workshop*, pages 491–498, Maryland, April 1993.

[18] W. Thompson and H. Pick Jr. Vision-based navigation. In *DARPA Image Understanding Workshop*, pages 149–152, January 1992.

[19] J. Wertz. *Spacecraft Attitude Determination and Control*, volume 73 of *Astrophysics and space science library*. Reidel, Boston, 1978.

[20] W. Whittaker, D. Bapna, M. W. Maimone, and E. Rollins. Atacama desert trek: A planetary analog field experiment. *Int. Symposium on Artificial Intelligence, Robotics and Automation for Space*, July 1997.

A Landmark-based Motion Planner for Rough Terrain Navigation

A. Hait   T. Siméon   M. Taïx

LAAS-CNRS
7, Avenue du Colonel-Roche
31077 Toulouse Cedex - France
{hait,nic,taix}@laas.fr

Abstract: *In this paper we describe a motion planner for a mobile robot on a natural terrain. This planner takes into account placement constraints of the robot on the terrain and landmark visibility. It is based on a two-step approach: during the first step, a global graph of subgoals is generated in order to guide the robot through the landmark visibility regions; the second step consists in planning local trajectories between the subgoals, satisfying the robot placement constraints.*

1. Introduction

Autonomous navigation on natural terrains is a complex and challenging problem with potential applications ranging from intervention robots in hazardous environments to planetary exploration. Mobility in outdoors environments has been demonstrated in several systems [16, 7, 8, 3].

The adaptive navigation approach currently developped at LAAS within the framework of the EDEN experiment [3] demonstrates fully autonomous navigation in a natural environment, gradually discovered by the robot. The approach combines various navigation modes (*reflex, 2d and 3d*) in order to adapt the robot behaviour to the complexity of the environment. The selection of the adequate mode is performed by a specific planning level, the navigation planner [11] which reasons on a global qualitative representation of the terrain built from the data acquired by the robot's sensors.

In this paper, we concentrate on the motion planning algorithms required for the *3D navigation mode* [13], selected by the navigation planner when very rugged terrain has to be crossed by the robot. On uneven or highly cluttered areas, the obstacle notion is closely linked with the constraints on the robot attitude, and therefore contrains the robot heading position. Planning a trajectory on such areas requires a detailed modeling of the terrain and also of the robot's locomotion system.

Several contributions recently addressed motion planning for a vehicle moving on a terrain [15, 9, 4, 2, 6]. In particular, the geometric planner we proposed in [15, 6] is based on a discrete search technique operating in the (x, y, θ) configuration space of the robot, and on the evaluation of elementary

Figure 1. Placement of an articulated chassis

feasible paths between two configurations. The overall efficiency of the approach was made possible by the use of fast algorithms for placing the robot onto the terrain (Figure 1) and checking the validity of such placements.

In the navigation experiments previously conducted with a real robot [13], motion control was limited to executing the paths returned by this planner by relying on odometry and inertial data. The unpredictable and cumulative errors generated by these sensors, especially significant on uneven and slippery areas, often caused important deviations leading to a lack of robustness at execution. To overcome this problem, the robot has to be equipped with environment sensors (eg. cameras) that can provide additionnal information by identifying appropriate features of the terrain, and allow to use sensor-based motion commands. Such primitives are more tolerant to errors than classical position-controlled primitives, but their feasability has to be checked in term of visibility of the landmarks along the trajectory.

The contribution of this paper is to propose a motion planning approach which considers a set of given landmarks (eg. terrain peaks [5]) in the terrain model, and computes a partitionning of the terrain into regions where particular landmarks are visible. The approach allows to produce trajectories that remain, whenever possible, inside these *visibility regions* where the robot can navigate with respect to the given landmarks using closed-loop primitives relying onto the sensor's data.

2. The Motion Planning Approach
2.1. Problem statement

We consider a geometric model of an articulated robot illustrated by Figure 1. The robot is composed of several axles linked by passive joints allowing to adapt its shape to the terrain relief. The terrain is described by surface patches defined from an elevation map in z associated with a regular grid in (x, y). The terrain model also contains a set of point landmarks corresponding to major terrain features that the robot should be able to track at execution. The robot motions are constrained by:

- **validity constraints** related to the feasibility of the motion (eg. stability of the vehicle, collision avoidance with the terrain, mechanical constraints).
- **visibility constraints** which traduce the ability of the robot to detect one or a set of landmarks from a given configuration.

While the first constraints need to be verified to guarantee the safeness of the motion, it would be too restrictive to only consider solutions satisfying the visibility constraint all along the path. Therefore, the planning approach should generate solutions alternating large portions verifying the visibility of the landmarks (sensor-based mode), and subpaths where the landmarks cannot be used at execution (position-controlled mode).

2.2. A two-step approach

In order to separate the integration of the two kinds of constraints, we propose a motion planner based on the following two-step approach:

- In the first step, subgoals are generated in order to guide the robot through landmark visibility regions computed from the terrain model and the landmark set. These subgoals and the start/final configurations are connected into a network of *possible* paths which minimize a cost reflecting the traversability of the terrain.
- In the second step, a motion planner based on [6] is used to transform the *possible paths* into feasible ones verifying the validity constraints.

The final trajectory is a sequence of trajectories from a subgoal to another one, alternately in and out of the visibility regions. We describe below the two steps of the approach respectively called **global planner** and **local planner**.

3. The Global Planner

3.1. Landmark visibility

The landmarks define a partition of the terrain into regions, called the **visibility regions**, where particular landmarks are visible. Consider a robot equipped with an omnidirectional sensor. A landmark $\mathcal{L}_i$ can be seen if the following constraints are respected:

- $\mathcal{L}_i$ is not hidden by a part of the terrain.
- the distance from the sensor to $\mathcal{L}_i$ is within some limit values.

For the first constraint, we use a hierarchical model of the terrain (see [15]) that allows an efficient collision checking: if the segment connecting the sensor center to the landmark does not collide with the terrain, then the landmark can be seen from this position. The test is performed for every position respecting the second constraint and the set of points obtained is the visibility region of the landmark.

For a given landmark $\mathcal{L}_i$, note that the visibility region may content several components. Each one is considered as a particular visibility region $\mathcal{R}_{ij}$.

3.2. Visibility graph

The visibility regions are then connected into a **visibility graph**. As illustrated by Figure 2, the graph has two type of edges. Some edges represent the possibility of a direct connection between two adjacent regions (ie. without crossing another region). These edges are first computed by a numerical propagation technique described in the next section. This propagation, issued from the boundary of the visibility regions, allows to associate to the edges a path minimizing a cost evaluated from the terrain shape. The endpoints of these paths define the nodes of the graph. They correspond to a set of subgoals located onto the boundary of the visibility regions. The graph is then completed by additionnal edges connecting the different subgoals generated in the same visibility region. The paths associated to these edges are also computed by a numerical propagation performed inside the visibility regions.

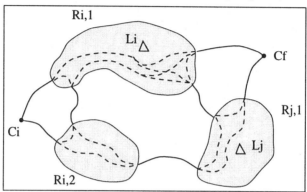

Figure 2. visibility regions and associated graph

3.3. The propagation algorithm

Let us now describe the propagation algorithm used to compute the visibility graph introduced in the previous section.

A **cost bitmap** is first computed by evaluating for each point of the terrain model, the slope and the roughness of a circular domain centered at this position, with a radius related to the size of the robot. It represents the difficulty for the robot to cross the domain.

The propagation consists in a wavefront expansion of a numerical potential obtained by integrating the cost across the bitmap, starting from each visibility region. The pixels belonging to the boundary of these regions are initialized to a null potential. At each iteration, the smallest potential pixel is expanded using an 8-neighbours propagation. Therefore, waves of increasing potential propagate from each region.

Whenever the waves issued from two regions meet together, an edge is created and the associated path is easily obtained by chaining back the pixels up to the associated region boundaries. The edge cost is the sum of the potentials computed at the meeting point. The propagation stops when all pixels have been expanded.

Path determination can take into account uncertainty growing outside the visibility regions by evaluating the maximum cost on a circular domain around the robot position, proportional to the uncertainty. This leads to a modification of the path that improve the robustness of the execution.

4. The Local Planner

The global planning step provides a sequence of subgoals between the initial and the goal configuration. The local planner is based on the algorithms described in [6]. Its role is to generate trajectories between the subgoals, while respecting the validity constraints (stability, non-collision, mechanical constraints).

The placement $\mathbf{p} = (\mathbf{q}, \mathbf{r}(\mathbf{q}))$ of the articulated robot (Figure 1) is defined by a configuration vector $\mathbf{q} = (x, y, \theta)$ specifying the horizontal position/heading, and by a vector $\mathbf{r}(\mathbf{q})$ of the joint parameters (the roll and pitch angles of the axles). For a given $\mathbf{q}$, the values of $\mathbf{r}(\mathbf{q})$ depend on the contact relations between the wheels and the terrain. Computing the robot placement is required for the evaluation of the validity constraints considered by the planner.

Planning is performed in two steps: a preprocessing step to compute robot placements on the terrain, and the planning step itself.

4.1. Preprocessing step

The preprocessing step is aimed at reducing the computational cost of the robot placement. It consists in slicing the orientation parameter, and in computing for each slice, two surfaces that characterize the evolution of the roll angle and elevation of a single axle, in function of its position onto the terrain. This step has only to be performed once, before the first call of the local planner.

4.2. Graph search

The planning step uses an approach similar to [1] in order to incrementally build a graph of discrete configurations that can be reached from the initial position by applying sequences of discrete controls during a short time interval. The arcs of the graph correspond to feasible portions of trajectory, computed for a given control. Only the arcs verifying the validity constraints are considered during the search.

The heuristic function used to efficiently guide the search is based on the potential bitmap computed by the global planner. It also allows to follow, whenever possible, the paths computed at the global level.

5. Results and Discussion

The motion planning approach has been implemented in C on a Silicon graphics indigo workstation with an R4000 processor. We describe below some simulation results.For these examples, the terrain is represented by a 114×179 elevation map (see Fig 3).

Several steps of the global planner are shown on Figure 4: the cost bitmap and potential bitmap computed from the terrain model and three landmarks (Fig. 4-a), and the obtained visibility graph (Fig 4-b). The landmarks visibility regions and the connecting paths are represented in dark on the figure.

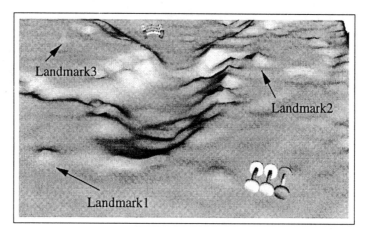

Figure 3. The terrain model, the selected landmarks and the initial/final configurations

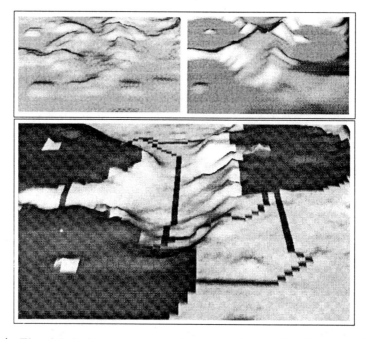

Figure 4. The global planning step. **a-** the cost and potential bitmaps and **b-** the computed visibility graph.

The final trajectory computed by the local planning step is displayed in Figure 5. One may note that a solution crossing the visibility regions computed for landmarks 3 and 1 would have been preferable for the visibility constraint. However, trajectories connecting the initial configuration to region 3 do not satisfy the validity constraints and the global planner had therefore to choose another path in the visiblity graph, leading to the final trajectory shown on

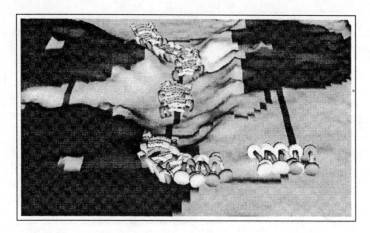

Figure 5. The final trajectory

	time (sec.)	
Preprocessing	cost bitmap	3.260
	quadtree	1.540
	slices	13.850
Global planning	visibility regions	0.500
	potential bitmap	2.470
	visibility graph	0.930
Local planning	graph searchs	2.000

Table 1. Some computation times

the figure.

Computation times are reported in the following table. In the *preprocessing* step, data structures are computed from the terrain and the robot models: the cost bitmap, the hierarchical terrain model (quadtree) used by the collision checker and the orientation slices used by the local planner.

In the next example, only one landmark has been selected (Fig 6). The first trajectory represents planning without landmarks [6]. The interest of the landmark-based approach is to allow the robot localisation in the landmark visibility region. Computation times are similar to the previous example.

The last example shows the influence of uncertainty on path determination. Figure 7 represents cost bitmap for different uncertainty values. In Figure 8, taking into account uncertainty modifies the path by moving away from hazardous regions.

In the presented approach, the planned trajectories pass through the landmark visibility regions allowing robot localisation. Integrating uncertainty in cost bitmap is a first step toward planning with uncertainty. This should be completed in a future work.

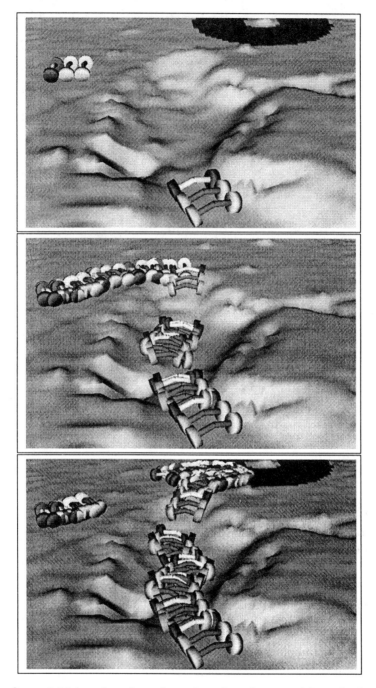

Figure 6. a-initial and goal configurations b-trajectory without landmark c-trajectory with landmark

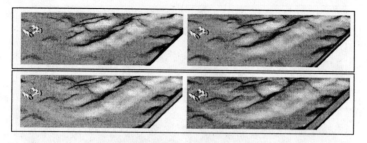

Figure 7. Cost bitmap with different uncertainty values

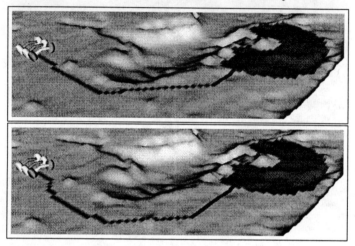

Figure 8. Path modification with uncertainty

6. Experiment

We are currently working on the integration of the presented motion planner in the EDEN experimental testbed. Experiments will be performed with the LAMA Marsokhod (Figure 9) equipped with a range finder and a stereovision system for terrain modelling and landmark recognition. The purpose is to demonstrate autonomous navigation of a robot on rough terrain.

The robot is given a global goal in an unknown environement and perceives incrementaly the natural environment. At each step, a perception of a portion of the environment is done. This sensor information allows terrain modeling [13] and landmark extraction [5].

To use the landmark-based motion planner approach it is necessary to define a goal in the current perception. The goal position is chosen in order to ensure a good matching between the different perceptions.

At this level the landmark-based motion planner is called with this goal and landmark position. During the excution, landmark visual servoing is used inside the visibility regions, and classical odometry method outside these regions. After the path execution, a new step is done as far the global goal is not reach.

Figure 9. The mobile robot Lama

References

[1] J. Barraquand and J.C. Latombe. On non-holonomic mobile robots and optimal maneuvering. *Revue d'Intelligence Artificielle*, 3(2), 1989.

[2] F. Ben Amar, P. Bidaud. Dynamic Analysis of Off-Road Vehicles. In *Fourth International Symposium on Experimental Robotics, Stanford (USA)*, July 1995.

[3] R. Chatila, S. Lacroix, T. Siméon, M. Herrb. Planetary exploration by a mobile robot: mission teleprogramming and autonomous navigation. In *Autonomous Robots Journal, Vol. 2, n° 4, p 333-344*, 1995.

[4] M. Cherif, C. Laugier. Dealing with Vehicle/Terrain Interactions when Planning the Motions of a Rover. In *IEEE International Conference on Intelligent Robots and Systems, Munich (Germany)*, September 1994.

[5] P. Fillatreau, M. Devy, R. Prajoux. Modelling of unstructured terrain and feature extraction using b-spline surfaces. In *International conference of Advanced Robotics, Tokyo (Japan)*, November 1993.

[6] A. Hait, T. Siméon. Motion planning on rough terrain for an articulated vehicle in presence of uncertainties. In *IEEE International Conference on Robots and Systems, Osaka (Japan)*, November 1996.

[7] M. Hebert. Pixel-based range processing for autonomous driving. In *IEEE International Conference on Robotics and Automation, San Diego (USA)*, 1994.

[8] E. Krotkov, M. Hebert and R. Simmons. Stereo Perception and Dead Reckoning for a Prototype Lunar Rover. *Autonomous Robots Journal, Vol. 2, n° 4, p 313-331*, 1995.

[9] T. Kubota, I. Nakatani and T. Yoshimitsu. Path Planning for Planetary Rover based on Traversability Probability. In *Seventh International Conference on Advanced Robotics, Sant Feliu (Spain)* September 1995.

[10] S. Lacroix, R. Chatila, S. Fleury, M. Herrb, T. Simeon. Autonomous navigation in outdoors environments: Adaptive approach and experiments. In *IEEE International Conference on Robotics and Automation, San Diego (USA)*, May 1994.

[11] S. Lacroix, R. Chatila. Motion and Perception Strategies for Outdoor Mobile Robot Navigation in Unknown Environments. In *Fourth International Symposium*

on *Experimental Robotics, Stanford (USA)*, July 1995.
[12] J.C Latombe. *Robot Motion Planning*. Kluwer Pub., 1990.
[13] F. Nashashibi, P. Fillatreau, B. Dacre-Wright, and T. Simeon. 3d autonomous navigation in a natural environment. In *IEEE International Conference on Robotics and Automation, San Diego (USA)*, May 1994.
[14] Z. Shiller and J.C. Chen. Optimal motion planning of autonomous vehicles in three dimensional terrains. In *IEEE International Conference on Robotics and Automation, Cincinnati (USA)*, 1990.
[15] T. Simeon and B. Dacre-Wright. A practical motion planner for all-terrain mobile robots. In *IEEE International Conference on Intelligent Robots and Systems, Yokohama (Japan)*, July 1993.
[16] C. Thorpe, M. Hebert, T. Kanade, and S. Shafer. Toward autonomous driving : the cmu navlab. part i : Perception. *IEEE Expert*, 6(4), August 1991.

Chapter 5

Heavy Robotic Systems

Field applications such as mining, cargo handling, construction and agriculture, offer an alternative view of the future of robotics science. Such applications are characterized by the use of high valued equipment in extremely harsh and demanding environments. On the positive side, the high value of equipment offers the opportunity for deploying sensor and computing technology whose cost would be prohibitive in any other application. Further, the harshness of the environment often provides its own economic arguments for developing and deploying robotic technology. Conversely, the size of machinery and the nature of the environment mean that considerable resources and skills must be employed in any experimental robotics program.

The paper by Salcudean et al. describes a position-based impedance controller for a teleoperated hydraulic excavator. A number of key issues in modeling, practical instrumentation and control of a hydraulic machine are addressed in the paper. Extensive results showing bucket contact forces in a ground smoothing application is presented. These particularly show some of the difficulties involved in instrumenting and controlling large hydraulic machinery in unknown environments. The use of a novel six-degree of freedom magnetically-levitated teleoperation master for controlling velocity and forces is also described.

Dissanayake et al. describe a semiautomatic container handling crane. The paper describes the modeling and instrumentation of a "Stewart platform" reeved container and crane. Instrumentation includes both rope tension and inertial measurements. Together with a stiffness model of the crane, this information is used to generate anti-sway control signals for the crane. Experimental work on a 1/15th scale model is described.

The paper by Durrant-Whyte et al. describes the development of high-integrity navigation systems for large outdoor mobile vehicles. It describes a navigation system structure which permits the analysis and quantification of sensor fault modes and overall system integrity. Results from a navigation system comprising dual GPS/IMU and MMW radar/encoder loops is described, demonstrating an ability to tolerate and recover from single sensor failures.

Corke describes a semi-automated 3500 tonne dragline. The paper describes the particular difficulties in modeling and controlling very large and poorly understood machines. The instrumentation employed obtains measurements of the state of the machine. The paper also describes the approach taken towards interfacing a partially automated machine with a human operator.

Evaluation of impedance and teleoperation control of a hydraulic mini-excavator

S.E. Salcudean, S. Tafazoli, K. Hashtrudi-Zaad and P.D. Lawrence
University of British Columbia
Vancouver, Canada
{*tims@ee.ubc.ca*}

C. Reboulet
ONERA-CERT
Toulouse, France

Abstract: A position-based impedance controller has been implemented on a mini-excavator. Its performance and an approach to evaluate its stability robustness for given environment impedances are discussed. A dual hybrid teleoperation controller is proposed for machine control. Issues of transparency are discussed.

1. Introduction

There are hundreds of thousands of excavator-based machines manufactured every year and widely used in the construction, forestry and mining industry. These are four-degree-of-freedom hydraulic arms as illustrated in Figure 1, having a cab "waist" rotation, and three links moving in the vertical plane, called, in order from the cab to the end-effector, the "boom", "stick" and "bucket". The controls and human interfaces of excavator-based machines are still rather primitive. Operators use joysticks or levers to control the extension of the individual cylinders, and not the bucket motion in task space by computer-coordinated cylinder control.

The need for coordinated cylinder control has been demonstrated in [1]. Machine modeling (arm dynamics and hydraulics) and control leading to accurate task-space motion have been discussed in [2]. Individual variable-displacement pump cylinder control has been used for fast, accurate and efficient coordinated motion of a CAT-325 machine [3].

Although the need for control of forces during excavation tasks has not been formally proven, the efficiency of such tasks depends on exerted forces and should be enhanced by force or impedance control of the excavation arm. Limited models of excavators and their interaction with the soil are presented in [4]. Kinematic and dynamic models for such machines assuming that the machine cylinders act as force sources are presented in [5, 6], with [6] adding simplified digging dynamics for digging simulations.

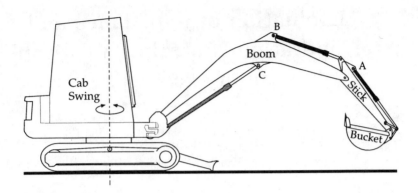

Figure 1. Mini-excavator Schematic.

Force-feedback teleoperated control of excavator-based machines has been reported in [7, 3]. The accomplishment of contact tasks relied entirely on the operator controlling the machine velocity via an active joystick, with stiffness modulated by end-point forces measured via hydraulic cylinder pressures. Since contact forces were controlled in closed loop only by the operator action and force measurements from differential pressure sensing had large errors due to cylinder friction, these results have not been entirely satisfactory.

The control of contact tasks could be improved by some form of impedance control [8] of the excavator arm. Impedance control of hydraulic robots has been presented before [?, ?, 9], but there has been no reported work of the impedance control of an excavator arm. A position-based impedance controller for an excavator arm and experimental results demonstrating impedance control of the stick cylinder are being presented in [10].

In this paper, issues of task-space stiffness/impedance control of an excavator are addressed and the problem of force-feedback teleoperation of excavato-based machines is revisited given this more versatile control mode. The paper is orgnized as follows: Section 2 briefly presents the mini-excavator used as an experimental base, Section 3 discusses issues of single-cylinder position control, task-space impedance control is presented in Section 4, while teleoperated control is presented in Section 5. Section 6 presents conclusions and future research directions.

2. Instrumented Mini-excavator

A Takeuchi TB035 mini-excavator has been used as a research platform for the experiments presented in this paper. The model and instrumentation for this machine can be found in [11].

Encoders have been installed to measure the cab-boom, boom-stick and stick-bucket angles. Load pins have been installed on the boom, stick and bucket cylinders and measure the boom cylinder force and the reaction forces of the stick and bucket cylinders. The pilot system for the main valves of the arm cylinders have been modified for computer control by using ON/OFF valves operated in differential pulsed-width modulation mode [12, 11]. The use of load-pins allows for much more accurate end-effector force-measurement than cylinder pressures, since joint friction is small and cylinder friction is substantial [13, 11]. A VME-based real-time system with the VxWorks operating system are being used to control the machine.

3. Single cylinder position control

Since the hydraulic cylinders behave like velocity sources, the range of attainable arm impedances is better when a position-based impedance control scheme is used [14]. For this purpose, inner-loop cylinder controllers were implemented with the goal of closely emulating velocity sources.

The controllers utilize compensation for the main spool dead-band, which, for safety reasons, is quite significant in all machines of these type. Proportional-derivative controllers with two sets of gains - one for cylinder extension and one for cylinder contraction - have been found to give satisfactory performance with a commanded to actual position transfer function described approximately by first order linear transfer functions of the form $1/(sT_i + 1)$, where the T_i's are of the order of 0.2 s. Let $l = [l_{boom}, l_{stick}, l_{bucket}]^T$, $l_d = [l_{boom\_d}, l_{stick\_d}, l_{bucket\_d}]^T$, and $P = diag\{P_1, P_2, P_3\}$. Experiments demonstrating impedance control of the stick cylinder are presented in [10].

4. Impedance Control

4.1. Impedance model

A desired task-space impedance of the following form is assumed:

$$f_0 - f_e = E_1 x + E_2(x - x_0) = (E_1 + E_2)x - E_2 x_0 \qquad (1)$$

where x_0 and x are the desired and actual bucket position, and f_0 and f_e are the desired and actual forces by the bucket on the environment. Typically $E_1 = E_1(s) = M_i s^2$ and $E_2 = E_2(s) = M_d s^2 + B_d s + K_d$, where s is the derivative or Laplace operator.

For a task-space environment described by $f_e = E_e x$, the proposed impedance control results in

$$(E_1 + E_2 + E_e)x = E_2 x_0 + f_0 \qquad (2)$$
$$(E_1 + E_2 + E_e)E_e^{-1} f_e = f_0 + E_2 x_0 , \qquad (3)$$

and x tracks x_0 if $\|E_2\| \gg \|E_1 + E_e\|$, while f_e tracks f_0 if $\|E_e\| \gg \|E_1 + E_2\|$.

4.2. Impedance Controller Design

An implementation of (1) can be realized using the linearized excavator arm dynamics and the hydraulics dynamics. The arm dynamics are given by

$$E_r x + J_r^{-T} \tau_g = J^{-T} f_c - f_e \qquad (4)$$

where $E_r(s) = M_r s^2$, M_r is the task-space excavator arm mass matrix, τ_g are the arm joint torques due to gravity, f_c are the applied cylinder forces, and

$$\dot{q} = J_c \dot{l}, \quad \dot{x} = J_r \dot{q}, \quad J = J_r J_c , \tag{5}$$

where q is the vector of joint angles. The hydraulics dynamics is described in task space as

$$x = J P J^{-1} x_d . \tag{6}$$

Let the impedance controller be given by

$$x_d = J \hat{P}^{-1} J^{-1} (E_1 + E_2 - E_r)^{-1} (E_2 x_0 + f_0 - J^{-T} f_c + J_r^{-T} \tau_g) , \tag{7}$$

or, equivalently,

$$l_d = \hat{P}^{-1} [J^T (E_1 + E_2 - E_r) J]^{-1} [J^T E_2 J l_0 + J^T f_0 - f_c + J_c \tau_g) , \tag{8}$$

where $\hat{P}^{-1}(s)$ is a stable approximation to the inverse of $P(s)$. With $G^{-1} = P\hat{P}^{-1}$, the following closed-loop impedance is obtained:

$$f_0 - f_e = [E_r + (E_1 + E_2 - E_r) J G J^{-1}] x - E_2 x_0 . \tag{9}$$

In the above, the Lapace variable s is used as the derivative operator and it is assumed that the arm configuration changes slowly enough for the Jacobian to be considered to be constant (so s and J commute).

In the impedance control law (7), the mass matrix term $E_r = s^2 M_r = s^2 J_r^{-T} D(q, \lambda) J_r^{-1}$ and the gravitational term $J_r^{-T} \tau_g(q, \lambda)$ can be evaluated using J_r, the arm mass matrix D, and the joint torques due to gravity τ_g, evaluated as functions of joint coordinates q and a set of inertial parameters λ. The parameters λ were previously identified using a least-squares fit of joint angle and cylinder force data [13, 11], while $D(q, \lambda)$ was obtained as a symbolic matrix function using Maple.

Note that the closed-loop dynamics (9) reduce to the desired impedance equation (1) when $G = I$. Also note that the control law is significantly simplified if $E_1 = E_r$ in the above. Since, typically, $E_2(s) = M_d s^2 + B_d s + K_d$, with all entries being positive definite, the intended task-space arm impedance will have the same or larger inertia, depending on whether $M_d = 0$ or > 0, so impact forces could not be reduced by this approach. An alternative for impact force reduction is to modify the force set point f_0 to $f_0 + E_f x$, where $E_f = M_f s^2 x$ is an inertia term (since E_f is not proper, a low-pass filter should be added to the inertial term). As long as G is close to the identity and $M_r + M_d - M_f > 0$, the system remains stable in spite of this positive feedback term.

4.3. Stability Analysis

Closed-loop system stability for a particular arm configuration and environment having dynamics $f_e = E_e x$ could be verified by determining whether

$$H = E_e - E_f + E_r + (E_1 + E_2 - E_r) J G J^{-1} \tag{10}$$

has a stable inverse using the multivariable Nyquist criterion. Guidelines for the choice of impedance parameters can be obtained by considering the scalar equivalent, with $P(s) = 1/(sT_1 + 1)$, $\hat{P}^{-1}(s) = (sT_1 + 1)/(sT + 1))$, $E_e(s) = m_e s^2 + b_e s + k_e$, $E_r(s) = m_r s^2$, $E_f(s) = m_f s^2$, $E_1(s) = m_r s^2$ and $E_2(s) = m_d s^2 + b_d s + k_d$, in which case equation (10) becomes

$$H = [(m_e + m_r - m_f)s^2 + b_e s + k_e] + [m_d s^2 + b_d s + k_d](sT + 1) . \quad (11)$$

A sufficient condition for stability is

$$\left| \frac{(m_e + m_r - m_f)s^2 + b_e s + k_e}{m_d s^2 + b_d s + k_d} \frac{1}{sT + 1} \right|_{s=j\omega} < 1, \quad \forall \, \omega . \quad (12)$$

As long as $m_e + m_r - m_f > 0$, for overdamped environments, choosing a critically damped E_2 is a sufficient condition for stability. For underdamped environments, the impedance parameters have to be selected for significant roll-off of E_2^{-1} before the resonant frequency of the environment.

4.4. Experimental Results

Experiments illustrating the effectiveness of task-space impedance control for a prototype leveling task are presented in this section. In such a task, the operator would move the bucket radially back-and-forth while exerting a normal force on the ground. Thus, the radial position R_t of the bucket tip, the bucket orientation α_t, and the vertical forces f_{ez} against the ground should be controlled. The impedance controller (8) was implemented with $\hat{P}^{-1} = I$ along the elevation axis Z_t, with

$$E_{1Z}(s) = M_r s^2 \quad \text{and} \quad E_{2Z}(s) = 400 s^2 + 10,000 s + 10,000 . \quad (13)$$

SI units are used throughout.

A piece of wood was laid on the ground in front of the excavator arm at an approximate elevation $Z_t = -1$ m. A desired trajectory as shown in dotted lines in Figure 2 was commanded. Only the bucket tip was in contact with the wood, in accordance with the kinematics and Jacobian calculations used in the controller. Figure 2 shows the bucket trajectory, Figure 3 shows the bucket forces, Figure 4 shows the cylinder extensions, and Figure 5 shows the cylinder forces. Position control results are shown on the left, impedance control results are shown on the right, and commanded trajectories are presented with dashed lines.

The experimental results show that in impedance mode, the bucket trajectory does comply to the environment contraint, transient forces are significantly lower, and steady-state contact forces tend to zero. By contrast, in position control, interaction forces are significantly higher. Note that because the arm does not comply to the constraint, the machine cab tilts up during position control, so the location readings of Figure 2 are in cab-frame, not ground frame.

The present impedance settings actually add to the robot mass by roughly 400 Kg. It is expected that the control law modification suggested in the previous sub-section will help reduce the level of the impact force on the machine.

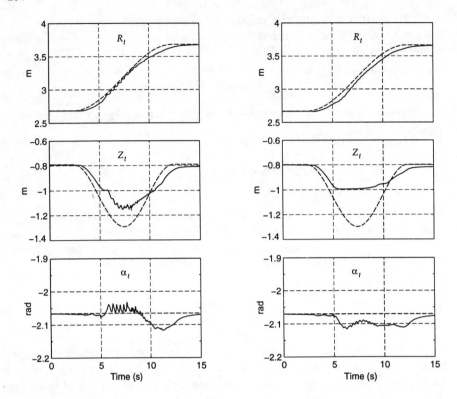

Figure 2. Position tracking in task-space under position and impedance control. Dashed lines show commanded positions.

5. Teleoperation

The existence of an effective impedance control law for the excavator arm allows more sophisticated teleoperation controllers to be implemented than the ones presented in [7, 3]. A four-channel teleoperation system is assumed as described schematically in Figure 6, where the achieved hydraulic arm impedance relationship (9) has been re-written as the "slave manipulator" dynamics

$$(Z_s + C_s)v_s = Z_d v_{s0} + f_{s0} - f_e , \qquad (14)$$

where $v_s = sx$, $v_{s0} = sx_0$, $f_{s0} = f_0$, $Z_s = E_r/s$, $Z_d = E_2/s$ and $C_s = (E_1 + E_2 - E_r)JGJ^{-1}/s$ is the compensator and the hydraulic dynamics.

The teleoperation master is assumed to be a force-source controlled (PD) mass as follows:

$$(Z_m + C_m)v_m = C_m v_{m0} + f_h + f_{m0} \qquad (15)$$

where $Z_m = M_m s^2$ is the master impedance, C_m is a position compensator, f_h is the hand force on the master, and f_{m0} is the master actuator force.

Force and position signals are communicated between the master and the slave, with $v_{s0} = C_1 v_m$, $f_{m0} = -C_2 f_e$, $f_{s0} = C_3 f_h$, $v_{m0} = -C_4 v_s$, and lead to

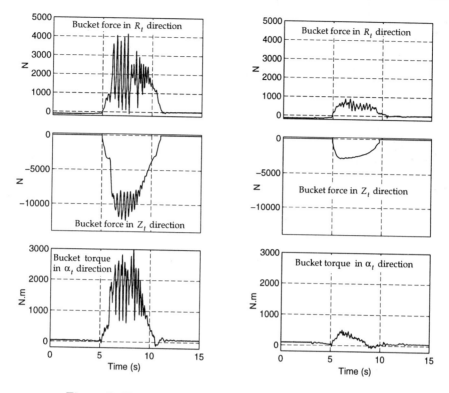

Figure 3. Trajectory forces (torques by the bucket tip).

the following teleoperation system dynamics:

$$\begin{array}{rcl} -C_m C_4 v_s + f_h & = & (Z_m + C_m)v_m + C_2 f_e \\ (Z_s + C_s)v_s - C_3 f_h & = & Z_d C_1 v_m - f_e \end{array} \quad (16)$$

5.1. Transparency and Dual Hybrid Teleoperation

Equation (16) can be solved for v_s and f_h in terms of v_m and f_e in hybrid matrix form [15, 16]:

$$\begin{bmatrix} f_h \\ -v_s \end{bmatrix} = \begin{bmatrix} Z_{m0} & G_f^{-1} \\ G_p & Z_{s0}^{-1} \end{bmatrix} \begin{bmatrix} v_m \\ f_e \end{bmatrix} . \quad (17)$$

If $f_e = Z_e v_s$, the impedance transmitted to the operator's hand $f_h = Z_{th} v_m$ is given by

$$\begin{align} Z_{th} &= Z_{m0} - G_f^{-1} Z_e (I + Z_{s0}^{-1} Z_e)^{-1} G_p \quad (18) \\ & Z_{m0} - G_f^{-1} (Z_e^{-1} + Z_{s0}^{-1})^{-1} G_p \quad (19) \end{align}$$

and, in terms of the parameters in (16), by

$$\begin{align} Z_{th} = &\, [I - (C_2 Z_e + C_m C_4)(Z_s + C_s + Z_e)^{-1} C_3]^{-1} \times \\ &\, [(C_2 Z_e + C_m C_4)(Z_s + C_s + Z_e)^{-1} Z_d C_1 + (Z_m + C_m)] . \quad (20) \end{align}$$

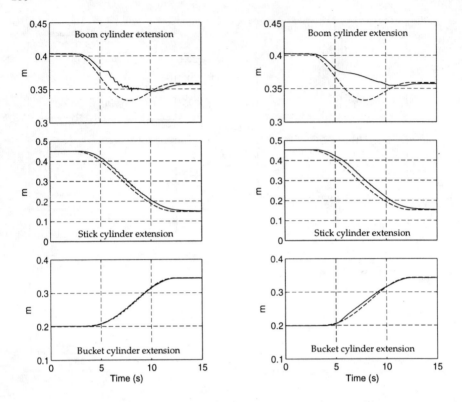

Figure 4. Cylinder extensions. Dashed lines show commanded extensions.

The teleoperation system is *transparent* if the slave follows the master, i.e., $G_p = I$ for position control and $G_p = I/s$ for rate control, and if Z_{th} is equal to Z_e for any environment impedance Z_e [15, 16, 17] (or, alternatively, $Z_{th} = Z_{t0} + Z_e$, where Z_{t0} is a "tool" impedance, usually taken to be Z_m [17]).

In special cases, such as identical master and slave dynamics, it is possible to design *fixed controlers* that provide *perfect* transparency [16], even when the slave manipulator is controlled by the master in velocity mode [17]. However, controller design is difficult (all teleoperation "channels", $C_m C_1$, C_2, C_3 and $Z_d C_4$ must be non-zero) and the stability robustness is quite poor. As an alternative, techniques using environment identification have been proposed [18] based on the architecture presented in [15]. Such schemes rely upon the identification of the environment impedance and its duplication at the master by adjusting C_m. At least with conventional identification approaches, it was found that environment identification converges slowly [18], has high sensitivity to delays, and therefore is unsuitable when the environment changes fast, as is the case when manipulating constrained objects.

For directions in which Z_e is known, the environment impedance does not need to be identified. In particular, in directions in which Z_e is known to be small (*e.g.*, free-motion), the master should act as a force source/position sensor and have low impedance, while the slave should behave as a position

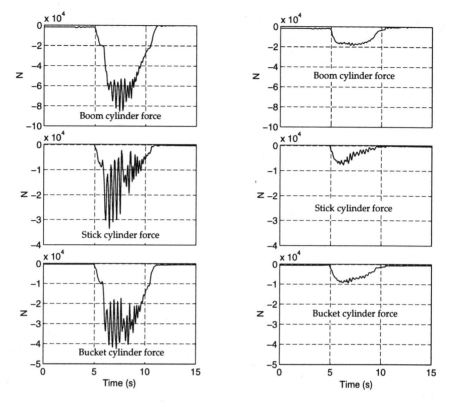

Figure 5. Net forces applied by the cylinders (load-pin reading minus gravity forces)

source/force sensor and have high impedance. Thus, in directions in which Z_e is small, positions are sent to the slave and forces are returned to the master, with C_1 and C_2 having unity transmission, and C_3, C_4 having zero transmission. The dual situation applies in directions in which Z_e is known to be large, (e.g., stiff contact or constraints). In those directions, the master should act as a force sensor/position source and have high impedance, with forces being sent to the slave and positions being returned to the master. Thus, in directions in which Z_e is large, C_1 and C_2 should have zero transmission, while C_3 and C_4 should be close to unity. From (20), it can be seen that the above insures that along very small or very large values of Z_e, the transmitted impedance equals that of the master $Z_m + C_m$, which can be set to the minimum or maximum achievable along required directions.

This concept of "dual hybrid teleoperation" has been introduced, studied and demonstrated experimentally in [19]. It has been shown that when the geometric constraints for a teleoperation task are known, the master and slave workspaces can be split into dual position-controlled and force-controlled subspaces, and information can be transmitted unilaterally in these orthogonal subspaces, while still providing useful kinesthetic feedback to the operator [19].

Consider the case of the leveling task discussed before. When the excavator

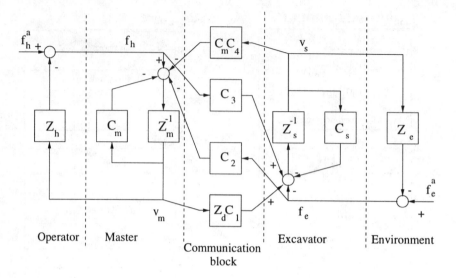

Figure 6. Four-channel teleoperation system.

bucket approaches the ground following the master position or velocity, it is controlled in position or rate mode. The master acts as a position sensor in the radial, axial and bucket orientation directions. Once contact of the bucket with the ground is detected, along the elevation axis the master impedance is set to a high value, while the excavator impedance is set to a low value. The master becomes a force sensor along this axis, with the sensed force being sent to the slave as an elevation axis force command.

5.2. Experimental results

A six-degree-of-freedom magnetically levitated wrist [20] has been used as a teleoperation master for controlling velocity and forces. The leveling experiment discussed above has been carried out, with the master and slave impedances being changed on contact detection as described above. The results will be reported in the final version of the paper.

6. Conclusion

Position-based impedance control of an excavator arm has been proposed, implemented and experimentally evaluated. In a prototype leveling task, it was shown that contact forces are substantially reduced, and the excavator arm stiffness is close to the one designed for. The use of impedance control to teleoperate the excavator in dual hybrid mode has been discussed.

This is the first time that the compliant control of an excavator arm has been reported. Applications of the technology are bound to follow, but there is much work yet to be done for the seamless integration of teleoperation with and without force feedback in the operation of such machines. This includes better position control of the indvidual cylinders, impedance identification for use in teleoperated control, and haptographical user interfaces to facilitate operation.

7. Acknowledgements

The authors wish to thank Icarus Chau for help with software development and Simon Bachmann for help with the hydraulics system. This work was has been supported by the Canadian NCE IRIS Project IS-4 and associated BC Infrastructure funds.

References

[1] U. Wallersteiner, P. Stager, and P.D. Lawrence, "A human factors evaluation of teleoperated hand controllers," in *International Symposium on Teleoperation and Control*, (Bristol, England), July 1988.

[2] N. Sepehri, P.D. Lawrence, F. Sassani and R. Frenette, "Resolved-mode teleoperated control of heavy-duty hydraulic machines," *Journal of Dynamic Systems, Measurement and Control*, vol. 116, pp. 232–240, June 1994.

[3] P.D.Lawrence, S.E. Salcudean, N. Sepehri, D. Chan, S. Bachmann, N. Parker, M. Zhu and R. Frenette, "Coordinated and force-feedback control of hydraulic excavators," in *Experimental Robotics IV, The Fourth International Symposium, Stanford Ca., June 30-July 2, 1995*, pp. 181–194, Springer Verlag, 1996.

[4] Alekseeva, T.V., Artem'ev, K.A., Bromberg, A.A., Voitsekhouskii, R.I., and Ul'yanov, N.A., *Machines for Earthmoving Work: Theory and Calculations*. New Delhi, India: Amerind Publishing Co., Pvt. Ltd., 1985.

[5] Vaha, P. K. and Skibniewski, M. J., "Dynamic Model of Excavator," *Journal of aerospace engineering.*, vol. 6, pp. 148–158, April 1993.

[6] Koivo, A J, Thoma, M, Kocaoglan, E, and Andrade-Cetto, J., "Modeling and control of excavator dynamics during digging operation," *Journal of aerospace engineering.*, vol. 9, no. 1, pp. 10–18, 1996.

[7] N.R. Parker, S.E. Salcudean, and P.D. Lawrence, "Application of Force Feedback to Heavy Duty Hydraulic Machines," in *Proceedings of the IEEE International Conference on Robotics and Automation*, (Atlanta, USA), pp. 375–381, May 2-6, 1993.

[8] N. Hogan, "Impedance control: An approach to manipulation, parts i-iii," *ASME J. of Dynamic Systems, Measurement, and Control*, vol. 107, pp. 1–23, March 1985.

[9] N. S. B. Heinrichs and A. Thornton-Trump, "Position-based impedance control of an industrial hydraulic manipulator," *IEEE Contr.Sys. Mag.*, vol. 17, pp. 46–52, February 1997.

[10] S.E. Salcudean, S. Tafazoli, P.D. Lawrence and I. Chau, "Impedance control of a teleoperated mini excavator," in *Proc. 1997 ICAR*, June 1997. to appear.

[11] S. Tafazoli, *Identification of frictional effects and structural dynamics for improved control of hydraulic manipulators*. PhD thesis, University of British Columbia, January 1997.

[12] S. Tafazoli, P. Peussa, P.D. Lawrence, S.E. Salcudean, and C.W. de Silva, "Differential pwm operated solenoid valves in the pilot stage of mini excavators: modeling and identification," in *Proc. ASME Int. Mech. Eng. Congr. and Expo., FPST-3*, (Atlanta, GA), pp. 93–99, November 1996.

[13] S. Tafazoli, P.D. Lawrence, S.E. Salcudean, D. Chan, S.Bachmann and C.W. de Silva, "Parameter estimation and friction analysis for a mini excavator," in *Proc. IEEE Int. Conf. on Robotics and Automation*, vol. 1, (Minneapolis, Minnesota), pp. 329–334, April 1996.

[14] D. Lawrence, "Impedance control stability properties in common implementations," in *Proc. IEEE Int. Conf. Rob. Aut.*, pp. 1185–1190, March 1988.

[15] B. Hannaford, "A Design Framework for Teleoperators with Kinesthetic Feedback," *IEEE Transactions on Robotics and Automation*, vol. RA-5, pp. 426–434, August 1989.

[16] D. A. Lawrence, "Designing Teleoperator Architecture for Transparency," in *Proceedings of the IEEE International Conference on Robotics and Automation*, (Nice, France), pp. 1406–1411, May 10-15 1992.

[17] M. Zhu and S. Salcudean, "Achieving transparency for teleoperator systems under position and rate control," in *Proceedings of the 1995 IEEE/RSJ International Conference on Intelligent Robots and Systems (IROS'95)*, (Pittsburgh, PA), August 5-9 1995.

[18] K. Hashtrudi Zaad and S.E. Salcudean, "Adaptive transparent impedance reflecting teleoperation," in *Proc. IEEE International Conference on Robotics and Automation*, (Minneapolis, Minnesota), pp. 1369–1374, April 22-28 1996.

[19] C. Reboulet, Y. Plihon and Y. Briere, "Interest of the dual hybrid control scheme for teleoperation with time delays," in *Experimental Robotics IV, The Fourth International Symposium, Stanford Ca., June 30-July 2, 1995*, pp. 498–506, Springer Verlag, 1996.

[20] S.E. Salcudean and N.M. Wong, "Coarse-fine motion coordination and control of a teleoperation system with magnetically levitated master and wrist," in *Third International Symposium on Experimental Robotics*, (Kyoto, Japan), Oct 28-30, 1993.

Control of Load Sway in Enhanced Container Handling Cranes

M.W.M.G. Dissanayake, J.W.R. Coates, D.C. Rye,
H.F. Durrant-Whyte, and M. A. Louda
Australian Centre for Field Robotics
Department of Mechanical and Mechatronic Engineering
The University of Sydney 2006 NSW, Australia.

Abstract: This paper describes the design, implementation and control of an enhanced container quay-crane. The new crane is based on a novel reeving arrangement which allows both fast and accurate gross motion as well as fine micropositioning. The paper describes the essential theory behind the mechanical design and the design of a controller that minimises load sway. The controller uses measurements of the rope tensions to provide artificial damping of the load. Experimental results from a $1/15^{th}$ scale model are presented. The control strategy was found to be extremely effective, damping spreader oscillations within two or three cycles.

1. Introduction

The box-rate achieved by quay-cranes is the single most important factor influencing the efficiency of a container port. Crane cycle times are limited by two main factors. First, containers may not always be aligned to allow easy pickup. Second, the load, suspended by ropes up to 50m long, has a tendency to sway during motion. These factors can reduce box-rates by about 50%. Considering that a typical vessel may require up to 1,000 container movements, the potential benefits of increasing box-rates can be substantial. A sustained small reduction in container positioning time can result in a terminal operator saving many millions of dollars per year. The work described in this paper aims to overcome both these problems by improving the reeving stiffness and damping, and by increasing the controlled degrees-of-freedom of a quay-crane to six.

Quay-cranes are used in ports to load and unload containers from ships. A typical quay-crane is shown in Figure 1. Containers are grappled by a "spreader" that is mounted on a "head-block". A set of wire ropes (the reeving) is used to hoist the head-block from an overhead rail-mounted trolley. The gantry on which the trolley runs can also be moved along an axis perpendicular to the trolley motion, giving the crane three translating degrees-of-freedom.

Operating a quay-crane is a demanding task that requires a highly skilled driver. One of the first steps towards achieving enhanced crane operation is gaining effective control of the spreader such that load sway is reduced, thereby increasing the potential operating speed of the crane. Current industrial practice in this area consists of relying on the skill of the crane driver to avoid load sway and of incorporating electrical or hydraulic anti-sway mechanisms

Figure 1. *Quay Cranes.*

into the reeving or control system. These devices can considerably increase the complexity of the crane, and many appear to be selected and evaluated on an ad-hoc basis. It is common for these anti-sway mechanisms to be switched off by crane drivers.

Many crane operators have found empirically that minor modifications to the reeving geometry can have a significant effect on the crane behaviour. In order to quantify these observations, it is important to model and to understand the dynamic behaviour of the crane, and to determine the key design parameters influencing the load sway. This understanding can, in turn, lead to improved reeving geometries and to control strategies that minimise load sway.

This paper examines the dynamic behaviour of a model quay-crane with a reeving arrangement based on the Stewart platform geometry [1]. The reeving used [2] is similar to that reported by Albus [3] and Dagalakis [4]. One of the advantages of this reeving arrangement is the ability to adjust both the spatial position and orientation of the load without moving the trolley or the gantry. This "micropositioning" can be achieved by operating independent hoist motors to change the lengths of the ropes carrying the head-block, or by other means. Furthermore, unlike a conventional four-rope reeving, the full constraint of the Stewart platform means that any load sway must result in elastic deformation[1] of the ropes. This effect substantially increases the stiffness of the crane, thereby reducing load sway and potentially increasing control bandwidth.

2. Reeving Geometry

In the reeving arrangement six ropes and a set of sheaves are used to suspend the head-block from a platform on a trolley that runs along a gantry beam. The sheaves are positioned at the apices of two horizontal trapezoids that are offset by rotation about a common vertical axis by 180°. The trapezoids are formed

---

[1] provided that the ropes do not lose tension

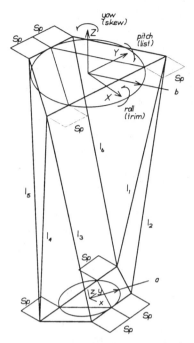

Figure 2. *Idealised modified Stewart platform reeving geometry.*

by inserting rectangular regions of identical width Sp into the centres of the two equilateral triangles in a Stewart platform. The reeving arrangement can only support a load without losing tension in the cables when the line of action of the load inertia force passes within the lower trapezoid. The allowable lateral eccentricity of the load can therefore be increased by increasing the width of the inserted rectangle. Figure 2 shows the idealised reeving, and defines parameters and frames of reference to be used in later sections.

The reeving arrangement in Figure 2 has the same geometric characteristics as the Stewart platform, yet provides a larger "footprint" on the container. Obtaining a large trapezoidal footprint is crucial, as it directly influences the crane's performance potential, particularly when the load is eccentric with respect to the spreader. The rope lengths are identical when the spreader is horizontal. Hoisting can therefore be achieved by using either one effective hoist drum or by using six separate hoist motors and drums. In contrast to conventional serial manipulators, the inverse kinematics for a modified Stewart platform can easily be obtained in closed form. Trigonometric equations for the six rope segment lengths required to achieve a desired spreader position and orientation can readily be derived.

3. Crane Control

3.1. Spreader Displacement from Rope Tensions.

Consider a small displacement $d\mathbf{p}$ of the spreader from the unloaded[2] position. In the world coordinates $F : \{t; X, Y, Z, \phi, \theta, \psi\}$ shown in Figure 2 this

---

[2] that is, the geometric position defined by the unstrained rope segment lengths $l_1, l_2, ..., l_6$

displacement can be expressed as

$$d\mathbf{p} = [dX, dY, dZ, d\phi, d\theta, d\psi]^T$$

It is assumed that the angular displacements in roll ϕ, pitch θ, and yaw ψ are small so that $d\mathbf{p}$ is a vector quantity, and the order of rotations is unimportant. The inverse kinematic equations may be written as a trigonometric function of the spreader position

$$\mathbf{l} = f(X, Y, Z, \phi, \theta, \psi) = f(\mathbf{p})$$

where the vector of rope segment lengths is

$$\mathbf{l} = [l_1, l_2, l_3, l_4, l_5, l_6]^T.$$

The perturbation in rope segment lengths required to effect $d\mathbf{p}$ is therefore

$$d\mathbf{l} = \mathbf{J}.d\mathbf{p} \qquad (1)$$

where $\mathbf{J} = \partial \mathbf{l}/\partial \mathbf{p}$ is the Jacobian which transforms from world co-ordinates to rope co-ordinates $L : \{t; l_1, l_2, l_3, l_4, l_5, l_6\}$. Note that here the Jacobian maps the world velocities $d\mathbf{p}$ into the "joint" velocities $d\mathbf{l}$ and is therefore the inverse of that typically defined in the robotics literature.

Now the rope tensions $\mathbf{T}$ are

$$\mathbf{T} = {}^L\mathbf{K}.d\mathbf{l} \qquad (2)$$

where ${}^L\mathbf{K} = diag[k_i], i = 1, 2, ..., 6$ is the 6×6 diagonal matrix whose elements represent the elastic stiffness $k_i = A_i E_i / L_i$ of each rope. Observe that ${}^L\mathbf{K}$ depends on the rope lengths, and hence on the height of the spreader. The quantity ${}^L\mathbf{K}$ may be regarded as a crane stiffness matrix in the rope co-ordinates L. Substituting equation (1) into (2) yields

$$\mathbf{T} = {}^L\mathbf{K}.\mathbf{J}.d\mathbf{p}$$

so that

$$d\mathbf{p} = \mathbf{J}^{-1}.diag(T_i/k_i), \qquad i = 1, 2, ..., 6. \qquad (3)$$

Note that $d\mathbf{p}$ is the spreader displacement required to cause the rope tensions to go from zero to T_i. If the crane design and installation is such that the rope lengths are equal, then equation (3) will give the spreader location with respect to a point directly below[3] the origin of the frame F even if the loading is eccentric. If the rope lengths are not equal to begin with, then the unloaded equilibrium position of the spreader will not be directly below the origin of F. The displacement $d\mathbf{p}$ is due to elastic deformation of the ropes under load, and is always measured from the unloaded equilibrium position of the spreader. We assume that the change in $\mathbf{J}$ as the ropes strain is insignificant.

---

[3] that is, along the negative Z axis

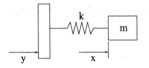

Figure 3. *Base-excited spring-mass system.*

3.2. Damping Augmentation

An uncontrolled quay crane has minimal damping, with the only significant damping forces arising from internal friction in the ropes, and from air resistance. Our objective is now to introduce sufficient damping that even an inexperienced crane driver may position the load rapidly and precisely. Let us develop a suitable control law by reference to the one degree-of-freedom base-excited spring mass system shown in Figure 3.

The equation of motion of this system is

$$m(\ddot{x} + \ddot{y}) + kx = 0$$

where x is the displacement of the load mass relative to the base, and m and k are the mass and spring stiffness. Let a position controller drive the base with motion y such that

$$\dot{y} = c\ddot{x}/m. \qquad (4)$$

With this control law, the equation of motion becomes

$$m\ddot{x} + c\dot{x} + kx = 0$$

and damping can be introduced by a suitable choice of c.

To implement this control law on the crane, we identify x with the six degrees of freedom $d\mathbf{p}$ in the previous section, and note that both eccentric loading and unequal rope lengths at no load will create a bias in $d\mathbf{p}$.

Observe from equation (4) that the control law integrates the measured relative displacement. Ideally, the controller will cause y to vanish as $t \to \infty$. In practice, however, any bias errors in the estimation of x will force the control variable y to diverge aperiodically.

Compensation is therefore introduced into the control law, to yield a transfer function

$$\frac{Y(s)}{X(s)} = \frac{c}{m} \frac{s}{(s+\alpha)^2} \qquad (5)$$

where α is chosen to lie well below the natural frequency of the system. The Bode diagram of this function rises at 20 dB/decade up to α and then falls at 20 dB/decade, so preserving the integrating effect at relevant frequencies. Zero gain at DC ensures that y vanishes as $t \to \infty$ even when there is a bias.

Full extension to six degrees-of-freedom is now possible. The equation of motion is

$$^F\mathbf{M}(\ddot{\mathbf{p}} + \mathbf{J}^{-1}.\ddot{\mathbf{l}}_d) + ^F\mathbf{K}.\mathbf{p} = 0$$

Figure 4. *The $1/15^{th}$ geometrical model quay crane showing the gantry beam, trolley, and trolley drive motor. The hoist ropes are at the lower left.*

where the displacement of the spreader is derived from $\mathbf{p} = \mathbf{J}^{-1}.diag[T_i/k_i]$, $\mathbf{l}_d$ are the commanded changes in the rope segment lengths, and $^F\mathbf{K} = \mathbf{J}^{-1}.{}^L\mathbf{K}.\mathbf{J}$ is the stiffness matrix of the crane in the world co-ordinates. We use a control law analogous to equation (4):

$$^F\mathbf{M}.\mathbf{J}^{-1}.\ddot{\mathbf{l}}_d = {}^F\mathbf{C}.\dot{\mathbf{p}}$$

resulting in a dynamic equation

$$^F\mathbf{M}.\ddot{\mathbf{p}} + {}^F\mathbf{C}.\dot{\mathbf{p}} + {}^F\mathbf{K}.\mathbf{p} = 0 \qquad (6)$$

which represents a damped multi degree-of-freedom system with a damping matrix $^F\mathbf{C}$. The control law now becomes

$$\ddot{\mathbf{l}}_d = \mathbf{J}.{}^F\mathbf{M}^{-1}.{}^F\mathbf{C}.\mathbf{p}.$$

As $\mathbf{p} = \mathbf{J}^{-1}.diag[T_i/k_i]$, and with the frequency compensation introduced in equation (5), we can express the control law in the Laplace transform domain as

$$\mathcal{L}(\mathbf{l}_d) = \mathbf{J}.{}^F\mathbf{M}^{-1}.{}^F\mathbf{C}.\mathbf{J}^{-1}.diag[T_i/k_i].\frac{s}{(s+\alpha)^2}. \qquad (7)$$

The term $\mathbf{l}_d$ is the change in the rope lengths, based on the rope tension measurements that encode the spreader displacement, that is required to add damping $^F\mathbf{C}$ to the system. Although one would usually choose $^F\mathbf{C}$ to be diagonal, the equation of motion (6) that results from the application of the control law (7) will be coupled, as $^F\mathbf{K}$ and $^F\mathbf{M}$ will generally not be diagonal.

4. Model Quay-Crane
4.1. Experimental Hardware

A $1/15^{th}$ geometric scale model of a quay-crane was built to verify the feasibility of mechanically constructing a crane with the proposed reeving arrangement.

Figure 5. *View of the model quay crane sheave platform on the trolley.*

Care was taken to preserve the major mechanical components, such as the beam, trolley, head block, sheaves and reeving geometry accurately to scale. A view of the model quay-crane can be seen in Figure 4.

Through-reeving is desirable for operational reasons on a quay crane. The through-reeved crane model is fitted with a partial-length rope tray to demonstrate that the rope catenary can be supported from the gantry beam. Catenary support is desirable to maintain the effective rope stiffness when micropositioning a load close to the limit of the crane's work space.

The six-rope reeving arrangement on the model is based on a modified Stewart platform, Figure 2, with the following geometrical parameters:

- lower circle radius $a = 73.5$ mm

- upper circle radius $b = 147.0$ mm

- spread distance $Sp = 73.0$ mm

Due to the relatively complicated reeving geometry and the need to provide the rope tray, the arrangement of sheaves on the trolley required a substantial design effort. Sheaves are placed in two horizontal layers on the trolley, as can be seen in Figure 5. The lower layer of sheaves lifts the ropes out of the rope tray, whilst sheaves in the upper layer redirect the ropes to the points on the trolley from which they drop to the head block.

The sheaves have been placed to be as near as possible to the ideal vertices in the modified Stewart platform geometry shown in Figure 2. In particular, the centres of all head block sheaves are located at (single sheaves) or adjacent to (double sheaves) the Stewart platform vertices. In the case of the trolley, the sheaves have been placed such that the mid point of the line joining the two "drop points" of a rope is close to the corresponding modified Stewart platform vertex. The load is a $1/15^{th}$ geometrical scale model of a 40 foot ISO cargo container, and is partially open to allow mass to be added. The operating mass of the container and head block is approximately 50 kg.

In the model, six remotely-mounted D.C. servo motors are used to drive the six steel ropes that position the head-block. The hoist motors are each rated at 90W continuous, and each drives a 50mm diameter hoist drum through a 35:1 reduction gearbox. The trolley is driven by an endless steel rope from a 50mm diameter drum. A 140 W DC servo motor drives the trolley drum through a precision 20:1 reduction gearbox. The available trolley travel distance is approximately 5.4 m. The trolley drive and hoist ropes are of 1.6 mm diameter multi-strand stainless steel construction. All drums are equipped with incremental optical encoders for measuring angular position, and with tachogenerators for velocity feedback to the amplifiers.

Any implementation of the control law depends on a knowledge of the location, attitude and the acceleration of the spreader and head-block assembly. For this purpose, the rope tensions were measured using strain-gauge tensometers mounted at the dead ends of the ropes.

4.2. Control Law Implementation

Each of the seven crane motors is driven in velocity mode by a servo amplifier that is rated at 25A peak. The system is controlled and co-ordinated by an IBM-compatible personal computer running the *chronos* [5] real time scheduler. A program written in *C* provides the user interface, computes inverse kinematics and implements seven independent PID position control loops running at 30 Hz to position the hoist and trolley motors. Two joysticks are used to provide trolley motion and hoist velocity commands, whereas micropositioning is presently commanded through keyboard inputs.

The control law used is equation (7), implemented for a Jacobian that is expressed as a function only of spreader height. That is, any micropositioning motion is neglected in the calculation of the Jacobian. For the purpose of exploring the effects of damping augmentation, the damping matrix $^F\mathbf{C}$ was set to $diag[c_X, 0, 0, 0, 0, c_\psi]$, so that damping was added only in the sway X and yaw ψ directions.

5. Experimental results.

Model construction and testing confirmed the mechanical feasibility of the reeving arrangement and allowed demonstration of the increased stiffness and the concept and efficacy of micropositioning.

A number of experiments were conducted to verify the effectiveness of the damping augmentation. The load was perturbed by hand, and the ensuing oscillations in rope tension were recorded. In particular, the response in sway (in the X direction) and in yaw (about the ψ axis) were studied, as these are known to be the most commonly excited in full-scale quay cranes. Figure 6 shows the cable tensions that arise in response to an initial load displacement of -300 mm in the X direction and -210 mm in the Y direction. Significant nonlinearity is evident as ropes 1 and 6 approach zero tension.

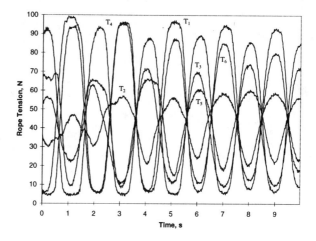

Figure 6. *Cable tensions for a load position perturbation in X, Y and ϕ.*

In Figure 7, the rope tensions **T** have been resolved into displacements $d\mathbf{p} = \mathbf{J}^{-1}.diag[T_i/k_i]$ in the directions of the world frame F. The presence of non-zero rope length adjustments $\mathbf{l}_d$ means that $d\mathbf{p}$ are the displacements relative to the spreader configuration after the ropes are lengthened by $\mathbf{l}_d$, assuming that **J** does not change significantly. Although $d\mathbf{p}$ shown in the figures is not an inertial displacement, it is representative of the load motion. The controller is active in this experiment, and it can be seen that the sway and yaw motions decay rapidly. In contrast, the Y motion which is not damping-augmented barely decays at all.

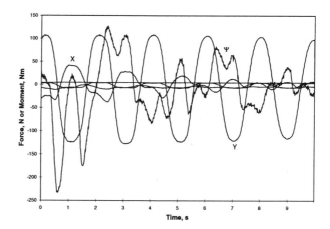

Figure 7. *Forces resolved in the directions of frame F. Since* **p** *is* $\mathbf{J}^{-1}.diag[T_i/k_i]$, *the forces are proportional to the load displacements in X, Y, Z, ϕ, θ, ψ.*

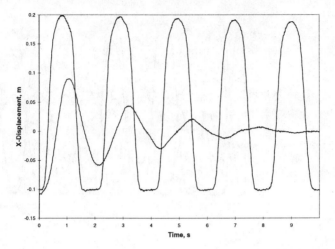

Figure 8. *The controller increases the damping ratio in sway from 0.002 to 0.12.*

Figures 8 and 9 allow direct evaluation of the controller's performance. The two traces in Figure 8 result from an initial load perturbation of -300 mm in the X direction. The efficacy of the controller is clearly demonstrated, with the damping ratio increasing from 0.002 to 0.12. Note that the modified Stewart platform reeving geometry has caused the natural frequency in sway to be increased from the "pendulum frequency" $\sqrt{g/l}$ of 0.29 Hz to a frequency of approximately 0.50 Hz. The controller causes the damped frequency to decrease to approximately 0.45 Hz. A similar effect is shown in Figure 9 for an initial yaw displacement of approximately $+37°$.

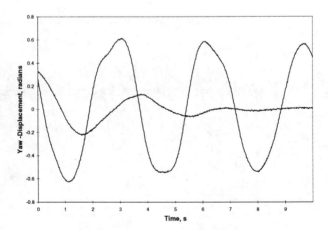

Figure 9. *The controller increases the damping ratio in yaw from 0.006 to 0.26.*

6. Future Work

The use of tensometers to estimate the position of the spreader is not feasible in the case of conventional cranes with four-rope reeving. Work is underway to mount a low-cost strap-down inertial platform on the head-block to augment the tensometer data and so obtain more accurate load position estimates. We also anticipate the opportunity to implement the control strategy described on a full-scale working quay crane.

An advanced driver-crane interface is also under construction. The objective of the user interface work is to allow a driver to operate the crane remotely. Driver input will be provided through a six degree-of-freedom input device such as the *Space Ball*. Cameras mounted on the crane structure, together with other sensors mounted on the spreader, will provide information about the location of the container. Direct range data will be obtained using an infra-red laser range-and-bearing sensor and a high-frequency sonar sensor. The laser sensor will be used to provide target bearing information while the sonar will be used to measure contact distances. Direct range data will be used to register visual information from cameras and to provide a qualitative numeric and visual display of the spreader location relative to the target container.

References

[1] Stewart D 1965 A platform with six degrees of freedom. *Proc. IME*. 180:371-386.
[2] Dissanayake MWMG, Rye DC, Durrant-Whyte HF Towards automatic container handling cranes. In: 1997 *Proc. IEEE Int Conf Robotics and Automation, Vol 1*, IEEE 1997.
[3] Albus J, Bostelman R, Dagalakis N 1993 The NIST ROBOCRANE. *J Robotic Systems*. 10:709-724
[4] Dagalakis NG, Albus JS, Wang B-L, Unger J, Lee JD 1989 Stiffness study of a parallel link robot crane for shipbuilding applications. *J Offshore Mechanics and Arctic Engng*. 111:183-193
[5] Auslander DM, Tham CT 1990 *Real-Time Software for Control*. Prentice Hall, New Jersey.

The Design of Ultra-High Integrity Navigation Systems for Large Autonomous Vehicles

Hugh Durrant-Whyte, Eduardo Nebot, Steve Scheding, Sala Sukkarieh, Steve Clark
Centre for Mining Technology and Equipment
Department of Mechanical and Mechatronic Engineering
University of Sydney, NSW 2006
Australia
[hugh,nebot,scheding,sala,clark]@tiny.me.su.oz.au

Abstract: This paper describes the design of a high integrity navigation system for use in large autonomous mobile vehicles. A frequency domain model of sensor contributions to navigation system performance is used to study the performance of a conventional navigation loop. On the basis of this, a new navigation system structure is introduced which is capable of detecting faults in any combination of navigation sensors. An example implementation of these principles is described which employs a twin GPS/IMU MMW radar/encoder navigation loop.

1. Introduction

This paper addresses the problem of developing ultra-high integrity navigation systems large outdoor autonomous guided vehicles (AGVs). Navigation, the ability to locate a vehicle in space, is a fundamental competence of any AGV system. To achieve any useful purpose, to guide itself from one place to another, an AGV must know where it is in relation to its environment. If this position information is inaccurate, unreliable, or in error, the AGV system will fail in a potentially catastrophic manner.

No system can be made 100% reliable; every sensor, all electronic components and any mechanical design will fail in some (possibly unknown) way, in some circumstance at some time. Consequently, navigation system design must always focus on integrity: if there is a failure it should be detected, identified and dealt with in a fail-safe manner. A system which is constructed from high reliability components may not itself have high integrity. However, a system constructed of relatively low reliability components may be designed to have high integrity.

The key to the development of a high-integrity, and thus useful, navigation system is through the use of a number of physically different sensors whose information is combined in such a way that all possible failures in all possible sensor combinations are detectable. This is the rational for the proposed design described in this paper.

This paper begins in Section 2 by describing and justifying an overall navigation system architecture which achieves this goal of high integrity. We describe how different sensor and system faults are detected and show that with the proposed design it is possible to quantify system integrity *a priori* in terms of fault rejection sensitivity. Section 3 describes one specific installation if this architecture which we are currently developing which employs both a DGPS/Inertial navigation loop together with a millimetre-wave radar/encoder navigation loop. Preliminary results of trials with these sensors are described are described in Section 4.

2 An Architecture for a High-Integrity Navigation System

The overall structure of the proposed navigation system is shown in Figure 1. Fundamentally, It consists of two physically independent "navigation loops", a third independent sensor that monitors differences between these two loops and an arbitration unit which compares and combines the results from these navigation different systems.

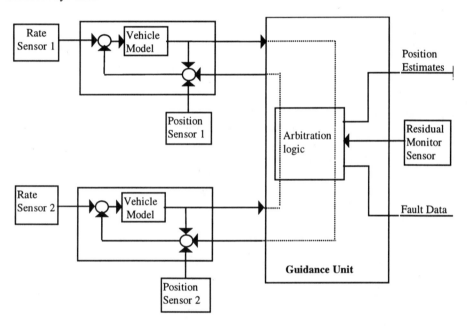

Figure 1: Proposed Navigation System Architecture

The essential principle encapsulated by this structure is the explicit recognition that all sensors fail at some point in time in a number of (possibly unknown) ways. The structure is designed such that any one failure in any one sensor can be detected and identified on-line. A failure in any two or more sensors may also be detected but not identified. The only way that this navigation structure will fail catastrophically is if any two sensors fail at the same time in exactly the same way. As the sensors are specifically chosen so that each operates on completely different physical principles, the probability of this common mode failure is as low as it can be. The aim of this proposed project is to design, implement, test and validate a navigation system based on these principles. The design and justification of this specific navigation system structure is now described in detail.

2.1 Navigation Loops

Any navigation system or navigation loop must contain at least i) a low frequency or position measurement sensor; ii) a high-frequency or rate measurement sensor; and iii) a model of the vehicle or process whose position and velocity is to be determined [7]. The reason for this is as follows:

A position measurement sensor is clearly essential to provide position estimates. In the position sensing process, the required position measurement is physically always a band-limited signal embedded in broad-band noise. Thus the position estimate must be constructed by passing this measurement through a low-pass filter (a smoother, weighted time average or a Kalman filter), with fixed cut-off frequency dictated by the position signal bandwidth. Beyond this cut-off frequency, position measurements are assumed dominated by signal noise. To provide position estimates beyond this cut-off, a sensor sensitive to high-frequency position changes must be employed. In particular, sensors that provide measurements of velocity or acceleration, once integrated to provide position, yield estimates of rapid changes in position. However, such sensors are not good at providing estimates of long term or slow changes in position as integration of the noise component in the signal leads to Brownian-like divergence of estimates. Thus the "low frequency" information provided by a position measurement sensor must be combined with "high frequency" information provided by a rate measurement sensor in a complimentary manner. Each sensor provides different information (they are not in any sense redundant), both sensors are necessary in any one navigation loop to provide essential vehicle navigation information.

In Figure 1, each navigation loop consists of this low-frequency/high frequency sensor combination arranged in a feedback mode (known as a feed-back complimentary filter [7]). Rate information is fed forward through the model of vehicle dynamics to provide an estimate of position. This estimate is compared with a position measurement and the error between the two is fed back to correct the rate sensor position estimates at source. This ensures that high frequency information from the rate sensor is fed straight through to the loop output, while low frequency information is corrected and filtered. The essential role of the vehicle model is now clear, it provides the means of relating rate information, obtained from wheel rotation and steer angle or from acceleration and rotation rates in an inertial package, to position information through a model of vehicle motion.

Thus, a position measurement sensor, a rate measurement sensor and a vehicle model are essential to any one navigation loop. Each is complimentary and there is little or now redundancy between components.

2.2 Arbitration and Residual Monitoring

The two navigation loops each provide estimates of vehicle location and heading derived from physically different sensor operating principles. The output from these two navigation loops may now be compared. If there is statistically no difference between the two estimates, then they may be combined to provide an improved overall estimate of vehicle location. If there is a significant difference between the two estimates a fault can be declared. If, at any one time, only one navigation loop is in fault, a comparison can be made with a third (residual monitoring sensor), to determine which loop is at fault and potentially why the fault has arisen. In the unlikely event that both loops fail simultaneously a fault can still be detected and declared on the basis of a difference in estimate output from each other and the residual sensor. However, in this case, it is not possible to identify the fault. The only case in which a fault can not be detected is if both navigation loops and the residual monitoring sensor all fail at exactly the same time in exactly the same way. Given the physical dissimilarity between the sensors employed in each loop, such a common mode failure will essentially never occur.

2.3 Quantifying Fault Rejection Capability

Each filter in each navigation loop acts as a low-pass filter with respect to position observations and as a high pass filter with respect to rate observations. The cut off frequency of the filter is determined by the relative noise levels in the two sensors but is otherwise always first order (with a 20db/decade roll-off) [8]. With this information it is not difficult to quantify fault rejection capability both in each loop and for the overall system. Consider the following four (exclusive and exhaustive) scenarios:

- **Rejection of low frequency faults (drift, bias, offset) in low frequency (position) sensors**: In any one navigation loop these faults are passed straight through the low-pass filter to the output and are thus undetectable. For the combined system, such faults are detected with a discrimination power equal to the ratio of the covariance in each loop; if the two loops are of essentially equal accuracy then the discrimination power is 3db.
- **Rejection of high-frequency faults (jump fault, mis-match, sudden failure) in low frequency sensors**: Such faults are naturally detected and attenuated by the low-pass characteristic of each navigation loop with a discrimination power that increases as 20db/decade past the filter cut-off frequency. Such faults are thus readily detected and rejected in any well design navigation loop.
- **Rejection of low-frequency faults (drift, bias offset) in high frequency (rate) sensors**: Again, such faults are naturally detected and attenuated by the high-pass (with respect to rate sensing) characteristic of each navigation loop with a discrimination power that decreases as 20db/decade up to the cut-off frequency.
- **Rejection of high-frequency faults (jump fault, sudden failure) in high-frequency (rate) sensors**: In any one navigation loop these faults are passed straight through the high-pass filter to the output and are thus undetectable. For the combined system, discrimination power is dependent on both the sensor

error rates and the vehicle model employed but is again, typically, 3db beyond the cut-off frequency.

An analysis of the fault discrimination capabilities of any given suite of navigation sensors is additionally a powerful tool in actually determining what combination of sensors as best fitted to a given navigation task.

3.0 The Design of a High-Integrity Navigation System

This section describes a current project to develop and demonstrate a high-integrity navigation system based on the principles outlined in Section 2.0 above. The system consists of two navigation loops, one based on the use of GPS a position sensor and an inertial measurement unit (IMU) as a rate sensor, the other based on millimetre-wave radar as a position sensor and encoders (wheel rate and steer angle) as rate sensors. This is shown in Figure 2. We now explain and justify the design of each proposed navigation loop.

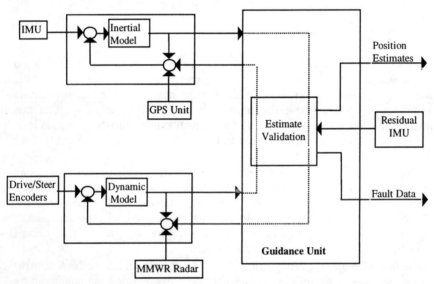

Figure 2: A High-Integrity Navigation System Design

3.1 The GPS/IMU Loop

The Global Positioning System (GPS) is now a widely used source of position measurement information in tasks such as vehicle tracking and dispatch, and surveying. The operating principles of GPS and also DGPS are well known and thus not discussed here [6].

GPS is relatively cheap to implement and is there for the using. It therefore makes sense to incorporate it in any proposed land vehicle navigation system. However, while GPS technology improves daily, as with any sensor it is important to understand the ways it may fail. Failures in GPS are predominantly caused by either constellation or local environment effects. Constellation faults, poor geometry, satellite availability, ephemeris error are generally abrupt and detectable with modern receivers. Environment effects, shadowing, EM interference and particularly multi-path propagation, are more problematic, leading to failures which may be both of abrupt and gradual (drift) types, many of which are unpredictable and undetectable. The probability is that these effects will be significant in relatively deep open-cut mines. It is a working hypothesis in this project that many of these effects are due to the scattering effects of operating at the relatively long wave-length of the GPS carrier. Consequently the second position sensor has been chosen to operate at a much shorter wave-length (see 3.2 below). Potential solutions such as the use of pseudo-lites have been rejected on the grounds of commonality with basic GPS. The results described in this paper use the Ashtec G12 DGPS unit with notional GDOP of approximately 0.5m. We are in the process of upgrading to the relatively new Novatel P2 DGPS system which incorporates carrier phase tracking to achieve a notional GDOP of 2cm.

In the design, position information from the GPS unit is put in feedback configuration with rate information from an inertial measurement unit (IMU) [1,2]. The unit employed is a Watson triaxial IMU system using vibrating structure gyros, solid state accelerometers, and two pendulum sensors to provide full three dimensional rate and attitude information. The use of the pendulum sensors is critical in providing accurate initial alignment information. Given correct initial alignment, attitude drift can be reduced to better than 1° per hour, and position drift to better than 0.5m per minute.

The "vehicle model" used in this navigation loop consists of a standard model of three-dimensional point kinematics relating accelerations and angular velocities to point position.

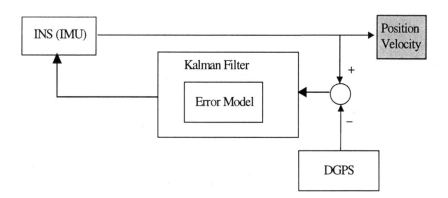

Figure 3: The GPS/IMU Indirect Feedback Filter

The GPS/IMU loop employs an indirect feedback filter as shown Figure 3 [7]. In this case high-frequency IMU information is still able to propagate through to the output without attenuation, and is compensated with low frequency corrections. The main feature of this filter is that the IMU may be reset from the filter estimates. This approach has the potential advantage of including other IMU internal parameters, such as platform misalignments, gyro and accelerometer bias, in the estimated state vector. The indicated position/velocity information derived from the INS is subtracted from the position/velocity GPS information. An error propagation model is used to fuse this observation and predict position/velocity and some parameters to correct the inertial measurement unit information. These parameters are fed back to the IMU unit.

3.2 The MMWR/Encoder Loop

Over the past five years, millimetre wave radar (MMWR) has found increasing use as automotive sensors, in guidance and collision detection tasks, as part of a number of international "intelligent vehicle highway" systems projects [4]. As an outdoor all weather navigation sensor, MMWR is almost ideal. The high operating frequency (77GHz) and the relatively short wave-length (4mm), provide a number of key operating characteristics. The radar itself can be made relatively small (100mm antenna is typical) while maintaining high levels of directional accuracy ($1°$ is common). In frequency modulation (FM) range measurement, very high swept band-widths (600MHz) can be obtained, translating in to very accurate range measurements (of order 10-25cm). MMWR has typical operating ranges of 2-300m and significantly is largely unaffected by heavy rain and dust. The MMWR unit used in this work is shown in Figure 4.

The MMWR/encoder loop is based on the same technology and principles previously employed by the authors in [3]. Position information is obtained by mechanically scanning the radar beam through $360°$, obtaining range and bearing measurements to a sparse number of special passive reflectors placed along the haul route. Position information from the MMWR is combined with encoder (wheel rotation rate and steer angle) information using a detailed kinematic vehicle model [5], together with a Kalman filter, in standard feedback configuration. The integrated MMWR/encoder system is capable of providing position accuracies of a few centimetres.

Figure 4: 77GHz MMW Radar Unit and Beacon used in Navigation System.

4.0 Preliminary Results

We focus here on describing results from the GPS/IMU system. The sensors were retrofitted in a Holden Ute vehicle. The car was driven in the neighbourhood of the University where buildings and other structures were present obstructing in some cases the direct path of the GPS satellites. Figure 5 shows the trajectory generated by the IMU alone, (black) and DGPS (grey). It can be seen that the error increases significantly after the first 500 meters. Figure 6 shows the estimates produced by the filter running on GPS data only. Figure 7 shows an enlargement of the position estimate after the first turn. In this case the DGPS generates a position solution which is affected by multipath. Since the filter is fusing all DGPS data the estimate becomes erroneous. Fault detection capabilities were included in the run shown in Figure 8. In Figure 9 it can clearly be seen that the fault is detected and the filter navigates with IMU alone during the period that the DGPS is in fault. There is no noticeable degrading of performance during the GPS fault. Finally Figure 10 shows the results of disconnecting the GPS after the first 500 meters. It can be seen that the attitude of the unit has been updated properly since the filter is able to predict the trajectory over this time with only small errors.

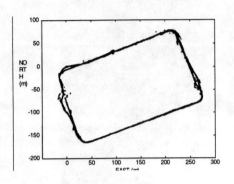

Figure 5: Trajectory without GPS   Figure 6 Trajectory with GPS

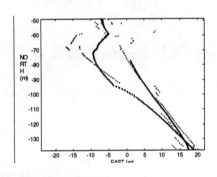 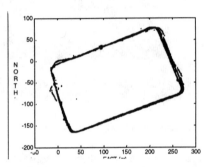

Figure 7: Enlarged Trajectory.   Figure 8: With GPS and Fault Detection

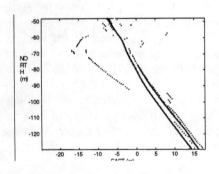 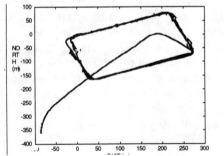

Figure 9. Enhanced GPS and FD   Figure 10. Disconnecting GPS

Bibliography

[1] B. Barshan and H.F. Durrant-Whyte, "A comparison of solid state gyroscopes", IEEE Trans. Instrumentation and Measurement, Vol 44, No 1, 1995.

[2] B. Barshan and H.F. Durrant-Whyte, "Inertial Sensing for Mobile Robotics", IEEE Trans. Robotics and Automation, Vol 11, No 3, 1995.

[3] H.F. Durrant-Whyte "An Autonomous Guided Vehicle for Cargo Handling Applications", Int. J. Robotics Research, 15(5), p407-441, 1996.

[4] M. Herbert, C. Thorpe and T. Stenz, (editors) "Autonomous Navigation Research at Carnegie Mellon", Kluwer Academic Publishers, 1997.

[5] S. Julier and Durrant-Whyte H.F. "Navigation and Parameter Estimation of High Speed Road Vehicles", Proc. IEEE Int. Conf. Robotics and Automation, 1995.

[6] A. Leick, "GPS Satellite Surveying, Second Edition", Wiley Interscience, 1995.

[7] P. Maybeck, "Stochastic Models, Estimation and Control Vol I", Academic Press 1979.

[8] E.M. Nebot, H.F. Durrant-Whyte, S. Scheding, "Frequency Domain Modeling of Aided GPS with application to High Speed Vehicle Navigation Systems", Proc. IEEE Int. Conf. Robotics and Automation, Alberquerque, April 1997.

[9] E.M. Nebot, H.F. Durrant-Whyte, S. Scheding, G. Dissanayake, "Slip Modeling and Aided Inertial Navigation of an LHD", Proc. IEEE Int. Conf. Robotics and Automation, Alberquerque, April 1997.

[10] A. Stevens, M. Stevens and H.F. Durrant-Whyte, "OxNav: Reliable Autonomous Navigation", Proc. IEEE Conf. Robotics and Automation, Japan, 1995.

Modeling and Control of a 3500 tonne Mining Robot

Peter I. Corke, Graeme J. Winstanley and Jonathan M. Roberts
CSIRO Division of Manufacturing Science & Technology
CRC for Mining Technology and Equipment
Pinjarra Hills, AUSTRALIA 4069.
http://www.cat.csiro.au/automation

Abstract: Draglines are extremely large machines that are widely used in open-cut coal mines for overburden stripping. Since 1994 we have been working toward the development of a computer control system capable of automatically driving a dragline for a large portion of its operating cycle. This has necessitated the development and experimental evaluation of sensor systems, machines models, closed-loop control controllers, and an operator interface. This paper describes our steps toward the goal through scale-model and full-scale field experimentation.

1. Introduction

A dragline, see Figure 1, comprises a rotating assembly that includes the 'house' (drive motors, controls, and operator cabin), tri-structure or mast, and boom[1]. The house rotates on a bearing surface on top of the 'tub' which sits on the ground. A large diameter ring gear is fixed to the tub and the house is rotated by a number of pinions driven by motors in the house. A walking dragline is able to drag the tub along the ground by means of large eccentrically driven 'walking shoes' at the side of the machine. The dragline has three driven mechanical degrees of freedom:
- the house and boom can *slew* with respect to the tub;
- the bucket can be *hoist*ed by a cable passing over sheaves at the tip of the boom;
- the bucket can be *drag*ged toward the house by a cable passing over sheaves at the base of the boom.

During digging the bucket motion is controlled using only the drag and hoist ropes. When the bucket is filled it is hoisted clear of the ground and swung to the dump position by slewing the house and boom. The drag and hoist drives now control the position of the bucket within a vertical plane that contains the centerline of the boom however the bucket is free to swing normal to that plane.

The configuration of the machine is given by

$$\mathbf{x} = [d\ h\ \phi\ \theta]$$

---
[1] Boom elevation angle is constant.

Figure 1. Annotated picture of the Bucyrus-Erie model 1370 dragline.

comprising respectively drag and hoist rope lengths, slew angle and swing angle, θ, of the bucket normal to the plane containing the boom and which moves like a pendulum

$$\ddot{\theta} = f(d, h, \phi, \theta)$$

which cannot be controlled directly. The control inputs are

$$\mathbf{u} = [\dot{d}\,\dot{h}\,\tau]$$

respectively the drag and hoist rope velocities and the slew torque. A good deal of operator skill is required to control, and exploit, the bucket's natural tendency to swing.

A complete cycle takes around 1 minute to complete, of which 80% is swinging the bucket through free space; it is this portion of the cycle that we are working to automate. The automation system will be activated by the operator once the bucket is filled, and it will automatically hoist, swing and dump the bucket (at an operator set reference point) before returning the bucket to a predefined digging point.

As can be seen from Figure 1 draglines are very large machines. They typically have boom lengths of over 100 m, weigh up to 5000 tonnes, and have bucket capacity of the order of 100 tonnes. Our concept is being prototyped on a Bucyrus-Erie 1370 dragline (BE1370) working at a mine located about 200 km from our laboratory in Brisbane. This paper describes ongoing work with that dragline, as well as an earlier phase of the experimental program which was conducted on a one-tenth scale 'model' dragline [1].

The motivation for this work is to improve the productivity of these very expensive machines ($US 40–80M). Current productivity is quite variable between operators and even the one operator at start and end of shift. It is estimated that increasing dragline productivity by 4% would save $US 2.4M/year per dragline. Our aim is for the bucket filling phase to be manually controlled, but that motion to the dump point and return, around 80% of the cycle time, will be automatic with the bucket following a defined trajectory through space.

Although superficially a dragline is quite different to a robot, due to its size and unusual actuation (by cables and winch drums), it is nevertheless useful to consider it as a potential robot — a 3DOF robot with one flexible link. With the addition of suitable sensors a dragline can be considered as a robot which can then perform controlled motion of its 'tool' along a defined trajectory.

The remainder of this paper is organised as follows. Section 2 describes the development of a sensor to determine the position of the dragline bucket in 3D space. Section 3 summarizes the electrical and mechanical dynamic modeling, parameter estimation and simulation results. Section 4 builds on the modeling work and covers control design, simulation and experimental results. Experimental data from a model and a production dragline will be used to illustrate the issues and the results achieved to date. Section 5 summarizes some relevant implementation details, in particular the novel operator interface. Finally Section 6 presents the conclusions.

2. Bucket position sensing

As already mentioned a good deal of operator skill is required to control the natural tendency of the bucket to swing. The motion of the dragline bucket with respect to the boom, $\theta(t)$, is like that of a pendulum, with the ropes[2] and aerodynamic forces providing damping. In order to control this motion, a critical part of the automation requirement, it is necessary to determine the bucket's position. The winch drums are fitted with encoders which can be used to obtain good estimates of the hoist and drag rope lengths, and these define a locus in space along which the bucket is free to swing. Some options to localize the bucket along that locus include:

1. fitting sensors such as accelerometers, GPS receivers or radio transponders to the bucket itself;

2. inferring bucket position from the torque exerted by the swinging load on the slew rotation axis;

3. using some non-contact sensor such as computer vision or radar.

The bucket's harsh interaction with the environment rules out the first option. The second option is motivated by examining the equations of motion, see equation (9). Our conclusion is that this effect is small and will be corrupted by machine non-idealities such as friction and backlash in the high-reduction slew gear train.

Our initial efforts[1] investigated the use of computer vision but it became clear that this is an extremely challenging application. The system must be extremely reliable and operate:

---

[2] these are dual 8 cm diameter multistranded steel cables.

Figure 2. BE1370 bucket images (a) raw bucket image. (b) edge image.

- 24 hours a day in all weather conditions (including tropical rain, fog and dust);
- at a rate of at least 3 Hz (a control constraint);
- with a large variation in scale as the bucket range varies from 5 to 100 m from the camera;
- with varying bucket orientation with respect to the camera;

While we were able to use computer vision for an early demonstration on the one-tenth scale dragline, see Section 4.1, this only operated reliably in good day light conditions. An extensive study using real image data[2] established that it was not practical to meet the constraints using computer vision. The deficiencies of the machine vision approach are best illustrated by the raw and edge bucket images from the BE1370 shown in Figure 2. Complicating factors include the problem of segmenting a bucket full of overburden against a background of overburden, moving camera and moving target, shadows, strong background texture and so on. Camera placement is also a non-trivial issue. We selected a view looking downward from the boom tip, since a horizontal view from the operators cabin has problems with the Sun shining directly into the camera at certain times of the day.

A workable alternative approach was inspired by video footage taken from our boom-tip camera panned out to view the horizon, see Figure 3(a). The hoist ropes are a very obvious feature, and the angle their centerline makes with respect to vertical is the swing angle of the bucket with respect to the boom plane. The concept was prototyped by a rope following algorithm that determined the location of the hoist ropes against the sky. The method was fast and accurate but required the background to be bright sky, and thus failed at night.

Our current approach is to use an infra-red scanning laser range finder (Sick PLS) mounted at the boom tip. This device measures range and bearing through 180° in a plane and has an angular resolution of 0.5° and a range accuracy of 2 cm. The sensor is mounted at the boom tip and its scanning

 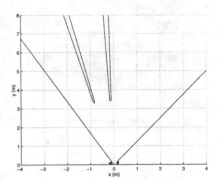

Figure 3. BE1370 hoist ropes (a) video image. (b) PLS range image.

plane is inclined at approximately 20° below the horizontal. In this position the only targets it will encounter are, in decreasing order of likelihood, the two hoist ropes, raindrops, insects, birds and dust, see for instance Figure 3(b). The sensor has been extensively tested under a range of conditions both in the laboratory and on the dragline. Figure 4 shows the PDF of target bearing and range occurrence for 1000 trials in normal daylight conditions and in heavy rain.

The two large spikes in the centre of the figures represent the dragline hoist ropes; the large spikes on the right are artifacts of the test range. The heavy rain introduces the small foreground spikes but the hoist ropes are still clearly discernible and can be reliably located with a tracking filter. Preliminary results (see Figure 5) show how the system is able to track the movement of the hoist ropes during machine operation. The rope bearing of Figure 5 is the mean bearing of the dragline's two hoist ropes.

The bucket sensing system combines the bearing information from redundant rangefinders, with the length of the hoist and drag ropes to determine the position of the dragline bucket in 3D space.

3. Electro-mechanical modeling and parameter estimation

This section summarizes our work on modeling and parameter estimation (see Corke et al.[3] for more details).

3.1. Electrical system

The dragline operator uses a combination of hand and foot controls in order to control the dragline. Hand-operated lever controls on either side of the operator's chair control drag (left hand) and hoist (right hand) rope speeds. Two mechanically coupled foot pedals control the torque applied to the slew motors; the right pedal slews right and the left pedal slews left. The outputs of the controls are command voltages in the range ±15 V to the drive control systems.

Each axis of the BE1370 is driven by four large (500 hp) separately-excited

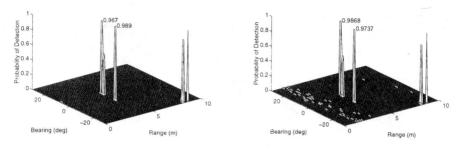

Figure 4. PDF of target bearing and range occurrence over 1000 trials (a) hoist ropes in daylight. (b) hoist ropes in heavy rain.

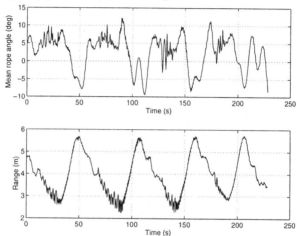

Figure 5. Sensed hoist rope motion during dragline operation.

DC traction motors which operate in a closed-loop configuration. The hoist and drag axes are speed controlled, while the slew axis is current controlled. The armature currents (up to 3000 A) are provided by Ward-Leonard electromechanical current amplifiers, see Figure 6. These comprise a generator driven at constant speed, ω_g, whose field current, i_{fg}, is is provided by the 'exciter', a solid-state controlled rectifier which functions as a voltage controlled current source. The demand input to the exciter comes from the operator's control via limiting and feedback circuitry. The generator terminal voltage

$$v_g = K_g i_{fg} \omega_g \qquad (1)$$

drives a current

$$i_a = (v_g - v_m)/2R_A \qquad (2)$$

through the motor where v_m is the motor back EMF and R_A is the armature resistance. The current gain of the Ward Leonard set on the BE 1370 is around 35 A/A. The back EMF

$$v_m = K_m(i_{fm})\omega_m$$

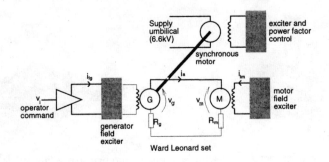

Figure 6. Ward-Leonard system.

is a function of both motor speed and motor field strength, i_{f_m}. For the BE1370 the motor fields of the hoist and drag drives can be weakened, reducing back EMF, and allowing higher motor speeds for paying out the ropes. Braking is achieved electrically by either regenerative braking (energy recovery) or 'plugging' (reverse motoring).

The resultant motor torque is

$$\tau = K_m(i_{fm})i_a \qquad (3)$$

where the torque constant, K_m, is a non-linear function of armature and field current.

3.2. Current and voltage control loops

Each motor is controlled by several feedback and limit loops. The hoist and drag axes have voltage (speed) feedback with a dynamic current limit, while the slew axis has current (torque) feedback with a dynamic voltage (speed) limit. All loops enforce a dynamic commutation power limit which is a function of the voltage and current product.

Current feedback is obtained from a current shunt in series with the generator via an isolation amplifier. Voltage feedback is obtained from a voltage divider across the generator and through a unity gain isolation amplifier.

Schematic diagrams for the dragline control system were analyzed symbolically to determine the volt-amp relationship which has been encapsulated in Simulink function.

The transfer function of the drag drive system has been identified from measured input-output data as

$$\frac{V(s)}{U(s)} = \frac{51}{[\zeta = 0.8,\, \omega_n = 1.3\,\text{rad/s}]}$$

and includes the lumped dynamics of the Ward-Leonard set and motor armature. Figure 7 shows the model response as well as measured data from the machine. The fit is excellent except for regions of high armature current (> 3000 A) where the current limit is enforced.

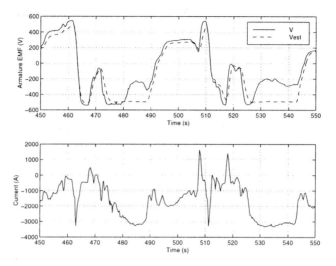

Figure 7. Data from BE1370 dragline. Top, plot of drag motor actual and model response to measured operator input versus time. Lower, plot of armature current versus time.

3.3. Mechanical model of bucket swing

When disengaged from the ground the bucket has four degrees of freedom; its spatial position and carry angle (the angle the bucket bottom makes with the horizontal plane). The bucket is under-actuated and the interaction between hoist, drag and carry angle is complex. In the work so far it has been sufficient to model the bucket as a point mass without concern for carry angle.

The equations of motion for the suspended bucket were derived using standard robotic techniques. A Lagrangian approach was used to determine the equations of motion. The two generalized coordinates are slew, ϕ, and swing, θ, and the generalized forces are slew torque and rolling friction, Q_1, and bucket drag, Q_2. Full details are given in Corke et al.[3].

The equations were derived symbolically using MAPLE, solved for $\ddot{\phi}$ and $\ddot{\theta}$, and the equations exported as 'C' which with some trivial wrapper functions became MATLAB 'MEX-file' source. The derived model is completely parameterized and the user provides a vector of specific dragline parameters such as house rotational inertia, boom length and angle above vertical, slew axis and bucket damping, and bucket mass. A number of coordinate frames were assigned and the kinematic equations derived.

3.4. Parameter estimation

As with conventional factory robots it is difficult to obtain information about motor parameters, inertia and friction. It is therefore necessary to apply parameter identification techniques to recorded machine input and output data. The remainder of this section applies this latter approach to the slew drive of the scale-model dragline, but the general approach could be applied to a well instrumented full-scale dragline, see for example Figure 7(b).

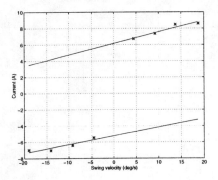 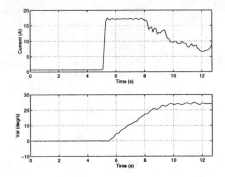

Figure 8. Model dragline experimental results. (a) Constant velocity tests. (b) 60% current step demand (15.3 A). Note that after 8 s the back EMF has risen sufficiently to decrease the armature current.

3.4.1. Friction identification

Friction was identified using a number of constant velocity tests and the average current and velocity points are plotted in Figure 8. This shows almost textbook characteristics of mixed Coulomb and viscous friction.

3.4.2. Motor electrical parameters

The velocity, current and terminal voltage during operation can be used to estimate the armature circuit resistance, R_a, and the motor back EMF constant, K_m[4]. Ignoring inductance, and assuming constant inertia, we can write the standard equations for a DC motor

$$v_a = K_m \omega + R_a i_a \tag{4}$$

which can be rewritten, for multiple voltage, speed and current measurements, in matrix form

$$\begin{bmatrix} v_{a_1} \\ \vdots \\ v_{a_N} \end{bmatrix} = \begin{bmatrix} \omega_1 & i_{a_1} \\ \vdots & \vdots \\ \omega_N & i_{a_N} \end{bmatrix} \begin{bmatrix} K_m \\ R_a \end{bmatrix}$$

which can be solved using least squares. For the model dragline the results of $R_a = 1.29\,\Omega$ (including wiring loss) and $K_m = 0.845\,\text{Vs/rad}$ compare well with the manufacturer's value of $0.5\,\Omega$ and $0.84\,\text{Nm/A}$.

3.4.3. House rotational inertia

Slew inertia was estimated by a current step-response test as shown in Figure 8 but rather than differentiate the velocity signal we chose instead to use a maximum likelihood procedure. The model was assumed to be of the form

$$J\ddot{\phi} + B\dot{\phi} + \tau_c(\dot{\phi}) = \tau_m$$

where ϕ is the slew angle, τ_m the applied torque, τ_c the Coulomb friction, J the total inertia and B the effective viscous friction coefficient. The inertia and viscous friction coefficient parameters were adjusted by the procedure in order to minimize the error between the actual and model response to the torque step. This produced an excellent fit and a viscous friction value close to that identified in section 3.4.1.

4. Bucket swing control

A common approach to non-linear system control, and the one used here, is to linearize the system dynamics about a given operating point. The linear form may be conveniently represented in statespace form as

$$\underline{\dot{x}} = \mathbf{A}\underline{x} + \mathbf{B}\underline{u} \qquad (5)$$
$$\underline{y} = \mathbf{C}\underline{x} + \mathbf{D}\underline{u} \qquad (6)$$

where $\underline{x} = [\phi \; \theta \; \dot{\phi} \; \dot{\theta}]$ and $\underline{u} = [h \; \dot{h} \; d \; \dot{d} \; \tau \; m_b]$.

To linearize the dragline slew axis and bucket dynamics about the steady-state, vertical, hoist rope configuration, we use the MATLAB function `linmod`, though the linearization could be achieved for exactly using the symbolic forms of the equations of motion of Section 3.3. Using parameters for the BE1370 with both ropes set at 60 m and with a full bucket, the linearized dynamics are

$$\mathbf{A} = \begin{bmatrix} 0 & 0 & 1 & 0 \\ 0 & 0 & 0 & 1 \\ 0 & -0.0237715 & 0 & 0 \\ 0 & -0.195456 & 0 & 0 \end{bmatrix} \qquad (7)$$

$$\mathbf{B} = \begin{bmatrix} 0 & 0 & 0 & 0 & 0 & 0 \\ 0 & 0 & 0 & 0 & 0 & 0 \\ 0 & 0 & 0 & 0 & 2.60417 \times 10^{-10} & 0 \\ 0 & 0 & 0 & 0 & 2.79806 \times 10^{-10} & 0 \end{bmatrix} \qquad (8)$$

The eigenevalues of $\mathbf{A}$ indicate the poles are at $0, 0, \pm 0.442j$ corresponding to undamped swing motion with a period of 14 s.

The state-space form is very sparse and can be written succinctly as

$$\ddot{\phi} = -0.0237715\theta + 2.60417 \times 10^{-10}\tau \qquad (9)$$
$$\ddot{\theta} = -0.195456\theta + 2.79806 \times 10^{-10}\tau \qquad (10)$$

which shows that the bucket swing angle, θ, weakly influences the motion of the house, ϕ — this is expected since the bucket inertia in this configuration is very signficant, around 30% of total rotational inertia. The bucket swing angle is only a function of itself and slew torque, not rope rates, and it is for this reason that a dragline's slew drive is current (or torque) controlled rather than velocity controlled.

Now that we have a linearized state-space representation we can apply any of a number of modern control design techniques. For instance, a linear-quadratic regulator (designed with $\mathbf{Q} = \text{diag}(1)$ and $\mathbf{R} = \text{diag}(1)$) yields the regulator

$$\tau = [3.1623 \; 2.6143 \; 10.8327 \; -5.3245]'\underline{x} \times 10^{-10} \qquad (11)$$

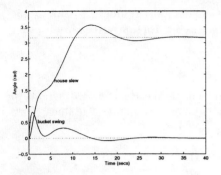
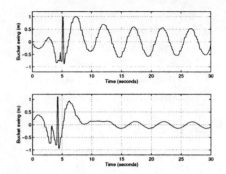

Figure 9. (a) Step torque response for BE1370 dragline linearized slew drive and bucket system. (b) Experimental results for model dragline. Top plot without vision control, bottom plot with vision control.

which shows that feedback of swing and swing rate is required to damp the bucket motion. This regulator is simply a proportional-derivative controller on slew and swing angles. A step response for this controller is shown in Figure 9.

Now that swing angle can be controlled it is possible to make the bucket follow an arbitrary trajectory through space. Standard robotic techniques can be used to compute the desired bucket position as a function of time, whilst avoiding obstacles, and meeting velocity and acceleration constraints imposed by the drive system. The axis control systems will then ensure that the bucket follows the desired spatial trajectory by providing appropriate setpoints for rope speeds and slew axis torque.

4.1. Control experiments

The closed-loop control experiments were conducted on the one-tenth scale model dragline, using a boom top mounted camera to provide estimates of the bucket centroid at 5 Hz[5]. Two types of experiment were conducted. The first was a cycle test which demonstrated many aspects of an automatic dumping cycle — coordinated slew, hoist and drag motion combined with visual swing control. The bucket moved cyclically between two predefined positions following a fifth-order polynomial path. Interestingly this smooth input tended not to greatly excite the bucket swinging mode.

The second experiment[5], less typical of normal operation, was a step response test. The boom slews between two defined positions with a bang-bang motor current trajectory. Figure 9 shows bucket motion in a step response test with and without the vision-based control system. The effect of vision-based control in rapidly damping out the major swing oscillation is very clear. The small residual oscillation is due to the poor low-speed motion control we could obtain with the model dragline which had particularly severe backlash (7 motor turns!) and considerable stiction.

5. Implementation details

Working on a production machine imposes considerable constraints such as limited access, minimal modification to existing electrical systems, and the need for an operator to regain control of the dragline at any time. These factors have considerably influenced our implementation which is described briefly in this section.

5.1. Operator interface

The operator's interface to the dragline is quite crude, and as mentioned in Section 3.1 comprises two joysticks and a pair of foot pedals. Initially we considerd using a space-ball type interface to directly demand the state of the suspended bucket with respect to a ground based coordinate frame, but this was perceived as too radical a change. The next option was to introduce a change-over switch so that the setpoint to the motor drive would be selected from the operator's control or the computer system, but this could introduce discontinuities at mode transitions.

This led to the development, not envisaged at project commencement, of a novel operator interface. The automation system 'drives' the dragline by physically moving the control joysticks and pedals; somewhat like an automotive cruise control or an aircraft auto-pilot.

Brushless DC servo motors and encoders have been fitted to each of the operator controls. The major design challenge was to select the means of coupling, tooth belts, and to choose the gear ratio so as to minimize the perceived motor cogging in the hand controls. The motors are servoed at a high-rate by dedicated hardware. The host monitors tracking error to determine if the operator is opposing the control, in which case the servo gains are ramped down to zero so as to smoothly transfer control of the dragline back to the operator.

Such an interface has the possibility, currently unexplored, to provide kinisthetic feedback so that, for example, the operator could 'feel' how hard the ground was during digging or how heavy the bucket was during lifting.

5.2. Computing and communications

The control computer is a VMEbus system comprising an MVME1603 PowerPC and Industry Pack (IP) digital and analog interfaces. Signals monitored include motor armature voltages and currents, regulator demands from the operator's controls, and encoders on the hoist and drag winch drums and the house slew drive. All signals are connected via isolation amplifiers and signal conditioning stages. The boom mounted rangefinders connect via an isolated RS422 serial network to an industrial PC that implements the tracking filter, and the results are communicated to the main processor via a local 10BaseT network. This network is in turn connected, via a 10BaseFL (fibre optic) network, to a radio LAN bridge (range up to 1 km) mounted on the dragline mast. The other end of the radio link is a caravan with an internal 10Base2/10BaseT network which serves as a mobile computing laboratory for the analysis of dragline data, and code development without requiring a physical presence on the dragline. All computers run under LynxOS, a real-time POSIX-compliant multitasking operating system.

6. Conclusions

This paper has discussed sensing, modelling, control and operator interface issues relevant to the automation of an electric walking dragline. Although these machines are quite different to what is normally considered a robot (in terms of physical scale, power, and method of actuation), the many techniques developed by the robotics community of the last 20 years are quite applicable to this application. The dragline can thus be transformed into a numerically controlled machine or robot. Critical to this transformation are sensing systems that are able to meet very stringent criteria, modeling of machine dynamics, parameter estimation and control design. Experimental data from a model and production dragline has been used to illustrate the issues and the results achieved to date. This paper has also described the evolution of our ideas over the course of the project; some due to the need for extreme reliability, such as the bucket sensing system, and some due to the needs of machine operators, such as the user interface. The approach has been verified experimentally on a one-tenth scale model dragline and the first automated swing of the production dragline is planned for the end of 1997. We see this current automation project as the first step toward eventual, fully automatic dragline operation.

Acknowledgements

The authors gratefully acknowledge the help of their colleagues Stuart Wolfe, David Hainsworth, Yunming Li, Stephen Nothdurft, Don Flynn, Ian Hutchinson, Peter Nicolay, Allan Boughen and Daniel Sweatman. The present work is funded by a consortium consisting of the Australian Coal Association Research Programme as project C5003, Pacific Coal Pty Ltd, BHP Australia Coal Pty Ltd and the Cooperative Research Centre for Mining Technology and Equipment (CMTE), a joint venture between AMIRA, CSIRO and the University of Queensland. ACIRL made the scale model dragline available for our closed-loop control experiments. Valuable inkind support has been provided by Bucyrus Erie Australia, Tritronics Pty Ltd, and Tarong Coal for allowing the automated swing system to be installed on their BE1370.

References

[1] D. W. Hainsworth, P. I. Corke, and G. J. Winstanley, "Location of a dragline bucket in space using machine vision techniques," in *Proc. Int. Conf. on Acoustics, Speech and Signal Processing (ICASSP-94)*, vol. 6, (Adelaide), pp. 161–164, Apr. 1994.

[2] G. Winstanley, P. Corke, and J. Roberts, "Dragline swing automation," in *Proc. IEEE Int. Conf. Robotics and Automation*, (Albuquerque, NM), pp. 1827–1832, 1997.

[3] P. I. Corke, G. Winstanley, and J. Roberts, "Dragline modelling and control," in *Proc. IEEE Int. Conf. Robotics and Automation*, (Albuquerque, NM), pp. 1657–1662, 1997.

[4] P. I. Corke, "In situ measurement of robot motor electrical constants," *Robotica*, vol. 14, no. 4, pp. 433–436, 1996.

[5] J. Roberts, P. Corke, and G. Winstanley, "Applications of closed-loop visual control to large mining machines," in *Video Proceedings of IEEE Int. Conf. on Robotics and Automation*, 1996.

Chapter 6

Non holonomic vehicles

Low and high level vehicle control depends on the vehicle structure. Planning and control execution of trajectories by non holonomic vehicles require the study of their kinematics model and environment conditions.

Autonomous maneuvering on non holonomic vehicles for moving in a structured dynamic environment is studied by Paromtchick, Garnier and Laugier. The problem faced up is the autonomous maneuvering for lane following, changing and parallel parking. Motion generation and vehicle control is split in three phases: localization, planning and execution. The vehicle kinematics determine the constraints to define possible trajectories based on the mission and range measurement of environmental objects.

Khatib, Jaouni, Chatila and Laumond proposed a dynamic path structure that they call a deformable structure to deal with environment changing conditions. The goal is to define a methodology linking planning and execution taking into consideration the vehicle structure. The work is an evolution of the method called elastic band that has been adapted from holonomic to non holonomic robots. The method considers distance to obstacles and uses convex potential functions to increase reactivity to deal with environment changes.

The paper of Lamiraux and Laumond addresses the problem of parametrizing a path with respect to time to generate a trajectory for a mobile robot towing a trailer. They present a numerical algorithm where they introduce the movement constraints for velocity and acceleration. The motion planning and control system is based on three main steps: non-holonomic path planning, trajectory computation that is the main objective of the pape,r and feedback control.

Autonomous maneuvers of a nonholonomic vehicle

I. E. Paromtchik, Ph. Garnier, and C. Laugier
INRIA Rhône-Alpes, GRAVIR
655, avenue de l'Europe, 38330 Montbonnot Saint Martin, France
E-mail: Igor.Paromtchik@inrialpes.fr

Abstract: Maneuvers of a nonholonomic vehicle in a structured dynamic environment are considered. The paper focuses on motion generation and control methods to autonomously perform lane following/changing and parallel parking maneuvers. Lane following/changing involves the tracking of a nominal trajectory in the traffic lane, and the generation and tracking of local lane-changing trajectories, for instance, for obstacle avoidance. Parallel parking involves a controlled sequence of motions, in order to localize a sufficient parking space, obtain a convenient start location for the vehicle and perform a parallel parking maneuver. The methods developed are tested on an automatic electric vehicle.

1. Introduction

The autonomous maneuvering of nonholonomic vehicles in dynamic environments is being studied by many research teams. The state-of-the-art of this domain reflects approaches of various complexity. A generalized approach involves planning a global path within an available map of the environment. Because of the computational costs, global planning is usually performed off-line. The subsequent following of the planned nominal trajectory involves reactive capabilities, in order to avoid collisions with unexpected obstacles. These two behaviors (trajectory following and obstacle avoidance) are in conflict, their simultaneous operation can lead to an oscillatory motion of the vehicle. However, if a nominal trajectory which is obstructed by an obstacle can be modified locally to avoid the obstacle and then return to the nominal trajectory, the oscillations can be eliminated. This issue is studied in the present paper using the example of lane following/changing maneuvers in a traffic environment.

A practical approach to motion generation and control for autonomous parallel parking in a traffic environment is also considered. Its key idea is to carry out a "Localization-Planning-Execution" cycle until a specified "parked" location of the vehicle relative to its environment is reached. The approach is based on range measurements to environmental objects around the vehicle. Feasible controls (steering angle and locomotion velocity) that correspond to a nominal trajectory leading to the "parked" location are planned and executed in real time. Once the motion has been carried out, the sensor data is used to decide whether the "parked" location has been reached and the parking maneuver is completed. This research work contributes to the French PRAXITELE programme that aims to develop a new urban transportation system based on a fleet of electric computer-driven vehicles [1]. At present, several experimental vehicles have been designed and their autonomous abilities are being developed.

A kinematic model of such a vehicle with front wheel steering is shown in Fig. 1. The vehicle's coordinates are denoted as a configuration $q = (x, y, \theta)^T$ relative to some reference coordinate system where $x = x(t)$ and $y = y(t)$ are the coordinates of the midpoint of the rear wheel axle, $\theta = \theta(t)$ is the orientation of the vehicle, and t is time. The motion of the vehicle is described by the equations

$$\begin{cases} \dot{x} = v \cos\phi \cos\theta, \\ \dot{y} = v \cos\phi \sin\theta, \\ \dot{\theta} = \frac{v}{L} \sin\phi, \end{cases} \qquad (1)$$

where $\phi = \phi(t)$ is the steering angle, $v = v(t)$ is the locomotion velocity of the midpoint of the front wheel axle, and L is the wheel base. The steering angle and locomotion velocity are two control commands (ϕ, v). Equations (1) correspond to a system with nonholonomic constraints because they involve the derivatives of the coordinates of the vehicle and are non-integrable [2]. Equations (1) are valid for a vehicle moving on flat ground with a pure rolling contact without slippage between the wheels and the ground. This purely kinematic

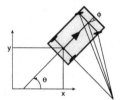

Figure 1. Kinematic model of a vehicle with front wheel steering

model of the vehicle is adequate to control low-speed motions, e.g. during parallel parking or lane following/changing in areas where only low-speed motions are allowed. For the high-speed motions, the dynamics of the vehicle must also be considered.

The notion "automatic vehicle" means that the vehicle is equiped with: (1) - a sensor unit to measure relative distances between the vehicle and environmental objects, (2) - a servo unit for low-level control of the steering angle and locomotion velocity, (3) - a control unit that processes data from the sensor and servo units and "drives" the vehicle by issuing appropriate servo commands. The sensor unit uses range sensors to measure relative distances between the vehicle and environmental objects. The servo unit consists of a steering wheel servo-system, a locomotion servo-system for forward and backward motions, and a brake servo-system to slow down and stop the vehicle. The microcomputer-based control unit monitors the current steering angle, locomotion velocity, travelled distance, coordinates of the vehicle and range data from the environment, calculates an appropriate local trajectory and issues the required servo commands.

2. Lane Following/Changing

Autonomous lane following is performed by tracking a nominal trajectory delivered by a global off-line path planner. In the case of unforeseen obstacles, the nominal trajectory is modified on-line, in order to avoid collisions. The modified trajectory has to satisfy temporal motion constraints and avoid collisions. In our earlier experiments, the trajectory following and obstacle avoidance behaviors were decoupled and considered independently, followed by a fuzzy behavior merging process. However, experiments showed that this produced oscillations of the effective motion of the vehicle during obstacle avoidance [3]. To remove these oscillations, a local trajectory is generated that avoids collisions with obstacles detected on the nominal trajectory. The local trajectory also allows the vehicle to catch up with the nominal trajectory (i.e. geometrical path and velocity profile along this path) after the obstacle avoidance. The major difference with the previous behavior-based approach is that the vehicle always follows a specific trajectory.

2.1. Lane Following

A method of trajectory following for a nonholonomic vehicle was described in [4]. This method guarantees the stable tracking of a feasible trajectory when the vehicle's control commands are:

$$\dot{\theta} = \dot{\theta}_{ref} + v_{R,ref}(k_y y_e + k_\theta \sin\theta_e), \qquad (2)$$

$$v_R = v_{R,ref} \cos\theta_e + k_x x_e, \qquad (3)$$

where $q_e = (x_e, y_e, \theta_e)^T$ represents the error configuration between the reference configuration q_{ref} and the current configuration q of the vehicle ($q_e = q_{ref} - q$), $\dot{\theta}_{ref}$ and $v_{R,ref}$ are the reference velocities, $v_R = v\cos\phi$ is the control command for the locomotion velocity of the midpoint of the rear wheel axle, k_x, k_y, k_θ are positive constants, and $\phi = \arctan\left(\frac{\dot{\theta} L}{v_{R,ref}}\right)$.

2.2. Lane Changing

Lane changing is carried out by generating and following a local trajectory. Such maneuvers are performed when the preplanned nominal trajectory would collide with an unforeseen obstacle. When an obstacle is detected, the nominal trajectory is translated to one side as shown in Fig. 2, in order to avoid collisions with the obstacle. The algorithm for collision avoidance involves the following iterations:

1. Generate a local trajectory which connects the nominal one with a collision-free local trajectory "parallel" to it (i.e. a parallel translation of the nominal trajectory).

2. Follow the local trajectory until the obstacle is overtaken.

3. Generate a local trajectory which connects the "parallel" trajectory with the nominal one.

4. Follow the local trajectory to catch up with the nominal one.

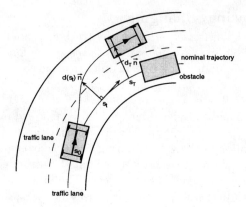

Figure 2. Translation of the nominal trajectory

A feasible trajectory for lane changing is obtained as a quintic polynomial

$$d(s) = d_T \left(10 \left(\frac{s}{s_T} \right)^3 - 15 \left(\frac{s}{s_T} \right)^4 + 6 \left(\frac{s}{s_T} \right)^5 \right), \quad (4)$$

where d_T is a distance between the two traffic lanes, s_T is a length of the nominal trajectory which is necessary to complete the lane changing maneuver, and $s = s_t$ is a length of the nominal trajectory since the start of the lane changing maneuver [5]. The distance d_T is supposed to be known. The minimal value of s_T is estimated as

$$s_{T,min} = \frac{\pi \sqrt{k\, d_T}}{2\, C_{max}}, \quad (5)$$

where C_{max} stands for the maximum allowed curvature:

$$C_{max} = \min \left\{ \frac{\tan(\phi_{max})}{L}, \frac{\gamma_{max}}{v_{R,ref}^2} \right\}, \quad (6)$$

γ_{max} is the maximum allowed lateral acceleration, and $k > 1$ is an empirical constant (e.g. $k = 1.17$ in our experiments).

When an obstacle is detected, a value $s_{T,min}$ is calculated according to (5) and compared with a distance between the vehicle and the obstacle. A decision is made, either to perform a lane changing maneuver or to slow down and possibly stop the vehicle. For the lane changing maneuver, the translation of the nominal trajectory is computed: at each instant t since the start of the maneuver, the reference position p_{ref} is translated along the vector $d(s_t).\vec{n}$ where $\vec{n}$ represents the unit normal vector to the velocity vector along the nominal trajectory. The reference orientation θ_{ref} is converted into $\theta_{ref} + \arctan\left(\frac{\partial d}{\partial s}(s_t) \right)$, and the reference velocity $v_{R,ref}$ is obtained as

$$v_{R,ref}(t) = \frac{dist(p_{ref}(t), p_{ref}(t + \Delta t))}{\Delta t}, \quad (7)$$

where $dist$ stands for the euclidean distance.

3. Parallel Parking

Autonomous parallel parking involves localizing a sufficient space (parking bay), obtaining a convenient start location for the vehicle relative to the bay, and performing a parallel parking maneuver. During localization the vehicle moves slowly along the traffic lane. Range data allows a local map of the environment alongside the vehicle to be built. Free spaces are detected, their borders are localized, and their orientation is calculated. The dimensions of the bay are compared with those of the vehicle and a decision on suitability for parking is made.

Drivers know from experience that before the parking maneuver starts, the vehicle must be oriented near parallel to the parking bay and it must also reach a convenient start position in front of the bay. A start location for parallel parking is shown in Fig. 3 where an automatic vehicle A1 is in a traffic lane. The parking lane with parked vehicles B1, B2 and a parking bay between them is on the right hand side of the vehicle A1. L1 and L2 are respectively the length and width of A1, and D1 and D2 are the distances available for longitudinal and lateral displacements of A1 within the bay. D3 and D4 are the longitudinal and lateral displacements of the corner A13 of A1 relative to the corner B24 of B2. The distances D1, D2, D3 and D4 are computed by the

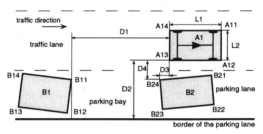

Figure 3. Start location for parallel parking

control unit from data obtained by the sensor and servo units. The control unit compares the length (D1-D3) and width (D2-D4) of the parking bay with the length L1 and width L2 of A1, where L1 and L2 include sufficient clearance for the vehicle to move around. If (D1-D3) > L1 and (D2-D4) > L2, the parking bay is sufficient for parallel parking.

During parallel parking, iterative low-speed backwards-and-forwards motions with coordinated control of the steering angle and locomotion velocity are performed to produce a lateral displacement of the vehicle into the parking bay. The number of such motions depends on the distances D1, D2, D3, D4 and the necessary parking "depth" which depends on the width L2 of the vehicle A1. The start and end orientations of the vehicle are the same for each iterative motion $i = 1, \ldots, N$.

For the i-th iterative motion (but omitting the index "i"), let the start coordinates of the vehicle be $x_0 = x(0)$, $y_0 = y(0)$, $\theta_0 = \theta(0)$ and the end coordinates be $x_T = x(T)$, $y_T = y(T)$, $\theta_T = \theta(T)$, where T is duration of the motion. The "parallel parking" condition means that

$$\theta_o - \delta_\theta < \theta_T < \theta_o + \delta_\theta, \tag{8}$$

where $\delta_\theta > 0$ is a small admissible error in orientation of the vehicle.

The following control commands of the steering angle ϕ and locomotion velocity v provide the parallel parking maneuver [6]:

$$\phi(t) = \phi_{max} \, k_\phi \, A(t), \quad 0 \le t \le T, \tag{9}$$

$$v(t) = v_{max} \, k_v \, B(t), \quad 0 \le t \le T, \tag{10}$$

where $\phi_{max} > 0$ and $v_{max} > 0$ are the admissible magnitudes of the steering angle and locomotion velocity respectively, $k_\phi = \pm 1$ corresponds to a right side (+1) or left side (−1) parking bay relative to the traffic lane, $k_v = \pm 1$ corresponds to forward (+1) or backward (−1) motion,

$$A(t) = \begin{cases} 1, & 0 \le t < t', \\ \cos \frac{\pi(t-t')}{T^*}, & t' \le t \le T - t', \\ -1, & T - t' < t \le T, \end{cases} \tag{11}$$

$$B(t) = 0.5 \left(1 - \cos \frac{4\pi t}{T}\right), \quad 0 \le t \le T, \tag{12}$$

where $t' = \frac{T-T^*}{2}$, $T^* < T$.

The commands (9) and (10) are open-loop in the (x, y, θ)-coordinates. The steering wheel servo-system and locomotion servo-system must execute the commands (9) and (10), in order to provide the desired (x, y)-path and orientation θ of the vehicle. The resulting accuracy of the motion in the (x, y, θ)-coordinates depends on the accuracy of these servo-systems. Possible errors are compensated by subsequent iterative motions.

For each pair of successive motions $(i, i+1)$, the coefficient k_v in (10) has to satisfy the equation $k_{v,i+1} = -k_{v,i}$ that alternates between forward and backward directions. Between successive motions, when the velocity is null, the steering wheels turn to the opposite side in order to obtain a suitable steering angle ϕ_{max} or $-\phi_{max}$ to start the next iterative motion.

In this way, the form of the commands (9) and (10) is defined by (11) and (12) respectively. In order to evaluate (9)-(12) for the parallel parking maneuver, the durations T^* and T, the magnitudes ϕ_{max} and v_{max} must be known.

The value of T^* is lower-bounded by the kinematic and dynamic constraints of the steering wheel servo-system. When the control command (9) is applied, the lower bound of T^* is

$$T^*_{min} = \pi \, \mathbf{max} \left\{ \frac{\phi_{max}}{\dot{\phi}_{max}}, \sqrt{\frac{\phi_{max}}{\ddot{\phi}_{max}}} \right\}, \tag{13}$$

where $\dot{\phi}_{max}$ and $\ddot{\phi}_{max}$ are the maximal admissible steering rate and acceleration respectively for the steering wheel servo-system. The value of T^*_{min} gives duration of the full turn of the steering wheels from $-\phi_{max}$ to ϕ_{max} or vice versa, i.e. one can choose $T^* = T^*_{min}$.

The value of T is lower-bounded by the constraints on the velocity v_{max} and acceleration $\dot{v}_{max}$ and by the condition $T^* < T$. When the control command (10) is applied, the lower bound of T is

$$T_{min} = \max\left\{\frac{2\pi\, v'(D1)}{\dot{v}_{max}},\, T^*\right\}, \qquad (14)$$

where the empirically-obtained function $v'(D1) \leq v_{max}$ serves to provide a smooth motion of the vehicle when the available distance D1 is small.

The computation of T and ϕ_{max} aims to obtain the maximal values such that the following "longitudinal" and "lateral" conditions are still satisfied:

$$|(x_T - x_0)\cos\theta_0 + (y_T - y_0)\sin\theta_0| < D1, \qquad (15)$$

$$|(x_0 - x_T)\sin\theta_0 + (y_T - y_0)\cos\theta_0| < D2. \qquad (16)$$

Using the maximal values of T and ϕ_{max} assures that the longitudinal and, especially, lateral displacement of the vehicle is maximal within the available free parking space. The computation is carried out on the basis of the model (1) when the commands (9) and (10) are applied. In this computation, the value of v_{max} must correspond to a safety requirement for parking maneuvers (e.g. $v_{max} = 0.75\, m/s$ was found empirically).

At each iteration i the parallel parking algorithm is summarized as follows:

1. Obtain available longitudinal and lateral displacements D1 and D2 respectively by processing the sensor data.

2. Search for maximal values T and ϕ_{max} by evaluating the model (1) with controls (9), (10) so that conditions (15), (16) are still satisfied.

3. Steer the vehicle by controls (9), (10) while processing the range data for collision avoidance.

4. Obtain the vehicle's location relative to environmental objects at the parking bay. If the "parked" location is reached, stop; else, go to step 1.

When the vehicle A1 moves backwards into the parking bay from the start location shown in Fig. 3, the corner A12 (front right corner of the vehicle) must not collide with the corner B24 (front left corner of the bay). The start location must ensure that the subsequent motions will be collision-free with objects limiting the bay. To obtain a convenient start location, the vehicle has to stop at a distance D3 that will ensure a desired minimal safety distance D5 between the vehicle and the nearest corner of the bay during the subsequent backward motion. The relation between the distances D1, D2, D3, D4 and D5 is described by a function

$$\mathcal{F}(D1, D2, D3, D4, D5) = 0. \qquad (17)$$

This function can not be expressed in closed form, but it can be estimated for a given type of vehicle by using the model (1) when the commands (9) and (10) are applied. The computations are carried out off-line and stored in a look-up table which is used on-line, to obtain an estimate of D3 corresponding to a desired minimal safety distance D5 for given D1, D2 and D4 [7].

When the necessary parking "depth" has been reached, some clearance between the vehicle and the parked ones is provided, i.e. the vehicle moves forwards or backwards so as to be in the middle of the parking bay between the two parked vehicles.

4. Experiments

The developed methods have been tested on an experimental automatic vehicle designed on the base of a LIGIER electric car. This is a four-wheeled vehicle with the front driven and steering wheels. The vehicle can either be driven as a car, or it can move autonomously. To allow autonomous motions, the vehicle is equiped with a control unit based on a Motorola VME162-CPU board and a transputer net. The sensor unit of the vehicle consists of ultrasonic range sensors (Polaroid 9000) and a linear CCD-camera. The steering wheel servo-system is equiped with a direct current motor and an optical encoder to measure the steering angle. The locomotion servo-system of the vehicle is equiped with 12 kW asynchronous motor and two optical encoders at the rear wheels to provide data on locomotion velocity of the vehicle. The vehicle also has an hydraulic braking servo-system. The developed steering and velocity control is implemented using ORCCAD software [8] running on a SUN workstation. The compiled code is transmitted via Ethernet to the VME162-CPU board.

An example of our experimental setup for lane following/changing on a circular road is shown in Fig. 4. The LIGIER vehicle has to follow a nominal trajectory along the circular traffic lane where another vehicle is moving at a lower velocity in front of LIGIER, as shown in Fig. 4a. When the obstacle is detected, a local trajectory for a lane change to the right is generated to avoid collisions, and LIGIER performs the lane changing maneuver, as illustrated in Fig.4b. Then, LIGIER moves in parallel to its nominal trajectory until the obstacle is overtaken. Further, a new local trajectory for a lane change to the left is generated, and LIGIER performs the lane changing maneuver to return to its nominal trajectory, as shown in Fig. 4c. Finally, LIGIER continues to follow its nominal trajectory, as illustrated in Fig. 4d.

An example of the control commands of the steering angle and locomotion velocity during the lane following/changing maneuvers on a circular road is shown in Fig. 5. The corresponding motion of the vehicle is depicted in Fig. 6 where the nominal circular trajectory and the local one are plotted. The vehicle performs a lane change to the right, moves in parallel to the nominal trajectory and performs a lane change to the left to catch up with its nominal trajectory. The locomotion velocity of the vehicle is increased when it moves along the local trajectory: as it is illustrated in Fig. 6, the duration of the motion along the local trajectory equals the duration of the motion along the nominal trajectory without the lane changing maneuvers.

Figure 4. Sequence of motions for lane following/changing on a circular road: a - following the nominal trajectory, b - lane changing to the right and overtaking, c - lane changing to the left, d - continuing with the nominal trajectory

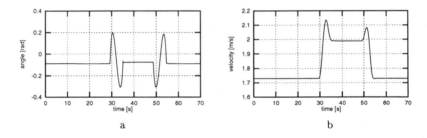

Figure 5. Control commands during lane following/changing: a - steering angle, b - locomotion velocity

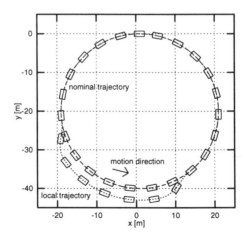

Figure 6. Lane following/changing on a circular road

An example of our experimental setup for autonomous parallel parking in a street is shown in Fig. 7. Autonomous parking can be carried out in an environment where there are moving obstacles, e.g. a pedestrian and another vehicle. As shown in Fig. 7a, the parking bay is in front of LIGIER at its right side between the two vehicles. Initially, LIGIER was driven to a position near the bay, the driver started the autonomous parking and left the vehicle. Then, LIGIER moves forwards autonomously in order to localize the parking bay, obtain a convenient start location and perform a parallel parking maneuver. When during this motion a pedestrian crosses the street in a dangerous proximity to the vehicle, as shown in Fig. 7a, this moving obstacle is detected, LIGIER slows down and stops to avoid the collision. When the way is free, LIGIER continues its forward motion. Range data is used to detect the parking bay. A decision to carry out the parking maneuver is made and a convenient start position for the initial backward movement is obtained, as shown in Fig. 7b. Then, LIGIER moves backwards into the bay, as shown in Fig. 7c. During this backward motion, the front human-driven vehicle starts to move backwards, reducing the length of the bay. The change in the environment is detected and taken into account. The range data shows that the necessary "depth" in the bay has not been reached, so further iterative motions are carried out until it has been reached. Then, LIGIER moves to the middle between the rear and front vehicles, as shown in Fig. 7d. The parallel parking maneuver is completed.

An example of the control commands (9) and (10) for parallel parking into a bay situated at the right side of the vehicle is shown in Fig. 8. The corresponding motion of the vehicle is depicted in Fig. 9 where the motion of the corners of the vehicle and the midpoint of the rear wheel axle is plotted. The available distances are D1=4.9 m, D2=2.7 m relative to the start location of the vehicle. The lateral distance D4=0.6 m was measured by the sensor unit. The longitudinal distance D3=0.8 m was estimated so as to ensure the minimal safety distance D5=0.2 m. In this case, five iterative motions are performed to park the vehicle. As seen in Fig. 8 and Fig. 9, the durations T of the iterative motions, magnitudes of the steering angle ϕ_{max} and locomotion velocity v_{max} correspond to the available displacements D1 and D2 within the parking bay (e.g. the values of T, ϕ_{max} and v_{max} differ for the first and last iterative motion).

The developed methods of motion generation and control for the lane following/changing and parallel parking maneuvers were tested. Because the vehicle is equiped with very simple ultrasonic sensors, only low-speed motions were allowed during the experiments. Also, small vertical objects such as posts can not be detected reliably. The execution of the maneuvers was found to be quite sensitive to the calibration of the steering wheel servo-system. To avoid accumulation of errors when computing the position and orientation of the vehicle during the lane following/changing maneuvers, landmarks were to be used. In the future, the experimental vehicle will be equipped with a more advanced sensor system.

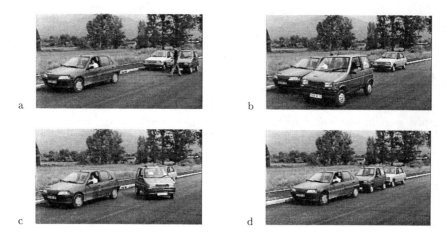

Figure 7. Sequence of motions for parallel parking: a - autonomous motion to localize a parking bay, b - obtaining a convenient start location, c - backward motion into the bay, d - parallel parking is completed

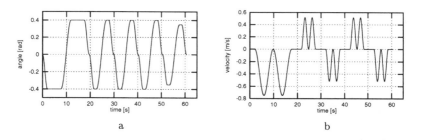

Figure 8. Control commands for parallel parking when backward and forward motions are performed: a - steering angle, b - locomotion velocity

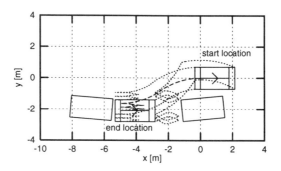

Figure 9. Parallel parking when backward and forward motions are performed

5. Conclusion

Motion generation and control methods to perform autonomous lane following/changing and parallel parking maneuvers were developed. Autonomous maneuvers were considered for a nonholonomic vehicle within a structured dynamic environment. The vehicle's constraints were taken into account to obtain feasible trajectories and control commands for the vehicle. The methods developed were implemented on an automatic electric vehicle and experimentally verified. The results obtained show the effectiveness of the developed methods of motion generation and control for autonomous maneuvers.

Acknowledgement

The authors would like to thank the members of the PRAXITELE and SHARP teams for their support during this work.

References

[1] Parent M, Daviet P 1996 Automated urban vehicles: towards a dual mode PRT (Personal Rapid Transit). *Proc. of the IEEE Int. Conf. on Robotics and Automation*, Minneapolis, USA, April 22-28, 1996, pp 3129-3134

[2] Latombe J-C 1991 *Robot Motion Planning.* Kluwer Academic Publishers. Boston, Dordrecht, London.

[3] Garnier Ph, Fraichard Th 1996 A fuzzy motion controller for a car-like vehicle. *Proc. of the IEEE/RSJ Int. Conf. on Intelligent Robots and Systems*, Osaka, Japan, November 4-8, 1996, pp 1171-1178

[4] Kanayama Y, Kimura Y, Miyazaki F, Noguchi T 1991 A stable tracking control method for a non-holonomic mobile robot. *IEEE/RSJ Int. Workshop on Intelligent Robots and Systems*, Osaka, Japan, November, 1991, pp 1236-1241

[5] Nelson W L 1989 Continuous curvature paths for autonomous vehicles. *Proc. of the IEEE Int. Conf. on Robotics and Automation*, Scottsdale, Arizona, USA, May, 1989, pp 1260-1264

[6] Paromtchik I E, Laugier C 1996 Motion generation and control for parking an autonomous vehicle. *Proc. of the IEEE Int. Conf. on Robotics and Automation*, Minneapolis, USA, April 22-28, 1996, pp 3117-3122

[7] Paromtchik I E, Laugier C 1996 Autonomous parallel parking of a nonholonomic vehicle. *Proc. of the IEEE Intelligent Vehicles Symp.*, Tokyo, Japan, September 19-20, 1996, pp 13-18

[8] Simon D, Espiau B, Castillo E, Kapellos K 1993 Computer-aided design of a generic robot controller handling reactivity and real-time control issues. *IEEE Trans. on Control Systems Technology*, December, 1993, pp 213-229

How to implement dynamic paths

M. Khatib, H. Jaouni, R. Chatila, J.-P. Laumond
LAAS-CNRS
7, Avenue du Colonel Roche, 31077 Toulouse Cedex 4 France

Abstract:

Sensor-based dynamic path modification is one of powerful issues to combine planning and reactive control. This paper treats problems related to flexible trajectories implementation and presents solutions to connect functional levels from the planning to the execution. A general dynamic path structure is proposed and discussed including execution needs and constraints. Implementation algorithms are then developed and presented for general metrics and execution contexts. Finally, solutions for car-like mobile robots are encapsulated in a complete system joining these different levels and mange there functional coherence.

1. Introduction

A robot should be able to plan its motion in order to reach its goal configuration as efficiently as possible. However, this motion planning being necessarily based on an environment model, new obstacles in the environment can make a planned path obsolete. Replanning a new path is in general costly, and unnecessary especially if the contingent obstacle can be avoided locally. In this case however, local obstacle avoidance must be guided to recover the planned path. A purely local avoidance may lose the connectivity of the planned path.

One approach to the problem of taking into account contingent obstacles is to consider the global path as being a deformable structure. The advantage of such an approach is also that it can deal with mobile obstacles in general, and the influence of local obstacle can also be considered consistently for the remaining of the path.

This approach was initially proposed and developed for holonomic robots by S. Quinlan and O. Khatib [5]. Their method, called the *elastic band*, consists in maintaining a sequence of connected local subsets of the free-space called *bubbles*. The resulting flexible path is then the concatenation of paths joining bubble centers.

Based on this work, we have adapted this method to car-like non-holonomic robots [3]. The main differences between the original method and our variant can be summed up in three points.

Firstly, the original method is mainly defined for holonomic robots, the contribution of the metric structure of feasible paths associated to the robot was not wholly specified. We use the metric space defined on feasible paths of non-holonomic robots to define bubbles as well as to specify the interactions between them.

The **second** point concerns the distance to obstacles. The original method uses the Euclidean distance, thus over-constraining the size of bubbles. We use as well a distance defined on feasible paths to obstacles. We have thus two alternatives and a treatment differentiating between mobile and immobile obstacles should be defined.

Lastly, band optimization and reactivity are completely based on convex potential functions, such as defined in [1], thus ensuring band stability without introducing restriction constraints as in the original method.

Dynamic path structure

As a result of our improvements as well as of the original concept of bubble band we define a dynamic path DP to be a connected sequence of primitives, each joining two configurations in free space. These primitives may be of several kinds such as geometric operations, sensor-based movements, ... Here we are interested in primitives which can be presented as subsets (bubbles) surrounding configurations (bubbles' centers) in a configuration space. A connected DP becomes a connected sequence of bubbles where the center of the first bubble is the start configuration and the center of the last bubble is the goal configuration. Such a DP is called a bubble band BB.

If a feasible trajectory can join the centers of any two connected bubbles with arbitrary start and finish speeds, in some fixed range; then the concatenation of such feasible trajectories along the BB is a feasible trajectory between the start and goal configurations.

A DP interacts with the surrounding environment via forces issued from potentials, function of the distance between adjacent configurations Dis_CC and the distance to obstacles Dis_CO; and applied to the configurations defining the DP. In the case of a BB these configurations are reduced to the centers of the bubbles.

A DP should always verify the following properties:

- It is collision-free;
- It is feasible;
- It converges asymptotically under some criterium of optimality in a static environment.

This definition will be used to create and maintain a permanent feasible and collision-free path joining given start and goal configurations. After defining such a path, we present also an execution system architecture (Kinan system) that exploits the dynamic structure of the path.

2. Solutions and implementation

Bubble bands BB have the advantages of simplifying the treatment of the free-space and of reducing the form of forces from functionals on continuous curves to functions on finite sequences of configurations. Thus the rest of this section will be consecrated to the definition and properties of BBs.

2.1. Bubble bands

As developed in [4], a bubble is the maximum local subset of the free-space around a given configuration of the robot which can be traveled in any direction without collision. To define a bubble more precisely, we use a metric space (a configuration space and a distance Dis_CC defined on it) based on elementary feasible paths, i.e. paths in the absence of obstacles, and a distance to obstacles function (Dis_CO) [3]. A bubble is thus characterized by its center, the configuration around which it is defined. Its radius is defined to be the distance to the obstacles. We note $B(\mathbf{p}, r)$ a bubble of center $\mathbf{p}$ and radius r.

A bubble band is a sequence of connected bubbles

$$\{B(\mathbf{p}_i, r_i) \mid i = 1 \ldots N_b\}$$

where for all $i = 1 \ldots N_b - 1$,

$$B(\mathbf{p}_i, r_i) \cap B(\mathbf{p}_{i+1}, r_{i+1}) \neq \phi$$

such that the first bubble of the sequence is centered on the start configuration of a path and the last bubble on the end configuration. Since the centers of two adjacent bubbles can be joined by an elementary feasible path, the global path between the start and end configurations is taken to be the concatenation of these elementary parts.

In the next subsection we shall discuss and the construction of a BB from an unnecessarily feasible path. Then the BB optimization and reactivity will be developed using potential functions.

2.1.1. Bubble band construction

Before constructing the BB, let us examine the behavior of individual bubbles.

Bubble classes Not all bubbles can be moved. The first bubble which is centered on the beginning of the path is usually considered to be immobile. We may also need to oblige the path to pass by a determined configuration. An immobile bubble is then centered on it. One special bubble is used to represent (track) the robot on the path and thus its movement is governed by that of the robot.

This leads us to distinguish between three classes of bubbles: mobile, immobile and track bubbles which are noted B_M, B_I and B_T respectively.

Bubble band construction Suppose we have a sequence of configurations $\{q_i, i = 1\ldots N_c\}$, which can be the result of a sequencer acting on some sort of path. The first bubble is created around the first configuration

$$B(\mathbf{p}_1, r_1) = B_I(\mathbf{q}_1, \text{Dis\_CO}(\mathbf{q}_1)).$$

It is an immobile bubble. Once a bubble is defined, e.g. $B(\mathbf{p}_i, r_i)$, a new mobile one is centered on the last configuration still belonging to it

$$B(\mathbf{p}_{i+1}, r_{i+1}) = B_M(\mathbf{q}_j, \text{Dis\_CO}(\mathbf{q}_j)),$$

where

$$\mathbf{q}_j \in B(\mathbf{p}_i, r_i) \quad \text{and} \quad \mathbf{q}_{j+1} \notin B(\mathbf{p}_i, r_i).$$

The recursion ends when the last configuration is reached

$$B(\mathbf{p}_{N_b}, r_{N_b}) = B_I(\mathbf{q}_{N_c}, \text{Dis\_CO}(\mathbf{q}_{N_c})).$$

The last bubble is also considered immobile.

This method guarantees the connectivity of the band and its presence in free space. Nevertheless, the overlapping of the bubbles may not be optimal. This redundancy is minimized once the creation-elimination method is applied.

A BB is created. From it a feasible path can be deduced. Already we have accomplished an important task. That of approximating any path by a feasible one. The next step will make this path dynamic.

2.1.2. Bubble band optimization and reactivity

As said earlier, the BB interacts with the environment via forces. Since the state of the band determines exactly the path associated to it, these forces are applied to the bubbles and thus to their centers. To guarantee the convergence of the band in a stabilized environment these forces are deduced from potential fields.

These forces are grouped in two sets: internal forces which tend to optimize the form of the band and external forces which act to avoid collision and guarantee the evolution in free space.

Internal forces These forces are local to the band. Each bubble exerts a force on its neighbors to keep them connected to it while preventing them from overlapping too much with it. But the most important force is that responsible for optimizing the form of the band. This force tries to align each bubble with its two adjacent ones.

It is shown in [3] that two bubbles $B(\mathbf{p}_i, r_i)$ and $B(\mathbf{p}_{i+1}, r_{i+1})$ are connected if and only if

$$r_i + r_{i+1} \leq \text{Dis\_CC}(\mathbf{p}_i, \mathbf{p}_{i+1}) - \epsilon_c$$

for all $\epsilon_c > 0$. On the other hand, two bubbles are said to overlap too much if

$$r_i + r_{i+1} \leq \text{Dis\_CC}(\mathbf{p}_i, \mathbf{p}_{i+1}) - \epsilon_o$$

for some $\epsilon_o > \epsilon_c$. Through a potential field deduced from these conditions

$$\mathcal{P}_f(\mathbf{p}_i) = \frac{K_f}{2} \Big(\text{Dis\_CC}(\mathbf{p}_i, \mathbf{p}_{i+1}) - (r_i + r_{i+1}) + \epsilon_c\Big)\Big(\text{Dis\_CC}(\mathbf{p}_i, \mathbf{p}_{i+1}) - (r_i + r_{i+1}) + \epsilon_o\Big),$$

the following force on $\mathbf{p}_i$ exerted by $\mathbf{p}_{i+1}$ is derived

$$\mathbf{F}_f(\mathbf{p}_i) = K_f \Big(\text{Dis\_CC}(\mathbf{p}_i, \mathbf{p}_{i+1}) - (r_i + r_{i+1}) + \frac{\epsilon_c + \epsilon_o}{2}\Big) \frac{\partial}{\partial \mathbf{p}_i} \text{Dis\_CC}(\mathbf{p}_i, \mathbf{p}_{i+1}).$$

The action of $\mathbf{p}_{i-1}$ on $\mathbf{p}_i$, noted $\mathbf{F}_b$, is just the symmetrical expression.

The last force acts to align a bubble with its neighbors. Consider $B(\mathbf{p}_{i-1}, r_{i-1})$, $B(\mathbf{p}_i, r_i)$ and $B(\mathbf{p}_{i+1}, r_{i+1})$ to be three consecutive bubbles. A string should attach $\mathbf{p}_i$ to a configuration on the segment $\overline{\mathbf{p}_{i-1}\mathbf{p}_{i+1}}$. This configuration is chosen to be the radius-weighted barycenter of the adjacent configurations. This induces the force

$$\mathbf{F}_c(\mathbf{p}_i) = K_c \left(\text{Dis\_CC}\left(\mathbf{p}_i, \frac{r_{i-1}\mathbf{p}_{i-1} + r_{i+1}\mathbf{p}_{i+1}}{r_{i-1} + r_{i+1}}\right)\right) \frac{\partial}{\partial \mathbf{p}_i} \text{Dis\_CC}\left(\mathbf{p}_i, \frac{r_{i-1}\mathbf{p}_{i-1} + r_{i+1}\mathbf{p}_{i+1}}{r_{i-1} + r_{i+1}}\right).$$

External forces External forces, whose resultant is noted $\mathbf{F}_e$, are due to obstacles. They can be deduced from potential functions such as defined in [1]. Such forces are expressed using a distance to obstacles. If the obstacles are immobile this distance is naturally Dis_CO. For a mobile obstacle, the minimum between the distance associated to it and Dis_CO may be used. This supposes that the robot knows what distance is associated to every mobile obstacle. This is not always true, especially for unknown obstacles. In this case the safest way is to use the Euclidean distance. To simplify this, we use Dis_CO in the expression of forces for immobile robots and the Euclidean distance, noted Dis_E, for mobile robots.

Bubbles movement Once all the forces are calculated, the band is modified by applying to each center the resultant of the forces acting on it. This induces a movement in the direction of the force. Here the passage between the force and the movement is discussed and then the criteria to be verified by the movement are presented.

Suppose the mass of the robot is a unity. This does not affect the generality of the method since the mass can always be taken account of through the force coefficients K_f, K_b, ... All the forces are derived from potentials and are thus conservative, whence it is sufficient to add some dissipative term to the resultant of the forces to guarantee the asymptotic convergence of the movement. The total force is then

$$\mathbf{F}(\mathbf{p}_i, \dot{\mathbf{p}}_i) = \mathbf{F}_f(\mathbf{p}_i) + \mathbf{F}_b(\mathbf{p}_i) + \mathbf{F}_c(\mathbf{p}_i) + \mathbf{F}_e(\mathbf{p}_i) - K_v \dot{\mathbf{p}}_i.$$

To translate this force into a displacement, we consider it constant for some interval dt. Thus

$$\Delta \mathbf{p}_i = dt \dot{\mathbf{p}}_i + \frac{dt^2}{2} \mathbf{F}(\mathbf{p}_i, \dot{\mathbf{p}}_i)$$

and
$$\Delta \dot{\mathbf{p}}_i = dt \mathbf{F}(\mathbf{p}_i, \dot{\mathbf{p}}_i).$$

Each bubble should now effectuate a displacement. Nevertheless, this displacement should guarantee some conditions. For the center of a bubble to remain always in the free space, the new center should remain in the old bubble. Displacements are thus truncated on the border of each bubble. Also a test of connexity should be done before and after the displacement. Bubbles may be inserted at these moments. At last redundant bubbles should be eliminated.

In short, the main loop is

For each bubble :
> Calculate resultant force.
> Test for connexity with previous and next bubbles and insert bubble if necessary.
> Effectuate displacement within the boundary of the bubble.
> Test for connexity with previous and next bubbles and insert bubble if necessary.
> If the bubble becomes redundant remove it from the band.

Freeze and advance Freeze_P The dynamic property of the band so convenient to react to the environment must be taken into account at the time of execution. A natural way to do so, while keeping as much of the path dynamic as possible, is to freeze a part of it; whence the following operation, which has the advantage of not inducing discontinuous effects on the forces applied on the band.

Suppose the existence of a bubble of class track whose center is the last configuration on the previously frozen part. Initially, it is centered on the start configuration. This bubble is immobile except when a part is being frozen. When a part is frozen on the band, all the bubbles it crosses are made immobile. The other bubbles behavior does not change except for the one next to the track bubble whose position is now on the end of the newly frozen part. It reacts now to the track bubble. Suppose this bubble is $B(\mathbf{p}_i, r_i)$ and call the track bubble $B(\mathbf{p}_t, r_t)$, then the force $\mathbf{F}_b$ due to the previous (track) bubble is modulated by the coefficient

$$\frac{\mathtt{Dis\_CC}(\mathbf{p}_i, \mathbf{p}_t)}{\mathtt{Dis\_CC}(\mathbf{p}_i, \mathbf{p}_{i-1})},$$

where $B(\mathbf{p}_{i-1}, r_{i-1})$ is the bubble before the track bubble and thus the last immobile bubble. Thus the modification of the force on $B(\mathbf{p}_i, r_i)$ is attenuated.

Discussion

As seen above, the BB is continuously modified under the action of forces applied to it. If these forces evolute smoothly with time, the behavior of the BB guarantees the three properties of DPs. Unfortunately, this is not always the case especially when a new obstacle is suddenly perceived that intersects the band or is too close to it. Such an event may produce a discontinuity in the band as well as the in forces applied to it.

Let us treat the discontinuity due to the presence of such an obstacle, the general problem being easy deduced from it afterwards. To smooth out its effects, the new obstacle is not taken into account directly but is approximated by a mobile obstacle moving towards it from a certain direction function of the position of the obstacle relative to the band.

This direction which is depends strongly on the obstacles modeling method, may be, for polygonal model, the direction of the barycenter of the obstacle relative to the nearest point to it on the path associated to the BB at the moment of perception of the new obstacle.

As for the general case, a discontinuity in the force may be smoothed by a continuous deformation from the previous value to the new by the bias of a monotonous differentiable function such as proposed in [1].

2.2. Execution controller (kinan)

Having constructed a dynamic path using the bubble band technique, we will discuss in this section the different aspects related to its execution. In fact, the BB is a dynamic system, evolving endlessly in function of environment changes. To execute the mobile output path, we have to satisfy these following constraints:

- The current executed configurations on the DP are, either dynamically joinable to the current robot state or immobile.
- Sensor-based path modification is coherent (correctly synchronized) with the current robot state in function of environment changes.
- Collision-free robot stop is guarantee, in case of path modification fail (obstacle avoidance fail).

Beforehand we suppose the existence of a parameterization unit able to compute the robot vector state profile on a given part of the path in function of time. Here the path is that issued from the corresponding metric space. On the other hand, a command module is supposed available to execute the parameterized part of the path. The role of the execution controller (kinan) is now to link the three modules (bubble band, parameterizer, and command module), while satisfying the above mentioned constraints.

The key idea of kinan system is based on the Freeze_P operator of the bubble band module. Within this operator, the BB is able to freeze a part, on the global path, defined by its length. Actually, the initial configuration of the frozen part is automatically considered as the center of the last immobile bubble. In this way, the global path is thus decomposed into two parts, the first is immobile and the second rest flexible. We can note that this operation guarantees the path continuity at the joint configuration. If we suppose that the parameterizer is able to compute the robot vector state profile between any two states, we will discuss now how to compute the part length to satisfy robot kinematic constraints, cycle time compatibility between modules and how to deal with the avoidance fail problem.

2.2.1. Partition

Suppose that T_b, T_p, T_c are respectively the cycle time of BB, parameterization and command modules. Where, T_b is the time necessary to update BB which includes distance computation, force computation, bubbles movement and connection check. T_p is the parameterization time of a part of the path and T_c is the robot command level period. In fact, it's easy to remark that T_b is considerably bigger than T_p and T_c. Moretheless, if we denote by T_u the necessary communication time to handle all data flow between modules updating thus a global cycle time of kinan system, we can deduce the first following temporal constraint:

$$T_{\text{part}} > T_b + T_p + T_c + T_u$$

where T_{part} is the needed time to execute the frozen part. This constraint expresses the minimum frozen part length, note ℓ_f, for a given nominal execution speed. On the other hand, given the maximum acceleration, the length ℓ_f should be enough for the robot to be able to stop in case of natural stop (cusp point, end of the path, ...) or an emergency. This is not all the truth, since a thought should be given to the rest of path. In fact, the robot should always move at the nominal desired speed that its path form allows. Then the remaining length on the path should be greater than the minimal length in which the robot can stop completely if it started this part with the nominal speed. If this is not verified the rest part of the path has to be frozen also.

Suppose now that the path is always available in a some memory zone, called poster, updated by the bubble band module. From now on when a part is said to be frozen, its length is determined satisfying the previous constraints and a **Freeze_P** request is sent to BB (figure 1). The BB updates then the flexible rest of path in its own poster.

Parameterization For a frozen part to be parameterized, its final speed should be determined. To do this, a forward look should be effectuated to test for a stop point (such as a cusp point for mobile robots) or the end of the path. In both cases the final speed is null. Otherwise it is the maximum speed variation that the parameterization algorithm can afford to reach the desired speed.

Command module In order to simplify communications handling operations we will suppose that the command module (say command) has the following behavior: the execution of the path is activated by a Track request. This last waits till one or more new parameterized parts are written in the execution controller poster (kinan poster).

The execution of the first parameterized part is started only if the last one ends with a null speed. While executing a part, the command updates robot state, part identifier, and especially the current execution time in its own poster. The execution continue while kinan poster is updated with new parameterized parts.

2.2.2. kinan execution algorithm

Having defined the different requests and operators, we will present here the general algorithm that manage the execution.

Initialization
    $s_i = 0$
    $v_i = 0$
-- *Execution time, initial speed and kinan poster are initialized.*
    Track
-- *A Track request is sent to the command.*
-- *The current execution time s_c is reset.*
Start
    While (Path $\neq \emptyset$)
        Part$_1$ = Freeze_P   -- *Freeze a part for the nominal execution.*
        Get v_f, s_f -- *v_f is the final speed and s_f is the expected final execution time.*
        Parameterize (Part$_1, v_i, v_n$)   -- *v_n is the nominal speed.*
        If $v_f \neq 0$ Then
            Part$_2$ = Freeze_P   -- *Freeze a part to stop.*
            Parameterize (Part$_1, v_n, 0$)   -- *Parameterize this part to stop.*
        End If
        Do
            Read s_c
        While $(s_i - s_c > \epsilon)$
        Write Part$_1$ and/or Part$_2$
        $v_i = v_f$
        $s_i = s_f$
    End While
End

The figure 1 shows the control architecture of Kinan system implementation.

Discussion
Related to the application presented in the next section, this algorithm presents one drawback which is mostly due to parameterization constraints. As will be seen later, limit speeds

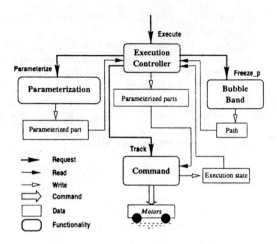

Figure 1. *Architecture of* Kinan *system*

on the ends of a part of a path are bounded by the parameterization module in function of the geometry of the part. Thus the initial speeds given to the parameterizer may be infeasible in some cases. To avoid this Path$_2$ can be parameterized before Path$_1$ with the nominal speed at the beginning and a stop at the end. The effective initial speed accorded by the parameterizer is then taken for the final speed of Path$_1$. This however speeds down the executor and should be avoided if possible. Nevertheless, we had to implement it on the car-like robot application presented next.

3. Application to car-like mobile robot

In this section we present the spesification of the general bubble band method to the case of the non-holonomic car-like robot with a minimum turning radius. First the associated metric space is discussed introducing thus the several primitives needed by the method. Then smoothing (which being so metric dependant was not incorporated in the genral method) and parametrization needed by execution are resumed. Having presented these tools, the general method can be automatically applied to this case and consequently is not detail in what follows.

3.1. Metric related methods

In this context, a configuration of the robot is the triplet consisting of its position and its orientation $(x, y, \theta)^T$ defined in the space $\mathcal{R}^2 \times \mathcal{S}^1$.

Distance configuration to configuration (Dis_CC) It has been proved in [6] that the shortest path between two configurations belongs to one of 48 simple sequences of at most five pieces, each piece being a straight line segment S or an arc of a circle C with at most two cusp points. The length of the shortest paths induces the configuration space $\mathcal{R}^2 \times \mathcal{S}^1$ with a metric structure compatible with the robot movement [2]. Figure 3.1-a. shows the projection of such a bubble on the movement plane. As expected positions at the side of the robot are farther than those on the movement axis which reflects the robot non-holonomic constraints.

Distance configuration to obstacles (Dis_CO) Based on [7], the problem of the distance between a robot configuration and a point or segment obstacle was solved [8]. This distance is noted Dis_CO : $(\mathcal{R}^2 \times \mathcal{S}^1) \times (\mathcal{R}^2 \times \mathcal{R}^2) \longrightarrow \mathcal{R}_+$.

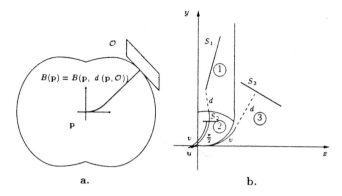

Figure 2. a. *Biggest collision-free bubble.* b. *Distance to segment in the different domains.*

Figure 3.1-b. shows the shortest paths from the origin of the configuration space to three segments presenting thus the three kinds of configuration to point shortest paths.

Gradient (Grad_Dis_CO) We have seen in section 2.1.2 that the path modification is achieved using potential fields which are functions of distance. The application of the $-\nabla$ operator to derive the force expressions from the previously defined potentials, brings us to calculate $\frac{\partial d}{\partial \mathbf{p}}$ where $\mathbf{p}$ is the configuration that varies. In fact, this last expression gives the direction of the force. Unlike the Euclidean distance, the car-like compatible distances, defined previously, have no normalized differentials. They contribute thus to the magnitude of the force. As well, they do not vary smoothly with $\mathbf{p}$, with consequent effects on stability.

	Path	$\dfrac{\partial d}{\partial x}$	$\dfrac{\partial d}{\partial y}$	$\dfrac{\partial d}{\partial \theta}$
①	$l_u^- l_{\frac{\pi}{2}}^+ s_d^+$	$\sin u$	$-\cos u$	$R \sin u - R$
②	$l_u^- l_v^+$	$\dfrac{\sin u}{\sin v}$	$-\dfrac{\cos u}{\cos v}$	$R\dfrac{\sin u}{\sin v} - R$
③	$l_v^+ s_d^+$	$-\cos v$	$-\sin v$	$R \cos v - R$

Table 1. *Derivatives of the distance from configuration to point.*

Table 1 shows derivatives of the distance Dis_CO which is here the distance from configuration to point. We note that the derivatives are continuous functions of x, y and θ almost everywhere except on the arc of center $(0, R)^T$ and of radius R from $(0,0)^T$ to $(R, R)^T$ separating the domains ② and ③. l is a left turn, r is a right turn and s is a straight line segment; The super-fixes + and − indicate forward and backward movements respectively and the indexes are the parameters, angles for turns and distances for segments (see fig. 3.1-b.).

Path between configurations (Path_CC) As a direct result of the fact that Dis_CC is a metric distance and of the deffinition of bubbles and of bubbles connectivity, the centers of two connected bubbles are joinable by a Reeds and Shepp path belonging to the union of both (operator Path_CC). This Reeds and Shepp (figure 3) path is intrinsicly parameterised by its length and thus a point on a path is easily defined by its metric distance from one

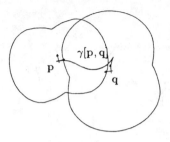

Figure 3. *Path between two consecutive centers*

end (Config_P).

Thus having defined the required primitives, the bubble band method described earlier is implemented and an example will be presented in the last section.

3.2. Smoothing and Parameterization

This paper is not only interested in generalising the bubble band and redefining it in the context of a metric space, it is also interested in presenting a complete system of dynamic path construction and execution. Before executing a path it sould be parameterized in order to respect the speed and acceleration limits of the robot. This being detailed in [3], will be recalled rapidly in this section.

Path smoothing As said earlier, the shortest path between two configurations is a sequence of straight line segments and arcs of circles. It is thus feasible by a car-like robot at the cost of stopping at each discontinuous switch of path curvature. To avoid this, the different parts of the path are approximated by Bezier curves smoothing the curvature at junctions. The advantages of using Bezier curves with respect to other technics (splines, clothoïdes, ...) are that they can be wholly define in a convex envelop as well as their derivatives. We can thus guarantee that the path executed lies in the union of the bubbles, thus in free-space. As well, the velocity and acceleration on these curves are bounded, for each belongs to a convex envelop of a finite number of constant vectors. As well they present a honogenuous representation of the several parts of the path (global parameterization).

Path parameterization Having now a smooth path (in fact, a part to parametrize fixed by the execution), a trajectory is defined on it, verifying linear and angular speed continuity and limit values, via a reparameterization. In our implementation we have opted for polynomial parameterization. Having to verify four constraints (the linear and angular speed limits at both ends of the path) parameterization by cubic polynomials was used. This imposes bounds on limit speeds in order to insure that the speeds does not exceed their maxima along the path. Thus the shape of the path influences indirectly the speed by which the robot follows it. The parameterizer then requires the initial and final requested speeds and returns besides the parameterization the initial and final effective speeds (An inconvenience to be handled by the executor as discussed above).

4. Experimentation

The car-like bubble band and the execution method were implemented on Hilare 2 robot and several experiments were effectuated to test and validate the method. Here we present a simple experiment to illustrate the behavior of the bubble band.

The first pavement of an arbitrary trajectory (figure 4-a.) are usually redundant. This is immediately cured by the application of internal forces which tend also to optimize the initial path.

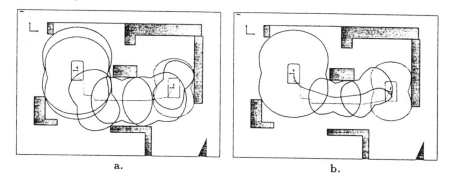

Figure 4. a. *Initial holonomic path.* b. *Application of internal and external forces and clearance from obstacles.*

Figure 4-b. shows the result of the application of internal and external forces on the band. The resultant path is not only geometrically feasible but it is also simpler than the initial one. In fact the bubble band tends to minimize a cost which is a complex combination of obstacle repulsion and path length. Thus the resultant path is usually simpler (at least shorter) and safer than the initial one.

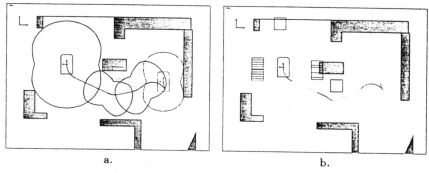

Figure 5. a. *Path modification.* b. *Execution path.*

An expected obstacles (the transparent square in the middle of figure 5-a.) modifies the band locally and this disturbance is propagated along the whole of the band losing influence as is get farther.

Figures 5-b. and 6-a. show the executed path and the execution trace respectively. An infinitismal cusp point may be perceived after passing the mobile obstacle enabling the robot to re-take the direction of the goal.

Finally, figure 6-b. presents the speeds, acceleration and curvature profiles of the execution.

In this experiment, Hilare 2 uses a predefined model of obstacles (grey polygons) and its own ultrasonic sensors to detect both modeled and unexpected obstacles. The robot runs under the control of kinan system and achieves real time performance at almost 0.5 meters per seconds.

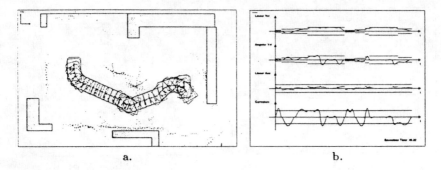

Figure 6. a. *Execution trace.* b. *Execution profile.*

5. Conclusion

The dynamic path technic presents a new methodology linking planning and execution. Thought reactivity combines the taking into account of global information to modify locally a trajectory and the propagation of such a modification to the totality of the trajectory. The supplementary advantage of this method is its capacity of adjust the trajectory to the environment diminishing thus the importance of planning to a simple topological guide. The experiments we have effectuated in this context proved both the great capacity of this approach. The bubble notion with the appropriate metric space enabled the elaboration of a unique robot-independent formulation.

The proposed execution system kinan enabled the coexistence of activities with different period cycles. The results obtained through experiments of the system confirm the necessity as well as the success of the proposed structure.

The cornerstone of this method is the definition of the distance function corresponding to the particular robot. In the car-like robot application, this distance can still be refined by taking into account the geometry of the robot by computing the real distance from the robot surface to obstacles.

On the other hand, in our experience, the output path is shown to be independent of the initial path pavement method. We still think that a theoretical proof is necessary to deduce the optimal character of the resultant **dynamic path** in presence of obstacles.

References

[1] M. Khatib. *Sensor-based motion control for mobile robots.* PhD thesis, LAAS-CNRS, Toulouse, France, December 1996. – Ref.: 96510.

[2] J.-P. Laumond and P. Souères. Metric induced by the shortest paths for a car-like mobile robot. In *IEEE International Workshop On Intelligent Robots and Systems, Yokohoma, (Japan)*, 1993.

[3] M. Khatib and H. Jaouni. Kinematics integrated in non-holonomic autonomous navigation. Technical Report 96346, LAAS-CNRS, September 1996.

[4] S. Quinlan. *Real-Time Collision-Free Path Modification.* PhD thesis, Stanford University, CS Departement, January 1995.

[5] S. Quinlan and O. Khatib. Elastic bands: connecting path planning and control. In *IEEE International Conference on Robotics and Automation, Atlanta, (USA)*, 1993.

[6] J.A. Reeds and L.A. Shepp. Optimal paths for a car that goes both forwards and backwards. *Pacific Journal Mathematics*, 145(2):367–393, 1990.

[7] P. Souères, J.-Y. Fourquet, and J.-P. Laumond. Set of reachable positions for a car. *IEEE Transactions on Automatic Control*, 39(8), August 1994.

[8] M. Vendittelli and J.-P. Laumond. Visible position for car-like robot amidst obstacles. In *Workshop on Algorithmic Foundations of Robotics, WAFR'96*, July 1996.

From paths to trajectories for multi-body mobile robots

F. Lamiraux and J.-P. Laumond
LAAS-CNRS
Toulouse, France
florent,jpl@laas.fr

Abstract: This paper adresses the problem of parametrizing with respect to time a feasible path for a mobile robot towing a trailer. We propose a robust approach, deriving from previous works on minimum time parametrization of a path for a manipulator. Numerical integrations of differential equations are replaced by piece-wise defined continuous functions. Our approach has been validated by experiments.

1. Introduction

This paper deals with a critical point in the classical chain of motion planning and control for mobile robots: how to transform a feasible path (i.e., that takes into account the kinematic constraints) into a feasible trajectory (i.e., that takes into account velocity and acceleration bounds).

Previous studies of this problem in mobile robotics provide dedicated solutions (e.g., [2, 1]) which usually consist in smoothing the planned path with canonical geometric curves (e.g., clothoids, involutes of a circle, splines) verifying good properties to transform the path into a trajectory.

Such a transformation appears more difficult if one considers multi-body mobile robots. On one hand, there is no canonical curves that could play for a multi-body mobile robot a role equivalent to clothoids for a simple two-driving mobile robot (e.g., the variation of the angle of a trailer pulled by a mobile robot along a clothoid cannot be computed analytically). On the other hand, even if such canonical curves existed, designing a collision-free path planner providing such curves would be a very difficult task.

Therefore, integrating general nonholonomic path planners into real experiments requires to face the problem of transforming *any* geometric feasible path into a feasible trajectory.

The solution presented in this paper is based on studies conducted in the framework of the manipulators. Most of the approaches are based on time-optimal control (see [3] for an overview). We propose a numerical version of the algorithm described in [7] and we apply it to a mobile robot pulling a trailer. The general algorithm sets numerical problems which are addressed in this paper. Moreover, while the original algorithm deals with bounds on accelerations, we take also into account bounds on the linear and angular velocities of the robot.

The presentation of the algorithm is introduced from its application to a mobile robot pulling a trailer. Nevertheless, its scope covers more general systems.

2. Problem statement

Figure 1 shows the mobile robot Hilare equipped with its trailer. The trailer is hooked up on the top of Hilare, on the vertical axis passing through the middle point (of coordinates (x,y)) of the driving wheel axle. The distance between this point and the middle point of the trailer axle is denoted by ℓ. The direction of the robot w.r.t. a given frame is denoted by θ. The direction of the trailer w.r.t. a robot centered frame is denoted by φ. The configuration space is $\mathbf{R}^2 \times (S^1)^2$.

From the control point of view, Hilare with its trailer corresponds to the following system:

$$\begin{pmatrix} \dot{x} \\ \dot{y} \\ \dot{\theta} \\ \dot{\varphi} \end{pmatrix} = \begin{pmatrix} \cos\theta \\ \sin\theta \\ 0 \\ -\frac{1}{\ell}\sin\varphi \end{pmatrix} v + \begin{pmatrix} 0 \\ 0 \\ 1 \\ 1 \end{pmatrix} \omega$$

where v and ω are respectively the linear and angular velocities of the robot. Both velocities and associated accelerations are submitted to the bounds v_{max}, $\dot{v}_{max}$, ω_{max} and $\dot{\omega}_{max}$ which are respectively $1m/s$, $1m/s^2$, $1rd/s$ and $\frac{3}{2}\pi rd/s^2$.

Figure 1. Hilare with its trailer

The motion planning and control system is composed of three main components:

- **Nonholonomic path planning**: The path planner is based on a two-step approach; a first collision-free geometric path is computed without taking into account the nonholonomic constraints; then this path is approximated by a sequence of feasible paths computed by a nonholonomic local path planner (see [6] for details).

- **Trajectory computation**: The nonholonomic path must be transformed into a trajectory; the current paper deals with this transformation.

- **Feedback control**: At the current level of the experimentation, our control law is very simple. It is sufficient to reach the experimental objectives presented below. When the robot goes forward, the trailer is not taken into account and we stabilize the robot according to the control law described in [4]. When the robot goes backward, we define a virtual robot which is symmetrical to the real robot with respect to the wheel axle of the trailer; then we apply the same control law to the virtual robot.

3. From path to trajectory
3.1. Constraints in the phase plane

Let us consider a feasible path computed by the nonholonomic path planner. This path is composed of a sequence of forward and backward sub-paths. Each of them can be processed independently. In the rest of the paper we will consider that the path corresponds to a single forward motion, i.e., we assume $0 \leq v$. Then, the constraints we are dealing with are: $0 \leq v \leq v_{max}$, $|\dot{v}| \leq \dot{v}_{max}$, $|\omega| \leq \omega_{max}$ and $|\dot{\omega}| \leq \dot{\omega}_{max}$, where v and ω are respectively the linear and angular velocities of the robot.

The path appears as a function f of real variable s from some closed interval $[s_0, s_{end}]$ to the configuration space. Transforming the path $f(s)$ into a trajectory consists in finding a real function $s(t)$ such that $f(s(t))$, where t represents time, is a feasible trajectory with respect to the velocity and acceleration constraints above. We denote $\dot{s}$ the time derivative of s.

Let us denote by $f_x(s), f_y(s), f_\theta(s)$ and $f_\varphi(s)$ the coordinates of the configuration $f(s)$ on the path f. We then get:

$$v = \left(\frac{df_x}{ds}(s)^2 + \frac{df_y}{ds}(s)^2\right)^{\frac{1}{2}} \dot{s} \text{ and } \omega = \frac{df_\theta}{ds}(s)\dot{s}$$

Notice that the constraints on the velocities and the accelerations do not involve $f_\varphi(s)$, i.e., the trailer.

By setting $\delta f_{(x,y)}(s) = \left(\frac{df_x}{ds}(s)^2 + \frac{df_y}{ds}(s)^2\right)^{\frac{1}{2}}$ and $\delta f_\theta(s) = \frac{df_\theta}{ds}(s)$, the linear and angular accelerations of the robot are:

$$\dot{v} = \delta f_{(x,y)}(s) \ddot{s} + \frac{d(\delta f_{(x,y)})}{ds}(s) \dot{s}^2 \text{ and } \dot{\omega} = \delta f_\theta(s) \ddot{s} + \frac{d(\delta f_\theta)}{ds}(s) \dot{s}^2 \quad (1)$$

Velocity constraints: The velocity constraints $0 \leq v \leq v_{max}$ and $|\omega| \leq \omega_{max}$ imply:

$$\dot{s} \leq \text{Inf}\left\{\frac{v_{max}}{\delta f_{(x,y)}(s)}, \frac{\omega_{max}}{|\delta f_\theta(s)|}\right\}$$

This constraint is represented in the *phase plane* $(s, \dot{s})$ by a curve that bounds from above the search space; this curve is called the *saturation curve*.

Acceleration constraints: From (1), $(s, \dot{s})$ being given in the phase plane, the acceleration constraints $|\dot{v}| \leq \dot{v}_{max}$, and $|\dot{\omega}| \leq \dot{\omega}_{max}$ impose $\ddot{s}$ to belong to two intervals. Then a necessary condition for $(s, \dot{s})$ to be admissible is that the two intervals intersect (in this case the intersection interval is denoted by $[\alpha(s, \dot{s}), \beta(s, \dot{s})]$). This condition is equivalent (after computation) to:

$$\dot{s}^2 \leq \frac{\dot{\omega}_{max}|\delta f_{(x,y)}(s)| + \dot{v}_{max}|\delta f_\theta(s)|}{|\Delta|}$$

with

$$\Delta = \delta f_\theta(s) \frac{d(\delta f_{(x,y)})}{ds}(s) - \delta f_{(x,y)}(s) \frac{d(\delta f_\theta)}{ds}(s)$$

In the phase plane the second member of the inequality gives rise to a curve called the *maximal velocity curve*.

Finally, the algorithm has to build a curve remaining in the part of the phase plane upper bounded by the lower envelop of the maximal velocity curve and the saturation curve.

3.2. Algorithm and its implementation

The algorithm proposed by [7] only takes into account the maximal velocity curve (i.e., the constraints on the accelerations). In its theoretical description, the algorithm builds integral curves corresponding to maximal or minimal accelerations. It also looks for characteristic points where some continuous function changes its sign.

In our algorithm we deal first with the acceleration constraints and then we introduce the velocity constraints.

3.2.1. Acceleration constraints

In this section, we don't take into account the constraints on the velocity of the robot (i.e., the saturation curve).

Slotine's algorithm: Let us denote by g the maximal velocity curve. The interval $[\alpha(s, \dot{s}), \beta(s, \dot{s})]$ is empty if $\dot{s} > g(s)$ and $\alpha(s, g(s)) = \beta(s, g(s))$ if $\dot{s} = g(s)$ and $\delta f_\theta(s) \neq 0$ (in our case, $\delta f_{(x,y)}$ never vanishes).

$$\kappa(s) = \frac{d\dot{s}}{ds} - \frac{dg}{ds}$$

is the difference between the slope of the phase plane trajectory at the maximal velocity curve and the slope of the maximal velocity curve. We say that $(s, \dot{s})$ is an *out-point* if $\kappa(s) > 0$ and an *in-point* if $\kappa(s) < 0$.

[7] define *characteristic points* as points where a phase plane trajectory can meet the maximum velocity curve without violating the acceleration constraints. These points are of three types:

1. *zero-inertia points* correspond to points where $\delta f_\theta(s) = 0$. There, the interval $[\alpha(s, g(s)), \beta(s, g(s))]$ is not reduced to a point and the slope of the maximal velocity curve is usually discontinuous.

2. *tangent points* correspond to points were $\kappa(s) = 0$. At these points a phase curve may be tangent to the maximal velocity curve if κ is decreasing at s.

3. *discontinuous points:* The functions $\frac{d\delta f_\theta(s)}{ds}$ and $\frac{d\delta f_{(x,y)}(s)}{ds}$ are not necessarily continuous. Thus the maximal velocity curve can be discontinuous.

Slotine's algorithm builds from each characteristic point a maximal acceleration curve forward by integrating the differential equation $\ddot{s} = \beta(s, \dot{s})$ and a minimal acceleration curve backward, by integrating $\ddot{s} = \alpha(s, \dot{s})$.

The solution is then the lower envelop of these curves, the maximal acceleration curve starting from $(s_{init}, 0)$ and the maximal deceleration curve ending at $(s_{end}, 0)$. See [7] for more details.

Numerical implementation: The numerical computation of maximal acceleration curves leads to discretized functions. If the derivatives of $\alpha(s, \dot{s})$ and $\beta(s, \dot{s})$ have no known bounds, the acceleration constraints can be violated (there is no way to avoid this problem by a numerical method).

Therefore the idea of our algorithm is to replace maximal acceleration curves by piecewise constant acceleration curves and to detect loops generated by the possibly miss of characteristic points.

The algorithm is composed of three main functions.

acceleration$(s_0, \dot{s}_0)$ Let $0 < \mu < 1/4$ be given. Starting from $(s_0, \dot{s}_0)$, we set $\ddot{s} = \mu\alpha(s_0, \dot{s}_0) + (1 - \mu)\beta(s_0, \dot{s}_0)$ and

$$\dot{s} = \sqrt{\dot{s}_0^2 + 2\ddot{s}(s - s_0)} \text{ while } \ddot{s} \in [2\mu\alpha(s, \dot{s}) + (1 - 2\mu)\beta(s, \dot{s}), \beta(s, \dot{s})]$$

This curve in the phase plane corresponds to a constant acceleration curve and thus is straightforwardly integrable. If s_1 is the first value for which $\ddot{s}$ leaves the interval, we set $\dot{s}_1 = \sqrt{\dot{s}_0^2 + 2\ddot{s}(s_1 - s_0)}$ and we define a new constant acceleration curve from $(s_1, \dot{s}_1)$. Then the operation is iteratively repeated until the generated curve reaches an out-point or a deceleration curve (the associated procedure returns either $OUT - POINT$ or $INTERSECTION$).

deceleration$(s_0, \dot{s}_0)$ is identically defined by replacing α by β and β by α and by going backward ($s_1 < s_0$). The curve is stopped as soon as it reaches an in-point or it enters the domain above an acceleration curve previously defined (the associated procedure returns $IN - POINT, INTERSECTION$ or $ABOVE - ACCEL$).

We notice that when μ tends to 0, these curves tend respectively to maximal and minimal acceleration curves. Their discretization is not fixed and depends on the variation of various coefficients. Moreover the validity of the acceleration along the curve can be verified with an as small as desired step size. Thus the choice of μ determines a compromise between curve optimality and their number of switching points.

characteristic $-$ point$(s_0, step)$ looks for a characteristic point from s_0. A step size being chosen, the function computes successively $\kappa(s_0 + k.step)$ for $k = 0, 1, ...$. When κ becomes negative, a dichotomy procedure is called to find the exact parameter of the sign change. The function stops when a characteristic point is found.

Algorithm:

$(s, \dot{s}) = (0, 0)$;
if $(s, \dot{s})$out-point
    next-curve $= DECEL$;
else
    next-curve $= ACCEL$;
while $(s < s_{end})$
    if next-curve $= ACCEL$
        end-cond-accel $=$ acceleration$(s, \dot{s})$
        if end-cond-accel $= INTERSECTION$
            add acceleration and deceleration curves to current phase curve;
            $(s, \dot{s}) =$ end of deceleration curve;
            next-curve $= ACCEL$;
        if end-cond-accel $= OUT - POINT$
            next-curve $= DECEL$;
    else
        $(s, \dot{s}) =$ characteristic $-$ point$(s, step)$
        end-cond-decel $=$ deceleration$(s, \dot{s})$
        if end-cond-decel $= INTERSECTION$
            add acceleration and deceleration curves to current phase curve;
            $(s, \dot{s}) =$ end of deceleration curve;
            next-curve $= ACCEL$;
        if end-cond-decel $= ABOVE - ACCEL$
            $(s, \dot{s}) =$ end of acceleration curve;
            next-curve $= ACCEL$;
        if end-cond-decel $= IN - POINT$
            $(s, \dot{s}) =$ end of acceleration curve;
            if end-cond-accel $= OUT - POINT$
                step $=$ step$/2$;
                next-curve $= DECEL$;
            else
                next-curve $= ACCEL$;
end while
deceleration$(s_{end}, 0)$

If the acceleration stops at an out-point and the next deceleration stops at an in-point, a characteristic point has been missed. Another one is looked for with a smaller step size. Indeed, if no bound on the derivative of κ is known, one never can be sure not to miss a characteristic point.

3.2.2. Velocity constraints

From now on we call the phase plane curve built in the former section the *acceleration phase curve*.

Taking into account the velocity constraints (i.e., remaining under the saturation curve) requires four routines:

followAccel follows the phase curve built in the former section until a velocity constraint is saturated. In this case the slope of the velocity constraint curve is computed. If it corresponds to a valid acceleration, followVelocityMax is called. Otherwise, deceleration is the next step of the algorithm.

followVelocityMax follows the velocity constraint curve until the resulting acceleration is not valid anymore, or until the velocity constraint curve cuts across the *acceleration phase curve*. In this last case, we call followAccel. If the acceleration is too big, acceleration is called. If the acceleration is too small, deceleration is called.

acceleration builds an acceleration curve as described in the former section. This function stops if it cuts across the *acceleration phase curve* or the velocity constraint curve. In the first case, followAccel is called. In the second case, deceleration is called if the slope of the velocity constraint curve is too small. Otherwise, followVelocityMax is called.

deceleration looks for the next *velocity characteristic point* (i.e. the slope of the velocity constraint curve corresponds to the minimal acceleration) and builds a deceleration curve backward from this point. Then followVelocityMax is called.

Once again characteristic points can be missed. This cases are detected and the step size of the characteristic point search is chosen smaller.

4. Results

In this section we give results from real experiments.

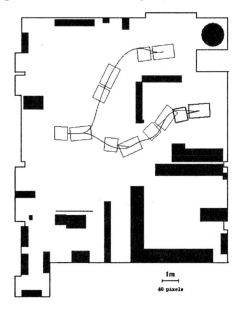

Figure 2. Path for a robot and its trailer

Picture 2 shows a path computed by a path planner for a robot towing a trailer [6]. The path is composed of two parts of opposite sense. These parts are dealt with independently.

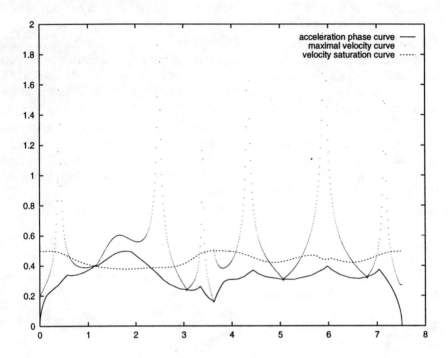

Figure 3. Acceleration constraints

Picture 3 shows the first computed curve obtained without taking into account the velocity constraints of the robot. Picture 4 shows the final curve taking into account all the constraints.

5. Conclusion

The problem of finding a convenient time parameterization of a path has been dealt with. This problem gave rise to theoretical solutions involving continuous functions and integral curves [7]. From an experimental point of view, the input of the time parameterization problem is the output of a path planner. Thus the algorithm has to run for a large range of inputs.

The critical point of the implementation revealed to be the uncontrolled variation of the coefficients $d\delta f_{(x,y)}/ds$ and $d\delta f_\theta/ds$. Thus a first constant step implementation led to failure.

That is the reason why we have chosen a less optimal but more robust implementation of Slotine's algorithm.

References

[1] S. Fleury, P. Souères, J.P. Laumond, R. Chatila, "Primitives for smoothing

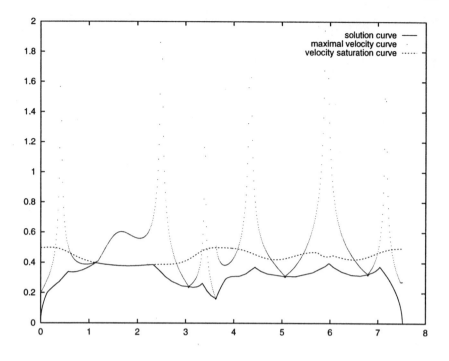

Figure 4. Velocity constraints

mobile robot trajectories", IEEE Transactions on Robotics and Automation, 11 (3), pp.441–448, 1995.

[2] Y. Kanayama and N. Miyake, "Trajectory generation for mobile robots," in *Robotics Research 3*, G. Giralt and O. Faugeras Eds, MIT Press, 1986.

[3] M. Renaud, J.Y Fourquet, "Time-optimal motions of robot manipulators including dynamics," in *Robotics Review 2*, O. Khatib, J.J. Craig, T. Lozano-Pérez Eds, MIT Press, 1992.

[4] C. Samson and K. Ait-Abderrahim, "Feedback Control of a Nonholonomic Wheeled Cart in Cartesian Space", IEEE International Conference on Robotics and Automation, pp 1136-1141, Sacramento, 1991.

[5] S. Sekhavat, J.P. Laumond, "Topological property of trajectories computed from sinusoidal inputs for nonholonomic chained form systems," IEEE International Conference on Robotics and Automation, Minneapolis (USA), 22-28 Avril 1996, pp.3383-3388

[6] S. Sekhavat, F. Lamiraux, J.P. Laumond, G. Bauzil, A. Ferrand, "Motion planning and control for Hilare pulling a trailer: experimental issues," IEEE International Conference on Robotics and Automation, Albuquerque (USA), 20-25 Avril 1997.

[7] J.J.E. Slotine and H.S. Yang, "Improving the efficiency of time-optimal path-following algorithms", *IEEE Transactions on Robotics and Automation*, 5 (1), pp 118-124, 1989.

Chapter 7

Legged locomotion

Legged locomotion gives place to many different problems, mainly planning the movement based on the robot structure, control of its execution and achieving an energy efficient walking. This chapter includes three papers dealing with three different walking robots having one, four or eight legs respectively.

Ahmadi and Buehler describe the experimental results of and electrically actuated one-leg hopping robot. With the aim to minimize energy consumption, mainly in vertical displacements that do not contribute directly to the effective mobility, a spring in series with the hip actuator is added to the hopping robot, ARL. To provide some compliance the control approach starts from the Raibert's idea of controlling hopping height, forward speed and body pitch, via three decoupled controllers and is addressed to drastically reduce energy consumption. The first results show the robustness of the approach and energy saving.

The work of Serna, Avelló, Briones and Bustamante deal with a wall-climbing robot conceived for inspection of nuclear power plants. The robot presented is the third generation of this pneumatic climbing robots family. The paper analyses the locomotive system, the pneumatic subsystem and the control. Some relevant performances are the excellent behavior when climbing vertical walls and the possibility to pass from horizontal to vertical surfaces.

The last paper, by Roßman and Pfeiffer deals with an eight-legged pipe crawling robot. In this case, the objective of the robotic system is its ability to move inside pipes. The effectiveness of the system relies on the flexible legs length and the capability to move inside tubes independently of their inclination (up to now, not curved pipes). The structure is based on two sets of four legs, each in star shape configuration. The paper describes the sensors and electronics and the control organization.

Preliminary Experiments with an Actively Tuned Passive Dynamic Running Robot

M. Ahmadi and M. Buehler

Centre for Intelligent Machines, Department of Mechanical Engineering
McGill University, Montréal, Québec, Canada, H3A 2A7

Abstract: This paper describes experiments with an electrically actuated one legged hopping robot, the ARL Monopod. While a spring-mass system, comprised of the leg spring and the body mass has previously provided the basic "passive dynamic" vertical motion, we have now added a spring in series with the hip actuator to use the same passive dynamic principle for the leg swing motion. This paper validates our previously proposed hip actuation controller, which actively tunes the leg swing motion during running to remain close to its passive dynamic response, and at the same time maintains balance and controls the robot's desired forward speed. Experimental data of stable running at $1m/s$ show a 75% reduction in hip actuation energy during flight, and a mechanical energy expenditure for the complete robot of only $64W$. The resulting specific resistance of the robot reduces to 0.36 at $1m/s$, which is about half the previously reported result for the robot without hip spring.

1. Introduction

Energy autonomy is a necessity for virtually any practical mobile robot. Achieving the required energy efficient running or walking in legged robots is particularly challenging, since much energy can be expended for internal motions, like vertical body motion, or swinging the legs. Since these motions do not contribute directly to mobility, the energy used to produce them should be minimized.

Just like a pendulum produces periodic motion efficiently by cycling kinetic and gravitational potential energy, the periodic internal motions needed in running can be produced efficiently as well, by cycling kinetic and spring potential energies. Many animals are superbly designed to exploit this principle, and manage to reduce the metabolic cost of running considerably by utilizing the elastic properties of their muscles, tendons, and bones [2].

In the area of robotic systems, it was Raibert [9] who pioneered the use of a system's unforced response (its "passive dynamics") to generate the gross motion for locomotion. To date, virtually all dynamically stable running robots utilize an actively tuned spring-mass system to provide the vertical robot motion [9, 8, 4, 6]. A beautiful example of just how far the passive dynamics approach can be carried is provided by McGeer [7] who built a two legged walking robot with knees, which was able to walk down shallow inclines, driven entirely by gravity without any actuation.

In [5] we show that at $1.2m/s$ the ARL Monopod uses about half of its total energy expenditure for swinging the leg. Much of this energy can be saved, since passive dynamics can provide the leg swing motion via a spring connecting the leg and the body at the hip joint as proposed in [10]. In our previous work [1], we presented a new control strategy to stabilize passive dynamic running of a one-legged robot. An alternative approach for controlling the same robot system with compliant leg and hip was proposed by François and Samson [3]. The controller is derived based on a simplified and linearized dynamic model of the robot, and was shown to be both stable and robust in simulations.

The current paper is devoted to the experimental validation of a controller [1] proposed earlier. In Sec. 2 the robot model and the design modifications to the ARL Monopod are described which give rise to the "ARL Monopod II." The hip and leg controllers are described in Sec. 3, together with the experimental results. Finally, Sec. 4 computes and discusses the energy expenditure as well as the specific resistance.

2. ARL Monopod II

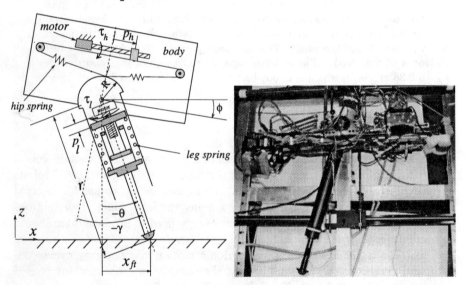

Figure 1. *A sketch of ARL Monopod II with important variables, and a photo.*

The ARL Monopod II is constraint to move in a vertical plane and consists of a body which is connected to a prismatic leg at the hip joint. A sketch of the mechanical system showing its degrees of freedom, together with a photo is presented in Fig. 1. The system has a total of seven degrees of freedom, including the leg length (r), leg actuator displacement (p_l), hip actuator displacement (p_h), leg angle (θ), and three degrees of freedom of the body ($x, z,$ and body pitch angle ϕ). Due to kinematic constraints, not all of them are free simultaneously. During stance there are five degrees of freedom $(\theta, \phi r, p_h, p_l)$, and during flight there are six $(x, z, \theta, \phi, p_h, p_l)$. The nominal values for some of the important parameters are also given in Table 1, followed by a table of indices.

J_b	body inertia	$0.8\ kgm^2$
J_l	leg inertia	$0.12 - 0.19\ kgm^2$
k_h	hip spring stiffness (linear)	$3800\ N/m$
k_l	leg spring stiffness	$6312\ N/m$
m_b	body mass	$11.5\ kg$
m_l	leg mass	$4.6\ kg$
r_0	leg length (when $p_l = 0$ in flight)	$0.64\ m$

Table 1. *Nomenclature and numerical settings.*

d	desired value	h	hip	f	flight
ll	lower leg	u	upper-leg	cg	center of gravity
$\hat{}$	amplitude	l	leg	lo	lift-off
$*$	passive dynamic	s	stance	td	touchdown
b	body	act	actuator	bu	body+upperleg

Table 2. *Description of indices.*

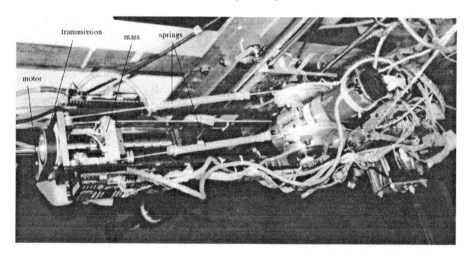

Figure 2. *A closeup of two of the four series hip compliances, inserted in the connecting strings between hip actuation ball screw transmission and leg.*

Both hip and leg actuators are connected in series with springs. The prismatic leg joint features a conventional helical compression spring. The hip has an extension spring made of natural rubber in the form of standard surgical tubing as shown in Fig. 2. Since the hip actuation mechanism employs Spectra strings, which transmit the linear ball screw motion via pulleys to the circular arc attached to the leg, cutting the strings and inserting the tubular spring offered an effective means of adding hip compliance. With four string locations where the springs could be added in parallel, the energy storage density requirements of the springs were drastically reduced as well. Even though the rubber springs show nonlinear behaviour at low and very large strains, we were able to operate in the linear region by suitable pre-loading, where the stiffness of each individual tubular spring is characterized by $\bar{\kappa}_{latex} = C_{latex}/l_{latex}$. l_{latex} is the length of the spring, and $C_{latex} = 51.5N$ is our springs' particular material constant.

3. Control

The basic control approach builds on Raibert's original idea of controlling hopping height, forward speed, and body pitch via three decoupled controllers [9]. Body pitch and hopping height are controlled during stance, while forward speed is controlled via touchdown foot placement during flight. After modifying Raibert's controller for the control of the passive dynamic hip oscillations, energy consumption was drastically reduced, while stable running was maintained as shown in Fig. 3.

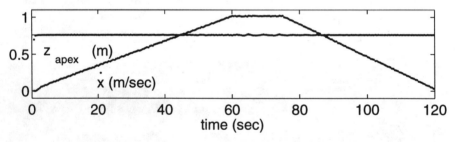

Figure 3. *Experimental data of controlled passive dynamic running. Robot speed $\dot{x}$ and hopping height z_{apex} are recorded during a two minute run. The robot follows a trapezoidal velocity profile from zero to $1m/s$. At the same time, the hopping height controller maintains closely the desired apex height.*

3.1. Hopping height control

The performance of the hopping height control is of critical importance to the success of the other controllers which are synchronized to the vertical motion. Any changes in hopping height will act as perturbations to the other controllers. Throughout the experiments, the robot is commanded to maintain a constant (desired) hopping height, $z_{apex,d}$. An energy based controller is used to accomplish this task. Note that the desired hopping height can be translated into a unique desired total vertical energy at apex, $E_{apex}(z_{apex,d})$. The objective of this controller is then to keep the *vertical total energy* of the robot,

$$E_v = E_{pot} + E_{kin,v} \qquad (1)$$

at the desired level, where E_{pot} is the gravitational and leg spring potential energy, and $E_{kin,v}$ the vertical component of the kinetic energy. Each step starts at apex and ends at the next apex. Let the vertical energy, E_v, at the

n_{th} apex be E^n_{apex} and let the n_{th} stance be occuring after the n_{th} apex. The difference between the desired energy and actual energy at the n_{th} apex is the amount of energy that we are adding to the system during the subsequent (the n_{th}) stance phase,

$$E^n_{act,d} = E^n_{apex} - E_{apex,d} + \hat{E}_{loss}.$$

$\hat{E}_{loss}$ is an estimate of the energy loss between two successive apexes. As an estimate we use the average of the past two losses,

$$\hat{E}_{loss} = \frac{1}{2}(E^{n-1}_{loss} + E^{n-2}_{loss})$$

where E^{n-1}_{loss} is the measured loss between $(n-1)_{th}$ and the n_{th} apex as,

$$E^{n-1}_{loss} = E^{n-1}_{apex} - E^n_{apex} + E^{n-1}_{act}$$

and E^{n-1}_{act} is the energy added by the leg actuator during the $(n-1)_{th}$ stance. It can be measured directly from the spring force and the actuator's linear velocity at the output of the leg ball screw transmission as

$$E_{act} = \int_{stance} F_l \dot{p}_l dt.$$

The remaining task is to add mechanical energy equal to $E^n_{act,d}$ to the vertical energy of the system during stance. A simple velocity controller can be used to control the injected energy by commanding the desired leg actuator velocity, $\dot{p}_{l,d}$ as a function of the continuously measured energy error,

$$\dot{p}_{l,d} = \kappa_{E1}(E_{act} - E^n_{act,d}).$$

However, this approach completely ignores the actuator dynamics. For example, at the beginning of stance this controller is likely to saturate the leg actuator. In addition, it injects most of the energy right after touchdown when the spring force is small. This results in high ball screw velocities, and as a result, higher damping losses than necessary and also high energy losses for accelerating and decelerating the ball screw. A better strategy is to add this energy when the spring is more compressed and eliminate actuator motion at touchdown and liftoff. To do this we multiply the above expression with a heuristic correction factor

$$\zeta = 1 - \frac{q_l}{q_{l,0}}$$

where q_l is the leg spring length and $q_{l,0}$ is the leg spring rest length. Thus ζ attains its maximum at maximum spring compression and vanishes at touchdown and liftoff.

A velocity feedback controller with feedforward compensation of the leg spring force, F_l, is used to follow the desired velocity,

$$\tau_l = -\kappa_{E2}(\dot{p}_l - \dot{p}_{l,d}) + r_l F_l,$$

where r_l is the leg ball screw lead in m/rad.

Fig. 4 shows experimental running data where the controller successfully injects various desired energy levels. Even when dynamic coupling with the forward motion exists, the controller performs well: Fig. 3 shows how the robot successfully maintains the desired hopping height even while the forward speed varies between zero and $1m/s$.

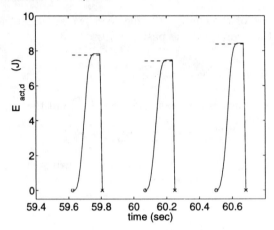

Figure 4. *The performance of the hopping height controller in adding the desired energy (dashed line) during three successive stance phases.*

Since the leg swing motion, described below, must be synchronized to the vertical motion, it is convenient to derive a scalar variable, termed "Locomotion Time," η, whose amplitude is independent of the hopping height. During flight we use the expression

$$\eta_f = \frac{-\dot{z}}{\dot{z}_{lo}}. \tag{2}$$

If we assume that the robot's mass is dominated by body mass and follows a ballistic motion during the flight phase, η_f is linear in time, since (2) is equivalent to

$$\eta_f = \frac{2}{T_f} t, \tag{3}$$

and $\eta_f \in [-1\ 1], \eta_f(t_{lo}) = -1, \ \eta_f(t_{td}) = 1$. During stance the locomotion time is simply a shifted and scaled time,

$$\eta_s = \frac{2}{T_s} t, \ \ \eta_s(t_{td}) = 1, \ \ \eta_s(t_{lo}) = -1,$$

where T_s is the previous stance time, used as a prediction of the current stance time. Note that for the definitions of the locomotion times, the time parameter is assumed to be zero at mid-phase.

3.2. Forward speed control

The basic idea behind the use of passive dynamic motion in the hip is illustrated in Fig. 5, and is presented in more detail in [1]. If we denote by x_{ft} the horizontal position of the toe with respect to the hip, a completely unforced, frictionless counter-oscillation of the leg and body coupled by the hip spring produces a sinusoidal response. It can now be seen that, with proper initial conditions and coordination with the vertical motion, one can assure that stance phase occurs during the period of approximately constant slope, equivalent to the robot forward speed. Thus the unforced response can provide the correct gross hip motion required for locomotion with minimal actuation, and thus energy consumption.

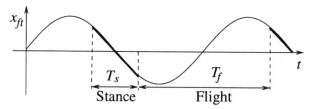

Figure 5. *Foot position with respect to hip, x_{ft}, produced by unforced (passive dynamic) leg swing motion, with linear approximations of stance phase (from [1])*

One complete vertical hopping cycle has to be synchronized with one complete leg swing cycle and where the robot touches ground with the proper leg touchdown angle. During flight, this can be achieved via

$$\theta_d(\eta_f) = \hat{\theta}_d \sin(\pi(1-\rho)\eta_f), \qquad (4)$$

where the duty factor, ρ, is defined as the ratio between stance time and one whole step duration,

$$\rho = \frac{T_s}{T_{step}},$$

and η_f is the flight locomotion time (3). At apex, $\theta_d(\eta_f = 0) = 0$, at liftoff $\theta_d(\eta_f = -1) = -\hat{\theta}_d \sin(\pi(1-\rho))$ and at touchdown $\theta_d(\eta_f = 1) = \hat{\theta}_d \sin(\pi(1-\rho))$. The duty factor, determined via the robot's design parameters, is selected such that the frequency of the desired leg angle trajectory (4), when synchronized to the vertical dynamics, can be the unforced response of the dynamical system consisting of the body and the leg, connected via the hip spring. The desired amplitude of the oscillation,

$$\hat{\theta}_d = \frac{\theta_{td,d}}{\sin(\pi(1-\rho))}$$

is determined by the desired leg touchdown angle, $\theta_{td,d}$, which in turn is determined by the robot's forward speed. Thus the overall leg angle controller, outlined in Fig. 6 determines the desired trajectory from the passive dynamics (4), whose amplitude is determined from the robot's forward speed (via the proper foot placement), and the resulting reference trajectory is then controlled via a model based leg angle controller. A more detailed description of these arguments is provided in [1].

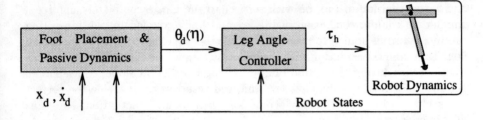

Figure 6. *The general outline of the control for the leg swing angle tracking in the flight phase*

The robot's forward speed is controlled during flight by servoing the leg to the proper touchdown angle. Raibert's foot placement algorithm [9] is used to find the desired foot position at touchdown,

$$x_{ft,td,d} = \frac{1}{2}\dot{x}T_s + \kappa_x(x - x_d) + \kappa_{\dot{x}}(\dot{x} - \dot{x}_d).$$

The first term on the RHS is the neutral foot position with respect to the hip, necessary to maintain the current forward speed. It uses the previous stance time as an estimate of the upcoming stance time. The other two terms are position and velocity error terms necessary for stable tracking. From geometry, the corresponding desired leg touchdown angle is

$$\theta_{td,d} = -\sin^{-1}\frac{x_{f,td,d}}{r}.$$

This completes the derivation of the desired leg angle trajectory derivation.

The tracking controller uses a simplified dynamic model. Assuming that the changes in the actuator position do not have significant dynamic effects except changing the leg inertia,

$$J_l(p_l) = J_{l,min} + (m_{ll} + m_{act})[(p_l + p_{l,0})^2 - p_{l,0}^2].$$

In addition, during the flight we neglect the friction forces acting on the robot through planarizer guides. We consider all the damping forces are small compared to inertial and spring forces. The resulting equations of motion in that case are

$$\begin{aligned} J_b\ddot{\phi} &= Rk_h\,[p_h + R(\theta - \phi)] \\ J_l\ddot{\theta} &= -Rk_h\,[p_h + R(\theta - \phi)] \end{aligned} \qquad (5)$$

and the coupled actuator dynamics is

$$\alpha_h\ddot{p}_h = \tau_h - rk_h[p_h + R(\theta - \phi)]. \qquad (6)$$

The above set of equations results in a system of six coupled first order differential equations. The control problem can be simplified considerably if we assume that the actuator bandwidth is much higher than that of the required leg oscillations. In that case, the control problem can be decoupled into two stages.

Actuator displacement control: The desired actuator displacement commanded by the either the leg angle controller during flight or the body pitch controller during stance is tracked using the model of the actuator dynamics (6). If the desired error dynamics for $e_{ph} = p_{h,d} - p_h$ are

$$\ddot{e}_{ph} + K_{v,f2}\dot{e}_{ph} + K_{p,f2}e_{ph} = 0,$$

the necessary controller is

$$\tau_h = \alpha_h \left[\ddot{p}_{h,d} + K_{v,f2}\dot{e}_{ph} + K_{p,f2}e_{ph}\right] + r_h k_h \left[p_h + R(\theta - \phi)\right] + r_h c_h \left[\dot{p}_h + R(\dot{\theta} - \dot{\phi})\right]. \tag{7}$$

The performance of the controller during a $1m/s$ run is shown in Fig. 7.

Leg Angle Control: Considering the hip actuator displacement p_h as the control input we develop a model based controller for leg angle tracking. The result is the desired hip actuator position which will be used in (7). If the desired final equation for the leg angle error, e_θ is

$$\ddot{e}_\theta + K_{v,f1}\dot{e}_\theta + K_{p,f1}e_\theta = 0,$$

then the input to the system (5), namely the desired hip actuator displacement, $p_{h,d}$ and its corresponding derivatives are

$$p_{h,d} = -R(\theta - \phi) - \frac{J_l}{Rk_h}\left[\ddot{\theta}_d + K_{v,f1}\dot{e}_\theta + K_{p,f1}e_\theta\right], \tag{8}$$

$$\dot{p}_{h,d} = -R(\dot{\theta} - \dot{\phi}) - \frac{J_l}{Rk_h}\left[\theta_d^{(3)} + K_{v,f1}\ddot{e}_\theta + K_{p,f1}\dot{e}_\theta\right],$$

$$\ddot{p}_{h,d} = -R(\ddot{\theta} - \ddot{\phi}) - \frac{J_l}{Rk_h}\left[\theta_d^{(4)} + K_{v,f1}e_\theta^{(3)} + K_{p,f1}\ddot{e}_\theta\right].$$

Note that we do not need to measure any second or higher order derivatives of states, since we can simply use terms from the equations of motion (5) and their derivatives. The performance of the leg angle controller during flight is shown in Fig. 7.

3.3. Body pitch control

Body pitch is controlled during stance in a similar fashion as is the leg angle during flight. To be properly synchronized with the stance time, the symmetric body pitch angle has to follow the trajectory

$$\phi_d(\eta_s) = \hat{\phi}_d \sin(\pi \rho \eta_s),$$

where the stance locomotion time η_s and the duty cycle ρ are as defined above. Note that this trajectory is derived during a ballistic flight of the system and simply specifies the body pitch trajectory during a portion of the cycle which now coincides with stance. In this case, the amplitude of the body oscillation and the leg oscillation are related as follows

$$\hat{\phi}_d = -J_b/J_l \hat{\theta}_d.$$

Since the amplitude of the leg oscillation, $\hat{\theta}_d$ has already been determined in the previous section, the desired body pitch trajectory is complete.

To track the desired body pitch trajectory, the same strategy as for the leg angle is used. However, the equations of motion during stance are different, and a simplified version we use is

$$\ddot{\phi} = Rk_h[p_h + R(\theta - \phi)] - m_b g z_{cg,b} \sin(\phi)/J_b \quad (9)$$

$$\ddot{r} = -M_{b,ul}[g\cos(\theta) + r\dot{\theta}^2] + k_l(r_0 - r + p_l)/M_{b,ul} \quad (10)$$

$$\ddot{\theta} = \left(-Rk_h[p_h + R(\theta - \phi)] - M_{bu}(r - z_{cg,bu})(2\dot{r}\dot{\theta} - rm_{b,ul}\sin\theta)\right)/(11)$$

$$[J_{l,cg} + M_{b,ul}(r - z_{cg,bu})]$$

where $r = p_l + q_l$ is the distance from the hip joint to toe.

The remainder of the controller development is analogous to above and is omitted here for brevity. The performance of this controller during a run is shown in Fig. 7.

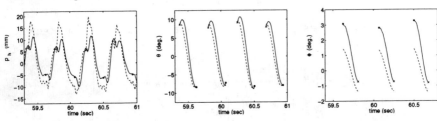

Figure 7. *The performance of the hip actuator controller (left), the leg angle controller during flight (center), and the body pitch controller during stance (right) during a 1m/s run, (dash: desired, solid: actual).*

4. Energetics

After having achieved stable running, as described in the previous section, we can now devote our attention to the primary motivation for the added hip compliance - the potential for energy savings. A common measure of energy expenditure is the integral of the absolute value of the instantaneous "shaft power,"

$$E_{shaft} = \int |\tau\omega| dt$$

where τ and ω are the torque and the angular velocity of the primary actuator, in our case, the leg and hip electric motors. Note that this measure does not include the motor efficiency of converting input electric power to shaft power, but does include the energy expenditure of moving the transmissions (in our case ball screws). We have summarized the energetics in Table 3 numerically, and graphically in Fig. 8, according to phases and motors. Compared to the energetics for running with a directly actuated hip [5], where the body pitch controller consumed $5J$ and the forward speed controller consumed $20J$ (at

1.2m/s), drastic reduction were achieved: The savings amount to approximately 75% for the forward speed controller and about 60% for the pitch angle controller.

	Stance Phase	Flight Phase
Leg Motor	Hopping Height: 12J	Leg Retraction: 10J
Hip Motor	Body Pitch: 2J	Forward Speed: 4J

Table 3. Energy cost of 28J per step divided by phase and actuator

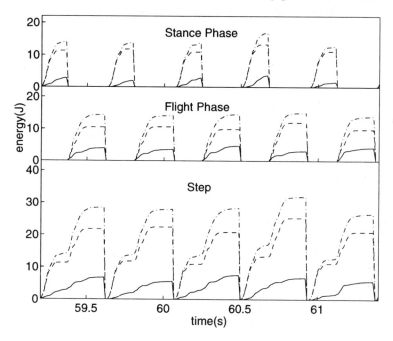

Figure 8. Experimental energetic data at 1m/s. (solid: hip motor, dashed: leg motor, dot-dash: total)

To compare this results to previous results we obtained in [5] we use the specific resistance, a standard measure of energy efficiency for mobile systems, which takes average power P, weight mg, and velocity into account,

$$\varepsilon(v) = \frac{P(v)}{mgv} = \frac{28J/hop \cdot 2.3 hops/s}{18kg \cdot 9.81m/s^2 \cdot 1.0m/s} = 0.36$$

The best value reported in [5] for the ARL Monopod with directly actuated hip was $\varepsilon(1.2m/s) = 0.7$. With on board batteries and controller, ARL Monopod II is heavier (to the body and leg mass in Table 1, we have to add the mass of the moving parts of the planarizer of 1.9kg, for a total mass of 18kg), was evaluated at a slower speed (1.0m/s vs. 1.2m/s), but with 28J uses approximately half the energy, and the specific resistance, as indicated above is only $\varepsilon(1m/s) = 0.36$.

Conclusion

This paper presented the first experimental results of controlled passive dynamic running of a one legged hopping robot with leg and hip compliance. The approach is very robust to deviations from the assumptions used to derive the controller. For example, it assumes a small duty cycle (ratio of stance to flight time), which was set to 1/6 in our previous simulations. However, in the experiments it was between 1/3 and 1/2, which clearly violates the straight line approximation shown in Fig. 5. In addition, the model based controllers rely on many parameters which are only approximately known. The energy savings obtained were dramatic and improved the energy efficiency of the robot by a factor of two.

References

[1] M. Ahmadi and M. Buehler. Stable Control of a Simulated One-Legged Running Robot with Hip and Leg Compliance. *IEEE Trans. Robotics and Automation*, 13(1):96–104, Feb 1997.

[2] R. McN. Alexander. Three Uses for Springs in Legged Locomotion. *Int. J. Robotics Research*, 9(2):53–61, 1990.

[3] C. Francois and C. Samson. Energy Efficient Control of Running Legged Robots. A Case Study: The Planar One Legged Hopper. Technical Report 3027, Unité de recherche INRIA, Sofia-Antipolis, France, 1996.

[4] P. Gregorio, M. Ahmadi, and M. Buehler. Experiments with an electrically actuated planar hopping robot. In T. Yoshikawa and F. Miyazaki, editors, *Experimental Robotics III*, pages 269–281. Springer-Verlag, 1994.

[5] P. Gregorio, M. Ahmadi, and M. Buehler. Design, Control and Energetics of an Electrically Actuated Legged Robot. *IEEE Trans. Systems, Man, and Cybernetics*, 27(4):(to appear), Aug 1997.

[6] A. Lebaudy, J. Prosser, and M. Kam. Control algorithms for a vertically-constrained one-legged hopping machine. In *Proc. IEEE Int. Conf. Decision and Control*, pages 2688–2693, 1993.

[7] T. McGeer. Passive dynamic walking. *Int. J. Robotics Research*, 9(2):62–82, 1990.

[8] K. V. Papantoniou. Electromechanical design for an electrically powered, actively balanced one leg planar robot. In *Proc. IEEE/RSJ Conf. Intelligent Systems and Robots*, pages 1553–1560, Osaka, Japan, 1991.

[9] M. H. Raibert. *Legged Robots That Balance*. MIT Press, Cambridge, MA, 1986.

[10] M. H. Raibert and C. M. Thompson. Passive dynamic running. In V. Hayward and O. Khatib, editors, *Experimental Robotics I*, pages 74–83. Springer-Verlag, NY, 1989.

ROBICEN: A pneumatic climbing robot for inspection of pipes and tanks

M. A. Serna and A. Avello
Escuela Superior de Ingenieros Industriales, University of Navarra
San Sebastián, SPAIN
maserna@ceit.es, alavello@ceit.es

L. Briones and P. Bustamante
CEIT
San Sebastián, SPAIN
lbriones@ceit.es, pbustamante@ceit.es

Abstract: ROBICEN is a family of three pneumatic wall-climbing robots designed for inspection of nuclear power plants. In particular, ROBICEN III is a very lightweight robot of only 2.5 Kg, capable of travelling at speeds of 0.3 m/s on flat walls and along large diameter cylindrical pipes. Its locomotive system is based on four compact suction cups, two fixed and two mobile, that allow the robot to move forward, backward and to rotate left and right. Its sensorial system, which is composed of four video cameras and an ultrasound sensor, allows the robot to take measurements of wall thickness. The robot is teleoperated from a PC computer through an RS-232 serial port.

1. Introduction

In the past two decades, advances in nuclear robotics have been boosted by several external factors. For example, as a consequence of the social mistrust produced by the well-know accidents of the Three-Mile-Island and Chernobyl, many national regulations have become stricter in matters of nuclear security. The need for remote manipulation techniques in the decommissioning and dismantling of nuclear facilities is also rapidly increasing as more and more nuclear plants reach their age limit or are closed by governments. Finally, many countries are following international recommendations to lower radiation dose limits on exposed personnel. These and other similar causes have increased the nuclear sector's awareness of the potential benefits of robotics.

Typically, nuclear robotics has focused on three main activities: decommissioning, inspection and power plant maintenance. In the decommissioning of radioactive facilities, remotely operated manipulators are used for cutting, grinding, grasping, cleaning, etc. in environments of medium and high radiation. Similar hazardous environments existing in fuel reprocessing plants and radioactive waste processing plants also require remote manipulation techniques. To help operators carry out complex and precise tasks, these remotely operated manipulators often use advanced technologies like stereo imaging, 3D virtual reconstruction of operation environments and force reflection.

Inspection robots in nuclear power plants are used to reduce the need for human intervention in surveillance and monitoring in environments of high temperature, humidity or radiation [1]. Inspection robots are not only able to take measurements but also to carry out more complex tasks like detection of water and steam leaks through noise and vision.

Finally, maintenance robots are used in normal power plant operation. Remotely operated devices provided with a dexterous manipulator arm reach the area of operation and perform tasks such as cleaning, welding, manipulating, painting, etc. in a way similar to a human arm.

Researchers have proposed a great variety of robot configurations and functional capabilities [2-8]. Wheeled robots and vehicles are the most common configurations for mobile robots. More recently, the idea of wall-climbing robots has been also presented in the literature [9-15].

2. The SRT project

In 1994, ENDESA and IBERDROLA launched the SRT project, whose aim has been to develop a set of robotic systems that meet the current and future requirements of Spanish nuclear power plants in inspection and manipulation.

Two large industrial companies, CASA and ENSA provide the project with technological expertise while three research institutions, CEIT, CIEMAT and the University of Cantabria, are in charge of advanced developments; UNESA acts as project coordinator.

Three different robotic devices have been developed for inspection tasks:
- ROBCAR, a track-suspended monitoring system robot for inspection of containment vessel internals.
- ANAES, a steam leak detector based on spectral analysis of noise.
- ROBICEN, a family of three pneumatic wall-climbing robots for inspection of tanks and large-diameter pipes.

A more detailed overview of the developments carried out in ROBICEN is given in the following sections.

3. ROBICEN I and II

ROBICEN I [16] was chronologically the first drive unit developed. Its locomotive system, which is shown in Figure 1, consists of three main rigid bodies with three corresponding suction cups underneath, one under each body, to provide adherence to walls. Four flexible pneumatic cylinders which simultaneously provide actuation force and structural support connect the distal bodies to the central body, establishng a flexible link between them. Four electrovalves located in the central body activate groups of two cylinders simultaneously to produce forward-backward translations and left-right rotations. Since the pneumatic cylinders are of single effect, eight traction springs are used to compress the actuators. Four sequences of cylinder activations, illustrated in Figure 2, are currently programmed in ROBICEN I to produce forward/backward motions and left/right rotations on both floor and walls.

Figure 1. ROBICEN I drive unit

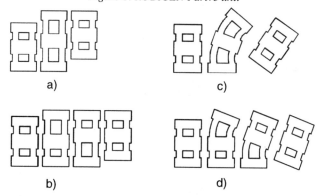

Figure 2. Motion sequences programmed in ROBICEN I: a) forward motion on floor, b) forward motion on wall, c) rotation on floor, d) rotation on wall.

ROBICEN II (shown in Figure 3) is also a fully pneumatic climbing robot which is provided with a mechanism that allows transitions from horizontal to vertical planes and vice versa by its own means. In addition, ROBICEN II is more stable and faster than the previous robot, partly due to its lower centre of mass. ROBICEN II is made up of two similar modules, one in front and the other at the rear, connected by a double four-bar linkage. When the linkage is actuated by two double-effect short-stroke cylinders, the front module rotates up to 90 degrees with respect to the rear one, thus allowing transitions from a horizontal to a vertical or oblique surface. Each module has two commercial pneumatic cylinders made of stainless steel and mounted in parallel to the robot's axis of motion. A linear guide mounted on each module provides structural stiffness and prevents the cylinders from receiving lateral forces. The simultaneous activation of both cylinders, which are hinged at both ends, produces forward motion, whilst the alternate activation of either cylinder produces rotation.

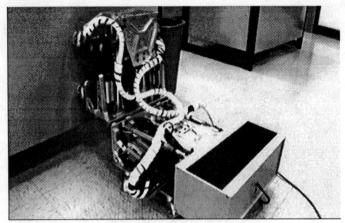

Figure 3. ROBICEN II in transition from a horizontal to a vertical surface.

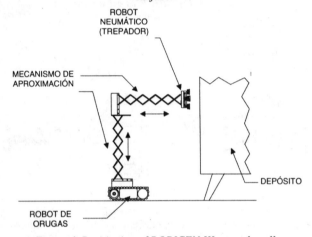

Figure 4. Positioning of ROBICEN III on tank wall.

4. ROBICEN III

ROBICEN III has been designed as a very lightweight and fast wall-climbing robot for inspection of tank and pipe walls in nuclear power plants. Its positioning on walls must be carried out by a manipulator arm carried by a remotely operated tracked vehicle, as shown in Figure 4. One of the first tasks preview for ROBICEN III is the inspection of a radwaste tank in C. N. Santa María de Garoña, Spain. The stainless steel tank has a 4 m. diameter, a height of 5 m., and is mounted on four steel legs. A radiation dose of around 1.2 Gy/h in contact precludes direct human inspection of the tank.

Table 1 shows a comparison between ROBICEN I, II and II.

	ROBICEN I	ROBICEN II	ROBICEN III
Max. dimensions (mm)	690x360x320	920x200x350	290x160x250
Mass (Kg)	22	24	2.5
Max. speed (m/s)	0.1	0.2	0.35
Air flow (Nlit/min)	1105	643	212
Suction cup area (cm^2)	377.5	272	62.5

Table 1. Comparison between ROBICEN I, II and III.

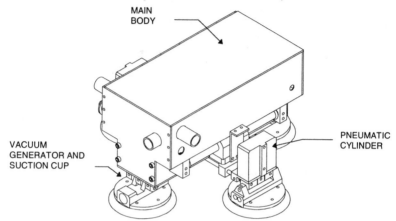

Figure 5. ROBICEN III wall-climbing robot.

4.1 Locomotive subsystem

Like ROBICEN I and II, this third version is a fully pneumatic robot with a locomotive subsystem based on suction cups. The drive unit of ROBICEN III, which is shown in Figure 5, is composed of a main body that accommodates electrovalves, vacuum sensors, cameras and control electronics. The joints and guides of the locomotive mechanism are fixed on this central body.

4.1.1 Locomotive mechanism

Two rodless pneumatic cylinders of 10 mm. diameter and 100 mm. stroke are mounted in parallel on each side of the main robot's body and work simultaneously as guides and as actuators (see Figure 6). Two additional transversal guides, mounted across the rodless cylinders with the aid of cylindrical joints, have one suction cup on each end.

To obtain a forward motion, the two lateral suction cups are simultaneously fixed to the wall whilst the two rodless cylinders are actuated from front to rear, producing a forward motion of the robot's body. Then, the two suctions cups fixed on the main body are simultaneously activated and fixed to the wall. Once the vacuum sensors detect that vacuum has been established, the lateral suction cups are released and the rodless cylinders are actuated to their forward end. By successively repeating this sequence, the forward motion of the robot, depicted in Figure 7, is produced. To produce a rotation, an analogous sequence of suction cup activations

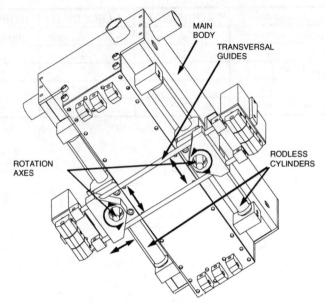

Figure 6. Bottom view of ROBICEN III showing the locomotive mechanism.

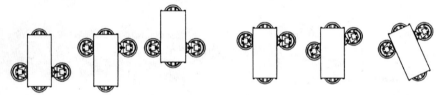

Figure 7. Forward motion of ROBICEN III   *Figure 8. Rotation of ROBICEN III*

must be produced, but this time the rodless cylinders must be actuated in opposite directions (see Figure 8).

4.1.2 Pneumatic subsystem

The pneumatic subsystem, whose circuit is depicted in Figure 9, is composed of the following elements:
- Two rodless cylinders of 10 mm. diameter and 100 mm. stroke made of aluminum and stainless steel, which are commanded by four miniaturised 3/2 electrovalves, produce the robot's motion.
- Four double-effect cylinders with short stroke, which are commanded by a 5/2 electrovalve, produce the up-down motion of the suction cups.
- Four embedded vacuum generators commanded by two 5/2 electrovalves.

Unlike the first two prototypes that used commercial pneumatic components for vacuum generation and suction cup actuation, in ROBICEN III a new design has been developed. In this design, the vacuum generator is embedded in the suction cup's body and the actuation cylinder is lighter more compact and more robust than current commercial versions. Actuator robustness is achieved mainly by two linear guides

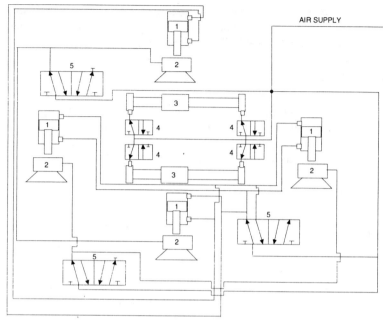

Figure 9. Pneumatic circuit of ROBICEN III: 1) Pneumatic cylinders 2) Vacuum generators and suction cups 3) Rodless cylinders 4) 3/2 electrovalves 5) 5/2 electrovalves.

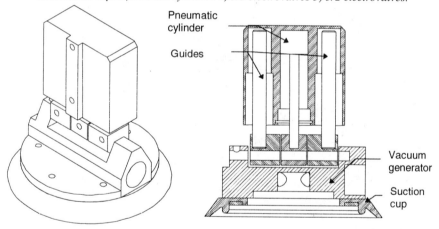

Figure 10. Vacuum subsystem with vacuum generator embedded in suction cup's body.

placed in parallel with the actuator's clynder. This novel approach obviates the need for a heavy commercial vacuum generator located in the main body and thus simultaneously reduces the robot's mass and lowers its centre of gravity. In addition, the existence of four independent vacuum generators makes the system more robust.

The vacuum subsystem shown in Figure 10, consists of three main components: pneumatic cylinder, vacuum generator and suction cup. The pneumatic cylinder that produces the up and down motion of the suction cup is reinforced with two guides to improve its lateral stiffness. The vacuum generator body is hinged to the pneumatic cylinder and guides so that the suction cup always orientates itself in a direction normal to the wall. Finally, the suction cup is made of silicon and has been designed to produce good adherence on flat walls with small to medium irregularities.

4.2 Control subsystem

The control subsystem of ROBICEN is composed of the onboard control card shown in Figure 11. The control card, which has been designed around the Intel 80535 processor, controls the electrovalves and the measurements of the sensors. Although all the electrovalves are of the on-off type, each rodless cylinder has a position sensor whose measurement is used in a feedback loop to control more precisely the robot's motions.

Each suction cup has a vacuum sensor whose measurement is used by the control algorithm to establish whether a step is successful. Whenever a step is considered unsuccessful, it is repeated with successively shorter strokes until sufficient vacuum is obtained. This simple strategy has proven to be very effective, allowing the robot to pass over normal weldings of steel tanks without compromising the robot's safety.

The robot is teleoperated from a PC that sends commands to the onboard controller card through an RS-232 serial port. Since onboard electronics can be damaged by radiation a new version of ROBICEN III with a minimum amount of embarked electronics is under development.

Figure 11. Control card of ROBICEN III

4.3 Sensorial subsystem

In addition to the position and vacuum sensors previously mentioned, ROBICEN III is currently provided with four pin-hole CCD video cameras and an ultrasound sensor to measure wall thickness. Other sensors can also be accommodated.

5. Conclusions

A family of three wall-climbing robots for inspection of nuclear power-plants has been presented. ROBICEN I, which is comparatively heavy and of limited mobility, has been used as a laboratory prototype in the feasibility study phase. ROBICEN II, which has the ability to pass from a horizontal to a vertical surface by its own means, is made of lighter materials and is a more sophisticated design. However, the ability to go from a horizontal to a vertical plane causes a significant increase in the robot's weight. Nevertheless, the overall performance has been satisfactory. Finally, ROBICEN III is a lightweight wall-climbing robot of only 2.5 Kg. with an original and efficient mechanical design. The excellent performance of ROBICEN III on vertical walls is partly due to the simplicity of its mechanical design and to the compact and robust design of its four independent vacuum generators embedded in the suction cup's body. ROBICEN III has been designed as a tank inspection robot and will soon be involved in the inspection of a highly radioactive radwaste tank.

Acknowledgements

The authors gratefully acknowledge the financial support of this project by Iberdrola and the leading role of Jesús Gómez Santamaría over the past two years.

References

[1] Gelhaus, F.E. and Roman, H.T. 1990. Robot applications in nuclear power plants. *Progress in nuclear energy, Vol. 23, N°1, pp. 1-33.*

[2] Nakayama, R., Kubo, K., Sato, K. and Taguchi, J. 1983. Development of nuclear power plant automated remote patrol system. *Proceedings IFAC-IFIP Symposium on Real-time digital control Applications, Vol.1, pp. 101-106.*

[3] Watanabe, A. and Kubo, K. 1988. Development of an automatic inspection robot for nuclear power plants. *Proceedings of IAEA international conference on man-machine in the nuclear industry, pp. 553-558.*

[4] Roman, H.T. and Marian, F.A. 1989. The year 2000 power plant. *IEEE Computer applications in power, Vol. 2, N° 2, April, pp. 19-21.*

[5] Roman, H.T. 1991. Robots cut risk and cost in nuclear power plants. *IEEE Computer applications in power, Vol. 4, N° 3, pp. 11-15.*

[6] Yamamoto, S. 1992. Development of inspection robot for nuclear power plant. *Proceedings of the 1992 IEEE International conference on robotics and automation, Vol. 2, pp. 1559-1566.*

[7] Kochan, A. 1992. Robots in the nuclear industry. *Industrial robots, Vol. 19, N° 2, pp. 15-17.*

[8] Schilling, R. 1992. Telerobots in the nuclear industry. *Industrial Robots, Vol. 19, N° 2, pp. 3-4 and 11-14.*

[9] Bahr, B., Lankarani, H.M. and Motavalli, S. 1991. Microcomputer-based control system for a surface-climbing robot. *Computers in engineering, Vol. 2, pp. 471-474.*

[10] Nishi, A. 1992. A biped walking robot capable of moving on a vertical wall. *Mechatronics, Vol. 2, N° 6, pp. 543-554.*

[11] Collie, A. 1992. Unusual robots. *Industrial robots, Vol. 19, N° 4, pp. 13-16.*

[12] Seward, D. 1992. Robots in construction. *Industrial robots, Vol. 19, N° 3, pp. 25-29.*

[13] Kerley, J.J., May, E. and Ecklund, W. 1992. The climbing crawling robot. *Industrial robots, Vol. 19, N° 4, pp. 32-34.*

[14] Ashikawa, M., Adachi, T., Katano, H., Miyashita, T., Matsui, S. and Kakikura, M. 1992. Development of wall coating removal robot. *Proceedings of 23rd. International symposium on industrial robots, October, pp. 709-714.*

[15] Gradetsky, V.G., Rachkov, M.Y. and Nandi, G.C. 1992. Vacuum pedipulators for climbing robots. *Proceedings of 23rd. International symposium on industrial robots, October, pp. 517-522.*

[16] Briones, L., Bustamante, P. and Serna, M.A. 1994. Wall-Climbing Robot for Inspection in Nuclear Power Plants. *1994 IEEE International Conference on Robotics and Automation, San Diego, Vol. 2, pp 1409-1414.*

Control of an Eight Legged Pipe Crawling Robot

Thomas Roßmann and Friedrich Pfeiffer
Lehrstuhl B für Mechanik, Technische Universität München
München, Germany
rossmann@lbm.mw.tu-muenchen.de

Abstract:

The project of an eight legged robot for pipe inspections is presented. The mechanical design, the functional principle and the control concept are described. The latter includes the task distribution to the hierarchically organized controller levels and the methods used for the control of the leg forces and robot motion.

1. Introduction

Tube systems differ in their pipe diameters, lengths, the mediums inside, the complexity of the tube arrangement etc.. Different kinds of robots have been developed for inspecting and repairing tubes from inside. They are driven by wheels or chains or they float with the medium. All types of robots have their specific difficulties, for example problems of traction or low flexibility and do not satisfy all requirements expected by the users. The aim of this project is the development of a robot moving forward by feet to study the possibilities and difficulties of legged locomotion in contrast to other systems. The higher flexibility of legs can be used to extend the possibilities of moving in tube systems. One of the requirements on the crawler is that it must be able to pass tubes independently of their inclination (from horizontal up to vertical pipes).

2. Design of the Robot
2.1. Mechanical Design

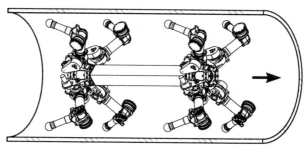

Figure 1. Pipe Crawling Robot

The robot shown in figure 1 has eight legs arranged like two stars. Four at a time are attached at the central body in a plane which contains the longitudinal axis of the crawler. These two planes are called leg planes. Each leg has two active joints, which are driven by DC-motors. Their axes of rotation are orthogonal to the leg planes. This provides each leg with a full planar mobility. The leg is mounted to the central body with an additional passive joint, which allows small compensating movements normal to the leg plane.

The crawler has a length of about $0.75m$ and is able to work in pipes with a diameter of $60 - 70cm$. In each of the eight legs, the lengths of the first[1] and the second[2] segments are $15cm$ and $16cm$, respectively. The highest possible torque of the hip joint is $78Nm$ short term and $40Nm$ permanently. The corresponding values of the knee are $78Nm$ and $20Nm$. In a stretched out position a leg is able to carry 6.5 times its own weight (less than $2kg$) permanently and 12 times for short time operations. Its mechanical design is based on a six legged walking machine [1][5]. The total weight of the crawler is about $20kg$ including the electronic parts. Details of the mechanical design can be found in [4].

2.2. Sensors and Electronic Equipment

The robot is controlled by five Siemens micro-controllers 80C167 CAN, which are installed on the crawler itself. One controller acts as a central unit. Each of the remaining four units controls two opposite legs. The controllers are able to communicate over a CAN bus system. Furthermore the CAN bus system is used to connect the controllers to an external PC, which allows the user to get informations about the system and to give simple control commands.

Each leg has two potentiometers to measure the angles of the active joints and two tachometer generators to measure the angular velocity of the motors. For measuring the contact forces to the pipe a special light weight sensor was developed, which is integrated in the last segment (see below). For future extensions the electronic architecture allows the implementation of further sensors. This can be necessary for the recognition of the tube geometry for example.

In figure 2 a cut though the force sensor in the outer leg segment can be seen. The sensor is based on strain gauges and consists of the following parts:

- The *measuring plate* with four strain gauges, which are arranged as a full bridge for the determination of the longitudinal forces.

- At the end of the *inner leg segment* the rubber ball which is used as foot is attached. It is held by the measuring plate on the one side and the membrane on the other side. On the inner leg segment four strain gauges are located as two half bridges[3] which are able to determine the bending torques of the "beam" in both directions. Furthermore the segment carries the entire sensor electronic[4] on four small printed circuit laminates (not

---

[1] distance between the two active joints (hip and knee)
[2] distance between knee joint and foot
[3] In figure 2 only two strain gauges can be seen. The other two are skewed with 90 degrees.
[4] power supply, amplifiers, DAC-converters for bridge balancing, devices for adjusting the sensor sensitivity

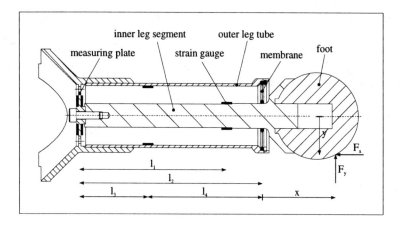

Figure 2. Force Sensor

shown in the figure).

- The *membrane* is needed to support the inner leg segment in the case of lateral forces. It is very important that it's longitudinal compliance is very high therewith the corresponding forces are not affected and can get to the measuring plate.

- The *outer leg tube* has as well as the inner leg segment two pairs of strain gauges for the bending torques. The tube guarantees a high lateral stiffness and load capacity in combination with a small own weight, which is a important aspect for walking machines.

The lateral forces and the bending torques caused by the lateral forces, respectively, are measured two times. Therefore it is possible to determine the forces independently of the actual contact point on the foot. The bending torques T (index olt for outer leg tube and index ils for inner leg segment) and the corresponding output of the sensor electronic U are given by the equations:

$$T_{olt} = \frac{l_4}{l_2}(F_x y - F_y x) - F_y l_4 \tag{1}$$

$$T_{ils} = \frac{l_1}{l_2}(F_x y - F_y x) \tag{2}$$

$$U_{olt} = k_{olt} T_{olt} \tag{3}$$

$$U_{ils} = k_{ils} T_{ils} \tag{4}$$

With a appropriate choice of the amplification factors k the lateral force can be given independently on x or y:

$$k_{ils} = \frac{l_4}{l_1} k_{olt} \rightarrow F_y = \frac{U_{ils} - U_{olt}}{k_{olt} l_4} \tag{5}$$

With the same method the lateral force in z-direction is determined.

As already mentioned the longitudinal force F_x is measured by the use of the measuring plate. It should be recognized, that all four strain gauges on the plate need the same sensitivity to keep the theoretical independence on an radial load. Caused by the high gradient of the elastic deformation and small errors in the strain gauge placements this can not be realized without further calibration. Therefore it is necessary to determine the sensitivity for each strain gauge separately and to adapt it accordingly.

3. Control of the Robot
3.1. Gait Pattern
The gait pattern is determined by the limited leg mobility. The two active joints provide the legs with a full planar mobility. Neglecting the small motions in the passive joints each leg with pipe contact reduces the degrees of freedom of the central body in the general case. With respect to the necessity to use two opposite legs at the same time, which allows to jam the crawler in the tube, it is useful for a maximum mobility to take the four legs of one leg plane at the same time. This provides the crawler with full mobility in this plane. An other advantage of this gait pattern is that no drilling friction is induced in the contact areas of the feet.

3.2. Task Distribution
The realization of the robot motion with respect to the requirements on stability, gait pattern, dynamics, and control can be achieved by a division of the whole problem in different subproblems. These subproblems differ in the their functionality and in their sphere of influence. Therefore, a control structure was chosen, which can be divided in two directions. The first direction is a function orientated division in coordination levels and levels, which have to realize the operations. The last ones are called operating levels. The second division is given through the concerned components. Therefore this division leads to central and local controller parts. The distribution of the different tasks to these four controller levels can be seen in figure 3.

- The *central coordination level* coordinates the phase characteristics of the two leg planes. Decisions on switching of the legs under load are made by this component. Furthermore, the problems which can only be mastered by a reaction of the whole robot should be solved in this level (e.g. the legs of one plane can not find contact in their whole working area).

- The *local coordination level* controls changes between the different leg phases. For a normal step the phases change in the following sequence: stance, protract, swing and retract. It also reacts to disturbances like avoiding small obstacles or finding no contact.

- The *central operating level* controls the position and the velocity of the central body. For this purpose a appropriate force distribution is given to the local operating levels. These commands must be created with respect to restrictions like satisfying the condition of sticking or the limitations of the electrical and mechanical components.

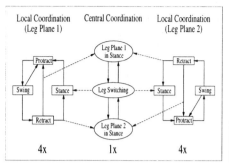 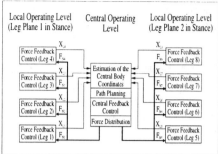

Figure 3. Level of Coordination and Operating Level

- The *local operating level* controls the applied forces during the contact phase and the motions of a single leg during the different air phases. In contrast to the last ones, which are really local problems, the leg forces are strongly coupled and therefore the force control can not be realized without a corresponding communication between the leg controllers.

4. Realization of the Controller Levels

4.1. Coordination Levels

The coordination levels are realized as finite state machines. A complex task is realized by a sequence of simple predefined behaviours (states). The switching between the different states is event related and also predefined. For example the local coordination level of one leg in the swing phase switches is the state retract in the case that a leg passes a limit in the front area. In the case a leg will touch an obstacle in the swing phase before it reaches this limit, it would switch in an other state, which realizes a behaviour for avoiding the obstacle. The different controller states of the coordination levels can be seen in figure 3. The arrows between the states indicate the state transition in the case of the corresponding events.

4.2. Operating Levels

The operating levels of the tube crawling robot consist of classical control elements. For the leg's position and velocity controllers, respectively, in non-contact phases, simple PID-controller can be used. This is possible, because the crawler motions are relative slow and the requirements on the accuracy of the leg motions are very low. Much more effort is required for the realization of the robot motion with respect to the conditions of sticking pipe contacts and preventing joints from overloading. This is caused by geometrical and kinetic nonlinearities of the system. It has many degrees of freedom, which are strongly coupled. Furthermore not all quantities necessary for controlling can be measured and therefore disturbance and state observers are needed. To get get a control design which corresponds to the described structure a cascade system was chosen. The inner control circuit of this system is a leg force controller, which is realized by the local controller level. The outer one controls the robot motion giving appropriate setpoints to the force controller. This corresponds to the central operating level.

4.2.1. Local Operating Level

Several control methods are tested for their suitability mainly for the force controller by using a simulation program for the system dynamics and a single leg test setup. The best results were archived by a controller which is based on the method of feedback linearization. In comparison to the others the control quality of this design was independent from the actual leg configuration. The disadvantage of this method is the more complicated and more complex structure, which requires more effort on communication between the different controllers as for example local PID-controllers [6].

To minimize the necessary computational power a simplified model was used for the controller design. In comparison to the model used for the simulation the following simplifications are done:

- Motions in the passive joints are not observable and not controllable by the legs of the corresponding leg plane. Therefore these motions must not be considered. This leads to a planar model with 11 degrees of freedom.

- The damping of the rubber balls (feet) is neglected.

- The masses of the segments are added to the central body and therefore the moments of inertia referred to the leg joints are decoupled from the central body coordinates. Caused of the light weight design the influence of this simplification is less than one per cent.

- It is assumed that the gear friction can completely be compensated by a disturbance observer and will not be considered further on.

Furthermore, the central body velocity and the actual direction of gravity are assumed to be known. In reality these variables must also be determined by an observer. This model yields the following equations of motion:

$$M_c \ddot{q}_c = A_c F + G \quad (6)$$
$$M_{li} \ddot{q}_{li} = J_{li}^T F_{li} - D \dot{q}_{li} + B U_i \forall i = 1, \cdots, 4 \quad (7)$$

Generally, the index c denotes the central body and li the i-th leg. M are the mass matrices, q the generalized coordinates. The Jacobian J_{li} is the derivative of the leg end point velocity w.r.t. $\dot{q}_{li}$. In the same way is J_{ci} the derivative w.r.t. $\dot{q}_c$ (later used). F is the vector of all leg forces and F_{li} that of the i-th leg. U_i are the motor inputs with the corresponding input matrix B, G the vector of the gravity forces in the plane and D characterizes the dependency of the motor torques on the angular velocities. The matrix A_c depends on q_{li} and q_c and describes the influence of the forces on the central body accelerations.

For the input-output-linearization the leg forces are taken as the system output. They can be calculated using the following equation:

$$F_{li} = C_{li}(x_{li} - x_{lif}) \quad (8)$$

x_{li}, x_{lif} are the coordinates of the leg end point and of the touching point at the tube. C_{li} represents the stiffness of the rubber ball.

Applying the control method yields the following equations for the closed inner control circuit. The desired linear dynamics is given through the matrices A_F and $A_{\dot{F}}$.

$$M_c \ddot{q}_c = A_c F + G \tag{9}$$
$$\ddot{F}_{li} = -A_{\dot{F}} \dot{F}_{li} - A_F F_{li} + A_F F_{li,nom} \quad \forall \ i = 1, \cdots, 4 \tag{10}$$

$F_{li,nom}$ is the new system input and contains the set points for the leg forces. An interesting property of the system is that the linear behaviour can be stabilized for any positive definite matrices C_{li} by consideration of some aspects in choosing A_F and $A_{\dot{F}}$. Caused by the uncertainness of the foot stiffness this property is an important advantage of this controller.

4.2.2. Central Operating Level

The remaining internal dynamics of the system correspond to the central body motion. The zero dynamics show a double integrating behaviour and are consequently not asymptotically stable. The task of the outer control circuit (central operating level) is to stabilize the system by giving the right setpoints $F_{li,nom}$. For that purpose, total forces and torques are determined to compensate gravity and to correct the robot position. Therefore the system is transformed into a form, in which the forces are split in active ones F_a[5] and passive ones F_p[6]. This transformation yields the following equations, if in (10) for both components of F_{li} the same decoupled dynamics is chosen:

$$M_c \ddot{q}_c = A_{c,a} F_a + G \tag{11}$$
$$\ddot{F}_a = -A_{\dot{F},a} \dot{F}_a - A_{F,a} F_a + A_{F,a} F_{a,nom} \quad \in \mathbb{R}^3 \tag{12}$$
$$\ddot{F}_p = -A_{\dot{F},p} \dot{F}_p - A_{F,p} F_p + A_{F,p} F_{p,nom} \quad \in \mathbb{R}^5 \tag{13}$$

It can be seen, that the passive forces are one sided decoupled and therefore stable in the sense of "bounded input, bounded output". They must not be considered further on. For the remaining system it is not difficult to determine the setpoints for $F_{a,nom}$ in that way that the central body motion can be stabilized.

4.2.3. Force Distribution

The three-dimensional vector of active forces $F_{a,nom}$ must be distributed to the four legs, i.e. to eight force components (vector F). This has to be done under consideration of the friction limits and without an overloading of the joint drives. Both can be influenced by an appropriate choice of the passive forces $F_{p,nom}$. Such kind of problems can be solved using an optimization or applying empiric rules. In comparison to the optimization, the last method has the advantage that less computational power is needed. The disadvantages are the higher load for the system components and the difficulty to find such

---

[5] they have influence on the cental body motion
[6] they are only used for spreading in the tube

rules for the general case[7]. Nevertheless caused by the reason, that the micro controllers in use have no good floating point performance this method has been implemented in a first step. In contrast, the connection to the external PC allows the implementation of real time optimizations for testing the suitability of different algorithms. Later of course this will be done using an on board PC. Simulations have shown that the possible payload for the pipe crawling robot can be enlarged significantly by minimizing the maximal joint torque T_j referred to the maximal possible torque $T_{j,max}$ of this joint:

$$\text{Max}(\frac{T_j}{T_{j,max}}) \to \text{Min} \tag{14}$$

$$\text{with} : \quad \boldsymbol{AF} = \boldsymbol{F}_{a,nom}; \quad \boldsymbol{GF}_a + \boldsymbol{g} \geq 0; \quad \boldsymbol{J}_l^T \boldsymbol{F} = \boldsymbol{T} \tag{15}$$

The first linear equality constraints of (15) guarantee that the leg forces are equivalent to the active forces $\boldsymbol{F}_{a,nom}$. The inequality constraints represent a linearized friction cone, which results in a friction pyramid and finally the last expression describes the correlation between leg forces and joint torques for the static case. This optimization problem can be transformed into the following linear optimization problem:

$$k \to \text{Max} \tag{16}$$

$$\text{with} : \quad \boldsymbol{AF}^* = \boldsymbol{F}_{a,nom}k; \quad \boldsymbol{GF}^* + \boldsymbol{g}k \geq 0; \quad \boldsymbol{J}_l^T \boldsymbol{F}^* = \boldsymbol{T}^* \tag{17}$$

$$-T_{j,max}/T_{max} \leq T_j^* \leq T_{j,max}/T_{max} \quad \forall \quad j = 1...8 \tag{18}$$

Thereby T_{max} is the over all maximal possible joint torque ($T_{max} = \text{Max}(T_{j,max})$). The solution of (14) results from the solution k_{max} of problem (16) by $\boldsymbol{F} = \frac{1}{k_{max}}\boldsymbol{F}^*$ and $\boldsymbol{T} = \frac{1}{k_{max}}\boldsymbol{T}^*$ respectively. It can be seen that k_{max} is the quotient of the normalized joint torques T_j^* and the real joint torques T_j. Therefore it will be called load factor. In addition to the payload the second big advantage of this cost function is that there are powerful algorithms solving linear programming problems, e.g. the simplex method. The only disadvantage of this method is, that the solutions of linear optimization problems show a non smooth behaviour, even in the case that the constraints are smooth. Such a behaviour has effects not acceptable for the force controller if the system stay in the vicinity of such a saltus and leads to the uselessness of linear optimizations. Nevertheless this disadvantage can be eliminated by a modification of the cost function:

$$\frac{f_Z(\sum |\Delta T_j^*|) + c_Z}{k + c_N} \to \text{Min} \tag{19}$$

f_Z, c_Z and c_N are constants, which can be used to adjust the behaviour of the solutions[8]. ΔT_j^* are the changes of the normalised joint torques corresponding to the last optimization result. These quantities in the numerator of the cost function prevent the solution of the optimization from jumping without a

---
[7] normal vectors and friction coefficients for every contact point are different
[8] for $f_Z = 0$, $c_Z = 1$ and $c_N \geq 0$ are (16) and (19) equivalent

significant change of the load factor k. The advantage of this kind of cost functions is that a modified simplex method can be used for solving the problem [3], which is as efficient as the simplex method for linear programming problems. The behaviour of both optimizations can be compared in figure 4, which shows the unsteadiness of the computed joint torques for the first cost function at a critical region and the corresponding results of the second one.

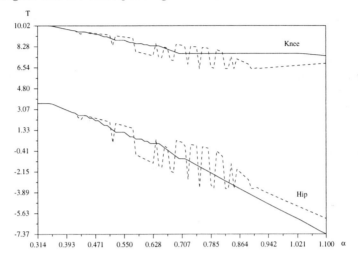

Figure 4. Optimization Results

4.2.4. Gravity and Velocity Observer

For controlling the robot motion it is necessary to know to crawlers velocity and angular velocity, respectively. Furthermore the dynamics of the closed loop system will have a much better behaviour if the effects of gravity in equation (11) are compensated by a feed forward control. Of course the crawler velocity as well as the robot position can be estimated from the corresponding values of the leg joints. But simulations have shown, that in contrast to the robot position the exactness of the estimated velocities is not sufficient all the time. This is caused by the compliance of the rubber balls, which are used as feet. In the case of fast changes of the leg forces (i.e. $100N$ in $0.1s$ by changing the active leg plane) the necessary relative velocity to deform the ball is about $50mm/s$ for a ball stiffness of $20N/mm$, and has therefore the same magnitude as the robot velocity. To reduce the error it is necessary to use a filter which is able to suppress the corresponding "oscillations" or to use a observer. Caused by the better performance and by the reason that an observer is already needed for the direction of gravity the second possibility was chosen. The observer design is based on the following equations:

$$\begin{pmatrix} \dot{q}_c \\ \ddot{q}_c \\ \dot{G} \end{pmatrix} = \underbrace{\begin{pmatrix} 0 & E & 0 \\ 0 & 0 & E \\ 0 & 0 & 0 \end{pmatrix}}_{=A_o} \begin{pmatrix} q_c \\ \dot{q}_c \\ G \end{pmatrix} + \begin{pmatrix} 0 \\ A_{c,a} F_a \\ 0 \end{pmatrix} \qquad (20)$$

Assuming that q_c can be estimated good enough (see above), $A_{c,a}F_a$ can be chosen as the measured output y. This is also the reason why $A_{c,a}F_a$ can be written as a known system input in equation (20). The usual observer approach with $\tilde{r}$ as the system error yields to the error differential equation

$$\dot{\tilde{r}} = A_o \tilde{r} - L(A_{c,a} F_a - \bar{A}_{c,a} \bar{F}_a), \qquad (21)$$

in which $\bar{A}_{c,a}\bar{F}_a$ corresponds to the output $\bar{y}$ of the observer. Equation (21) can be linearized and for $\tilde{r} = 0$ and for the resulting system a observer design is possible for which global stability can be shown. Therefore the nonlinear system is also stable at least near the linearization point. The quality of the designed observer can be seen in figure 5, which shows the comparison of the observed and the "real" crawler velocity of a simulation.

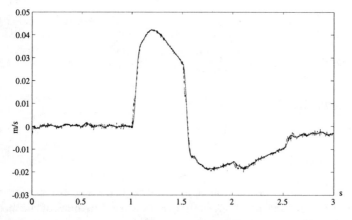

Figure 5. Observed Crawler Velocity

5. Dynamics

A simulation program, which includes all the relevant properties of the robot, was developed. By means of this program it is possible to get informations about the system behaviour and to determine the motor power reserves. Since the elastic eigenfrequencies of the system parts are very high, a modelling as a rigid body system is favourable. The system parts are the central body, the rotors of the motors, the shafts of the gears, and the segments of the legs. Different to industrial robots, the stiffness of the gears is negligible for the system behaviour. The reasons are the extreme light weight design, the very short lever arms and the small moments of inertia of the segments. The friction of the used Harmonic Drive Gears, which strongly depends on the torque, has great influence on the control and on the loads of the motors (coulomb friction in meshing). For consideration of this effect in the simulation, "normal torques" are established to calculate tangential friction torques that act against the direction of the rotation. To include sticking without load (effects like No-Load Starting Torque and No-Load Back Driving Torque) an initial tension λ_0 of the gears is introduced. For sticking under load the transmitted torques are added to the initial tensions. The dynamics are described by the following equations:

$$M\ddot{q} = h + W_N\lambda_N + W_H\lambda_H + H_R(\lambda_{Na} + \lambda_0) \tag{22}$$
$$\ddot{g}_N = W_N^T\ddot{q} + \tilde{w}_N = 0 \tag{23}$$
$$\ddot{g}_H = W_H^T\ddot{q} + \tilde{w}_H = 0 \tag{24}$$

with:
$$H_{Ri} = -\mu_i w_{Ti}\text{sign}(\dot{g}_{Ti}) \tag{25}$$
$$\lambda_{Nai} = |\lambda_{Ni}| \tag{26}$$

M denotes the mass matrix, h is the vector of gyroscopic forces, active moments and active forces. The vector of the generalized coordinates q (its dimension is 62) contains the six degrees of freedom of the central body, the angles of the leg joints and the degrees of freedom of the motors and the gears. λ_N is the vector of the normal torques (see above) and λ_H the vector of the constraining torques caused by sticking of the gears. Corresponding to this, the constraints are denoted g_N and g_H with the Jacobian matrices W_N and W_H. Different to usual stick-slip problems, the normal torques can have positive and negative values and therefore they correspond to the normal forces of a bilateral guiding device. The friction torques of the rotating gears are determined by the absolute values of the normal torques, the initial tensions and the friction coefficients μ_i. They are projected into the generalized sliding directions through w_{Ti}. As usual, the passing from sticking to slipping is detected by reaching the stick limit $\lambda_{Ti} = \mu(\lambda_{Ni} + \lambda_{0i})$ and from slipping to sticking with the kinematic condition $\dot{g}_{Ti} = 0$. In order to determine the accelerations, the equations are transformed into a linear complementary problem, which can be solved with the Lemke algorithm [2].

The results of a simulation of the whole robot crawling in a vertical tube are presented in figure 6. The robot walks about $0.5m$ and makes several steps. In the upper left graph the robot position in direction of the tube axis is shown and in the upper right one you can see the corresponding velocity. In the graphs in the second row shows the normal force (left) and the the tangential forces (right) of the first leg. The steps can easily be detected in the force diagrams. The peaks in the velocity diagram occur during switching of touching legs.

6. Conclusion

Before the presented algorithms were implemented on the robot, they were tested in the simulation program and a single leg test setup. Both of them

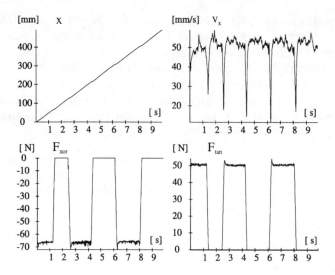

Figure 6. Simulation of the Whole Robot

are also used for testing the mechanical design. At the test setup the leg is mounted on a fixed frame and can walk on a conveyor-belt, which is motor driven and can be run with different velocities. As well as the simulations of the whole system, the measurements at the test setup yielded the desired results. Furthermore with the help of the test setup, it was possible to prove the quality of the simulation model. Caused by this procedure the putting into operation of the crawler brought no further problems and it worked at the first go. The crawler is now able to walk through straight pipes independently of the position inside the tube or the inclination of the tube (from horizontal up to vertical pipes). Future work aims mainly at the extension of the controller levels for walking through curves and other more demanding tube geometries. This refers mainly to strategies of finding appropriate contact points and the detection of the actual tube geometry.

References

[1] Eltze, J.: Biologisch orientierte Entwicklung einer sechsbeinigen Laufmaschine, no. 110 in Fortschrittsberichte VDI, Reihe 17, VDI-Verlag, Düsseldorf, 1994.
[2] Glocker, C.; Pfeiffer, F.: Multiple Impacts wich Friction in Rigid Multibody Systems, Nonlinear Dynamics, Kluwer Academic Publishers, (1996).
[3] GROSCHE, G.; ZIEGLER, V.; ZIEGLER, D.; ZEIDLER, E., eds., Teubner-Taschenbuch der Mathematik Teil II, B. G. Teubner, Stuttgart, Leipzig, 1995.
[4] Herrndobler, M.: Entwicklung eines Rohrkrabblers mit vollständigen Detailkonstruktionen, Master's thesis, Lehrstuhl B für Mechanik, TU München, 1994.
[5] Pfeiffer, F.; Eltze, J.; Weidemann, H.-J.: The TUM-Walking Machine, Intelligent Automation and Soft Computing, 1 (1995), pp. 307–323.
[6] Roßmann, T.; Pfeiffer, F.: Contol and Design of a Pipe Crawling Robot, Proc. of the 13th World Congress, of Automatic Control, I. F., ed., San Francisco, USA, 1996.

Chapter 8

Sensor Data Fusion

Autonomous navigation usually requires the fusion of multiple sensor data to extract the required world information to move safely and efficiently. In this chapter new approaches for world modeling, world interaction and vehicle positioning are presented. An additional and different kind of application requiring also sensor fusion, deals with automation of paper recycling.

Henriksen, Ravn and Andersen study how to improve the performance of autonomous vehicles interacting with the surrounding environment. The work is based on the "sens-act" paradigm, fusing data from three kind of sensors: vision, sonars and mechanical probing. They propose an architecture based on three subcomponents: perception, world modeling and actuation, where the perception unit is not necessarily passive. They experiment with such a system through indoor environments, navigating and doing some actions such as door opening. The mechanical probing device or active antenna, a one d.o.f. arm, operates either as a sensor or as a manipulator.

A quite different problem is treated by Faibish, Bacakoglu and Goldenberg. They develop a robotic system used for detection, sorting and grading of paper objects from visual features. The system consists of the sensing devices: vision and US, and the robot arm with a vacuum gripper. Vision is used for object segmentation, features extraction and color detection (to detect the non written area). The experimentation has been addressed to define the optimal parameter set to automate the process of paper recycling.

An algorithm for dynamic map building based on a 2D range finder is proposed by Vandorpe, Van Brussel, De Schutter, Xu and Moreas. The method takes into account uncertainties of the parameters before matching corresponding primitives. The estimation of a primitive is improved during a continuous observation. The experimentation is carried out on the mobile robot LiAs. Experiments in a extremely cluttered environment produced good maps built from geometrical primitives, line segments and clusters.

Hanebeck and Schmidt develop a new approach to estimate mobile robot posture. It consists on an integrated localization system for initializing and continuously updating the robot posture. The limited capabilities of the simple sensors used is compensated by their data processing and intelligent strategies. Efficient closed-form solutions are derived for calculating the robot posture based on both angle and distance measurements. After its initialization, the robot starts moving and its posture prediction is estimated based on incremental sensor data. Since statistical data fusion does not apply, set-theoretical data fusion is utilized. Experimental results are shown from the navigation tests performed on the mobile robot ROMAN.

Autonomous Vehicle Interaction with In-door Environments

Lars Henriksen, Ole Ravn and Nils A. Andersen

Department of Automation, Technical University of Denmark,
Building 326, DK-2800 Lyngby, Denmark,
E-Mail: or@iau.dtu.dk

Abstract

Abstract: The paper describes the possibilities of enhancing the performance of autonomous vehicles by interacting with the surrounding environment. Benefits and drawbacks of the use of different sensors and strategies for integrating the acquired knowledge are examined in an example case of a 'through partly closed door' navigation task for an AGV. This illustrates the increased ability of sensors to acquire knowledge by actively doing explorative actions. A combined mechanical actuation and probing device accentuates the sense-act paradigm.

The theory and algorithms used are described in detail, and extensive experiments with an implementation of the 'through door' navigation algorithm on our non-holonomic test-bed Autonomous Guided Vehicle are documented.

1. Introduction

Autonomous Guided Vehicles (AGVs) have been research topics for several years but still only few commercial vehicles are available that solve real-life problems. However there seem to be a large potential for developing vehicles with such capabilities. If available they will certainly be quite useful in a wide variety of tasks like e.g. transport in warehouses and hospitals, autonomous cleaning, etc.

An important factor regarding functional robotic systems including mobile platforms is the interaction between vehicles and the surrounding environment. To use the robot or mobile platform for any real task other than merely avoiding obstacles interaction with the enviro. nent is needed. Here focus is put on intelligent interaction between the vehicle and environment through the use of multiple sensors, sensor fusion techniques, a 'se e-act' paradigm for solving complex tasks and finally the integration of these into manoeuvres.

Especially the 'sense-act' paradigm and the combination of actuators and sensors in one (logical or actual) device is interesting as it enables interaction at a lower cost and at less mechanical complexity. In this paper the existence of basic locomotion and perceptual capabilities such as trajectory following [Ravn et.al, 1995], vision, pan/tilt, sonar, odometry etc. is assumed.

The paper is organized as follows: Section 2 describes some extensions to well known architectures in the area of perception enabling a structured approach to active sensing. Active sensing in turn is useful when interacting with the environment in a intelligent fashion. Three sensor types (vision, sonar and mechanical probing) are described in relation to the use of active sensing and benefits and drawback are outlined. Furthermore the combination of the sensors for enhanced performance is described. In section 3 the general concepts introduced in section 2 are illustrated through an example. The example used is an AGV opening a partly open door using a low-DOF manipulator and several sensors. Section 4 concludes the experiences obtained.

2. Interaction with the environment

Much previous work with vehicles such as a typical AGV have focused on obstacle avoidance to such a degree that vehicles seem almost mysophobic. All surrounding environment is perceived as obstacles which should be avoided by a safe (and large) margin. Our approach attempts to utilize and interact with the environment to enhance the vehicle performance especially in man-made environments. The perspective is to be able to use the vehicle with high performance in man-made environments with only small adaptations made to the environment. The example considered in this paper is navigation through a possibly closed doorway. It is possible to consider equipping all doors with automatic door-openers, but this is rather costly and inflexible. A vehicle with the capability of detecting the door and open it if needed would be more flexible and economically attractive.

When faced with a task involving interaction between the environment and the autonomous system, considerations be made about how to achieve the task in the most reliable, flexible and cost-efficient way. In the example introduced later of opening a partly closed door it could be considered equipping the door with an automatic door opener and having the autonomous system communicate with this. This is a rather simple solution when only a few doors are considered but the flexibility is low as only certain doors can be opened. A balance between flexibility and reliability should in each case based on the domain knowledge about the task. At the lower single sensor level the modifications to the environment could enable or robustify the algorithms used to yield better performance at a low cost in flexibility. This is utilized in the experiment section.

Architectures like RCS shown in figure 1 and other hierarchical architectures for autonomous systems are based on the three subcomponents perception, world modelling and actuation. These components are then grouped into hierarchical layers with respect to complexity.

The key observation is that for instance the perception unit is seen as a passive unit with respect to the environment. We propose to view the perception unit as a possibly active unit where the unit itself can contain an actuation unit. Doing this provides several advantages. A single sense (or perception) operation can contain several actuation components and this is done without making the central architecture of the autonomous system more complex.

Many sensors can successfully be mapped into this paradigm. Just to mention a few, vision and image processing is normally perceived as a passive act but if the preceding steps of finding the optimal view and maybe changing lighting conditions is taken into account the complete perception step clearly contains active parts.

During the last years several investigations of different sensors for autonomous vehicles have been made and attention to the area of fusing the information has also

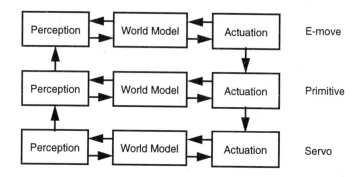

Figure 1. Simplified view of the RCS-3 Architecture

been considerable. We have based our approach on the following sensor types, integrating the data and utilizing the sensors in an active fashion to get the needed data:

2.1. Vision

As reported in [Andersen et al., 1991] vision used as a sensor has several benefits. It is multi-purpose and can be reconfigured to measure different aspects on-line. It is non-intrusive and is remote sensing by nature. However special measures have to be taken in order to cope with the limited accuracy of the measurements and the large amounts of information.

When used in a practical applications a priori domain knowledge about the environment and the task as well as modifying the environment can be used to limit the computational burden and to robustify the measurements. Adding physical markers to the environment is sometimes useful but in some situations it is not feasible. Active signalling using structured light is one way of achieving markers without physically adding elements to the environment. Also a large amount of domain knowledge helps to improve performance, [Natagani, Yuta, 1995].

2.2. Sonar

Sonar is interesting due to low cost and physical complexity, but the low reliability of the range readings makes the use for detection and location purposes difficult. Compensation for this can be attempted by the moving the vehicle, verifying data and modifying the environment by use of acoustic reflectors mounted on the object measured. Still performance is reported not to be sufficient for industrial environments, [Budenske, Gini 1994].

2.3. Active antennas

Although most frequently reported as pure sensing devices antennas can be used both as probing devices in the sensing phase and as simple manipulators in the acting phase. This feature can reduce the complexity of the system with equivalent performance to systems with separate sensing and actuating devices. [Kaneko *et. al.*, 1995] describes how to find the location of the contact point between a flexible antenna and an object in 3D. However to get a reasonable precision a special probing strategy is required which includes motion of the vehicle. The use of contact sensors and compliance in the device can be used to measure the mechanical compliance of the environment giving valuable information. Proximity sensors mounted on the probing device can yield more precise information about the exact location than achievable using for instance vision.

3. Experiments

The example exercise is a drive-through-door task for an AGV (see figure 2). It is assumed that the AGV is led to the door by higher levels of navigational subsystems including planners, dead-reckoning, vision etc. When the through-door task is initiated the AGV is thus positioned in front of door from a few meters away and facing it with a direction approximately normal to the door opening. The state of the door is not known.

The strategy for solving the problem is a three step procedure: First the location of the door is found and the degree of openness is estimated. If the door is not sufficiently open the door is manipulated using the 1-DOF manipulator. Finally the door opening is traversed.

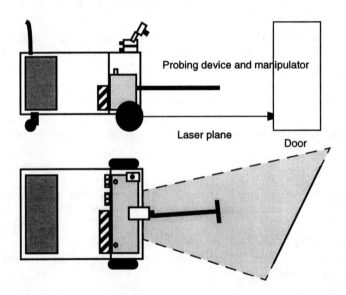

Figure 2. Experimental setup.

3.1. Vision sensor

The task of the vision subsystem is twofold. First the position of the door should be detected and the degree of openness should be estimated. Second the it should guide manipulator to the point of attack by estimating the 3D point of the door edge.

The combination of a laser and a camera possesses several advantages over a line scanning laser ranging system [Henriksen and Krotkov, 1997] which generate the same 1D type of data. First it is possible to correlate range readings with vision features providing a basis for more complex algorithms. Second, it is a less expensive alternative to laser ranging as most mobile robots are already equipped with a vision system. Third, the combination is mechanically less complex as it contains no moving parts making the system potentially more robust. Fourth, it builds in a level of redundancy as sensing can be switched to pure vision if the laser should fail.

The virtual sensor consists of two parts; a CCD camera and a line laser. The laser is a solid state line generator projecting a red laser plane towards the door. In the current configuration the laser plane is parallel to the floor at a height of approx. 10 cm above it. The camera is a commercial 512*512 pixel grey level type which is mounted approximately 50 cm above the laser, pointing 20 degrees down. The FOV of the laser and the camera is approximately the same. By mounting the laser a little behind the camera it is ensured that the laser will illuminate the entire width of the image. This configuration gives a 3D resolution under 1 cm 2 meters from the door. Ranges up to 6 meters can be extracted.

3.1.1. Door localization

With the laser turned off the on-board computer acquires an image. The laser is switched on and another image is captured and subtracted from the first image. This data is available every 120ms. The difference image of structured light is then thresholded using adaptive threshold levels leaving the signature of the laser line on the door and door frame. Missing pixels in the line due to possible harsh lighting conditions is reconstructed and the position of the line is extracted.

In detecting the door and finding the door edge it is assumed that the door is left ajar. This results in steps in the laser profile at the door edge and at the door frame which is detected by differencing and thresholding the line positions. If the door is open more than say 10 degrees the laser line disappears through the opening into the next room leaving an easily detectable gap in the laser line. The edges of this gap are treated the same way as the differenced and thresholded line positions.

Using the geometry between the laser plane and the camera the door edge is converted from image coordinates to 3D coordinates in the local AGV coordinate system. This 3D coordinate is used to suggest a locomotion command to bring the vehicle into a position where the 1-DOF arm is capable of opening the door (in collaboration with the locomotion system of the AGV). Also the arm is commanded to a position where it is ready to grip the door edge. See figure 3.

If the width of the door is known the 3D location of the hinges can be computed using the door edge and a point on the laser line on the door. Together with the door frame edge the degree of openness of door can be estimated. This is used to detect if the door is sufficiently open to grasp door or if the opening is sufficiently large for a traverse without the need for an opening procedure. Also the global position of the vehicle relative to the door opening can be computed using the hinges and the door frame enabling the AGV to position itself optimally in front of the door for opening

Figure 3. The laser illuminated door.

and traverse. Detection of doors opening inwards or outwards is detected by the position of the door edge relative to the door frame edges.

3.1.2. Limitations of the vision system

To maintain the integrity of the vehicle it must be prevented from acting on erroneous data. This is procured though various checks of various potential hazardous situations described below.

Correct estimation of the range requires that enough light be reflected back to the camera to generate a line point [Henriksen, 1996]. In the general case, this requires that the illuminated surface be diffusely reflective. If the reflection is specular, the light is reflected away from the camera resulting in no line point. If the surface is not reflective but absorptive, as in the case of dark panels or books, then the weak return signal may also cause detection to fail (no reading).

The range of door angles detectable is limited by the effective angle of incidence. If the door is open to a degree where it points almost towards the AGV the intensity of the incident light is too low to generate a sufficiently strong reflection. This situation is potentially hazardous as the door is not visible and thus appears open. However if presuming that the door has to be open to a degree where the back side of the door is visible before commencing a traverse this situation of invisible door/door at critical angle is recognized and an exception is signalled to probe the door with the manipulator.

Line detection can also be corrupted by the lighting conditions. In very bright light some areas of the image may saturate resulting in local zero-difference which again leaves no laser points for the line finder. If not in a saturated area the AGC of the camera reduces the to a level where the laser line in the difference image is of the same order of magnitude as the pixel noise. In this poor signal-to-noise ratio the threshold algorithm is more likely to encounter difficulties in finding a suitable threshold level which again may confound the subsequent operation of the line finder.

The signal to noise ratio may also be impoverished by unstable grabbing of the image. If the frame synchronization pulse is detected imprecisely the entire frame will be shifted sideways. When subtracting two such images of different synchronization vertical lines appear in the difference image distracting the line finder.

Finally the system assesses whether the density of line data is sufficient. If this is not the case corrective action is necessary. For some of the very low level problems, like shifted frame synchronization appropriate actions can be directly associated with the problem. In the case of shifted frame synchronization problem, a new difference image is obtained to see if the problem was just a result of spurious unfavourable conditions. For other problems, such as unfortunate angle of incidence on door, different actions can be taken involving other systems of the AGV (like relocating the vehicle to a more favourable position or probing the door with the manipulator). In this situation an exception is signalled.

3.1.3. Performance of vision

The position estimation algorithms and acquisition checks described above have been implemented and tested in a large number of experiments on our AGV. The tests were conducted in a typical lab setting at the Institute of Automation.

In terms of detecting erroneous 3D points the performance is excellent. In 1300 runs the system issued *no* commands at all that could endanger the integrity of the vehicle. The vision subsystem led the vehicle to grasp the door edge consistently and reliably with a safe margin to obstacles (door and door frame). In figure 4 the range of door angles that can be detected by the vision system is depicted.

If the vision acquisition metrics signal that no data is available other subsystems of the AGV may be employed to resolve the problem. This could be e.g. mechanical probing devices or audible sensors or the locomotion subsystem. Few false rejections of good data are encountered, mainly due to shifting frame synchronization pulses. These kinds of unnecessary rejections of the door are intermittent and are self-correcting in the next attempt.

The 3D position absolute accuracy is within 5 cm precision when no calibration information about the internal or external camera parameters or laser parameters is used (image centre, focal length., camera roll, pan, tilt, laser plane height, roll, tilt). It is expected that precision can be increased considerably using the calibrated values. See figure 5. In terms of repeatability the precision is under 1cm, see table (1).

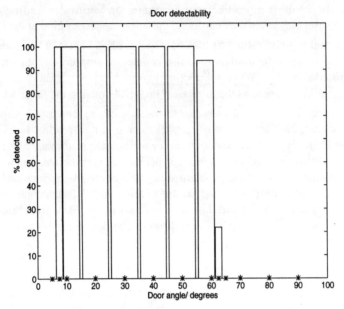

Figure 4. Availability of door edge coordinates.

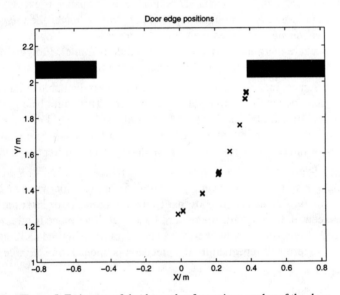

Figure 5. Estimates of the door edge for various angles of the door.

Table 1: Repeatability of door edge coordinates.

Angle (degrees)	Mean x / y (m)	Standard deviation x / y (mm)
5	- / -	- / -
7.5	0.3778 / 1.9403	1.5176 / 2.5728
10	0.3704 / 1.9051	1.0126 / 1.4647
20	0.3403 / 1.7579	0.8561 / 1.5870
30	0.2815 / 1.6134	0.8838 / 0.3000
40	0.2210 / 1.4952	2.4388 / 4.9548
50	0.1245 / 1.3781	0.6450 / 1.8054
60	0.0142 / 1.2809	0.4514 / 1.3761
62.5	-0.0148 / 1.2637	0.8768 / 1.5412
65, 70, 80, 90	- / -	- / -

It is noted that the maximum standard deviation peaks around 40 degrees. This is a result of the laser line starting to illuminate the width of the door causing the edge detector to find positions ranging from the front edge to the rear edge of the door. For angles higher than 40 degrees the rear edge is found consistently. This is appropriate as the interesting feature is the point of attack of the manipulator rather than a fixed point on the door.

The cycle time is currently about 3.2 sec. on the (MC68020 based) image processing computer used in the test configuration. This includes no effort for optimizing the algorithms and it is expected that the speed can be increased considerably without significant loss of detection reliability by streamlining the code. Furthermore the time consumption can be reduced additionally by establishing more realistic illumination as the system uses more time to cope with harsh lighting conditions.

3.2. Mechanical probing device

In this work the AGV is equipped with a 1-DOF arm, which is able to rotate in a horizontal plane in front of the AGV. The arm is telescopic with the outer part spring activated to give passive compliance in the length direction. The arm is operated with a DC-motor with a tachometer and the position of the arm is measured with a potentiometer. The end of the arm is equipped with a contact sensor. The arm controller has two modes, position control and impedance control. The 1-DOF manipulator of the AGV is used both for actuation and sensing.

When the manipulator is used for sensing it is run in two modes:

1. Panning the arm at constant speed while monitoring the position error and torque of the arm.
2. Moving the AGV while monitoring the contact sensor and the position error and torque estimate of the arm.

Combination of these modes enables the AGV to determine the position and size of an obstacle in arm plane in front of the AGV.

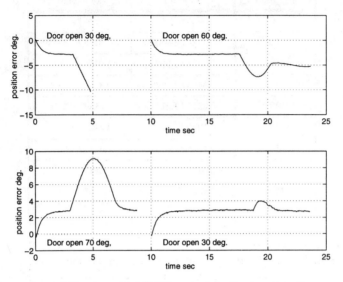

Figure 6. Panning with 1-DOF manipulator for door detection.

Figure 6 shows 4 detection pans. The arm is panned with a constant speed of 5 degrees/sec and the position error is monitored as a function of time. The upper plot shows 2 pans from right to left. In the first the door is opened 30 degrees and after the initial acceleration to a steady state error of approx. 3 degrees the door is hit and the error shows a clear rise. In the second pan the door is opened 60 degrees and starts to move when hit by the arm which leeds to a rise in the error to accelerate the door followed by a fall when the door is moving.

The lower plot shows two pans from left to right. Because of the panning direction the arm will always be able to move the door (in the closing direction), why the two pans exhibit the same characteristics. As seen from the plots the impact position of the arm is easily detected.

3.3. Results

Execution of the open-door task is solved combining the methods described above and extensive experiments have been carried out in the laboratory verifying the approach. The test bed autonomous vehicle is described in more detail in [Ravn and Andersen, 1993] and the algorithm is described in the following.

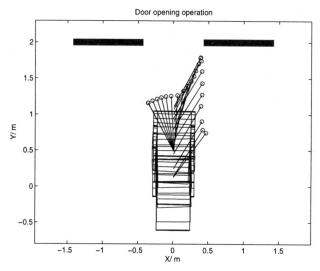

Figure 7. The AGV opening a door.

The laser detection system returns four different scenarios:

3.3.1. Door 'half open':

The laser system is unable to detect the exact position of the door edge but returns the position of the doorframe.

1. Move the end of the arm well within the door edge, using distance estimate of door frame and standard door width measure.
2. Detect door impact using detection panning.
3. Push door open using impedance control, monitoring arm position to detect proper opening.

3.3.2. Normal door opening:

The laser system returns the exact position of the door edge.

1. Move the end of the arm to the correct side of the door edge.
2. Detect door impact using detection panning.
3. Push door with arm using impedance control, monitoring arm position. If arm does not move, drive the AGV slowly backwards until arm movement starts.
4. Push door open using impedance control, monitoring arm position to detect proper opening.

3.3.3. Door nearly closed:

The laser system returns the best estimate of door edge.

1. Move the AGV with the arm to the correct side of the expected door edge. Move until contact sensor is activated.
2. Detect door impact using detection panning.
3. Proceed as in 'normal door opening'.

3.3.4. No vision information:

The laser system is unable to give any information about the door.
1. Pan in front of the AGV. If no detections move a little forward and repeat pan.
2. Move arm away from door move a little forward.
3. Detect door impact using detection panning.
4. Calculate door opening direction using the two door detections.
5. Move AGV to 'open position and proceed as 'normal' door opening.

4. Conclusion

The aim of this work has been to explore and integrate different sensors and algorithms for sensor fusion as well as the 'sense-act' paradigm. The procedures developed has been successfully tested on the test bed autonomous vehicle. Also it is shown how the interaction with the environment facilitates the sensing task yielding improved overall performance. Finally it is exemplified that the door-opening-task can be solved using simple means with convincing performance.

5. References

Albus, James S. (1988). System Description and Design for Multiple Autonomous Undersea Vehicles. *Technical report 1251*, National Institute of Standards and Technology.

Andersen, N., Ravn, O., and Sørensen, A. (1991). Real-time vision based control of servomechanical systems. In *Proceeding of the 2nd International Symposium on Experimental Robotics — ISER'91*.

Andersen, Nils and Henriksen, Lars and Ravn, Ole (1995). Visual Positioning and Docking of Non-holonomic Vehicles. In proceedings Fourth International Symposium on Experimental Robotics 1995.

Budenske, John, Gini, Maria (1994). Why Is It So Difficult For a Robot To Pass Through a Doorway Using Ultrasonic Sensors?. In *Proceedings of IEEE International Conference on Robotics and Automation 1994*.

Henriksen, Lars and Krotkov, Eric (1997). Natural Terrain Hazard Detection with a Laser Rangefinder. *In Proceedings of IEEE International Conference on Robotics and Automation 1997 - ICRA '97*.

Henriksen, Lars (1996). Laser Ranger Based Hazard Detection. *Technical report CMU-RI-TR-96-32*. Carnegie Mellon University.

Kaneko, Makoto and Kanayama, Naoki and Tsuji, Toshio (1995). 3-D Active Antenna for Contact Sensing. In *proceedings for IEEE International Conference on Robotics and Automation 1995*.

Ravn, O. and Andersen, N. A. (1993). A test bed for experiments with intelligent vehicles. In *Proceedings of the First Workshop on Intelligent Autonomous Vehicles*, Southampton, England.

Ravn, O., Andersen, N. A., and Sørensen, A. T. (1993). Auto-calibrations in automation systems using vision. In *Proceedings of the Third International Symposium on Experimental Robotics*.

An Experimental System for Automated Paper Recycling

S. Faibish, H. Bacakoglu and A.A. Goldenberg
Robotics and Automation Laboratory,
Department of Mechanical and Industrial Engineering,
University of Toronto
Toronto, Canada
faibish@me.utoronto.ca

Abstract: This paper presents the experimental aspects of a robotic system used for detection, sorting and grading of paper objects from visual features. The experimental system is used as an automated paper recycling system. The system uses simplified stereo vision to compute position estimate of unknown objects. A vector of geometrical and textural features is used for sorting the paper, according to its grade. The grading process uses additional contact sensors for refining grading from visual features. Supervised learning is used for training the decision system to separate paper from other objects. The experimental aspects of the training process are presented in detail. A special vacuum gripper was designed for gripping paper objects detected by the vision system. The experimental results prove the efficiency of the proposed techniques.

1. Introduction

The basic principle of the proposed system is the use vision and ultrasonic sensors for detecting and grading paper objects. In order to enable real-time operation of the vision system special image processing techniques are used. The image frame of the workspace sampled by a pair of CCD cameras is divided into sub-frames. The images of the empty sub-frames are stored in memory and used as reference images. The detection and grading algorithm computes the difference between the sampled images of the sub-frames and the reference sub-frames corresponding to each camera. The segmentation of the objects is performed using the difference images of the sub-frames. An adaptive threshold process is used for segmentation and detection of paper objects. Only sub-frames containing new objects are analyzed reducing the computation time.

The computation scheme used has several advantages including: (i) only image sub-frames containing objects of interest are analyzed; (ii) smaller images are sampled and processed; (iii) there is no need for matching the two images of an object (from each camera) for the stereo vision algorithms; and (iv) the efficiency and computation speed of the image processing algorithms are improved. Two issues related to 3D visual sensing from stereo pairs are addressed in this paper. First issue is concerned with the image processing techniques used for enabling real-time 3D position estimates, using photogrammetry methods, as applied to moving objects of unknown structure. Second, the fusion of color images and ultrasonic data is used for detecting and grading, with high confidence levels, small paper objects. A special vacuum gripper was designed for grasping paper objects detected by the vision system.

Determination of objects' position in a dynamic factory environment, like material handling or mechanical assembly, is a major task in robot vision. The 3D position of an object can be found by monocular or stereo vision. Monocular vision has been reported to attract much attention, but it is assumed that object dimensions in the scene are known a priori [6]. In our particular application we have used calibrated stereo vision since the object dimensions are not known. Experimental results show that stereo vision compensates for some of the uncertainties that may occur from a single camera. Possible sources of such uncertainties are poor lightening, noise in the image and camera calibration errors.

A two-step camera calibration method is used to obtain our imaging model [1]. After the calibration is completed for each camera, the location of the point, in world coordinates can be obtained by extracting its image projections with sub-pixel accuracy (see [7]). The coordinates of the point are transformed to the robot coordinate system by a rotational and a translational transformation, accurately measured.

In this paper we concentrate our attention to the experimental issues connected to the detection of paper objects. The main problem of detecting paper from other objects is that the visual properties of paper objects are almost identical to properties of other materials.

2. Brief Description of the System

In order to proof the feasibility of the proposed concept a demonstration system was build. The principle block diagram of the demonstration system is presented in Figure 1.

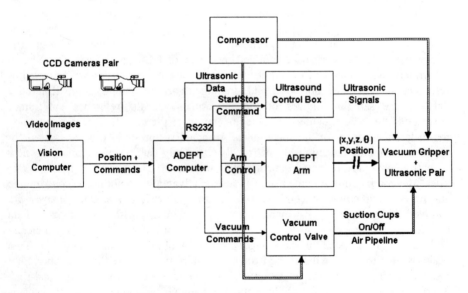

Figure 1: Schematic block diagram of the system

The experimental setup included 4 components: (i) a vision system; (ii) an ultrasonic sensor system; (iii) a robot arm; and (iv) a vacuum gripper. A brief description of the components is presented in the next paragraphs.

2.1. The vision system

Consists in a pair of color CCD cameras with mono-focal lens and focal length of 4.8 mm, and two uniform illumination sources placed over the scene. All the vision components are placed on a tripod pedestal. The video outputs of both cameras are digitized into 24 bit/pixel RGB data by a PC Pentium 133 MHz vision computer. The vision software was implemented in Matlab interfaced with the image grabbing board driver using a Dynamic Data Exchange mechanism. The control commands generated by the vision computer are sent to the robot arm using a RS232 serial port.

2.2. The ultrasonic sensor

Consists in a pair of emitter/receiver probes, containing low frequency resonant crystals and an electronic controller box. The ultrasonic probes are mounted on the vacuum gripper and the control box is mounted on the arm. The sensor measures the time delay of the sound traveling through a tested object. A similar sensor is used for detection of cavities inside telephone poles.

2.3. The Robot Arm

The robot arm used, had to be powerful enough to raise the vacuum gripper, the ultrasonic probes, and a maximum load of 3 kg. The most adequate robot configuration for manipulation of large loads is a SCARA type robot. An ADEPT SCARA manipulator with four degrees of freedom, able to carry loads up to 5.5 kg, was used. The ADEPT computer was connected by 2 serial ports, to the vision computer and to the ultrasonic sensor.

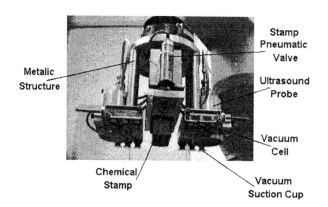

Figure 2: The Vacuum Gripper View

2.4. The vacuum gripper

Was designed and built in the laboratory and consists in three components: a vacuum gripper, 2 ultrasonic probes and an aluminum structure (see Figure 2). The vacuum gripper consists in four vacuum cells each containing a group of three suction cups. Each vacuum cell is connected to an independent valve that disconnects the airflow in case of vacuum lost enabling independent action of each cell. The vacuum generator and the vacuum valve are mounted on the second link of the arm. The gripper is able to hold a 50 paper sheets pile and up to 3 kg weight load.

3. Overview of Algorithmic Modules

In this section we present the main algorithmic modules of the system. The algorithmic modules are: Vision module, Robot control module and Data fusion and decision module.

3.1. Vision Module

This section describes the up-to-date progress in the vision algorithms used for paper recycling. The reader is assumed to have a basic knowledge of mid-level vision and pattern recognition.

3.1.1. Image Acquisition

The vision system acquires images through two color CCD cameras. The system is capable of acquiring 8-bits gray level image as well as 24-bits color RGB images. The imaging system is modeled using a pin hole camera model as in Figure 3. In this model a point (X, Y, Z) in the world coordinate system (WCS) is projected to the image point (x_D, y_D) in the image coordinate system (ICS) and identical to Camera Coordinate System (CCS). The lens is characterized by its effective focal length λ.

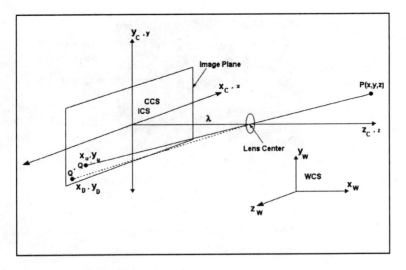

Figure 3: Pin hole camera model

3.1.2. Calibration

The issue of camera calibration in the context of machine vision is to determine: the *intrinsic parameters* which give information about the optical and geometrical camera characteristics such as focal length, scale factors, lens distortion and the intersection point of the camera axis with the image plane and the *extrinsic parameters* which give information about the position and orientation of the camera frame relative to a world coordinate system such as rotation and translation.

To obtain these parameters, calibration points with known world and image coordinates are used. Each camera is calibrated independently. The calibration target is a 50 cm cube that has 45 white circles on black background as shown in Figure 4. The matching is carried out between the centers of these circles and their projections on the image plane.

Figure 4: The Calibration Target

3.1.3. Image processing

This is the most important step of processing large data sets to make the results more suitable for classification than the original data. The process does not increase the information content, but it does increase the dynamic range of the data. There are two approaches for image processing: *spatial domain processing* and *frequency domain processing*. We have worked with spatial domain method since it is faster and the pixels are manipulated directly.

3.1.4. Object Segmentation

A spatially dependent object segmentation method is used within each cell on the conveyor. The difference image between, the sampled image of a sub-frame (Figure 5b) and the reference image of the sub-frame (Figure 5a) is computed and then filtered using a median filter to remove white noise. Linear gamma correction is used to enhance the image. Since we deal with objects with unknown gray level properties, an adaptive threshold technique is used to segment the object from background. Because of its speed and robustness, we chose Otsu's Method [5] to convert the gray level image to binary image. This binary image is then used to extract the pixels belonging to the object (Figure 5c). The histogram of the filtered difference image shows good segmentation properties (Figure 5d).

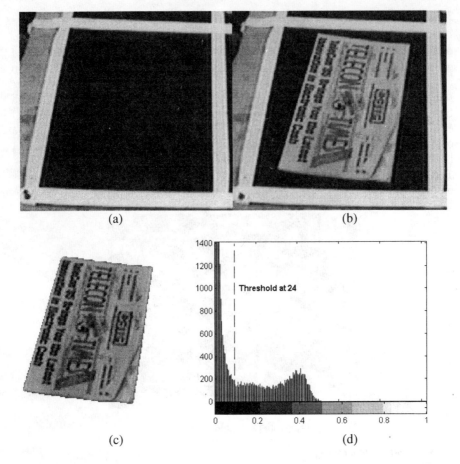

Figure 5: Empty frame (a), frame with object (b), segmented object (c), histogram (d)

3.1.5. Photogrammetry

A two camera stereo vision system is used for obtaining the location of objects with respect to the robot. To obtain the location of an image point (x_1, y_1) in the world coordinates, the same point has also to be extracted from the other image with the coordinates (x_2, y_2). In order to avoid model matching, the centroid of the object in each image is used. The centroid algorithm has also facilitated to work with sub-pixel accuracy. Finally, the four degrees of information obtained from the two cameras is used to calculate the world coordinates (X, Y, Z) via pseudo-inverse least squares solution. At the end of this stage the orientation of the object is computed using Hough transform techniques [4].

3.1.6. Features Extraction

The feature extractor uses two different types of features to describe the structural properties: *shape descriptors* and *textural features*. The shape descriptors used are: *circularity, area, perimeter* and *area-to-perimeter ratio*. There are many textural features. We used a set of 7 textural features including: *uniformity of the textural energy, entropy of the image, contrast of the gray level variation, inverse difference moment of the image, correlation of the co-occurrence matrix, homogeneity of the texture* and *cluster tendency of the texture*.

3.1.7. Classification

In general pattern classification techniques are grouped into two: *parametric* and *non-parametric techniques*. In this research a non-parametric pattern classification was used. Several non-parametric classifiers have been tested, including: *Fisher's linear classifier* [3], *Nearest-Neighbor classifier, Condensed Nearest-Neighbor classifier* and *Perceptron classifier*. Among these classifiers Fisher's method gave the most accurate results (see Figure 6). One important thing to consider here is the linearly separate behavior of the patterns. Non-linearity represents a major research challenge.

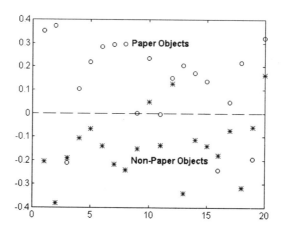

Figure 6: Fisher Linear Classifier, `o'=Paper, `*'= Non-Paper, - - - = Paper/Non-paper border

3.2. Robot control module

In order to implement the data fusion and manipulation of the objects an ADEPT robot arm was used. The arm was controlled from the ADEPT computer. The control tasks includes the next functions: Trajectory planning for point to point movements, Motion control for minimum trajectory tracking errors, Ultrasonic sensor contact force control and Vacuum gripper (x,y) position control.

3.3. Ultrasonic sensor control

The contact ultrasonic sensor is used for grading paper objects detected by the vision system with low confidence. The ultrasonic sensor measures the time delay between the emission and detection of the returned signal traveling through the tested material. In order to facilitate accurate measurements two requirements must be fulfilled: the contact force of both ultrasonic probes is identical and equal to the set value, and secondly that the entire contact area of both ultrasonic probes touches the tested objects.

3.4. Vacuum gripper control

The vacuum gripper consists in four vacuum cells each containing 3 suction cups and was designed such that each cell will be able to grip independently from the other cells. This design was used in order to enable gripping of planar objects of different sizes independently. In order to grip an object the vision system computes the location, orientation and size of the object in ADEPT coordinate system. Depending on the estimated size of the object the ADEPT computer decide to grip it using four, two or one cell and compute the equivalent position and orientation of the gripper such that the maximum number of vacuum cells will touch the object.

3.5. Data fusion and decision module

The decision algorithm is based on a multi-level active sensing process. The active sensing includes action and sensing tasks. The actions are performed by the ADEPT arm and the sensing is performed by the Stereo Vision and the Ultrasonic sensor. The action tasks are: Gripping, Dropping, Touching Object, Out-of-Scene, and the sensing tasks are: Detecting Visual Features, Detecting Color of Objects, Measuring Ultrasonic Delay. A similar data fusion paradigm is used in Automatic Target Detection (ATR) systems [8].

The data fusion and decision algorithm may be described as a feedback control process including: Perception, Response Action (feedback signal) and Decision. This closed loop system combines perceptions and actions in order to make an assertion and to validate the hypothesis. The first stage of the algorithm is the *Sensor Image Processing* of the stereo vision sensor. At the end of this stage the objects of interest are detected, separated from the background, and matched in both stereo images. The *Geometrical Parameters* of the objects of interest, in robot task space, are computed. Furthermore, additional *Algorithms* are used to estimate the *Shape* and *Texture Features* of the detected objects. The features are analyzed and the object is classified using a *Continuous Inference Scheme*. A decision is taken concerning the *Actions* to be taken in order to clarify the behavior type. A specific action is associated to each behavior type. At the end of the process a final decision about the type and grade of an object is taken.

4. Experimental Results

In this section are discussed the implementation aspects of the computation algorithms including parameter tuning and robustness. Experimental results are also presented and analyzed in order to improve the data fusion, decision and control algorithms for future implementations.

4.1. Parameters Tuning

The performance of the implementation in software is influenced by the parameters of the sensing scheme. The relevant parameters are: *Object detection threshold* used for detection of an object in a sub-frame; *Object content threshold* used for finding the pixels belonging to the interior of an object; *Color threshold* used for detecting the area of the un-printed region of a paper object; *Noise threshold* used for selecting a noise free area of the image of an object; and

Ultrasonic threshold used for paper/non-paper object discrimination from ultrasonic measurements.

4.1.1. Object detection threshold

This parameter is used for detection of an object in a sub-frame. After computing the difference between the sampled image and reference image of a sub-frame we have to detect the presence of an object. The positive detection is obtained when the maximum absolute value of the difference is greater than this threshold level. If the threshold is too small the difference generated by the noise, caused by changes in illumination for example, can be higher than the threshold and object detection occurs. This is a false alarm and will increase the processing time of the whole scene. On the other hand if the threshold level is too high dark colored objects will remain undetected. In this case, the undetected objects will remain in the scene and the paper object detection rate of the system will be poor. Fortunately a unique value complying with all the tested objects and illumination levels can be found. This last remark is true only in cases when the illumination of the scene is constant and uniform, as in our case. The optimal value obtained from experiments was 0.3 (77/256).

4.1.2. Object content threshold

The parameter is used for finding the pixels belonging to the interior of an object. After detecting the presence of a new object inside a sub-frame all the pixels belonging to that object must be found. This parameter is critical when analyzing printed paper objects. If the color of the printing is very dark his color will be very close to the color of the background and the absolute value of the difference will be smaller than the value of the threshold. A special case is the one of non-uniform illuminated objects with multiple facets. This case is very similar to the case of printed paper where low illumination level is interpreted as a dark spot. If threshold's value is too small, large amounts of noise will be inserted in the content of the object leading to mismatching between the stereo images resulting in large position errors. When threshold's value is too high a large number of pixels belonging to the object will be missed resulting in erroneous texture features detection and to wrong classification. This problem is reduced in cases of convex objects when the missing internal pixel may be filled using a region growing algorithm applied to convex hulls. Unfortunately the value of the threshold depends on the color of the object. An adaptation scheme was used for tuning this parameter. The adaptation law depends on the statistical properties of the object, such as: mean, variance and peak values for all the pixels belonging to an object. The threshold values were changed in the range 0.2 to 0.61.

4.1.3. Color threshold

This parameter is used for detecting the area of the un-printed region of a paper object. The selection of the parameter influences the accuracy of geometrical features of an object. The optimal value (0.35) was selected using the histogram of the internal pixels of the object. This value was suitable for all the tested objects.

4.1.4. Noise threshold

The parameter is used for selecting noise free area of the image of an object. The value of this parameter depends on the average color of the outer object area inside a sub-frame. A secondary parameter used in the same context is the level of the background noise. If the background noise is under this level the confidence of the geometrical features is 1. Otherwise the confidence is proportional to the ratio of the noise level and the color of the background. The main source of background noise, in our case was the change in the ambient illumination and this parameter takes care of the illumination noise.

4.1.5. Ultrasonic threshold

This parameter is used for detection of paper objects from non-paper objects. The optimal value was selected based on experiments after testing all the objects used as the learning set for the Vision system. The selected value was in the range 60-80 nano-seconds. Lower values correspond to metal objects, and higher values correspond to wood or plastic.

4.1.6. Features weighting coefficients

These weighting coefficients multiply the textural and geometrical features to form a cost index. Their values are obtained using a linear supervised learning scheme based on Fisher's classifier. The values of the weighting parameters are the coefficients defining a hyper-plane in features space. Figure 7 shows the linear classifier is not an optimal solution and non-linear hyper-surfaces must be found for improving the paper objects classification specially for grading paper based on visual features. The figure presents a 2D features space (for simplicity) and shows the linear and non-linear solutions. The linear solution leaves paper objects classified as non-paper and the other way. Of course the linear classifier is wrong.

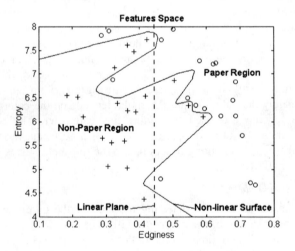

Figure 7: Linear and non-linear classifiers

4.2. Experiments Results

The purposes of the experiments were to: (i) find the optimal parameter set for best detection and grading performance, (ii) evaluate the performance of the system, (iii) test the robustness of the algorithms when applied to real objects picked from Municipal Solid Waste stream, and (iv) prove the feasibility of the proposed detection and grading method.

The results of the experiments were very useful for solving few image processing problems and tuning the parameters for best results. The problems of image processing were: non-uniform illumination, segmentation of dark objects with low reflectance, such as clothes, and detection of the bounds of the sub-frames.

4.2.1. Illumination

This problem appeared in cases of dark colored objects for which a lack of illumination led to misdetection of the presence of the object. The solution was to improve the illumination by adding light sources. The illumination was increased and was made uniform such that the top surface of the objects received maximum illumination while the side surfaces were darker. In this way only the pixels from the top surface passed the adaptive threshold level and were detected as object pixels by both cameras. This improved also the matching of the object from both images and as a result the position estimation errors decreased.

4.2.2. Detection of bounds

The second problem was the detection of the bounds of the sub-frames, i.e. the delimitation between the internal and external pixels to a sub-frame. The problem was amplified by the large difference between the black color of the frame background and the white color of the separation strips. The color of the background was selected black in order to improve the contrast of paper object to the background. Due illumination changes, image noise, and pixel quantization two problems occurred: false detection of objects and wrong selection of object pixels. The solution was to define a mask for each sub-frame defining the interior pixels and eliminating the boundary pixels.

At the end of the image processing stage a set of 50 images of segmented objects was defined and used as the learning set for the vision system. The selection of the learning set depends on user's choice, and the user is not always able to estimate textural features. The false detection rate of paper objects was of 8% and the false alarm rate was 2%. An example of the image processing results is the object presented in Figure 5. The resulting The object detection threshold was 24 while the best fitting threshold value was 30 (the first local minim in the histogram of Figure 5d). Using this value all the pixels belonging to the object were detected with no false pixels. The position estimation errors were (0.012, 0.02, 0.002) meters and 2 deg. Those values are enough for the robot to place the ultrasonic sensor in the correct measuring location and orientation. Additional ultrasonic tests were needed for sorting. The result of the ultrasonic sensing was 69 nano-seconds which is in the range of paper objects.

5. Conclusions and Remarks

As a conclusion of the experimental results two types of improvements appear to be critical to the performance of an effective industrial automated paper recycling system: (i) the sensing needed for detection and grading; and (ii) the efficient manipulation of the paper objects.

The vision system is the core sensing device for both detecting and grading paper objects. The image processing algorithms are robust to small changes in illumination and color balance, but the confidence of the detection is reduced, due to high image noise levels. The adaptive threshold mechanism needs to be improved using additional parameters, and not only object color as is done now. Improved algorithms should be used for automatic tuning of the image processing parameters.

The stereo vision, based on phtogrammetry, proved to be precise enough for the needs of such a system, but it is too slow (80 milli-seconds/sub-frame) for industrial applications. The computation time was acceptable but 90% of the time was spent on the 3D position computations.

The additional ultrasonic sensor used in the experiments was an off-the-shelf item and was not tuned for this specific use. Even so, the ultrasonic sensing functioned well. The detection of paper objects was correct in most of the cases (90%) and in all the experiments no non-paper object was detected as paper.

The manipulation of paper objects using the vacuum gripper was efficient. The suction power was enough to grip 50 paper sheets at a time. The gripper was able to pick small as well as large objects including books, wood, metal, plastic and glass objects.

It is clear that, in order to cope with large amounts of objects, as needed by the industry, contact manipulation and sensing needs to be eliminated. The learning based vision sensing alone, will be able to detect paper and non-paper objects and also separate the different grades of paper using non-linear classification algorithms.

References

[1] H. Bacakoglu, Calibration of Stereo Vision Pairs, M.Sc Thesis, University of Waterloo, 1995.
[2] S. Faibish and I. Moscovitz,"A New Closed-Loop Non-Linear Filter Design", in Proc. of the 1-st European Control Conf., Grenoble, France, July, 1991.
[3] R. A. Fisher, "The use of multiple measurements in taxanomic problems", Ann. Eugenics, vol. 7, Part II, 1950.
[4] J. Illingworth and J. Kittler,"A survey of the Hough Transform", J. of Computer Vision, Graphics, and Image Processing, Vol. 44, pp. 87-116, 1988.
[5] N. Otsu, "A threshold selection method from gray level histograms", IEEE Trans. Syst. Man Cybern., vol. SMC-9, no. 1, pp. 62-66, 1979.
[6] Y. C. Shiu and S. Ahmad,"Calibration of wrist-mounted robotic sensors by solving homogeneous transform equations of the form AX=XB," IEEE Trans. on Robotics and Automation, Vol. 5, No. 1, pp.16-29, February 1989.
[7] M. A. Sid-Ahmed and M. T. Borai,"Dual Camera Calibration for 3-D Machine Vision Metrology, " IEEE Trans. on Instrumentation and Measurement, Vol. 39, No. 3, pp. 512-516, June 1990.
[8] J. A. Stover, D. L. Hall and R. E. Gibson, "A Fuzzy-Logic Architecture for Autonomous Multisensor Data Fusion", IEEE Trans. on Ind. Electronics, vol. 43, no. 3, pp. 403-410, June 1996.

Positioning of the Mobile Robot LiAS with Line Segments Extracted from 2D Range Finder Data using Total Least Squares

J. Vandorpe, H. Van Brussel, J. De Schutter, H. Xu, R. Moreas
Division PMA, Department of Mechanical Engineering, Faculty of Engineering,
Katholieke Universiteit Leuven, Celestijnenlaan 300B, 3001 Heverlee, Belgium
Tel: 32-16-322480 Fax: 32-16-322987
Email: jurgen.vandorpe@mech.kuleuven.ac.be

Abstract: In this paper, a new algorithm is described for dynamic map building with geometrical primitives for a mobile robot. The dynamic map is built up using a 2D range finder mounted on the mobile robot LiAS[1] which is navigating in the environment. The dynamic map is used for both planning and localisation purposes. The map is composed of line segments and circles. For the extraction of line segments, a total least squares algorithm is used. The parameters describing the geometrical primitives are provided with uncertainties which are used in the matching phase and which are necessary if the map is used for localisation. This paper describes in detail how the uncertainty on the robot position and the uncertainty on a single range measurement leads to the uncertainty on the parameters of a geometrical primitive. Promising experimental results obtained by the algorithm in real unstructured environments are presented.

1. Introduction

An accurate map of the working environment is necessary at two levels of the control hierarchy of an integrated mobile robot. Intelligent real-time planners [1-2] need an up-to-date representation of the mobile robot's surroundings to find a path to the goal in a partly known environment. The map is also often used by a position estimation module where the perceived primitives in the dynamic map are compared to a prior known world model [3-4].

A widespread map building method is the occupancy grid based representation. The concept was introduced by Moravec and Elfes in the 80's [5-6]. Occupancy grids represent the environment as a two-dimensional array of cells and proved to be very simple and quite useful for obstacle avoidance and planning purposes.

With the introduction of (relatively) low-cost and high-accuracy 2D range finders the occupancy grid seems no longer the best solution. Because the latest generation of range finders has a range of up to 50 meters and sample frequencies up to 1 kHz, the number of cells of which the confidence value has to be changed, is too large. The main drawback of grid-based world models however, is probably the fact that it is impossible to use them directly for position estimation. Therefore, map building gradually geared up towards algorithms using geometrical primitives.

---

[1] Leuven Intelligent Autonomous System

Several researchers built line segment maps directly from the range data of 2D range finders. In [7] a method is presented where a set of short line segments approximate the shape of almost any kind of environment. A similar method was described in [8].

The here proposed method is different and more general because it takes into account the uncertainties of the parameters before matching corresponding primitives. Doing so, the estimate of a certain primitive keeps improving while it is observed continuously. The resulting map contains a number of line segments each represented by a set of parameters with corresponding uncertainties. A major drawback of a map composed out of line segments only, is that small objects like table legs or other narrow poles are not modelled. This can cause serious problems if the map is to be used by a path-planning algorithm. Therefore, we propose a map built up by two different geometrical primitives. The first primitive is the line segment which is used to model all objects with a width exceeding 30 cm. Measured points which lie close to each other but do not pass the criteria for line extraction are called clusters and are represented by circles.

2. The Range Finder and the Mobile Robot LiAS

Fig. 1 shows the PMA mobile robot LiAS. The vehicle is equipped with several sensors. Encoders and gyroscopes are used for deadreckoning. For perception, LiAS is equipped with ultrasonic sensors, a tri-aural acoustic radar and a new laser range

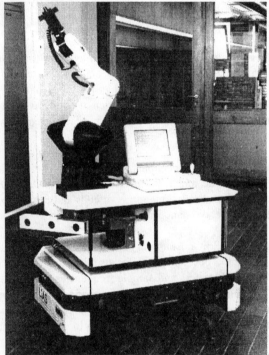

Figure 1: The mobile robot LiAS

finder which is used for the map building experiments described in this paper. A powerful on-board transputer system with 8 Transputer cards takes care of different modular tasks which are performed in parallel.

The 2D range finder is a commercial eye-safe laser scanner (PLS-scanner from Sick Optic Electronic). It has a scanning angle of 180 degrees with an angular resolution of 0.5 degrees and a range accuracy of less than 5 cm over a total range of 50 m. The scanning time for a complete scan of 360 range measurements takes about 600 msec.

3. Deadreckoning

Since the robot position has a direct influence on the parameters of the geometrical primitives, a good position estimate is necessary. In [4], a sensor fusion algorithm for position estimation of LiAS was described. A Kalman filter combines the data from the encoders and gyroscopes. The result is an accurate estimate of the robots pose vector $X_r = [x_r \ y_r \ \theta_r]^T$ in a global reference frame and a corresponding covariance matrix C_{Xr}.

4. The Geometrical Primitives

4.1. Introduction

Fig. 2 shows the representation of the two basic primitives. Line segments are represented in a global frame {G} by:

- the parameters $_G\rho$ and $_G\theta$ from following line equation:

$$_Gy \sin {_G\theta} + {_Gx} \cos {_G\theta} - {_G\rho} = 0, \quad (1)$$

- their corresponding variances $\sigma^2_{_G\rho}, \sigma^2_{_G\theta}$ and $\sigma_{_G\rho _G\theta}$ represented by covariance matrix $_GC_{line}$, and
- the begin and end points $(_Gx_b, _Gy_b)$ and $(_Gx_e, _Gy_e)$ expressed in global cartesian coordinates.

Clusters are represented by the global cartesian coordinates of the centre point $(_Gx_c, _Gy_c)$ and by the radius R of the circle.

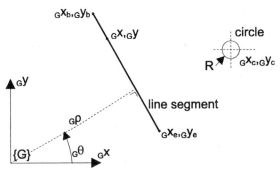

Figure 2: Parameters of a line segment and a circle in a global frame

The next sections will describe how these parameters are calculated from the range measurements and the position estimate of the mobile robot in a global frame.

4.2. Conversion of measured points to local cartesian coordinates

The line-segment parameters in a local frame {L} are calculated from measurements in cartesian coordinates. The range finder provides the range r and the angle α of a measured point. The measurement vector $S = [r\ \alpha]^T$ is relative to the local reference frame of the range finder.

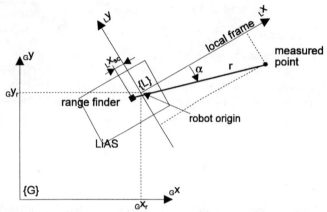

Figure 3: Conversion of measurements to local cartesian coordinates

The measured points are converted to cartesian coordinates $_LX_i = [_Lx_i\ _Ly_i]^T$ in a local reference frame {L}, located at the robot's origin:

$$_Lx_i = {_Lx_{sc}} + r\cos\alpha, \tag{2}$$
$$_Ly_i = {_Ly_{sc}} + r\sin\alpha, \tag{3}$$

where $[_Lx_{sc}\ _Ly_{sc}]^T$ denotes the relative position of the scanner on the robot. In a more general form, equations (2-3) can be written as:

$$_LX_i = f(S). \tag{4}$$

To obtain the uncertainties on the parameters of a line $(_L\rho, {_L\theta})$, the uncertainties on each measured point $_LX_i$ belonging to that line, have to be calculated. The uncertainty of a measured point $_LX_i$ is represented by a covariance matrix $_LC_{Xi}$ and depends on the uncertainty of the measurement (r,α) which is represented by the covariance matrix C_S. The values of C_S are obtained from the sensor specifications and therefore C_S is a diagonal matrix since the outputs of the sensor are not correlated. $_LC_X$ is calculated using following expression:

$$C_X = G\,C_S\,G^T, \tag{7}$$

where G represents the Jacobian of f to S.

4.3. Extraction of line segments

In this section, the measured points $_LX_i$ in the local frame are used for the extraction of lines with parameters $(_L\rho, {_L\theta})$ in local coordinates. The line-fitting algorithm uses a total least squares algorithm also called orthogonal regression. In a second step, the local parameters are converted to global parameters $(_G\rho, _G\theta)$. Finally, the uncertainties corresponding to $(_G\rho, _G\theta)$ are determined.

Criteria for line extraction

To add a newly measured point $_LX_i = [_Lx_i \; _Ly_i]^T$ to the current line, three conditions have to be satisfied:
- the distance of the new point to the previous point must be smaller than a given threshold,
- the distance of the new point to the current line must be smaller than a given threshold,
- the angular difference $\Delta\alpha$ between the new angle α_i and the angle α_{i-1} of the previously measured point must be smaller than a certain threshold.

Orthogonal regression in the local reference frame

The problem of fitting a line through a series of points is also known as linear regression. Linear regression can be solved by the classical least squares (LS) approach [11]. However, the classical LS-algorithm assumes error-free abscissas (or ordinates) and seeks to minimise the deviations of the ordinates (or abscissas) to the fitted line. Since the measured points have errors in all directions, total least squares (TLS) is more appropriate. The total least squares approach minimises the squares of the deviations of the points orthogonal to the fitted line. Therefore, the TLS-method is also called orthogonal regression.

The basic TLS equations to find the line parameters (a,b) from equation $y = ax+b$, can be found in [11]. The resulting solution is a closed function of all the measured points. For the here presented problem, the 'closed form' aspect is very important because the solution must be partially derived to all the measured points. The partial derivatives are used to calculate the uncertainties on the line parameters. Since we use ρ, θ- parameters to represent a line, the TLS solution is derived for our specific case.

The orthogonal distance d_i of point $(_Lx_i, _Ly_i)$ to a line is given by:

$$d_i = {_Ly_i} \sin {_L\theta} + {_Lx_i} \cos {_L\theta} - {_L\rho} . \tag{8}$$

The parameters $(_L\rho, _L\theta)$ of a line which fits through a series of points $_LX_i$, using the total least squares algorithm, are determined by minimising the following criterion:

$$SUM = \sum_{i=1}^{n} \left({_Lx_i} \cos {_L\theta} + {_Ly_i} \sin {_L\theta} - {_L\rho} \right)^2 \tag{9}$$

At the extremal points of equation (9), the partial derivatives to $_L\rho$ and $_L\theta$ must equal zero:

$$\frac{\partial (SUM)}{\partial {_L\rho}} = 0, \quad \frac{\partial (SUM)}{\partial {_L\theta}} = 0 . \tag{10}$$

Before solving (10), the regression parameters R_x, R_y, R_{xx}, R_{yy} and R_{xy} are defined by:

$$R_x = \sum_{i=1}^{n} {}_L x_i, \quad R_y = \sum_{i=1}^{n} {}_L y_i,$$

$$R_{xx} = \sum_{i=1}^{n} {}_L x_i^2, \quad R_{yy} = \sum_{i=1}^{n} {}_L y_i^2, \quad R_{xy} = \sum_{i=1}^{n} {}_L x_i {}_L y_i. \tag{12}$$

Equations (10) can be written as:

$$_L \rho = \frac{R_x \cos {}_L \theta + R_y \sin {}_L \theta}{n}, \tag{13}$$

$$0 = (R_{yy} - R_{xx}) \sin {}_L \theta \cos {}_L \theta + R_{xy} (\cos^2 {}_L \theta - \sin^2 {}_L \theta)$$
$$+ (R_x \sin {}_L \theta - R_y \cos {}_L \theta) {}_L \rho. \tag{14}$$

Substitution of (13) into (14) leads to a quadratic equation which can be readily solved providing $_L\theta$ and $_L\rho$.

Conversion of the line parameters ($_L\rho, _L\theta$) to the global frame

The robot position $_G P = [_G x_r \; _G y_r \; _G \theta_r]^T$ is taken into account for the conversion of the local parameters ($_L\rho, _L\theta$) to the global parameters ($_G\rho, _G\theta$) (Fig. 4):

$$_G\rho = {}_G y_r \sin {}_G\theta + {}_G x_r \cos {}_G\theta + \text{sgn}(_L\rho), \tag{15}$$
$$_G\theta = {}_G\theta_r + {}_L\theta + \delta\pi. \tag{16}$$

The sign of $_L\rho$ in (18) is chosen so that $_G\rho$ is always positive. If the sign of $_G\rho$ is negative then δ equals 1, otherwise, δ equals 0.

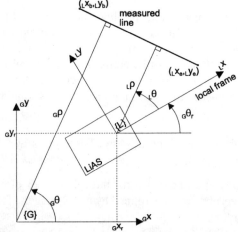

Figure 4: *conversion of a line segment to the global frame*

The parameters ($_G\rho, _G\theta$) describe the equation of a boundless line. To obtain a line segment, the coordinates ($_G x_b, _G y_b$) and ($_G x_e, _G y_e$) of the begin and end point are determined.

Determination of the uncertainties on $_G\rho$ and $_G\theta$

Similar as for the calculation of $_G\rho$ and $_G\theta$, a two-step method is followed to calculate the covariance matrix $_GC_{line}$ of the line, expressed in the global reference frame. First the variances $\sigma^2_{_L\rho}, \sigma^2_{_L\theta}$ and $\sigma_{_L\rho _L\theta}$ on $_L\rho$ and $_L\theta$ in a local reference frame are calculated from the uncertainties on all the points $_LX_i$ belonging to the extracted line. In the second step the variances of the line parameters in global coordinates are determined.

Step 1:
It was shown above that the parameters $_L\rho$ and $_L\theta$, are a function of all the measured points $_LX_i$ belonging to the line:

$$_L\rho = f_1((_Lx_1, {_Ly_1}), (_Lx_2, {_Ly_2}),...,(_Lx_n, {_Ly_n})), \quad (17)$$
$$_L\theta = f_2((_Lx_1, {_Ly_1}), (_Lx_2, {_Ly_2}),...,(_Lx_n, {_Ly_n})). \quad (18)$$

f_1 and f_2 are given by equations (12) and (17).

The covariance matrix $_LC_{line}$ corresponding to the line, expressed in the local frame, is derived from (17-18):

$$_LC_{line} = \sum_{i=1}^{n} J_i {_LC_{Xi}} J_i^T, \quad (19)$$

where $_LC_{Xi}$ is the covariance matrix of point $_LX_i$, and J_i is the Jacobian of f_1 and f_2 to $_LX_i$. The elements of J_i are closed-form analytical functions.

Step 2:
The global line parameters $_G\theta$ and $_G\rho$ are a function of the local parameters $_L\theta$ and $_L\rho$ and the robot position $_GP = [_Gx_r\ {_Gy_r}\ {_G\theta_r}]^T$. Their relationship was described above by equations (15-16).

The covariance matrix $_GC_{line}$ of the global line parameters is given by:

$$_GC_{line} = F\ {_LC_{line}}\ F^T + G\ {_GC_P}\ G^T, \quad (20)$$

where $_GC_P$ is the covariance matrix of the pose vector $_GP$, F is the Jacobian of equations (15-16) to $_L\theta$ and $_L\rho$, and G is the Jacobian of (15-16) to $_Gx_r$, $_Gy_r$ and $_G\theta_r$.

4.4. Extraction of clusters

As in the extraction of line segments, intermediate parameters are first calculated in the local robot frame. Next, the local parameters are transformed to global parameters, taking the robot position into account. If at least two consecutive points $_LX_i$ (in local frame) do not pass the criteria to be added to a line, it will be checked if these points form a cluster and can be represented by a circle. The only condition which has to be verified to add a new point $_LX_i$ to a current cluster, is that the distance of the new point to the centre of the circle must be smaller than a certain threshold.

The coordinates of the centre point ($_Lx_c, _Ly_c$) in the local frame are calculated by following expressions:

$$_Lx_c = R_x/n, \quad _Ly_c = R_y/n, \tag{21}$$

where R_x and R_y are given by (11) and n is the number of cluster points. The radius R is calculated from:

$$R = \sqrt{\sigma_{x_c}^2 + \sigma_{y_c}^2}. \tag{22}$$

Similar as with lines, the local parameters are transformed to global parameters, taking into account the robot position.

5. Matching

5.1. Matching of line segments

After the extraction of a new line, the world model has to be updated. Two cases can be distinguished: (i) the new line matches to a line in the model and is used to obtain a better estimate; (ii) the new line does not match to a line in the model and is transferred to the model directly.

Several criteria have to be checked to make a distinction between the two cases. One of them is a statistical test, involving the covariance matrices of both lines. Given the new line represented by $_GL_n$ and covariance matrix $_GC_n$ and the corresponding line in the model given by $_GL_m$ and covariance matrix $_GC_m$, a scalar normalised distance, called *Mahalanobis* distance, is calculated using following expression:

$$\textit{Mahalanobis}(_GL_n, {_GC_n}, {_GL_m}, {_GC_m}) = (_GL_n - {_GL_m})^T \cdot (_GC_n + {_GC_m})^{-1} \cdot (_GL_n - {_GL_m}) \tag{23}$$

The normalised Mahalanobis distance is compared to a threshold to determine the condition to distinguish matching lines. Finally, two matching lines are combined in a static Kalman filter to calculate a better estimate.

5.2. Matching of circles

Only the distance between the centre points of both circles is checked to decide if a new circle corresponds to a circle already in the world model. Similar as in the case with line segments, matching circles are used to calculate a new estimate.

6. Dynamic Map Building: Deleting of Primitives

Since the final aim is to build a dynamic map, objects which have been removed from the real world, have to be removed from the world model as well. When a new line is added to the world model or a new estimate of an existing line in the world model has been made, an algorithm checks if any primitives have to be deleted or modified. A triangular area (called wipe-triangle) between the scanner and the line segment is used to determine these primitives. The deleting algorithm was described earlier in [9].

7. Experimental Results

7.1. Map building

Fig. 5 shows a world model representing the PMA-laboratory. Although the environment was extremely cluttered, only 52 line segments and 24 circles were needed to build the world model. The uncertainty of the lines depends on the number of measurement points and on the number of times the line has been improved with a new matching line. In Table 1 shows a number of line segments from Fig. 5 with their parameters and corresponding uncertainties. The uncertainty on the parameters of line segment 3 is quite high because the line was built using only 3 points.

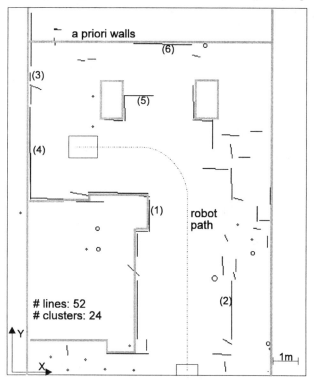

Figure 5: Resulting world model

nr.	times detected	ρ cm	θ deg	σ_ρ cm	σ_θ deg
1	13	897	89	99	0.64
2	12	1412	90	41	0.35
3	7	153	91	1109	4.5
4	10	156	90	9	0.08
5	37	1714	0	28	0.27
6	38	2046	1	33	0.17

Table 1: Uncertainties on lines from Figure 5

7.2. Positioning with line segments

7.2.1. Kalman filter with geometrical constraints

In [3], a high-performance positioning module for the PMA mobile robot LiAS was described. Geometrical primitives are matched with primitives from a known prior model. The Kalman filter based positioning module is very flexible and general since it uses different sensory data to provide a better position estimate. Moreover, the positioning module can deal with partial data which by itself cannot lead to a complete estimate of the position. The positioning module uses only external sensor information when and if it is available because the enhanced deadreckoning with encoders and gyroscopes enables the robot to navigate for long distances when no external data is available. Partial sensor data are formulated as geometrical constraints and are applied in a Kalman filter for the positioning of the mobile robot LiAS.

Generally the constraint equation is represented by:

$$G(P,S) = 0, \tag{24}$$

where $P = [x_r\ y_r\ \theta_r]^T$ denotes the robot position in the global frame, and S is the perception vector which includes the sensor measurement and the position information of the beacon in general.

7.2.2. Position correction using line segments

Natural landmarks such as line segments are used in the position estimation module. In this method, complete geometrical primitives are matched with primitives in a prior model. Unlike methods where single measurements are matched, matching with the wrong primitive is avoided. Wrong matching is quite serious since it can lead to a complete loss of position. To distinguish lines which match with the prior model, the Mahalanobis distance (described in Section 5.1) is used.

In section 4.3, each newly detected line segment is represented by the parameters $_GL_n = [_G\rho_n\ _G\theta_n]^T$ in a global reference frame and its corresponding covariance matrix $_GC_n$. The parameters α and d represent the new line in a local robot frame. These parameters are the local intermediate line parameters which were calculated in section 4.3 ($[\alpha\ d]^T = [_L\rho\ _L\theta]^T$).

The intermediate parameters $[\alpha\ d]^T$ are accompanied by the corresponding uncertainties $\sigma_{\alpha\alpha}$, σ_{dd} and $\sigma_{\alpha d}$. As can be seen in Fig. 6, two geometrical constraint equations can be derived:

$$G_1(P,S) = {_G\theta} - \alpha - {_G\theta_r} = 0 \text{ or } \pi, \tag{25}$$
$$G_2(P,S) = {_Gy_r} \sin {_G\theta} + {_Gx_r} \cos {_G\theta} - {_G\rho} + \text{sgn}(d) = 0, \tag{26}$$

where $P = [_Gx_r\ _Gy_r\ _G\theta_r]^T$ is the position of the robot in the global frame. The complete measurement vector is given by $S = [_G\rho\ _G\theta\ d\ \alpha]^T$. The sign of d in equation (25) and the result of equation (26) (π or 0) are dependent on the robot position with respect to the line ('line between robot and origin' or 'robot between line and origin').

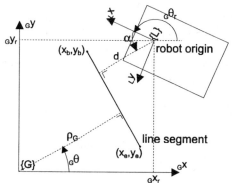

Figure 6: Mobile robot and measured line segment

Fig. 7 shows a test trajectory with a total length of 35 meters. During the execution of the trajectory (speed = 35 cm/sec), 7 prior known line segments and 2 clusters were used to update the position. The uncertainties at the different positions in the trajectory are shown in Table 2. Note that the line segments correspond very well with the prior model. Naturally, this is due to the fact that the uncertainty on the robot position is kept small ($\sigma_x, \sigma_y < 3$ cm) with the external corrections during the complete robot motion.

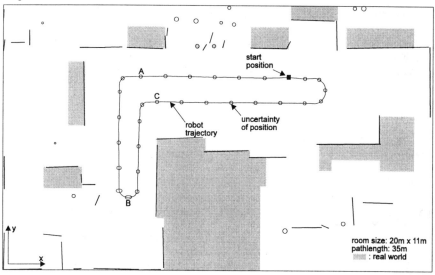

Figure 7: Execution of a trajectory with Kalman filter and natural beacons

Position	σ_x (cm)	σ_y (cm)	σ_θ (deg)
Startpos	3.00	3.00	1.00
A	2.23	2.19	1.26
B	4.47	2.44	1.30
C	2.35	2.34	0.61
Endpos	2.41	2.39	0.87

Table 2: Uncertainties at some positions from Figure 7

8. Conclusions

This paper describes good results of a map building algorithm with a 2D range finder. The parameters on the geometrical primitives are provided with uncertainties depending on the uncertainty of the robot position estimate and the uncertainties of all the measurements leading to this primitive. The quality of the world model allows the mobile robot to use the world model for localisation purposes. The localisation module compares the perceived primitives to a known a-priori model using the corresponding uncertainties to estimate the robot position. The position estimation is based on a 'constraint' Kalman filtering technique and was already described in [3]. Finally, the world map was also used by the reactive planner described earlier in [2].

Acknowledgements

This work was sponsored by the Flemish Institute for Scientific Research (IWT) and by the Belgian Program on Inter-university Attraction Poles initiated by the Belgian State -Prime Minister's Office- Science Policy Programming. The scientific responsibility is assumed by its authors.

References

[1] Vandorpe J., Van Brussel H., 'A Reflexive Navigation Algorithm for an Autonomous Mobile Robot', Proc. Int Conf. on Multisensor Fusion and Integration of Intelligent Systems, 1995, pp. 251-259.
[2] Vandorpe J., Van Brussel H., Xu H., 'LiAS: A Navigation Architecture for an Intelligent Mobile Robot System', Special Issue of IEEE Multisensor Fusion and Integration for Intelligent Systems in the Transactions on Industrial Electronics, June, 1996, pp. 432-440.
[3] Vandorpe J., Van Brussel H., Xu. H, Aertbelien E., 'Positioning of the Mobile Robot LiAS using Natural Landmarks and a 2D Range Finder', Proc. Int. Conf Multisensor Fusion and Integration for Intelligent Systems, 1996, pp. 257-264.
[4] Xu H., Van Brussel H., De Schutter J., Vandorpe J. 'Sensor Fusion and Positioning of the Mobile Robot LiAS', Proc. Int. Conf. Intelligent Autonomous Systems 4, 1995, pp. 246-253.
[5] Elfes A., 'Using Occupancy Grids for Mobile Robot Perception and Navigation', IEEE, Computer, June 1989, pp. 46-57.
[6] Moravec H.P., 'Sensor Fusion in Certainty Grids for Mobile Robots', AI Magazine 9 (2) 1988, 61-74.
[7] Gonzales J., Ollero A., Reina A., 'Map Building for a Mobile Robot with a 2D Laser Rangefinder', Int. Conf. Rob. & Autom., 1994, 1904-1909.
[8] Fortarezza G., Oriolo G., Ulivi G., Vendittelli M., 'A Mobile Robot Localisation Method for Incremented Map Building and Navigation', Proc. of the 3rd Int. Symp. in Intell. Robotic Systems, 1995, 57-65.
[9] Vandorpe J., Van Brussel H., Xu H., 'Exact Dynamic Map Building for a Mobile Robot using Geometrical Primitives Produced by a 2D Range Finder', Proc. Int. Conf. on Robotics and Automation, Minneapolis, April 1996, pp. 901-909.
[10] Vandorpe J., 'Navigation Techniques for the Mobile Robot LiAS', PhD. dissertation 97D3, Katholieke Universiteit Leuven, Dept. Werktuigkunde, PMA, Jan. 1997, ISBN/1996/7515/46.
[11] Hald A., 'Statistical Theory with Engineering Applications', John Wiley & Sons Inc., New York, 1952.

Mobile Robot Localization Based on Efficient Processing of Sensor Data and Set–theoretic State Estimation

U. D. Hanebeck, G. Schmidt
Institute of Automatic Control Engineering (LSR)
Technical University of Munich
80290 München, GERMANY
e-mail: {hnb,gs}@lsr.e-technik.tu-muenchen.de

Abstract: This paper summarizes a new approach for mobile robot localization based upon data from an onboard multi sensor system. Initialization of the robot posture as well as recursive in–motion posture estimation is considered. Accuracy, robustness, and long term stability of the proposed approach is demonstrated by means of long range experiments with path lengths of more than one kilometer and robot velocities of up to 1 m/sec.

1. Introduction

A new approach is introduced to estimate the absolute posture, i.e., position and orientation, of a fast mobile robot on a planar surface. The estimation procedure is based upon data from an onboard multi sensor system and comprises two phases: 1. Determining the initial robot posture and 2. recursive updating of the robot posture during fast motion.

This work was motivated by the need for closed–loop navigation of mobile service robots equipped with cheap and simple sensors. These sensors just provide sparse data, suffer from production tolerances and missing calibration. The paper focuses on two different types of sensors: Angle measurement systems and range sensor arrays with an arbitrary geometry. Their limited perception capabilities are compensated by intelligent sensing strategies. As a result, environmental features can be sensed with high sampling rates. Hence, stochastic errors become negligible in comparison with systematic uncertainties. Since statistical data fusion methods are not appropriate in this case, set–theoretic data fusion is employed. For that purpose, a new solution framework for ellipsoidal sets has been developed, which leads to nonlinear estimators, that allow the uncomplicated treatment of systematic and correlated uncertainties.

To increase localization accuracy, efficient closed–form solutions are derived for calculating the robot posture based on both angle and distance measurements. These solutions include the consideration of both measurement errors and landmark position uncertainties as an integral part.

Accuracy, robustness, and long term stability of the proposed approach is demonstrated by means of a prototype implementation used for navigating

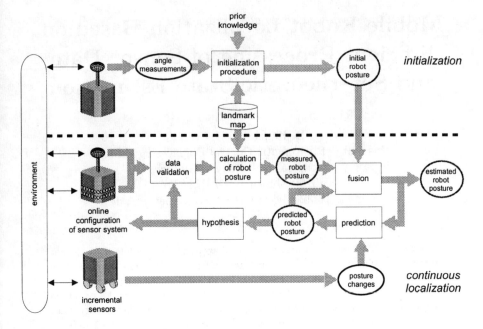

Figure 1. Overview of the proposed localization scheme.

the omnidirectional mobile service robot ROMAN on a day–by–day basis. For obstacle detection and localization purposes, ROMAN is equipped with a multi sonar system. Furthermore, it employs an onboard laser–based angle measurement system to determine the azimuth angles to known artificial landmarks. Experiments include floor–inspection missions with complete floor coverage. In this context, non–stop long range missions with path lengths of more than one kilometer and robot velocities of up to 1 m/sec have been performed.

2. Problem Formulation and Solution Approach

A mobile robot could in principle be localized by dead–reckoning, i.e., by propagating a given initial posture with data from incremental sensors like odometer or gyroscopes. However, due to systematic and correlated uncertainties in conjunction with the required integration steps, dead–reckoning suffers from accumulating errors. Thus, its accuracy is sufficient for small path lengths only.

To keep the localization accuracy within specified bounds though, additional geometry sensors are required for repetitively updating the robot posture with respect to certain environmental features. In many applications, closed–loop navigation is desired even for high–speed maneuvers, which necessitates high update rates.

Furthermore, the measurements performed by the robot are uncertain. For the sensors considered, these uncertainties contain systematic and strongly correlated components. In addition, the landmark positions are assumed to be known, but uncertain.

The localization procedure should produce precise estimates of the robot posture despite these problems. In addition, a realistic quantification of the resulting posture uncertainty should be provided. This is not only a prerequisite for reliable operation of internal localization mechanisms like landmark prediction or data validation. It also allows to assess the maneuverability of the robot in narrow environments.

An overview of the proposed localization system is given in Fig. 1. It comprises two phases: 1. The initialization phase for determining the initial robot posture with little prior knowledge. 2. The continuous localization phase for updating the robot posture during motion.

One of the unique features of this approach is the rigorous modeling of uncertainties by just specifying amplitude bounds. This is advantageous, when detailed stochastic models are not appropriate, not available or too complicated. As a result, rather than point estimates, the algorithm propagates all feasible states that are compatible with the a priori amplitude bounds for landmark position uncertainties and measurement noise. The inherently high complexity of this approach is reduced by approximating the sets of feasible states with ellipsoidal sets. A new formalism for ellipsoid calculus yields simple and computationally attractive algorithms for prediction and fusion.

3. Initialization of Robot Posture

At the beginning of a mission, only little prior knowledge about the robot posture is available. A set of angle measurements is collected while the robot is motionless. For these measurements, the initialization procedure determines the unknown association with landmarks. The following difficulties arise, cf. Fig. 2:

- The landmarks are uncoded, i.e., indistinguishable.

- Erroneous measurements occur, i.e., some angles are caused by unmodeled objects.

- The environment is in general non–convex. Thus, only a subset of all landmarks is visible at a time. In addition, some landmarks are occluded by dynamic objects like humans or other robots.

Many authors assume a perfectly known initial posture, thus eliminating the need for an initialization procedure. In [16], angle–based initialization procedures have been introduced without considering erroneous measurements.

Here, a simple and efficient initialization procedure has been developed [7], which is based on an interpretation tree and a new linear solution for converting angle measurements to robot postures. It copes with the above mentioned difficulties and, in addition, makes use of prior knowledge.

4. In–Motion Localization

Once the robot posture is initialized, the robot may begin its mission. To ensure accurate and stable in–motion robot localization despite the limited

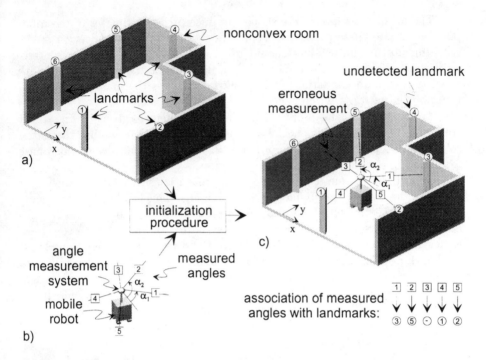

Figure 2. Example for robot posture initialization based on angle measurements: a) Landmark map. b) Measured angles. c) Resulting association.

perception capabilities of the considered sensors, both spatial and temporal continuity conditions are exploited.

For that purpose, incremental sensors are used to predict the robot posture and to postulate landmark hypotheses. Based on these hypotheses, the sensor system is configured online, i.e., sensing and processing power are focused on useful landmarks at a very early stage. In addition, measurement hypotheses are derived for validating actual measurements. To calculate the robot posture based on measured data, new closed–form solutions have been developed. The measured robot posture is fused with the predicted posture in a set–theoretic setting. The result then serves as the basis for the next prediction step.

4.1. Prediction of Robot Posture Based on Incremental Sensor Data

In a set–theoretic setting, each prediction step consists of calculating the set of all feasible postures given the current set of estimated postures and the set of posture changes. Since the result is not an ellipsoid, an ellipsoid containing all feasible postures is used instead. Details on how to efficiently calculate a minimum volume bounding ellipsoid are given in [4].

It is important to notice, that set–theoretic prediction does not require the estimation and the posture changes to be uncorrelated. Hence, the set of predicted postures represents an upper bound for correlated or even systematic uncertainties.

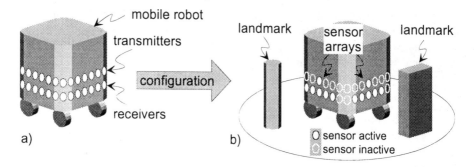

Figure 3. Online configuration of multi sonar system. a) The available transmitters and receivers. b) Forming sensor arrays for landmark tracking.

4.2. Online Configuration of the Sensor System

The set of predicted robot postures is used together with the landmark map to select useful landmarks and predict their positions with respect to the robot. Subsequently the sensor system is configured in such a way as to focus on these expected landmarks.

In case of the multi sonar system, the closest transmitter is selected to illuminate the respective landmark and several receivers are combined to form a receiving array, Fig. 3. Since configuration is performed at a high rate, these sensing arrays track the associated landmarks. Further exploiting the given prior knowledge about the landmark positions allows the use of high–frequency sampling schemes [3].

4.3. Calculation of Robot Posture Based on Angle and Distance Measurements

The straightforward calculation of the robot posture based on angle or distance measurements leads to nonlinear systems of equations, that can only be solved iteratively. Here, however, efficient closed–form solutions have been developed, which solve these problems exactly, i.e, are *no* approximations.

4.3.1. Angle Measurements

Several solutions for calculating the posture given measured angles to a set of known landmarks have been reported. For the case of three landmarks, a classical closed–form solution exists, which has been applied in [9, 16]. However, uncertainties in angles measurements or landmark positions have a strong impact on localization accuracy. Thus, it is advantageous to include more than three landmarks. For the case of $N > 3$ landmarks, a heuristic procedure has been introduced in [17] for selecting the best triple, which is then used for localization. Some authors average all possible triple solutions, which is computationally complex. In [18], an iterative solution is used, which is not well suited for real time applications. A closed–form solution, which consists of $N(N-1)$ equations, but does not consider uncertainties, is given in [1].

Here, a new algorithm has been developed for calculating the posture of an observer based on N angle measurements. It consists of only $N-1$ linear equations for the observer position and N linear equations for the orientation. The posture is calculated as the least–squares solution of the system of linear equations, which includes the consideration of errors in landmark positions and angle measurements as an integral part.

4.3.2. Distance Measurements

For obstacle detection and robot self–localization with a multi sonar system, the first step is to locate objects (obstacles and landmarks) based on range measurements collected with hull–mounted sensor arrays of arbitrary geometry. Real objects are assumed to be composed from either point or plane reflector primitives. The required ranges from the signal transmitter to the receiver arrays via the objects are determined via time–of–flight measurements.

For one signal emitter and a linear array of receivers, closed–form solutions are given in [11, 14]. For the two–dimensional case, multi frequency holography may be applied where intersecting the elliptic curves defining the object locations is done numerically on a grid. This would result in a high computational load in three dimensions and in addition, grid–based schemes are not very efficient for the case of few distinct reflectors. Analytic solutions are often based on a far–field approximation, that may not be accurate enough for certain applications. On the other hand, the straightforward least–squares solution of the exact analytic measurement equations requires a costly iterative minimization.

Here, *efficient* algorithms for locating reflective objects with an *arbitrary array geometry* have been developed [6]. The non–linear measurement equations are transformed in such a way, that a least–squares solution for the object locations may be calculated in closed form. A priori knowledge about the measurement uncertainties is modeled in a set–theoretic framework, i.e., ranging errors are assumed to be unknown but bounded in amplitude. This is a useful alternative to stochastic models when coping with real–world phenomena like correlated noise, unknown statistics, systematic errors, and errors which are not mutually independent for different sensors. Based on this uncertainty model, an error propagation analysis is performed. Weighting matrices for the least–squares estimator are derived, that are optimal, when the ranging errors are within practical limits. For discrimination of the two types of reflective objects and for discarding erroneous measurements, a set–theoretic hypothesis test is provided.

4.4. Set–theoretic Fusion of Measurement and Prediction

Each measurement defines a set of feasible robot postures. This set may be fused with the set of predicted states. If the two sets possess common points, set–theoretic fusion consists of calculating their intersection. Since the intersection of measurement set and prediction ellipsoid is not in general an ellipsoidal set, a bounding ellipsoid is calculated.

The calculation of the minimum volume bounding ellipsoid requires numerical optimization. First results for simple suboptimal bounding by convex

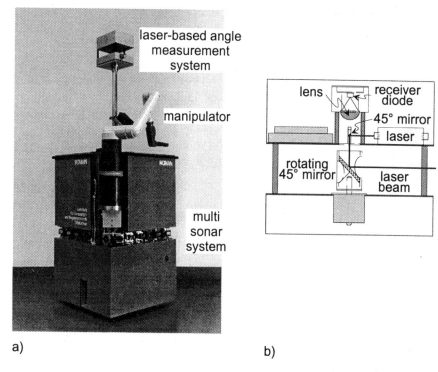

Figure 4. a) Service robot ROMAN. b) Laser–based angle measurement system.

combination of measurement and prediction set are given in [15]. For MISO[1]–systems, i.e., one–dimensional observations, closed–form solutions for calculating bounding ellipsoids have been developed in [2]. In [13], these procedures have been proposed for robotic applications.

Here, a new fusion formalism for the case of MIMO[2]–systems, i.e., vector observations, has been derived, which generalizes the approach in [13].

5. Experimental Validation
5.1. Prototype Implementation

Accuracy, robustness, and long term stability of the proposed approach is demonstrated by means of a prototype implementation used for navigating the omnidirectional mobile service robot ROMAN on a day–by–day basis. ROMAN is a full–scale robot adapted to indoor requirements: width 0.63 m x depth 0.64 m x height 1.85 m, cf. Fig. 4 a). Maximum velocity is about 2 m/sec; the robot's weight is 260 kg.

Dead–reckoning is based on the robot's odometer and a simple gyroscope. For obstacle detection/localization and self–localization, ROMAN is equipped with a multi sonar system. Furthermore, an onboard laser–based

---

[1] Multiple Input Single Output
[2] Multiple Input Multiple Output

angle measurement system determines the azimuth angles to known artificial landmarks.

5.1.1. Odometer and Gyroscope

Since ROMAN is an omnidirectional mobile robot, it is almost impossible to provide dedicated measurement wheels. Thus, the odometer is based on the drive wheels. Due to a short wheel base, imperfect wheel coordination, and uncertain wheel/floor contact points, the odometer is reliable for small path lengths only.

To support the wheel–based odometer, a simple gyroscope is employed, which suffers from a slowly time–varying unknown offset and a time–varying scale factor.

5.1.2. Multi Sonar System

ROMAN is equipped with a multi sonar system for

- detection and localization of unknown objects,
- self–localization with respect to known natural landmarks.

The system comprises 24 piezo-ceramic emitter and receiver elements, which are mounted in a horizontal plane around the mobile platform. Range measurements are performed in pulse–echo mode at a carrier frequency of 40 kHz. Maximum range is about 4 m for most reflector types.

Simple analog threshold circuits are used for detecting the returned echos. This leads to amplitude dependent ranging errors, which are deterministic and correlated for different sensors.

5.1.3. Laser–Based Angle Measurement System

Besides the multi sonar system, ROMAN is equipped with an onboard laser–based angle measurement system to

1. support localization in unstructured environment,
2. provide a reference system for evaluation purposes.

The principle of operation is shown in Fig. 4 b): An eye–safe laser beam scans the environment in a horizontal plane and determines the azimuth angles to known artificial landmarks, i.e., retro–reflecting tape strips attached to the walls. The landmarks are not coded and the system does not provide distance information. 20 horizontal 360° scans per second are performed; absolute measurement accuracy is about $0.02°$.

A map contains the nominal positions of the identical landmarks. Landmark positions have been acquired by the robot itself during an exploration trip. During the map building process, unavoidable uncertainties occur, which are estimated and additionally stored in the map.

5.2. Long Range Experiment: Floor Inspection

To demonstrate the relevance of the proposed localization system in real world applications, long range floor inspection missions with path lengths of more than one kilometer have been performed.

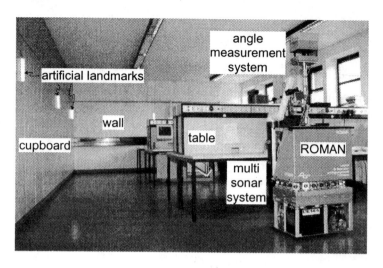

Figure 5. Robot workspace containing artificial landmarks and natural landmarks like ⋄ walls, ⋄ cupboards, and ⋄ tables.

By splitting up the localization system into two parallel and independent subsystems, the performance of the proposed approach for different combinations of geometry sensors with incremental sensors can be compared online. The posture estimation based on ultrasonic sensor data is used for navigation purposes, while the estimation based on laser data is used as an independent reference. In real applications, of course, all available sensor data is used for navigation purposes.

A part of the robot workspace is shown in Fig. 5. Artificial landmarks are required for the laser–based angle measurement system; natural landmarks like walls, cupboards, and tables are used by the multi sonar system. The landmark map employed is shown in Fig. 6.

During the inspection mission, the robot follows the pre–planned path shown in Fig. 6 b), which is generated automatically by the procedure described in [8]. The unlimited long–term stability of the proposed localization system is underlined by initializing the robot posture only once and then repetitively traveling along the path without stopping. After 12 loops, the total distance traveled is about 1056 m; the total time is 46 min. The number of $180°$ turns is 132 and the number of $90°$ turns is 72. The maximum speed is 1000 mm/sec; the average speed is about 400 mm/sec, since the robot automatically slows down during sharp turns.

The resulting posture estimates based on laser data are plotted in Fig. 7 a). These estimates are used for reference purposes only and are *not* available for robot navigation. In this experiment, posture estimates based on ultrasonic sensor data are used for navigation purposes, Fig. 7 b). When the robot crosses the highlighted areas, not enough natural landmarks are available for sonar–based navigation. Hence, the localization accuracy decreases. However, outside these areas, the sonar–based posture estimate is not significantly

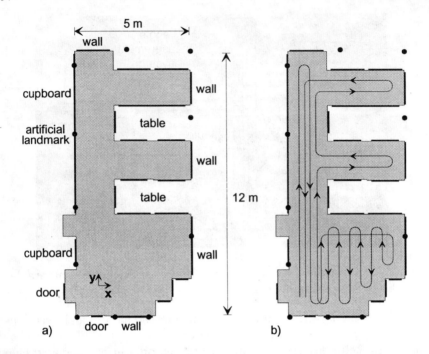

Figure 6. a) Landmark map. b) Pre–planned path for floor inspection mission with full floor coverage.

worse compared to the laser–based posture estimate. The maximum absolute deviation between true and estimated robot posture is about ±5 cm and ±1° for laser–based localization and about ±10 cm and ±2° for sonar–based localization.

For comparison purposes, the posture changes supplied by odometer and gyroscope have been integrated and plotted in Fig. 7 c). Obviously, dead–reckoning suffers from strongly correlated accumulating errors.

6. Conclusions

A key component for mobile service robots has been introduced, i.e., an integrated localization system for initializing and continuously updating the robot posture. It is based on an onboard multi sensor system comprising simple and cheap standard sensors. The limited perception capabilities of these sensors are, however, compensated by intelligent sensing and processing strategies, that focus the available sensing and processing power on useful landmarks.

Furthermore, efficient closed–form solutions are used for calculating the robot posture based on the sensor data. Measurement errors and landmark position uncertainties are considered as an integral part to increase localization accuracy.

The appropriate treatment of the predominant systematic and strongly correlated uncertainties is guaranteed by the use of a new set–theoretic estimation framework.

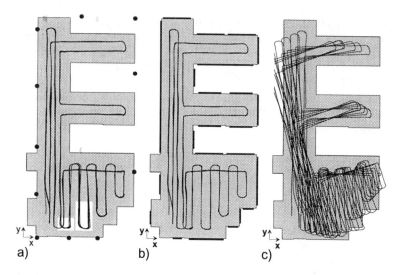

Figure 7. a) Posture estimate based on laser data for 12 loops. b) Posture estimate based on ultrasonic sensor data for 12 loops. c) Pure dead–reckoning, i.e., integrating posture changes supplied by odometer and gyroscope.

The proposed approach is very general and may thus be used with any robot equipped with an onboard sensor system. In this paper, the effectiveness of the new localization system has been demonstrated by navigating an omnidirectional service robot on long range missions through an indoor environment.

The performance of the proposed fusion scheme has been evaluated for different combinations of geometry sensors and incremental sensors. The maximum absolute deviation was found to be about ± 5 cm and $\pm 1^o$ for angle–based localization and about ± 10 cm and $\pm 2^o$ for range–based localization.

In this application, set–theoretic estimation concepts produced good results, because deterministic uncertainties were predominant. In cases where stochastic errors play the major role, stochastic filters are of course the appropriate tool. However, when both error sources occur simultaneously, the application of a new filter type could enhance estimation quality [5]. The so called SSI–filter (Statistical and Set–theoretic Information Filter) combines set–theoretic and statistical estimation concepts and also contains them as border cases.

References

[1] Betke M, Gurvits L 1997 Mobile Robot Localization Using Landmarks. *IEEE Transactions on Robotics and Automation*, Vol 13, No 2, pp 251–263

[2] Deller J R, Nayeri M, Odeh S F 1993 Least–Squares Identification with Error Bounds for Real–Time Signal Processing and Control. *Proceedings of the IEEE*, Vol 81, No 6, pp 815–849

[3] Hanebeck U D, Schmidt G 1995 A New High Performance Multisonar System for Fast Mobile Robot Applications. In: Gräfe V (ed) *Intelligent Robots and*

Systems 1994 (IROS'94), Elsevier Science, Amsterdam, pp 1–14

[4] Hanebeck U D, Schmidt G 1996 Set–theoretic Localization of Fast Mobile Robots Using an Angle Measurement Technique. In *Proceedings of the 1996 IEEE International Conference on Robotics and Automation, Minneapolis, MN*, Vol 2, pp 1387–1394, 1996.

[5] Hanebeck U D, Horn J, Schmidt G 1996 On Combining Set–theoretic and Bayesian Estimation. In *Proceedings of the 1996 IEEE International Conference on Robotics and Automation, Minneapolis, MN*, Vol 4, pp 3081–3086, 1996.

[6] Hanebeck U D, Schmidt G 1996 Closed–Form Elliptic Location with an Arbitrary Array Topology. In *Proceedings of the 1996 IEEE International Conference on Acoustics, Speech, and Signal Processing, Atlanta, GA*, Vol 6, pp 3070–3073

[7] Hanebeck U D, Schmidt G 1996 Localization of Fast Mobile Robots Using an Advanced Angle–Measurement Technique. *IFAC Control Engineering Practice*, Vol 4, No 8, pp 1109–1118, 1996.

[8] Hofner C, Schmidt G (1995) Path Planning and Guidance Techniques for an Autonomous Mobile Cleaning Robot. In: Gräfe V (ed) *Intelligent Robots and Systems 1994 (IROS'94)*, Elsevier Science, Amsterdam, pp 241–257

[9] Krotkov E 1989 Mobile Robot Localization Using A Single Image. *Proceedings of the 1989 IEEE International Conference on Robotics and Automation, Scottsdale, AZ*, pp 978–983

[10] Leonard J J, Durrant-Whyte H F 1991 Mobile Robot Localization by Tracking Geometric Beacons. *IEEE Transactions on Robotics and Automation*, Vol 7, No 3, pp 376–382

[11] Peremans H, Audenaert K, Van Campenhout J 1993 A High–Resolution Sensor Based on Tri–Aural Perception. *IEEE Transactions on Robotics and Automation*, Vol 9, No 1, pp 36–48

[12] Rao B S Y, Durrant-Whyte H F, Sheen J A 1993 A Fully Decentralized Multi–Sensor System for Tracking and Surveillance. *The International Journal of Robotic Research*, Vol 12, No 1, pp 20–44

[13] Sabater A, Thomas F 1991 Set Membership Approach to the Propagation of Uncertain Geometric Information. In *Proceedings of the 1991 IEEE International Conference on Robotics and Automation, Sacramento, CA*, pp 2718–2723

[14] Sabatini A M, Di Benedetto O 1994 Towards a Robust Methodology for Mobile Robot Localisation Using Sonar. In *Proceedings of the 1994 IEEE International Conference on Robotics and Automation, San Diego, CA*, pp 3142–3147

[15] Schweppe F C 1973 *Uncertain Dynamic Systems*. Prentice–Hall

[16] Sugihara K 1988 Some Location Problems for Robot Navigation Using a Single Camera. *Computer Vision, Graphics, and Image Processing*, Vol 42, pp 112–129

[17] Sutherland K T, Thompson W B 1994 Localizing in Unstructured Environments: Dealing with the Errors. *IEEE Transactions on Robotics and Automation*, Vol 10, No 6, pp 740–754

[18] Yagi Y, Nishizawa Y, Yachida M 1995 Map–Based Navigation for a Mobile Robot with Omnidirectional Image Sensor COPIS. *IEEE Transactions on Robotics and Automation*, Vol 11, No 5, pp 634–648

Chapter 9

Modeling and Design

In this chapter different modeling techniques for robot design: arm, wrist or specific positioning system are presented. The model analysis provides relevant information for the manipulator design and evaluation, and parameter determination and component design.

Sabin et al. present the i.ARES project, consisting of an autonomous-on-ground vehicle (Rover), fully sensorized and with a high decisional intelligent level, and they describe in more detail the manipulator. The objective of the manipulator is to collect samples, measure soil resistance and perform visual inspection. The specifications of i.ARES are, first to be able to operate within a given volume which depends on the attachment point to the rover, second it has to be able to withstand an external force of 15N and external torque of 1Nm.

A new method of modeling nonlinear friction as a function of both position and velocity is introduced by Popovic and Goldenberg. The modeling technique uses spectral analysis to describe non-linear friction in complex mechanisms. The method is experimental in its different steps, setup, static friction and sliding friction. A DFT-based friction model that considers static, sliding, Coulomb, and negative damping frictions and asymmetry effect is described. This model has the advantages over the friction model of first, a higher accuracy, second, friction is described as a function of position and velocity and third, the accuracy can be adjusted.

Canfield and Reinholtz present their invention of a three d.o.f. device suited either as a robotic wrist or a platform manipulator. It is called carpal wrist due to its similarity to the human wrist. The wrist structure is the one of a parallel robot. Its kinematics position analysis show that real time control is possible, velocity analysis shows that the wrist is free of singularities, and the dynamic force analysis provide a design tool for actuator selection. Dexterity is evaluated using several criteria.

A second parallel architecture is presented by Merlet, a 3 d.o.f. positioning system. Mips, a mini-positioning system can have applications in the medical field or in inspection. For most of these systems these number of d.o.f. are usually enough. Miniaturization of robotic structures is not a problem of scale alone, but many other questions enter in the problem. For that reason the mechanical and sensing structure is described, and analysis is done on the kind of suitable actuators: piezoelectric, SMA, magnetic actuation or polymers.

i.ARES MANIPULATION SUBSYSTEM

Fernández Sabin, Mayora Kepa, Basurko Jon

IKERLAN

Mondragón (Spain)

email: iares@ikerlan.es

Gómez-Elvira Javier

INTA

Torrejón de Ardoz (Spain)

email: gomezej@inta.es

García Ricardo, González Carlos, Selaya Javier

IAI

Arganda del Rey (Spain)

email: gonzález@iai.csic.es

Abstract: This paper deals with the Manipulation Subsystem of the i.ARES Project demonstration vehicle (rover) designed for moon exploration purposes. The subsystem is a very light robotics structure with six degrees of freedom and a sensorized gripper (vision & force sensors). It includes the control electronics hardware and the software required for autonomous functions and for on-ground teleguiding.

1. Introduction

The i.ARES Project is an EUREKA Programme developed with the collaboration of French, Hungarian, Russian and Spanish Companies, under the technical management of CNES (France). The aim of the Project is to develop an Autonomous On-Ground Demonstrator Vehicle (Rover) fully sensorised and equipped with a Manipulation Arm. Project environment specifications are directed to a potential Moon Exploration. The high degree of autonomy required to execute the system tasks needs a powerful on-board decisional system obliging the use of intelligent components. The Teleguiding possibility forces a high sensorization (cameras, sensors, etc.) to allow the commanding from an external on-ground operator.

The Manipulation Subsystem is under the responsibility of an Spanish Consortium composed by three Companies: IKERLAN, INTA and IAI. This paper explains the Spanish contribution to the project: The design, manufacturing and testing of the Manipulation Arm with a Dextrous End-Effector, the developing of the arm commanding software, the on-board software to allow autonomous operations (vision and manipulation commanding and control) and the on-ground software for Teleguiding.

2. i.ARES Manipulation Specifications and Functions

The IARES Manipulator is the key element of the i.ARES Rover since this is the component that interacts with the soil environment (manipulating objects and allocating and managing sensors) and its characteristics have an important effect on the Rover performance. It must comply to the following Moon Scenarios missions:

- **Capture non-structured samples** of 10 to 50 mm side & 0.5 kilograms, with force interaction.
- **Measure soil resistance** to detect soft and dangerous surfaces for the vehicle.
- **Perform Visual Inspection** of the vehicle itself.

The Manipulator Arm operates in three ways: under on-ground human operator control (Teleguided Mode with images and force values sent also during the grasping), automatically adapting a pre-programmed conditioned sequence to the real environmental conditions of the task (Autonomous Mode) or based on advance interactive autonomy (Teleoperated Mode).

The Manipulator is designed to work under on-ground conditions, but is representative in terms of functions, mass and volume of a moon qualified model. The components used to design and manufacture the hardware components (arm structure, end-effector and electronics) are of industrial or military origin, but can be changed for adequate space qualified components or, in the case of the electronics, could be qualified under the right moon conditions. The Manipulation Subsystem has a high degree of integration incorporating low mass actuators, position, force and vision sensors and all the required electronics in a very small volume.

3. i.ARES Manipulation Technical Description

The dexterity needed to comply to previous specifications requires a low mass and a much smaller and compact design. It should integrate mechanical parts and control electronics to command the moving and grasping activities taking into account the vision & force sensors information fully monitoring both status and operations. A summary of the parts and most important functions achieved by the Manipulation Subsystem are:

- The **Arm Structure** which is a open chain six degrees of freedom mechanism of specific low mass design. It holds an **Dextrous End-Effector** to grasp the samples allowing force interaction.
- The **Electronics** for the Arm movements control, based on a powerful T805 transputer architecture of high reliability in a low mass and volume compact design
- The **Software of Manipulation** including the Arm Control Software, running in the transputer board to command movements, the On-board Software, running in the Rover main computer to manage the Manipulation tasks

execution in Autonomous Mode, and the Soil - Segment Software, in the on-ground main computer to manage the Manipulation tasks execution in Teleguided Mode.

4. i.ARES Manipulation Mechanical Description

The iARES specification, from a mechanical point of view, has two main requirements: 1) the arm must be able to work inside a workspace volume which depends on the attachment point to the rover, being approximately a semicilynder of 0.8 m radius, and 2) it has to be able to withstand a external force of 15 N and external torque of 1 Nm.

4.1 Arm Mechanical Structure

The main problem to design the i.ARES manipulator arm is to cope with two main constraints, namely: only commercial equipment have to be implemented in the robot, and space engineering criteria must be taken into account throughout all the project stages. Both are difficult to combine and have a strong impact on the final prototype. Besides, the harness must be routed along the inside of the arm so as to enable an easier installation of a potential thermal blanket allowing axes rotation. The structure must be as light as possible and the robustness and lubrication have to be representative of a space model, etc.

After a trade off between different potential configurations, it was selected a classical one with six degree of freedom, two in the shoulder, one in the elbow and three more in the wrist as depicted in *Figure 1*.

The length between axes 2 and 3 is 0.367 m and 0.508 m from 3 to the end-effector interface.

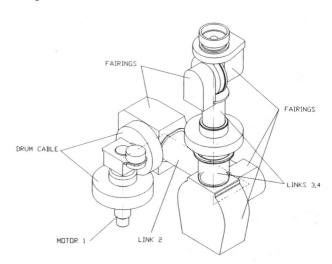

Taking into account the lightness requirement a scanning of the DC motor market was done trying to find the ones with a best mass/torque ratio. Direct drive, hollow shaft motors (to ease the harness routing), brushless DC motor, etc. were analysed,

Figure 1. IARES manipulator configuration

keeping at least a DC motor with Harmonic Drive® reductor implemented inside the set. An additional reduction was necessary in some joints to fulfil the torque and speed requirements. The actual motorization is shown in the table 1.

The propioceptive sensorization is based on encoders linked to the motor shafts. They have a resolution of 1000 ppr (joints 1 to 5) and 500 ppr for the motor 6. In order to have an absolute reference in the manipulator, a Precision Rotative Transducer (potentiometer type sensors) from Sfernice is placed at the output of each joint. These transducers have a linearity of 0.025% for the first three and 0.05% for numbers 4 and 5. The last joint has only a microswitch reference.

As previously mentioned, the lightness was a strong requirement which led to implement as many non metallic elements as possible, taking into account that the used materials had to be space qualified.

Joint	Motor	Reducer	Total Reduction	Rated torque (Nm)	Rate speed (rpm)
1	RH-8-3006	HFUC 20 50	5000	100	0.6
2	RH-8-3006	HFUC 25 120	120000	240	0.025
3	RH-8-3006	HFUC 20 80	8000	160	0.0375
4	RH-11-3001		100	3.9	3.9
5	RH-8-1502	Bevel gear 5:1	500	8	3.
6	RH-5-5502	Bevel gear 6:1	600	1.74	9.2

Tabla 1. Joint data.

Following these criteria, CFRP has been used to build up the links, the fairings and the cable drums. Some pieces of these drums have also been manufactured with DELRIN.

Figure 2. Joint 4, 5 and 6

The harness was a cornerstone of the design due to the fact that the motoreducers selected were not of the hollow shaft type, and the cables had enough protection from the outside environment (space requirement). The selected solution was to design a special device called "cable drums" (see Figure 1) which allow the cable to turn in a controlled way minimising the torque induced by them. A special test with a mock up of a drum was carried out to get confidence in the design, verifying that the cable rolled up correctly inside the element.

In Figure 2, the joints 4, 5 and 6 can be seen without the harness and attached to a special bracket. The pictures highlights the CFRP elements, links 3 and 4 and the fairings of joint 5.

4.2 End-Effector Mechanical Structure

The End-Effector of the arm is made of aluminium in order to save mass (actual mass is less than 1 kg). The EE external envelope, including the fingers, has a volume of 164 mm (length) by 64 mm square side, excluding the camera. It contains all the electronics to control force sensors and to command the electric actuator for the three fingers grasping and includes a camera and a laser light.

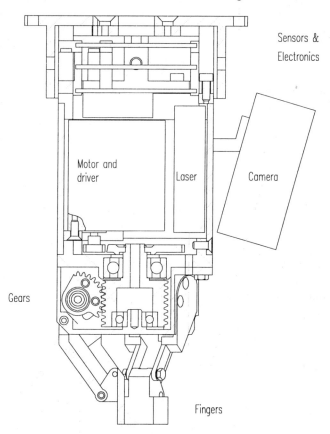

Figure 3 End-Effector

The Manipulator End-Effector has been designed with an specific mechanical structure which contains force sensors, a vision equipment (a camera and a laser emitter), an electric actuator, the grasping mechanism and the electronics required to assure the subsystem maximum performance. The following EE Modules of figure 3 are described from the upper part (close to the Manipulator extreme flange) to the lower (close to the samples):

1- <u>Manipulator Interface Module</u> allows the EE Mechanical & Electrical link to the IARES Manipulator Arm. EE mechanical interface is obtained by means of screws and pins. Electrical Interface requires a harness internally routed with eleven wires for the EE internal components (six for power supply and five for a serial link communication) and a harness routed externally with six wires for the MVS (four for camera and laser power supply and two for the images transmission). This bracket has a cup-shaped structure that limits the F/T Sensor deformation to avoid any damage in sensor structure.

2- <u>Force/Torque Sensor Module</u>, able to calculate three Forces and three Torque, with a range of ± 20 N and ± 1.5 Nm. F/T Sensor is a mechanical structure based on several flexible beams, with some strain gauges bonded in order to detect the deformation and calculate the forces. It could contain some electronics. The three electronics boards foreseen for the subsystem are: gauges connection board with the Wheatstone bridges and compensation resistors, analog hardware board for the signals conditioning, amplification and filtering and digital hardware board that should include A/D converter, a microcontroller, RAM and EEPROM memory, digital Inputs/Outputs and a Serial Link for the EE communication to an external Controller.

3- <u>Intermediate Module</u> containing the DC Motor with the required spur gear Reducer; the MVS and some electronics boards. This module also contains the analog board to drive the DC Motor. This bracket allows the MVS Mechanical & Electrical link to the EE structure. MVS components are attached by means of screws and pins to the bracket internal or lateral flanges.

4- <u>Grasping Mechanism</u> that interacts with the objects. It will be configured with worm gear reducer for the rotation of the three fingers able to catch the samples. It will also include the sensors to measure the Grasping Force (able only to measure the Force in the fingers attaching direction- 25 N range) and a connector with a minimum of three pins to allow the Soil Resistance Measuring Tool connection (not required in the i.ARES specifications).

After comparing different design concepts simplicity and optimum performance were given priority in order to maximise reliability.

5. i.ARES Manipulation Software Description

Three level of software are defined in the I.ARES Manipulation Subsystem: The **Soil-Segment Software** running under an on-ground operator, the **On-Board Software** running in the main computer of the vehicle and the **Software of Arm & End-Effector Movements Control** (running in the Manipulator TRAM Module and in the End-Effector computer) commanding straight line movements of the Arm and managing force sensor of the End-Effector.

5.1 On-Ground Segment Software

The **Soil- Segment Software** (in the on-ground main computer) manages the Manipulation Subsystem tasks execution in Teleoperated and Teleguided Modes under the control of an on-ground human operator.

Ground Segment Commands & Emergency Buttons

The ground segment carries out the control of the rover and its built-in arm. As far as this publication is concerned we are dealing mainly with the Spanish group developments: this software starts working after the rover has been stopped and the command given to initiate an arm procedure. We have considered that the ground segment is made up by a computer with a screen and a mouse. No other hardware has been considered necessary. The program we are speaking about is given control by a higher level program and will release control to it upon its termination.

The ground segment of I-Ares is the device that issues commands to the flight segment. These commands can be sorted in four great groups: Automatic Sample Taking, Teleprogrammed Sample Taking, Soil resistance measuring and self inspection. The program has to be capable of continuous display of the actual arm state and of issuing commands to halt any action the rover might be executing at a given moment.

Emergency buttons are assumed to be needed urgently and so, they are located in a way that permits them to be pressed without any delays. They are the Pause, Continue, Stop, Next, Power Off, Power On, Laser Off and Laser ON buttons. Last four buttons are self explaining. Pause button stops the actual action permitting to continue it with the Continue button, or the next button if it is desired to skip the action. The stop button automatically skips to the next command.

Status Lights & Command Execution

The screen displays continuously five lights that show the state of important variables of the process. They are INIT, POWER, DEPLOY, CALIBRATION, EXECUTION which show respectively whether the initialisation has been accomplished, the arm has been powered, deployed, is calibrating or executing some command.

The arm has the following modes of Operation, each of which has its own control screen: Automatic Mode Sample Collection, Teleprogrammed Mode Sample Collection, Teleguided Mode, Soil Measuring Automatic Mode, Soil Measuring Teleprogrammed Mode, Self Inspect Mode

The **Automatic Mode** commands are executed through a similar interface: The screen shows one of the pictures taken by the arm camera, and provides with the means to change it by another. Then a list of points is made, either for sample taking or for hardness measuring. The list is later sent to the rover for execution.

The **Teleprogrammed Mode** commands are in fact a way to defer totally the decisions of sample taking to the human operator. In this mode the operator sends the individual command to the flight module without comment. In order to permit it the operator has access to all the elementary commands of the arm and selects them at will. These commands include the capability to take and visualise pictures.

Self Inspect Mode

The self inspect mode is a semiautomatic procedure. Upon entering on the self inspect mode a list of points is requested from the main control computer. This list is interpreted as a series of points through which the arm is to pass. At the end a picture is taken.

5.2 On-Board Power-PC Software

The **On-board Software** (running in a PowerPC, the computer of the Rover) commands the transputer node, performing image sensor treatment and managing the Manipulation Subsystem tasks execution in Autonomous Mode. This is the most important piece of software of the manipulator, due to the fact that it must be able to decide under unknown conditions.

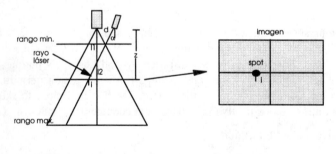

$z = l1 + l2 = d/tg\ \alpha + i\ tg\ \alpha$

Figure 4: *Camera and laser light spot*

First function of this is related to the Manipulator Vision Subsystem: the camera assures image acquisition of the soil working area with enough resolution to allow the samples identification and the laser allows the calculation of the soil true distance, considering target distances from 30 to 80 cm. The selection of a CCD camera with 40° by 31° angular aperture and F = 6 mm allows 0.84x1.14 mm per pixel of resolution at 80cm height (9 pixels for a sample of 10mm side, that is good enough for the human operator identification). The laser sensor requires a minimum power of 25 mW, and to belong to class IIIA it is necessary a pulse activation during a maximum of 90 milliseconds. The calculation of the soil distance (figure 4) is based on the position of the laser spot in the image by using the formula of figure 4.

Second and most important function is to obtain the sample contour and, combined with the data of a 3D analysis of the sample using laser sensor, to calculate the right end-effector position and orientation for grasping purposes. After sample grasping it will be left in a box and mass is measured. Force sensor at end-effector level is controlled and the reaction under potential collisions or measurement of ground resistance are possible.

5.3 Control Transputer & End-Effector Software

The **Software of Arm Movements Control** runs in the TRAM Module, based on a T805 Transputer processor using six LM628 microcontrollers to perform servo joints movements. The software developed is based on the ACT-CONTROL library that provides functions to generate & control different kind of trajectories:

- Control the manipulator movements commanding straight line movements in camera, tool or laser co-ordinates.
- Receive orders and send information from and to the PowerPC.
- Command the End-Effector and manage force data.

The software works in a synchronous way with real time events hardware dependent. Motions are specified in joint position and velocity or using different Cartesian co-ordinates. The motion computer generates a trapeze that controls position, velocity and acceleration every sampling period of 8 to 10 milliseconds.

EE digital board microcontroller is a powerful SIEMENS 80C196 that permits a complex software implementation to manage and command the dextrous functions assigned to the EE. This module has four main functions:

- Co-ordination Service: It has the Initialisation Procedures, the Autotest and the Supervision activities.
- Communication Service: It offers the protocols for two serial lines: RS422 and RS232 (only for testing phases).
- Sensor Signals Treatment: It contains the offsets and calibration matrix to calculate F_{xyz} and T_{xyz}.
- Grasping Commands: It has the programs for the gripper opening and closing functions.

The RS422 communication link to the MANI controller allows the "commands" reception and the "status information" sending, including synchronous Fxyz and Txyz data, minimising wires number.

6. i.ARES Manipulation Electronics Description

6.1 Control Transputer Electronics

The electronics defined for the arm actuators control is a powerful architecture to cope with the control of the manipulator. The selected architecture is based on the use of a **Transputer Module (TRAMs)** containing a processor in which one or several processes can run. These processes are loaded on the transputer from the host, after that the transputer network runs autonomously, communicating with the host only when required.

The **TRAM Module** is complemented with two additional boards: The **Control Board** based on specially dedicated peripheral microcontrollers (one per actuator of the arm - motor and encoder), which permits the Manipulation Main Processor (a T805 transputer) devote its CPU time for other tasks (i.e., communication with other subsystems, such as the gripper controller). It contains the required circuitry for the Control loops (high and local levels). For *High Level Position Feedback* a rotatory potentiometer per joint is used and for safety reasons three brakes are foreseen. The other specific board is the **Power Board**, developed to drive the actuators through PWM signals (the microcontrollers output). It provides the power for the different arm devices, i. e. the Motors, the Brakes, the End-Effector and the Sensors (camera, laser, gripper, etc.)

Hence, the external interfaces provided are:

- **Parallel inputs/outputs**, to/from the LM's microcontrollers.
- **Serial lines (RS422)**: Mobile Robot Main Processor and Gripper Controller.
- **Analog to Digital Conversor**, to read the positions obtained by the potentiometers.
- **Auxiliary digital inputs/outputs**, for additional devices, such as motor brakes.

6.2 End-Effector Hardware

End-Effector electronics is included in the EE mechanical structure and has four boards of 50 by 50 mm side. F/T Sensor structure contains three electronics boards linked through connectors (figure 5) and the intermediate module contains the motor driver, the fourth hardware board. Boards are from the upper to the lower:

- Connection Board, very close to the horizontal bars strain gauges. It supports interface harness soldering points and contains the strain gauges connections, the Gauges Wheatstone Bridges and the compensation resistors. It has an independent four legs support to the sensor structure.

- Analog Board for the conditioning, amplification and filtering of the signals coming from the Wheatstone Bridges (six channels).
- Digital Board that includes the A/D converter for the bridges analog signals, a microcontroller SIEMENS 80C196, RAM and EEPROM memory, digital Inputs/Outputs and a Serial Link. It is in charge of manage F/T Sensor, Driver of the Grasping Motor, High Level Communication (Serial Link) and the Grasping Force Sensors (not implemented in this model). This board has a connector for the motor driver board.

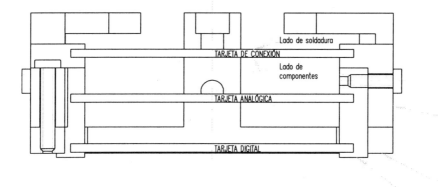

Figure 5: End-Effector Boards.

Analog and Digital boards are linked through four guiding pins that allows its modular insertion to the connection board and an easy disassembly for programming or adjusting purposes.

Driver board is connected to the Digital board (see figure 3) and commands DC Motor rotation through PWM output signals. It reads motor current and allows to control the power consumption and the grasping force.

7. References List

[1] Maurette, Michael. Planetary Rover Operations. 4th ESA Workshop "ASTRA 96"

[2] Fernández, Sabin. Iares Manipulator Dexterous End_Effector. 4th ESA Workshop "ASTRA 96"

[3] Selaya, Javier. Iares Manipulator Control Hardware. 4th ESA Workshop "ASTRA 96"

Modeling of Nonlinear Friction in Complex Mechanisms Using Spectral Analysis*

M.R. Popović
AlliedSignal Aerospace Canada
Etobicoke, Canada
e-mail: milos.popovic@alliedsignal.com

A.A. Goldenberg
Robotics and Automation Laboratory
University of Toronto
Toronto, Canada
e-mail: golden@me.utoronto.ca

Abstract

In this paper a new method of modeling nonlinear friction as a function of both position and velocity is introduced. The proposed empirical model uses spectral analysis to identify the main sources of position dependent friction in complex mechanisms. In addition, the model describes the contribution to the overall friction of every moving part of the mechanism. In the paper, the proposed spectral-based modeling technique is used to describe static friction and sliding friction (which includes negative damping friction) of the first joint of PUMA 560 robot. This is the first friction model truly capable of describing nonlinear friction in complex mechanical systems.

1. Introduction

A number of control strategies have been developed to overcome problems caused by the nonlinear friction (stick-slip effect) [2]. The most relevant strategies presented in the literature are: friction compensation using dither signal [1]; friction compensation through position or force control using high gain feedback [4]; adaptive feedback friction compensation [3]; robust nonlinear friction compensation [12]; model-based feedforward friction compensation [1, 5, 10]; and precise positioning using pulse control [7, 9, 14]. An essential part in the use of the above compensation methods is friction modeling. Friction models are either used as a part of the controller [1, 3, 5, 7, 9, 10, 14], or to verify stability of the proposed methods [1, 3, 4, 12]. Nearly thirty different friction models have been proposed so far [10]. Among them, the most frequently used models are Tustin's [1], Canudas de Wit's [3], Armstrong's [1], Hess-Soom's [6]

---
*Supported by a grant from MRCO, Ontario, Canada.

and Haessig-Friedland's [5]. Common to all these friction models is that they were initially designed to describe nonlinear friction in simple cases such as sliding of two flat surfaces or rolling of perfectly rounded body on a flat surface. Later on, the authors used the same models to describe nonlinear friction in complex mechanisms. As complex mechanisms may have numerous sliding and rolling parts, these simple models cannot fully describe the complexity of the friction phenomena generated by such mechanisms. Additionally, except for Armstrong's [1], Dahl's [1] and Haessig-Friedland's [5] models all other models describe friction as a function of velocity only, eventhough friction depends on both, position and velocity. As a result, none of the existing friction models leads to completely satisfactory method of describing nonlinear friction in complex mechanism.

In this paper a new modeling technique that uses spectral analysis to describe nonlinear friction in complex mechanisms as a function of both position and velocity is proposed. To the best of authors' knowledge this is the first modeling strategy specifically designed to describe the contribution to the overall friction of every mechanical part of a complex mechanism. In the paper the spectral-based modeling technique is used to experimentally obtain the model of the static friction[1] and the sliding friction[2] of the first joint of PUMA 560 robot. Although in this paper the results are presented for the PUMA 560 experimental setup the proposed modeling strategy can be applied to a variety of different mechanisms.

The organization of the paper is as follows. Section 2 describes the experimental setup. Static friction and sliding friction experiments are presented in Section 3. The spectral-based diagnostic method which determines sources of friction in complex mechanisms is introduced in Section 4. In Section 5 the diagnostic method is used to model friction of the first joint of the PUMA 560 robot. Conclusions are given in Section 6.

2. Experimental Setup

In the paper, nonlinear friction is investigated using the first joint of the PUMA 560 robot [13] (see Figure 1). The first joint was selected for experiments because it represents a fairly complex mechanism whose performance is not influenced by gravity.

The first joint of the PUMA 560 robot is actuated with a permanent magnet DC-motor. Position and velocity measurements of the first joint are obtained from the two encoders mounted on the motor shaft (see Figure 1). The built-in Unimate encoder, whose resolution is 62610 encoder increments per joint revolution [13] (1000 encoder increments per encoder's shaft revolution), is used to measure position. A Canon Laser Rotary encoder, whose resolution is 20285640 encoder increments per joint revolution [8] (324000 encoder increments per encoder's shaft revolution), was added to measure velocity.

The first joint is controlled with a modified Unimate controller [13] which

---

[1] Static friction is the torque (force) necessary to initiate motion of a mechanism from rest.
[2] Sliding friction is the friction torque (force) generated by a mechanism during motion. It consists of Coulomb friction, viscous friction and negative damping friction.

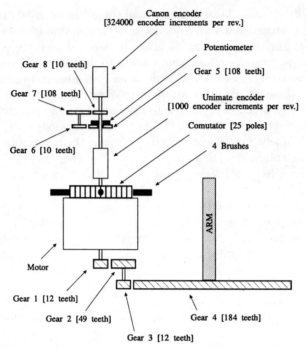

Figure 1. Schematic diagram of the first joint of the PUMA 560 robot

allows both open-loop and closed-loop control of the mechanism. This feature enabled us to carry out experiments using open-loop control scheme, and at the same time, perform positioning and warming-up[3] procedures using closed-loop control. The control program was written in C++, and the sampling period is 1.5 msec.

3. Experiments

In literature [1, 11] it was already established that static friction and sliding friction are repeatable and position dependent. Therefore, in this paper we shell only describe static friction and sliding friction experiments required to generate the overall friction model of a complex mechanism.

3.1. Static Friction Experiments

Static friction was measured by conducting a number of break-away experiments, as suggested by Armstrong [1]. First, the warm-up procedure was executed. Second, the arm was positioned at a starting point, and the controller was switched to the open-loop configuration. Then, the motor current was increased until static friction of the first joint was overcome (break-away). As soon as the arm overcame static friction (represented by a move of at least 10 encoder increments) the motor current was brought to zero and the arm was left to come to rest. During this "relaxation" period the initial position

---

[3] In order to eliminate the influence of the dwell time effect [1, 9], prior to every experiment, the arm was "warmed-up" by 6 cycles of vigorous motion throughout the workspace.

of the arm and the break-away current were recorded. After the arm came to rest a new experiment was initiated. These experiments were carried out for a large range of desired arm positions. The experiments were conducted for two different arm configurations and two different directions of rotation of the first joint. Ten experiments were conducted for each combination of joint rotation and arm configuration (in total 40 experiments).

The static friction characteristics from the same set were processed to obtain average static friction characteristics for each direction of rotation. Figure 2.a shows one such average static friction characteristic calculated for a set of ten static friction curves obtained for the vertical arm configuration and a clockwise direction of rotation of the first joint.

3.2. Sliding Friction Experiments

Sliding friction was studied by maintaining the angular velocity of the experimental mechanism constant and by measuring friction as a function of position. In this way a number of sliding friction versus position characteristics were obtained for different constant angular velocities. These sliding friction characteristics together with the static friction characteristics obtained in Subsection 3.1 were later used to generate a complete friction model of the experimental mechanism.

The sliding friction experiments were conducted as follows. First, the warm-up procedure was performed. Second, the robot arm was positioned at a predetermined starting point and the controller was switched to track a constant angular velocity $\dot{\theta}_d$. Throughout the experiment the friction torque and the Unimate encoder readings were recorded. Since the PID controller tracks the desired velocity with an error, it was decided to record friction torque and the corresponding position only when the angular velocity of the mechanism $\dot{\theta}$ was in the range of ±1% from the desired angular velocity $\dot{\theta}_d$. The experiments conducted have shown that no matter how many times the controller was set to track the desired velocity $\dot{\theta}_d$ it was not possible to obtain sliding friction measurements for more than 40% of positions of interest. In order to obtain sliding friction measurements for all the desired positions it was necessary to command the controller to track each of the following velocities: $0.90 * \dot{\theta}_d, 0.91 * \dot{\theta}_d, 0.92 * \dot{\theta}_d, \ldots 1.10 * \dot{\theta}_d$; at least 10 times (in total 210 experiments). This enabled the sliding friction for the desired velocity $\dot{\theta}_d$ to be measured at all positions of interest. In addition, for every position at least 6 (on average 20) sliding friction readings were obtained. The different friction readings for the same position were averaged and these averaged values were sorted out according to their positions. For the few positions in which sliding friction was not recorded the cubic spline interpolation was used to generate the corresponding friction values from the available average sliding friction data.

This type of experiment was conducted for the following set of desired angular velocities: 0.005, 0.008, 0.01, 0.03, 0.05, 0.08, 0.09, 0.1, 0.2, 0.3, 0.4 and 0.5 rad/sec. The experiments were conducted for the same arm configurations and the same directions of rotation of the first joint as the experiments in Subsection 3.1. In total 10080 experiments were conducted. Due to space limitations only few average sliding friction characteristics obtained for the vertical arm configuration and a clockwise direction of rotation of the first joint (similar to that in Subsection 3.1) are presented in Figures 2.b-f.

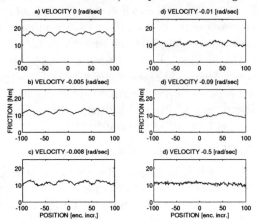

Figure 2. Six enlarged average friction characteristics as a function of position obtained for constant angular velocities

3.3. Discussion

A comparison of static friction and sliding friction characteristics obtained in Subsections 3.1 and 3.2 shows that these characteristics have similar positional dependency, in particular the characteristics obtained for lower angular velocities. It is also important to mention that as the velocity increases, the fluctuation in friction decreases reaching a minimum for angular velocities of 0.4 and 0.5 rad/sec. The reason for the decrease in fluctuation as the velocity increases can be explained as follows. During the static friction regime all moving parts of the mechanism are motionless and asperities of their contact surfaces are all bonded. Once static friction is overcome (breakaway) and the asperity contacts are broken, all sliding and rolling parts of the mechanism are in the boundary lubrication regime. As a result, tips of the asperities are establishing and breaking contacts very fast causing random friction and significant friction fluctuations (see Figure 2.a). Also during this phase, non-smooth transitions caused by the gears are more visible because the velocity of the mechanism is not large enough to build up a fluid film between the gears' contact surfaces. As the velocity increases, the lubricant spreads between the contact surfaces and the tips of the asperities are not in contact as much as before causing a drop in the overall friction and a reduction in friction fluctuations. During this transition period friction fluctuations caused by the random friction and non-smooth transitions are decreasing but are still significant, as shown in Figure 2. Once all the contact surfaces are fully lubricated, friction fluctuations caused by random friction and non-smooth transitions significantly decrease and the average friction reaches its minimum (angular velocities in the range from -0.08 to -0.09 rad/sec). As the velocity further increases the average friction augments as a result of the increase in viscous friction, while the friction fluctuations caused by random friction and non-smooth transitions further decrease making the friction characteristic smoother, as shown in Figures 2.e - 2.f. These findings fully confirm previously published results in [1, 10, 11].

The overall average friction as a function of angular velocity was further calculated from the average friction versus position characteristics obtained for angular velocities 0, 0.005, 0.008, 0.01, 0.03, 0.05, 0.08, 0.09, 0.1, 0.2, 0.3, 0.4 and 0.5 rad/sec. This was done by further averaging average friction characteristics obtained for different angular velocities (friction characteristics shown in Figure 2) and arranging the obtained values in an increasing order depending on their angular velocity. As a result, the overall average friction torque as a function of angular velocity was obtained for the experimental mechanism, as shown in Figure 3. Note that in Figure 3 the second through to the fifth average friction values represent the negative damping phenomenon. Therefore, Figures 2.b - 2.d are the first sliding friction versus position characteristics ever recorded for negative damping velocities. Also these characteristics are the first to show that negative damping friction is position dependent.

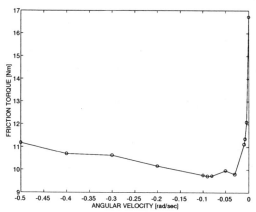

Figure 3. Overall average friction as a function of angular velocity for a clockwise direction of rotation of the first PUMA 560 joint

The experiments also confirmed the fact that friction is strongly influenced by two phenomena: **Asymmetry** and **Dwell time** effects [1, 10]. Asymmetry is a phenomenon that results in dissimilar friction characteristics for opposite motion directions [1]. In this paper problems caused by asymmetry were overcome by introducing distinct friction models for each direction of motion [1, 11]. Dwell time is a period of time during which friction characteristics of the mechanism significantly change because the mechanism is motionless [1, 11]. In our experiments, a "warming-up" procedure was used to eliminate the influence of dwell time effect.

4. Diagnostic of Friction in Complex Mechanisms

In this section the diagnostic procedure that was developed to generate the overall friction model is introduced. The diagnostic procedure starts by applying the DFT analysis on the average static friction and sliding friction

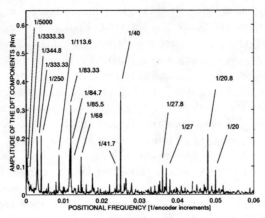

Figure 4. DFT components of the static friction characteristic

characteristics obtained in Section 3 (see Figure 2). The average characteristics are used in the analysis because averaging preserves the repeatable part of the measured phenomenon and reduces noise and random signal effects. As a first step in performing the spectral analysis, the range of positions of the analyzed signal has to be selected. The range determines the minimum "positional" frequency that can be observed using the DFT analysis [2] (also called "basic frequency"). The working range of the first joint of the PUMA robot is equivalent to 55600 encoder increments, which corresponds to 320° of the first joint rotation. Therefore, the smallest frequency we could select as the basic frequency is $1/55600$ (encoder increments)$^{-1}$. As we are unable to measure friction of the first joint for the entire range of encoder positions (55600 encoder increments) and as our position encoder has a resolution of 1000 encoder increments we have decided to use the range of 10000 encoder increments in the spectral analysis. Thus we have selected the frequency $1/10000$ (encoder increments)$^{-1}$, which is 10 times lower than the frequency of the position encoder, to be the basic frequency. With this basic frequency it is possible to observe practically all phenomena of interest except for those generated by the rotation of gear 4 and those generated by the arm rotation. The reason is that the phenomena generated by the arm and gear 4 rotations have a frequency of $1/62610$ (encoder increments)$^{-1}$, which is lower than the minimum measurable frequency $1/55600$ (encoder increments)$^{-1}$. The frequency $1/10000$ (encoder increments)$^{-1}$ can be accepted as a sufficiently small to be used as basic frequency for the purpose of this study. This choice of basic frequency significantly simplified the analysis and the process of associating Fourier components with mechanical parts which are generating them. Nonetheless, it is possible to use some other frequency as the basic one. After the characteristics with the proper range are obtained they are analysed using the DFT method. The results of one such analysis conducted for the static friction characteristic shown in Figure 2.a are given in Figure 4.

Since there are almost 10000 different DFT components (each DFT component is defined by its amplitude A, phase B and period C [2]) obtained for each analysed characteristic it is impractical to study all of them. Hence, it was decided to investigate only those DFT components that are considered significant. For this purpose it was decided that the DFT components whose amplitudes are greater than or equal to 0.5% of the zero DFT component (a mean value of the measured characteristics) are considered significant. The DFT components which do not satisfy this requirement are discarded as insignificant. Other criteria for selecting significant components can be used depending how refined the DFT analysis should be. For the purpose of this study the proposed component significance criterion provided us with more than satisfying accuracy. This will become clear in the next section where the overall fiction model is introduced.

Note that the DFT components obtained from the spectral analysis are not given in the time domain, instead they are given in the "position" domain. In other words, the DFT components are not functions of frequency (Hz), but functions of reciprocal value of a number of encoder increments, or (encoder increments)$^{-1}$. As a result, each DFT component in Figure 4 corresponds to a phenomenon that repeats itself every n number of encoder increments, where n = (value on the x-axis)$^{-1}$. For example, the component marked "1/83.33" in Figure 4 corresponds to a phenomenon that repeats itself every 83.33 encoder increments. In order to explain the relationships between each phenomenon and a DFT component in Figure 4, a detailed description of the first PUMA joint was required (see Figure 1) as well as the number of encoder increments the Unimate encoder generates during one motor revolution (see Section 2). Following are the examples how it was determined which parts of the experimental mechanism are responsible for the existence of significant DFT components given in Figure 4[4]:

- **Frequency 0.004 (encoder increments)$^{-1}$.** If one commutator pole is not aligned properly with the neighboring commutator poles, it rubs the brushes of the DC motor more or less than the other poles, causing fluctuations in friction. In between two rubbings of a non-aligned pole with two adjacent brushes, exactly 250 encoder increments are counted. This means that the non-aligned commutator pole generates a fluctuation in friction with a frequency of $1/250 = 0.004$ (encoder increments)$^{-1}$. Hence, sliding of the brushes on the commutator is associated with the DFT component 0.004 (encoder increments)$^{-1}$.

- **Frequency 0.012 (encoder increments)$^{-1}$.** One rotation of gear 1 is equivalent to 1000 encoder increments. Therefore, a part of gear's circumference which belongs to a single tooth of gear 1 is equivalent to $1000/12 = 83.33$ encoder increments. This means that between engagements of two neighboring teeth of gears 1 and 2, 83.33 encoder increments are counted. In other words, friction fluctuation which is generated by engaging and disengaging of gears 1 and 2 has a frequency of $1/83.33 = 0.012$

---

[4] A detail explanation how different parts of the experimental mechanism generate significant DFT components is provided in [10].

(encoder increments)$^{-1}$. Hence, meshing of gears 1 and 2 is considered responsible for the existence of this DFT component.

- **Frequency 0.024 (encoder increments)$^{-1}$.** Frequency 0.024 (encoder incre- ments)$^{-1}$ is the closest frequency obtained to the frequency of the second harmonic of frequency $1/83.33 = 0.012$ (encoder increments)$^{-1}$, which is $1/41.66 = 0.024$ (encoder increments)$^{-1}$. Hence, meshing of gears 1 and 2 generates this DFT component.

If these findings are arranged in a table, a one-to-one map between the significant DFT components and the phenomena responsible for their existence is obtained (see Table 1). From Table 1 it could be concluded that the major sources of friction fluctuation for the first PUMA joint are: meshing of gears 1 and 2; meshing of gears 3 and 4; rotation of gear 6 and sliding of the brushes on the commutator. Once this list is obtained the diagnostic procedure is completed.

j	$1/C_j$ (enc. incr.)$^{-1}$	C_j enc. incr.	Sources of the DFT components
0	0	∞	Mean friction
1	0.0001	10000	Gear 6 and basic frequency
2	0.0002	5000	Eccentricity of gears 6 and 7
3	0.0003	3333.33	Gear 6 and basic frequency
4	0.0029	344.82	Meshing of gears 3 & 4
5	0.003	333.33	Motor, motor shaft, encoders, gears 1 and 5
6	0.004	250	Brushes and commutator
7	0.0059	169.49	Meshing of gears 3 & 4
8	0.008	125	Brushes and commutator
9	0.0088	113.63	Meshing of gears 3 & 4
10	0.0117	85.47	Meshing of gears 3 & 4
11	0.0118	84.74	Meshing of gears 3 & 4
12	0.012	83.33	Meshing of gears 1 & 2
13	0.0147	68.03	Meshing of gears 3 & 4
14	0.0176	56.82	Meshing of gears 3 & 4
15	0.024	41.67	Meshing of gears 1 & 2
16	0.025	40	Brushes and commutator
17	0.0265	37.74	Meshing of gears 3 & 4
18	0.036	27.78	Meshing of gears 1 & 2
19	0.037	27.03	Rotation of gear 6
20	0.048	20.83	Meshing of gears 1 & 2
21	0.072	13.89	Meshing of gears 1 & 2

Table 1. Frequencies and periods of the significant DFT components obtained for $\nu = 4$

The diagnostic method outlined in this section is the simplest approach to determine which mechanical parts of the rotary mechanism are causing positional dependency of friction. Using this diagnostic method one can determine which mechanical parts need to be redesigned in order to minimize the positional dependency of friction. Although this method does not provide us with a specific measure of contribution of every part of the mechanism to the overall mean friction, it can be used as an indicator which parts are most likely to be the main sources of average friction.

5. Complete Friction Model

In this section a DFT-based friction model that describes static friction and sliding friction as a function of both position and velocity is described.

As a first step in developing the model, a distinction between the DFT components obtained for different average friction characteristics had to be made. In order to do that the following notation was introduced. All DFT components A_j, B_j and C_j obtained for the friction characteristic with angular velocity $\dot{\theta}_i$, were labelled as $A_j(\dot{\theta}_i)$, $B_j(\dot{\theta}_i)$ and $C_j(\dot{\theta}_i)$ respectively, where $j = 0, \ldots, J_i$ denotes a number of significant DFT components obtained for the average friction characteristic with the angular velocity $\dot{\theta}_i$, and $i = 1, \ldots, I$ denotes a number of analysed friction characteristics. Also, angular velocities $\dot{\theta}_i$ were arranged in such a way that the property $|\dot{\theta}_i| \leq |\dot{\theta}_{i+1}|$ holds. As a result, significant DFT components obtained for the static friction characteristic were labelled as $A_j(\dot{\theta}_1)$, $B_j(\dot{\theta}_1)$ and $C_j(\dot{\theta}_1)$, where $\dot{\theta}_1 = 0$ rad/sec. Note that components $B_0(\dot{\theta}_i)$ and $C_0(\dot{\theta}_i)$ (where $i = 1, \ldots, I$) are always equal to 0.

After the parameters of the DFT components were labelled according to which friction characteristic they belong to, they were further sorted out in the following manner. First, the second significance criterion was introduced. This criterion says that the DFT component is significant for all analysed characteristics if it is significant (its amplitude is greater than or equal to 0.5% of the amplitude of the zero DFT component) for at least ν number of analysed friction characteristics. The simplest choice of ν is 1. This means that the DFT component which is found to be significant for at least one analysed friction characteristic is significant for all other friction characteristics. In this case, for $\nu = 1$, 77 different significant DFT components were obtained, which were too many to be handled by the MATLAB software used in this analysis. Instead $\nu = 4$ had to be used, which means that a DFT component was significant only if it was significant for at least 4 different analysed characteristics. As a result, 22 different significant DFT components and their corresponding amplitudes, phases and periods were obtained (see Table 1).

Once all of the significant DFT components were obtained, all $A_j(\dot{\theta}_i)$ parameters with the same j index were grouped in descending order according to their angular velocity $\dot{\theta}_i$. Then, functions $A_j(\dot{\theta})$ which represent parameters A_j as functions of angular velocity $\dot{\theta}$ were obtained by interpolating values of $A_j(\dot{\theta}_i)$ on the interval $[\dot{\theta}_1, \dot{\theta}_I]$ (see Figure 5). In this particular case a third order spline interpolation was used to interpolate values of $A_j(\dot{\theta}_i)$ on the interval [0 rad/sec, -0.5 rad/sec], where the interpolation step was -0.001 rad/sec. The same procedure was repeated for parameters $B_j(\dot{\theta}_i)$ and the corresponding $B_j(\dot{\theta})$ functions were obtained (see Figure 5). If functions $A_j(\dot{\theta})$ and $B_j(\dot{\theta})$ and the corresponding parameter C_j which represent the amplitude, the phase angle and the period of the DFT component j, respectively, are substituted in the Fourier series the following function is obtained:

$$\tau_f(\theta, \dot{\theta}) = A_0(\dot{\theta}) + \frac{\pi}{2} \sum_{j=1}^{J} A_j(\dot{\theta}) \sin(\frac{2\pi}{C_j} - B_j(\dot{\theta})) \qquad (1)$$

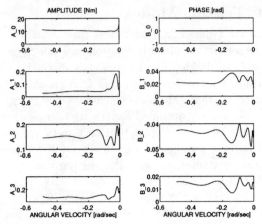

Figure 5. 4 out of 22 significant DFT components given as a function of angular velocity obtained from both static friction and sliding friction characteristics

where θ denotes the absolute angular position of the mechanism and $\dot{\theta}$ denotes mechanism's angular velocity ($\dot{\theta} \in [\dot{\theta}_1, \dot{\theta}_I]$ where $I = 13$, $\dot{\theta}_1 = 0$ rad/sec and $\dot{\theta}_I = -0.5$ rad/sec). Function (1) represents an experimentally obtained model of the static friction and the sliding friction as a function of both angular position and angular velocity. This model also describes the negative damping friction which is included in the model as a part of the sliding friction.

The friction model obtained by substituting functions $A_j(\dot{\theta})$ and $B_j(\dot{\theta})$ (see Figure 5) into equation (1) describes the friction of the experimental setup for angular positions from 5000 (25.9714°) to -4999 (−31.5198°) encoder increments and angular velocities in the range from 0 rad/sec to -0.5 rad/sec. In other words this model describes friction for 10000 different angular positions and 501 different angular velocities, which is equivalent to 40.08 MB of data. As a result, the obtained friction model cannot be presented graphically for all of the above positions and velocities using the available MATLAB software. Instead, a small part of the model is presented graphically in Figure 6. The characteristic shown in Figure 6 describes friction for angular positions from -2480 to -2500 encoder increments and for angular velocities from 0 rad/sec to -0.2 rad/sec. If it is taken into consideration that this characteristic describes less than 0.1% of the friction characteristic function (1) models, the extent of information in the proposed model can be appreciated.

6. Conclusions

In this paper a diagnostic method, which could be used to determine sources of friction in complex mechanical devices, is proposed. This diagnostic method uses spectral analysis to identify which mechanical parts of a mechanism generate friction and the relative contribution of each part of the mechanism to the positional dependency of the overall friction. This diagnostic method can be used as a powerful tool to identify which mechanical parts of a complex mechanism need to be redesigned in order to minimize positional dependency of friction.

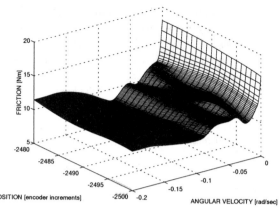

Figure 6. Modeled friction characteristics given as a function of position and velocity

The proposed diagnostic method can be used to develop a complete friction model of any complex mechanism. In the paper it was demonstrated how the diagnostic method can be used to model both static friction and sliding friction in a complex mechanism as a function of position and velocity. The proposed DFT model describes: static friction, viscous friction, Coulomb friction, negative damping friction and asymmetry effect, as functions of both position and velocity. These five phenomena are the most significant of all the nonlinear friction phenomena. The proposed model does not describe time lag, history effect, dwell time, Dahl effect and load dependency. Time lag and Dahl effect were not modeled because it was not possible to observe them with the chosen experimental setup. History effect was not modeled because it is not completely understood and an adequate method to describe it does not exist. Dwell time effect was not modeled because it was eliminated using the warm-up procedure. In order to address load dependency experiments were conducted using different loads and it was observed that the average friction was constant while the position dependent friction component varied with load. The reason why the average friction was constant is the fact that the mechanism's transmission elements and roller bearings were preloaded [14]. As for the position dependent component it was not possible to determine how the load affected it and for this reason load dependency was not modeled.

The proposed empirical model developed using the DFT modeling technique has a number of advantages over existing friction models. First, the accuracy of the proposed model is higher than the accuracy of any other existing friction model. This can be explained by the fact that the proposed model describes all repeatable friction phenomena, and does not filter out any significant friction component, except for the random friction. Second, this model describes friction as a function of both position and velocity. Third, the accuracy of the model could be easily adjusted by increasing or decreasing the number of DFT components used in the model. In particular, the proposed friction model with the chosen significance criteria models about 50-60%

of the position dependent friction. By including more DFT components this result could be improved by up to 70%. Beyond this accuracy it is believed that it is impossible to further improve the accuracy of the model because of the random friction component, which should not be modeled since it changes constantly as a result of wear. Fourth, this is the only generic model that can truly describe nonlinear friction in complex mechanisms. As a generic model, it can be used to compensate for friction as a part of a model-based feedforward friction compensator, or to simulate adverse effects of nonlinear friction on the performance of complex mechanisms and to identify major sources of friction in mechanisms.

References

[1] Armstrong B., <u>Control of Machines with Friction</u>, Kluwer Academic Publishers, Boston, USA, 1991.

[2] Bloomfield P., <u>Fourier Analysis of Time Series - An Introduction</u>, John Wiley and Sons, New York, 1976.

[3] Canudas de Wit C., <u>Adaptive Control for Partially Known Systems</u>, Elsevier, Boston, USA, 1988.

[4] Dupont P.E., "Avoiding Stick-Slip in Position and Force Control Through Feedback", *Proc. IEEE Int. Conf. on Robotics and Automation*, 1991, pp.1470-1475.

[5] Haessig D.A., and Friedland B., "On the Modelling and Simulation of Friction", *J. of Dynamic Systems, Measurement, and Control*, Vol. 113, September 1991, pp.354-362.

[6] Hess D.P. and Soom A., "Friction at a Lubricated Line Contact Operating at Oscillating Sliding Velocities", *J. of Tribology*, Vol.122(1), 1990, pp.147-152.

[7] Higuchi T. and Hojjat Y., "Application of Electromagnetic Impulsive Force to Precise Positioning", *IFAC 10th Triennial World Congress*, 1987, pp.283-288.

[8] Nishimura T., and Ishizuka K., <u>Canon Laser Rotary Encoders</u>, Motion, Canon Inc., Tokyo, September/October 1986.

[9] Popović M.R., Gorinevsky D.M., and Goldenberg A.A., "Fuzzy Logic Controller for Accurate Positioning of Direct-Drive Mechanism Using Force Pulses", *Proc. of IEEE Int. Conf. on Robotics and Automation*, Vol.1, 1995, pp.1166-1171.

[10] Popović M.R., <u>Friction Modeling and Control</u>, PhD Thesis, University of Toronto, Toronto, Canada, 1996.

[11] Popović M.R., and Goldenberg A.A., "Friction Diagnostics and Modeling Using DFT Analysis", *to appear in Proc. of IEEE Int. Conf. on Robotics and Automation*, 1997.

[12] Southward S.C., Radcliffe C.J., and MacCluer C.R., "Robust Nonlinear Stick-Slip Friction Compensation", *ASME Winter Annual Meeting*, 1990, ASME paper No. 90-WM/DAC-8, 7 pages.

[13] *Unimate PUMA Robot Manual 398H*, Unimation Robotics, A Condec Company, Connecticut, USA, 1980.

[14] Yang S. and Tomizuka M., "Adaptive Pulse Width Control for Precise Positioning Under the Influence of Stiction and Coulomb Friction", *Trans. ASME J. Dynam. Syst. Meas. and Control*, September 1988, Vol. 110, No3, pp.221-227.

Development of the Carpal Robotic Wrist

Stephen L. Canfield
Tennessee Technological University
Cookeville TN, 38501-5014, USA
slc3675@tntech.edu

Charles F. Reinholtz,
Virginia Polytechnic Institute & State University
Blacksburg, VA 24061-0238, USA
creinhol@vt.edu

Abstract: The manipulator described in this paper is a novel three degree-of-freedom device that is ideally suited as a robotic wrist or platform manipulator. Because of its similarity to the human wrist, the invention has been named the "Carpal Wrist." Much like its natural counterpart, the Carpal Wrist has eight primary links, corresponding to the eight carpal bones of the human wrist, a parallel actuation scheme, similar to the flexor and extensor carpi muscles along the forearm, and an open interior passage, which forms a protected tunnel for routing hoses and electrical cables, much like the well-known carpal tunnel. The Carpal Wrist also has the significant advantages of possessing closed form forward and inverse kinematic solutions and a large, dexterous workspace that is free of interior singularities (either considered separately or as part of a manipulator arm). As a result of its symmetric parallel architecture, the Wrist can handle a large payload capacity and can easily be adapted to a variety of actuation schemes. One embodiment of the Carpal Wrist is shown in Fig. 1.

1. Introduction

While parallel-architecture manipulators have long been recognized for their high-rigidity and large payload-to-weight properties, few have been developed for application, primarily because of complications in kinematic and dynamic modeling. The mathematical model of any manipulator must be developed in order to allow the necessary motion control of the device. The mathematical model provides a mapping from the input space (called joint space) to the output space (called tool space) of the manipulator. Given a desired task in terms of motion of the robot tool, the mathematical model determines the required motor input parameters. Advanced manipulator performance through automatic control becomes possible when the model includes inertial or dynamic effects of the manipulator and tool. The research leading to the development of the Carpal Wrist is significant because it presents a complete kinematic and dynamic model of a parallel-architecture manipulator, and thus may provide significant improvement over current serial robot technology.

This paper will briefly describe the theoretical and mechanical development of the Carpal Wrist. Some of the anticipated advantages and disadvantages will be discussed. A general model of the Wrist will then be presented that describes the kinematic geometry of the device. From this model, the kinematic and dynamic analyses necessary for implementation will be demonstrated. Finally, results of the theoretical model will be reviewed and their application demonstrated in design of a prototype Wrist.

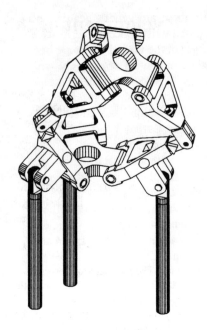

Figure 1: The Carpal Wrist

1.1 Background of the Carpal Wrist

The Carpal Wrist concept evolved from work in spatial parallel mechanisms, particularly a parallel, all-revolute constant velocity coupling [1,2]. The need for improved static and dynamic robot performance and the versatility of parallel mechanisms combined to give rise to the Carpal Wrist concept. The Carpal Wrist has a spatial, parallel-architecture consisting of three symmetric 5-revolute chains connecting the ground (base) to output (distal) plate (Fig. 2). This concept provides several inherent design advantages, including:

- A high strength-to-weight ratio, high rigidity, and improved dynamic characteristics
- Durability, stemming from its all-revolute design
- Structural symmetry that provides advantages over many proposed spatial parallel devices including closed-form solutions to the positional kinematics
- The possibility of either rotary actuation or linear actuation
- All actuated members are directly linked to the base, hence all input actuation is relative to ground
- A protected, enclosed central passageway
- A large, singularity free workspace

While these are general features of the proposed wrist concept, the relative advantages of the Carpal Wrist are best understood in context of existing wrist technology. A comparison of some key application requirements is made on a point-by-point basis in Table I. Note that the Carpal Wrist has many of the advantages of general parallel devices without sacrificing the kinematic simplicity and workspace size and dexterity of serial wrists. While the Wrist concept is not without disadvantages, it does have sufficient merit to warrant detailed kinematic investigation.

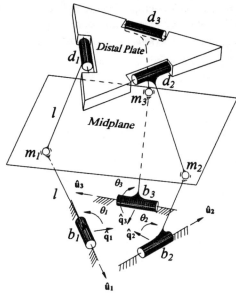

Figure 2: Kinematic Diagram of the Carpal Wrist

TABLE I: Features of Alternative Manipulator Structures

Attributes	Serial Wrists	Parallel Wrists	CARPAL Wrist
Positional Kinematics	Closed-form for many common arrangements	Implicit, usually requires numerical solution	Closed-form
Workspace / Dexterity	**Roll-Pitch-Roll:** Interior Singularities, large Workspace **Roll-Pitch-Yaw:** Boundary Singularities, Smaller Workspace	Typically limited	Large, singularity-free workspace, boundary singularities only
Strength-to-Weight ratio	Low	High	High
Actuation	Remotely located actuators Complex transmission required, Control axes relative to previous axis	Centrally-located linear actuators	Remotely located actuators; Control all axes relative to ground
Routing Tooling Requirements	Generally external routing	Internal routing possible	Internal, protected-tunnel routing

The Carpal Wrist has broad potential for commercial application. Forming this concept into a valid working model has involved a significant amount of research and development. The four areas of primary research are:
- Positional Kinematic Analysis
- Instantaneous Kinematic Analysis and Control
- Dynamic Force Analysis
- Prototype Design, Implementation & Testing

These issues have been addressed and validated. Closed-form positional kinematic solutions have been developed allowing real-time robot control [3]. A study of the instantaneous kinematics demonstrated the Wrist to be free of singularities [4], and to have a high degree of dexterity when compared to other conventional robotic wrists [5]. A dynamic model of the Wrist was developed, providing a design tool for actuator selection under high-speed operation [6], and a simulation environment to design high-level control systems [3]. The theoretical work was applied, tested and verified through the design and fabrication of a Wrist prototype [7]. Additionally, this research has led to general modeling and analysis techniques with applications in areas such as parallel manipulators as well as manipulator kinematics and dynamics.

2. Kinematic Position Analysis

Kinematic position analysis is the first step in creating a complete mathematical model of the new manipulator architecture. This analysis begins with developing a general descriptive model that describes the kinematic geometry of the device. From this geometric model, a mathematical function relating the input to output position parameters is formed. The goal of this function is to describe the forward and inverse position relationships in closed-form.

A schematic diagram illustrating the geometry of the Carpal Wrist is shown in Fig. 2. It consists of two rigid plates, referred to as the basal and distal plates, coupled together by three five-revolute (RRRRR) chains, with the central three revolutes intersecting at a common point. As a conceptual simplification, the central three intersecting revolutes can be replace with a spheric joint. However, it is considered that this all-revolute design will lead to increased reliability and superior precision. Although other possible geometries exist, this model will specifically address configurations in which the revolute axes of the basal and distal plates form equilateral triangles with base dimension, b, and which have six connecting arms of equal length, l. This symmetric construction is specifically chosen to result in a closed-form kinematic analysis. By actuating the three basal revolute joints, θ_1, θ_2, and θ_3, the motion of the distal plate may be controlled relative to the basal plate. This wrist is capable of producing pitch and yaw rotations in excess of 180 degrees combined with a third translational or "plunging" motion. The plunging motion is defined as an extension along the normal axis of the distal plate. A fourth degree-of-freedom may be added in the form of a revolute joint on the distal plate to accommodate tasks requiring general orientations of the end-effector. Note that adding roll motion in this manner results in an overall structure that is no longer parallel, but rather a hybrid parallel/serial device.

Forward kinematic analysis of the Carpal Wrist addresses the problem of finding the output orientation of the distal plate (the tool attaches to the distal plate and assumes its orientation), given the three base leg inputs, θ_1, θ_2, θ_3. This forward kinematic problem is solved using the midplane symmetry of the manipulator. Since the input base leg angles are given, and all the manipulator geometry is known, the location of the three points at the ends of the base legs are determined and used to define the midplane. Since the desired assembly mode of this wrist is one which provides symmetry about the midplane, the basal plate can be reflected to define the distal plate and thus locate the position and orientation of the Wrist output. This yields a closed-form solution to the forward kinematic analysis.

The inverse kinematic problem solves the required input angles for the base legs given an output position and orientation (pose) of the distal plate. Note that in formulating this problem, the output pose must be consistent with the mechanism's available pose freedom. This consistent pose is most easily specified as the pointing direction of the distal plane, and its plunge distance, or distance from Wrist center. Using the specified output pose, the distal plate is defined in terms of its three distal nodes. Then, vectors connecting the distal and basal nodes are generated, and the midplane is formed at the bisecting points of these three vectors. The position of the three midplane nodes are found as the intersection of the circle formed by each leg and the midplane. Finally, the angle of each base leg is determined, giving the required inputs for inverse control. Note that in this inverse solution, multiple solutions exist, 2^3 or 8 to be exact. This arises in the two possible solutions that occur for each of the three leg branches. These multiple solutions are called closures and in the geometry of the Wrist, the outer closure is chosen for all solutions. Also note that this solution proceeds in a step-by-step progression, with a fixed number of operations at each and is therefore closed-form. This provides the necessary tools for real-time control of the Carpal Wrist.

The workspace of the Wrist is shown in Fig. 3 as a series of surfaces, each representing a different fixed plunge value. The size of the available workspace is a function of the base-to-leg length ratio, $R_b = b/l$, and the plunge distance-to-leg length ratio, $R_d = p_d/l$ [1]. For any fixed R_d, the Wrist sweeps through a spherically constrained workspace. A series of these spherically constrained workspace plots are shown for various ratios, $R_d = 0.7467$ to 0.8667 while the ratio R_b is held fixed at 0.6667. The total workspace volume may be visualized as the sum of all the individual fixed plunge ratio plots. These dimensionless parameters, in particular $R_b = b/l$, provide design variables to control the size of available workspace. For example, decreasing the R_b ratio increases the theoretical workspace size.

3. Instantaneous Kinematic Analysis

The instantaneous kinematics or velocity analysis of the Wrist relates the output tool velocity to the input angular leg velocities. This analysis is position dependent, and assumes apriori knowledge of the position information. From the position analysis, the kinematic transformation between joint space and tool space can be represented as:

$$x = f(\theta) \quad (1)$$

Instantaneous or velocity analysis follows directly from the position analysis. Here, the input velocity vector, ω is mapped into the output space velocity vector, v, by the matrix, J called the Jacobian of the manipulator:

$$v = J\omega. \quad (2)$$

This matrix equation demonstrates the linear relationship between the input and output velocities. Since the vector function f from the position analysis is available, although implicit, velocity analysis results can be obtained from direct differentiation to obtain the Jacobian matrix.

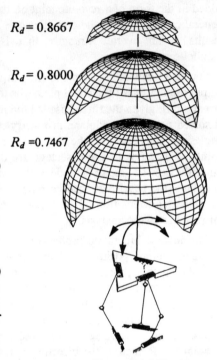

Fig. 3: Theoretical Workspace

The system Jacobian plays a central role in the kinematics of the Wrist. As shown in the equation above, it is used to define the velocity relationship of the actuated input velocities to the output or tool velocities. It is also important in quantifying the performance of the Wrist. Two key performance measures, the size and continuity of the available (singularity-free) workspace, and the manipulability or dexterity measure throughout the workspace, are contained in the algebra of the Jacobian matrix.

Once the Carpal Wrist's Jacobian is derived, the performance measures of the new device can be evaluated. Results from these evaluations play an important role in comparing the new robotic wrist to current wrist technology.

3.1. Closed-Form Jacobian

The Jacobian of the Carpal Wrist architecture is generated from the forward kinematic position equations. Partial differentiation of the three outputs with respect to the input joint angles generate the elements of J:

$$J = \begin{bmatrix} \partial \hat{N}_x / \partial \theta_1 & \partial \hat{N}_x / \partial \theta_2 & \partial \hat{N}_x / \partial \theta_3 \\ \partial \hat{N}_y / \partial \theta_1 & \partial \hat{N}_y / \partial \theta_2 & \partial \hat{N}_y / \partial \theta_3 \\ \partial p_d / \partial \theta_1 & \partial p_d / \partial \theta_2 & \partial p_d / \partial \theta_3 \end{bmatrix} \qquad (3)$$

where N_x, N_y are components of the symmetric midplane unit normal vector, and p_d is the plunge distance. The derivatives in this matrix have been determined in closed-form [4] resulting in a closed-form expression for the Jacobian. This Jacobian gives the output velocity of the Wrist as:

$$\begin{Bmatrix} \dot{\hat{N}}_x \\ \dot{\hat{N}}_y \\ \dot{p}_d \end{Bmatrix} = [J] \begin{Bmatrix} \dot{\theta}_1 \\ \dot{\theta}_2 \\ \dot{\theta}_3 \end{Bmatrix}. \qquad (4)$$

3.2. Singularity Analysis

It has been suggested that the Carpal Wrist contains a large, singularity-free workspace based on observation of the prototype model. From the Jacobian previously developed, an equation for the determinant is derived as a function of the three input angles, θ_i, $i = 1$ to 3:

$$\det[J] = \begin{pmatrix} \left(\partial \hat{N}_x / \partial \theta_1 * \partial \hat{N}_y / \partial \theta_2 * \partial p_d / \partial \theta_3 + \partial \hat{N}_x / \partial \theta_2 * \partial \hat{N}_{y2} / \partial \theta_3 * \partial p_d / \partial \theta_1 + \partial \hat{N}_x / \partial \theta_3 * \partial \hat{N}_y / \partial \theta_1 * \partial p_d / \partial \theta_2 \right) \\ - \left(\partial \hat{N}_x / \partial \theta_3 * \partial \hat{N}_y / \partial \theta_2 * \partial p_d / \partial \theta_1 + \partial \hat{N}_x / \partial \theta_1 * \partial \hat{N}_{y2} / \partial \theta_3 * \partial p_d / \partial \theta_2 + \partial \hat{N}_x / \partial \theta_2 * \partial \hat{N}_y / \partial \theta_1 * \partial p_d / \partial \theta_3 \right) \end{pmatrix} \qquad (5)$$

Singular positions can be found by setting this determinant equal to zero and solving for θ_1, θ_2, and θ_3. A search for zeros of the determinant reveals two infinities of singular positions, i.e., the singular input values form a surface. For the Wrist to be singularity free inside its workspace, this surface of singular positions must coincide or be outside the workspace boundary (the workspace boundary is defined by closure in the kinematic position equations). This was demonstrated by evaluating det[J] for singular input values and then comparing these with kinematic closure (which is indicated by the inability of the mechanism to assemble which occurs when the discriminant goes to zero in the position solution) at these same values. The results from this test illustrate that, as a singular position is approached (det[J] $\to 0$), the workspace boundary is also approached (*discriminant* $\to 0$).

3.3. Dexterity Results

The dexterity of the Carpal Wrist has been characterized using several quantitative definitions of dexterity. One index, presented by Soper et al., [8], defines dexterity as the relative stretching between the input and output velocity vectors:

$$D = \frac{\|v\|}{\|\omega\|} = \frac{1}{\mu} \qquad (6)$$

where D is the Wrist dexterity, v is the output velocity vector, ω the input velocity vector, and μ the mechanical advantage. Expanding this definition for the square Carpal Wrist Jacobian gives an explicit equation for D:

$$D = \sqrt{v^T v \left(v^T (J^{-1})^T J^{-1} v\right)^{-1}}. \qquad (7)$$

Note that dexterity definitions are functions of workspace position. Overall manipulator dexterity has been demonstrated to be best defined as the relative uniformity in dexterity over the manipulator workspace [5]. This measure, called the dexterous measure, is characterized by a low standard deviation in dexterity over the workspace and represents the most desirable manipulator behavior. The dexterous measure of the Carpal Wrist is compared to two other orienting devices, a serial pitch-yaw wrist and a five-bar pointing device, over a solid, hemispherical workspace. The results of this comparison are shown in Table II, demonstrating the Carpal Wrist to have the lowest standard deviation for the chosen dexterity definition, and therefore has the greatest uniformity in dexterity.

Table II: Results of Dexterous Measure Evaluation

Wrist Type	σ(dexterity)
Carpal Wrist	0.0375
Serial Pitch-Yaw	0.2047
Pointing Five-Bar	0.1249

5. Dynamic Force Analysis

The kinematic analysis of the Carpal Wrist is complimented by characterizing the dynamics of its parallel structure. Previous static analysis has verified that the Carpal wrist exhibits the improved force bearing capacity of parallel devices [7]. Here, the parallel structure is shown to be particularly advantageous when considered dynamically, due to its light-weight structure and multiple load-bearing members. The dynamic equations of motion are derived in closed-form by direct application of Lagrange's equations to the kinematic model. The model assumes a massive tool and includes all gravitational, inertial, and gyroscopic effects. The equations of motion provide closed-form evaluation of the actuation moments based on general tool trajectories.

The dynamics of the Carpal Wrist are derived using Langrange's approach, with the generalized coordinate system chosen as the input joint angles and the kinematics developed in canonical form. The Lagrangian, expressed as a function of generalized coordinates, depends on the manipulator energy state, i.e., the potential and kinetic energy. Therefore, calculating the Lagrangian requires position and velocity information. Since closed-form solutions for both the position and velocity analyses of the Carpal Wrist have been found, the Lagrangian may be expressed in closed-form. The Lagrangian formulation is advantageous since it removes internal constraint forces from the equations of motion and allows the choice of generalized coordinates, in this case the joint space coordinate system. Lagrange multipliers can be introduced as desired to inspect stress in the manipulator links due to the dynamic forces.

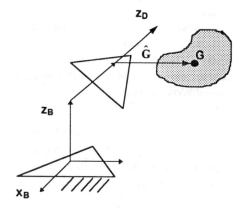

Figure 4: Wrist Mounted Tool

In developing the equations of motion, tensor subscript notation is applied throughout using the standard Einstein summation convention. This notation is useful as the derivatives of the velocity equations will result in third order tensors.

5.1. Wrist Dynamics

Figure 4 shows a tool, modeled as a lumped mass, mounted on the Wrist. Position and velocity information of the Wrist output and tool are known from the kinematic model, which provides a mapping between the input space joint angles (joint space) and the output space (tool space) This mapping is represented in both forward and inverse form through the functions f_1 and f_2, and J:

$$(\alpha,\beta,p) = f_1(\theta_1,\theta_2,\theta_3), \quad (\theta_1,\theta_2,\theta_3) = f_2(\alpha,\beta,p), \quad (\dot{\alpha},\dot{\beta},\dot{p}) = J'(\dot{\theta}_1,\dot{\theta}_2,\dot{\theta}_3) \quad (8)$$

where:
α, β, and p represent the output coordinates,
θ_1-θ_3 represent the input space coordinates,
f_1, f_2 represent the forward and inverse kinematic mappings respectively, and
J' represents the Jacobian, mapping input to output velocities.

With position and velocity information known, the energy of the Wrist and tool system can be determined, expressed in the Lagrangian. Lagrange's equations are then written and expanded to solve explicitly for both the required input motor torques and for the joint space accelerations. The resulting equations of motion [6]:

$$M_I = \sum_m \left\{ m \left[v_i \left(\frac{\partial^2 v_i}{\partial \theta_m \partial \theta_I} \dot{\theta}_m + \frac{\partial^2 v_i}{\partial \theta_m \partial \theta_I} \ddot{\theta}_m \right) + \frac{\partial v_i}{\partial \theta_I} \left(\frac{\partial v_i}{\partial \theta_m} \dot{\theta}_m + \frac{\partial v_i}{\partial \theta_m} \ddot{\theta}_m \right) \right] + \left[\frac{\partial^2 \omega_i}{\partial \theta_m \partial \theta_I} \dot{\theta}_m + \frac{\partial^2 \omega_i}{\partial \theta_m \partial \theta_I} \ddot{\theta}_m \right] (R_{ij} I_{jk} \omega_k) \right.$$
$$+ \frac{\partial \omega_i}{\partial \theta_I} \left(\frac{\partial R_{ij}}{\partial \theta_m} \dot{\theta}_m I_{jk} \omega_k + R_{ij} I_{jk} \left(\frac{\partial \omega_k}{\partial \theta_m} \dot{\theta}_m + \frac{\partial \omega_k}{\partial \theta_m} \ddot{\theta}_m \right) \right) - m v_i \frac{\partial v_i}{\partial \theta_I} + \frac{\partial \omega_i}{\partial \theta_I} R_{ij} I_{jk} \omega_k + \tfrac{1}{2} \omega_i \frac{\partial R_{ij}}{\partial \theta_I} I_{jk} \omega_k$$
$$\left. + mg \left[\frac{\partial p}{\partial \theta_I} (1 + R_{33}) + p \frac{\partial R_{3j}}{\partial \theta_I} + \frac{\partial R_{3j}}{\partial \theta_I} G_j \right] \right\}$$

(9)

where:
v is the velocity of the mass center of the output,
ω is the angular velocity of the output,
R_{ij} is the rotation matrix between the distal and basal plates,
I_{jk} is the moment of inertia tensor of the tool expressed in the tool frame, and
G_i is the vector locating point G with respect to the center of the distal plate expressed in the tool frame (Fig. 4).

5.2. Application to the Carpal Wrist Prototype

To verify the equations of motion, a dynamic force analysis of the prototype Carpal Wrist was performed using the dynamic model created above, and then comparing the results with experimental dynamic force data. A dynamic force measurement system was incorporated into the Carpal Wrist prototype carrying a generic, cylindrical-mass payload. This dynamic measurement system, shown in Fig. 5, records forces in the actuator rods that connect the stepper motors to the driven base legs of the Wrist. Axial forces in the actuator rods are measured with four strain gages, arranged in a full bridge with two longitudinal gages and two Poisson's gages. The gage output is acquired and recorded through a *Metrabyte DAS-16* data acquisition board. From the acquired gage voltage, force in the actuator rods and finally a measure of actuator moments are calculated. This measure is then compared to results from the equations of motion simulating the prototype Wrist move.

The prototype performed a longitudinal move, a longitudinal slice over the hemispherical workspace sweeping from one extreme of deflection (90 deg.) to the other. The path velocity and acceleration profile were matched to the upper limit of capacity of the stepper motors driving the prototype. This move begins with a constant acceleration reaching maximum velocity in approximately .1875 seconds, moving across the majority of the workspace at constant velocity, and finally then decelerating in 0.1875 seconds. The time of the entire move was three seconds. Results of the required input motor moments from the dynamic model are shown at the top of Fig. 6 while results of the experimental data are shown at the bottom. Comparing the model and experimental data

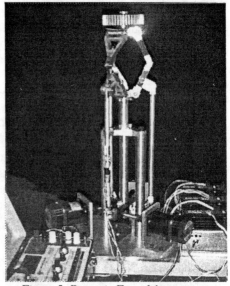

Figure 5: Dynamic Force Measurement System

demonstrates similarity in the required input motor forces, verifying the dynamic model. The primary difference in the theoretical and experimental results are the absence of peaks of required force shown in the theoretical model but not the experimental results. This difference is accounted for in several factors. Primarily, effects of friction, ignored in the dynamic model, had a significant effect in the prototype at the start and end of the move. Additionally, the path profile assumed for the theoretical model assumed a constant acceleration. However, the inability of the motors to meet the peak force would alter the actual acceleration slightly, instead maintaining a level input motor force demonstrated in the experimental results.

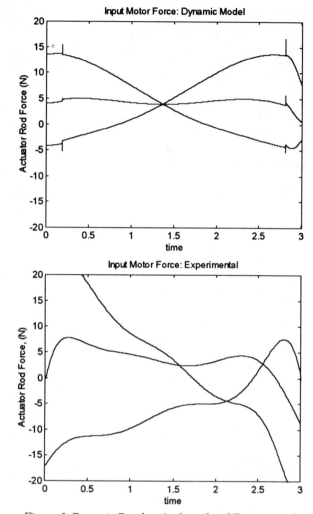

Figure 6: Dynamic Results: Analytical and Experimental

6. Carpal Wrist Prototype

The theoretical evolution of the Carpal Wrist concept has resulted in a complete mathematical model representing the kinematics and dynamics. The validity of the concept as well as the model are demonstrated by constructing the kinematic model into a physically viable prototype, and finally, a working device that can meet the demands of the industrial user. A proof-of-concept, working model of the Carpal Wrist was designed and developed, shown in Fig. 7. The mechanical design of this prototype is presented in detail by Ganino [7].

7. Results and Conclusions

This paper has presented the theoretical and prototype development of a new, parallel-architecture robotic wrist. The goal of the theoretical modeling has been to create the tools necessary to form the Carpal Wrist concept into a working model for industrial application. A summary of the primary areas of model development has been presented. Closed-form positional kinematic solutions have been developed allowing real-time manipulator control. The instantaneous kinematic analysis has demonstrated the Wrist to be free of singularities and to have a high degree of dexterity when compared to other conventional robotic wrists. Finally, the dynamic model has been developed providing a design tool for actuator selection under high-speed operation, and a simulation environment to design high-level control systems. Resulting from this work, a proof of concept prototype has been designed and is currently being demonstrated to industry for future application.

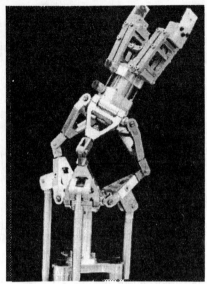

Figure 7: Prototype Carpal Wrist

8. Bibliography

[1] Canfield, S. L., and Reinholtz, C. F., 1995, "Development of an All-Revolute-Jointed Constant-Velocity Coupling," *Journal of Applied Mechanisms & Robotics*, Vol. 2, No. 3, July, pp. 13-19.
[2] Canfield, S. L., Salerno, R. J., and Reinholtz, C. F., 1995, "Design of an All-Revolute, Linkage-Type, Constant-Velocity Coupling," *SAE International Off-Highway & Powerplant Congress & Exposition*, September 11-13, 1995, Milwaukee, Wisconsin, ISSN 0148-7191, Paper no. 952133.
[3] Canfield, S. L., 1997, "Development of the Carpal Robotic Wrist," Ph.D. Thesis, Virginia Polytechnic Institute & State University, Blacksburg, VA.
[4] Canfield, S. L., Ganino, A. J., Salerno, R. J., and Reinholtz, C. F., 1996, "Singularity and Dexterity Analysis of the Carpal Wrist," *Proceedings of the 1996 ASME Design Engineering Technical Conferences*, Irvine, CA, August 18-21, 96-DETC/MECH-1156.
[5] Canfield, S. L., Soper, R. R., and Reinholtz, C. F., 1997, "Uniformity as the Guide to Evaluation of Dexterous Manipulator Workspaces," *1997 ASME Design Automation Conference*, Sacramento, CA, September 15-20, DETC97/DAC-3969.
[6] Canfield, S. L., Soper, R. R., and Reinholtz, C. F., 1997, "Dynamic Force Analysis of the Carpal Wrist," *1997 ASME Design Automation Conference*, Sacramento, CA, September 15-20, DETC97/DAC-3970.
[7] Ganino, A. J., 1996, "Mechanical Design of the Carpal Wrist: A Parallel-Actuated, Singularity-Free Robotic Wrist," Master's Thesis, Virginia Polytechnic Institute & State University, Blacksburg, VA.
[8] Soper, R. R., Canfield, S. L., Reinholtz, C. F., and Mook, D., 1997, "New Matrix-Theory-Based Definitions for Manipulator Dexterity," *1997 ASME Design Automation Conference*, Sacrament, CA, September 15-20.

First experiments with MIPS 1 (Mini In-Parallel Positioning System)

J-P. Merlet

INRIA, BP 93 06902 Sophia-Antipolis Cedex, France

Abstract: We present preliminary results of the design of a mini in-parallel 3-d.o.f. positioning system called MIPS. MIPS degrees of freedom are one translation and two orientations, which are obtained by the motions of linear magnetic actuators acting within a special in-parallel mechanical architecture. Its overall width will be about 1cm for a length of about 3cm. MIPS should be useful in medical applications and in inspection tasks.

1. Introduction

Mini positioning system have drawn a lot of interest in the recent past, especially for medical applications and inspection tasks. Building a miniature robot is not only a scaling problem. Indeed a simple scaling of a mechanical architecture like a serial robot performing well in a macro-environment will not work at a miniature scale as the friction forces (decreasing slowly with the size) will quickly take over the inertial forces (decreasing as the square of the size). Consequently a working mechanical device at small scale should rely for its motion on the deformation of its geometrical structure instead on the more classical concept of relative motion of its links. Parallel architectures rely exactly on this geometrical deformation concept and have been indeed used at different scale with good result: huge size flight simulator, large-scale positioning system, medium size robot and mini positioning system [1],[8],[13]. In these examples the height of the robot ranges from several meters to two centimeters although basically the same mechanical architecture is used. Furthermore parallel manipulators are very efficient as source of force as all the actuators are directly delivering their forces to the mobile platform (this may be seen if we consider the robot ratio L/M where L is the nominal load and M the mass of the robot: this ratio for serial arm will be at best 0.1 for 6 d.o.f. robot while this ratio may exceed 20 for parallel robots).

Broadly parallel manipulators can be divided in two classes: the first class have actuators in the legs connecting the base to the moving platform while the second class uses grounded actuators. The classical Gough-platform is an example of the first class while the HEXA robot [14] and the prototype we have presented in [11] are examples of the second class. For a miniature robot with drastic constraints on the size locating actuators in the leg is not a good solution as the interference between the legs will greatly decrease the useful workspace of the system. So a manipulator of the second class will be more appropriate.

For the applications we are considering MIPS will act as an active head mounted on some other positioning system. For example MIPS could be mounted as the end of an endoscope for insuring the fine positioning of surgical tools (as the devices proposed by Sturges [16], Wendlandt [18], which use cables and winches as actuators or the pressure-driven devices of Grundfest, Burdick [3] and Treat [17]) or as an inspection head for a mobile platform inspecting pipes. Therefore the overall mobility of the system will be insured both by the degrees of freedom of the support robot and by those of the head. A careful analysis of medical applications has shown that most of the tasks could be performed with a 3 d.o.f. robot having both translation and orientation capabilities. We will see however that MIPS may be reconfigurable to provide various combinations of d.o.f. The necessary range of motion should be in the vicinity of 5 mm for the translation part and ± 15 degrees for the orientation. The available force at the center of the platform should be around 0.15 N, but the robot should be able to withstand a larger force (especially during the travel to the point of interest). Our objectives are:

- an overall diameter of the robot in the range of 1cm

- a minimum height of the robot so that it can be used even in curved pipes

- a low stiffness: this is an element of safety especially for medical applications. But the robot should be able to withstand large forces in some configuration

- an autonomous robot with respect to the actuation: indeed an external actuation through flexible wire power transmission is a problem as soon as the length of the wire become important

- a modular robot from the control and power view point. In some mode the system should be able to perform some fixed motion autonomously without relying on any connection with the external world.

- a modular robot with respect to the provided d.o.f.: by a simple change in the mechanical architecture the operator may construct a robot with different d.o.f..

2. Mechanical architecture

Among all the possible 3-d.o.f. parallel robot one of the most promising structure has been proposed by Lee [8] which has also been used as the wrist of the ARTISAN robot of Khatib [7]. This structure is presented in figure 1. In this system each leg is connected to the base with a revolute joint and to the platform with an universal joint. A linear actuator in the leg enables to change the leg length and 3 d.o.f. of the platform could be controlled by changing the three leg lengths: a translation along the vertical direction and two orientations. We have decided to use this idea but without having the actuator in the legs: instead we use the principle presented in [11], where the joints close to the base are moving along a vertical direction while the legs have a fixed length. This

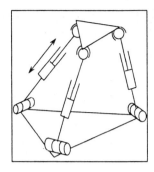

Figure 1. Lee 3-d.o.f. robot

enable to use very thin legs (leading to a very low mass of the moving elements of the robot) and therefore to decrease the risk of interference between the legs while enabling to keep the overall diameter of the robot quite small. The new architecture is presented in figure 2. This architecture is modular: for example

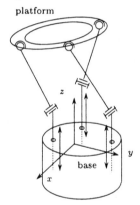

Figure 2. The mechanical architecture of the proposed robot

by changing the axis of the revolute joints we may modify the rotation axis of the platform or by connecting the platform with a rigid link fixed on the base with a ball-and-socket joint we will get a wrist with 3 rotational d.o.f.

In our basic design the revolute joint axis of leg 2 and 3 is the x axis while the joint axis of leg 1 is the y axis. With this disposition the platform may perform a translation along the z axis and rotations around the x, y axis. Note that with this architecture simple motion can be performed with very simple control laws for the actuators. For example:

- a similar periodic inputs on the three actuators will create an up and down motion of the platform

- a periodic input on leg 1 and a similar input with a phase shift of 180 degree on leg 2 and 3 will create a rotation of the platform around the y axis

- a periodic input on leg 2 and a similar input with a phase shift of 180 degree on leg 3 will create a rotation of the platform around the x axis

2.1. Kinematics

A reference frame $O, (\mathbf{x}, \mathbf{y}, \mathbf{z})$ is defined. Similariy we define a moving frame $C, (\mathbf{x_r}, \mathbf{y_r}, \mathbf{z_r})$ where C is an arbitrary point of the moving platform. The location of the moving platform will be defined by the coordinates of C in the reference frame and by a rotation matrix R relating the two frames (a vector whose components is expressed in the moving frame will be denoted by a superscript ^{r}).

Let define A_i as the connecting point of leg i to the revolute joint, B_i its connecting point to the universal joint on the moving platform, A'_i a reference point on the linear actuator axis. The axis of the revolute joint is $\mathbf{n_i}$, the leg length l_i and λ_i will be the height of A_i with respect to A'_i.

In order to solve the inverse kinematics consider the constraint equation on the leg length:
$$\|\mathbf{A_i B_i}\| = l_i$$
or:
$$\| - \lambda_i \mathbf{z} + \mathbf{A'_i O} + \mathbf{OC} + R\mathbf{CB_i^r}\| = l_i$$
squaring the previous equation leads to:
$$\lambda_i^2 - 2\lambda_i \mathbf{z}.(\mathbf{A'_i O} + \mathbf{OC} + R\mathbf{CB_i^r}) + \|\mathbf{A'_i O} + \mathbf{OC} + R\mathbf{CB_i^r}\|^2 - l_i^2 = 0 \quad (1)$$

Hence the articular coordinates are obtained as solution of a second order equation. One solution will have a height greater than B_i and will be discarded. As expected not all possible motions can be performed as we have the constraint equation:
$$\mathbf{A_i B_i}.\mathbf{n_i} = 0$$

The direct kinematics problem may be reduced to the direct kinematics of Lee prototype. For given heights of the actuator there may be up to 16 different configurations for the moving platform. The configurations can be determined by solving a 16 order polynomial in the tangent of the half angle of one of the revolute joint [10].

A software tool has been designed to simulate the motion of the robot, estimate its workspace and find the optimal design according to the task at hand.

3. Actuation

According to our basic design the three linear actuators will be disposed side by side in the main body of the robot. For a miniature device the actuation scheme may be: electrical motor, piezo-electric linear actuators, wires actuation, shape memory alloy (SMA), polymeric gels and films, or magnetic actuation. Although small electric motors (with a diameter of 8mm) are available we have not retained this solution. Indeed they need a reduction gear and a mechanism to transform the rotary motion into linear motion. Their diameters are such that they have to be put on top of each other in the body, leading to a robot with an important height.

Piezo-electric linear actuators are interesting for their low weight and high force and have been used in the past for micro parallel robot [1] or serial robot [2],[5]. Their main drawbacks are their rather limited range of motion (even with stacked actuators the range of motion is about 2 mm) and their cost.

Wire cable robots have also been used in the past [16],[18]. They have the advantage of compacity as soon as the winches are external to the robot (internal winches lead to the use of electrical motors) but one our objective prohibit the use of external power transmission.

SMA [9] have also been considered: by changing their temperature they have the property to come back to a given geometry. They have the advantages of compacity and low weight but their control is far from simple and it is difficult to predict their behavior in a surrounding where the temperature may be varying. Furthermore their bandwidth is usually very low.

Some polymers like the perfluorosulfonic acid polymer are able to bend when a low voltage is applied onto its surface [4],[15]. The amplitude of bending is usually very low (less than 0.1 mm) and consequently these actuators will need to be stacked to get the necessary range of motion.

Magnetic actuation is interesting and has been considered in the past [6],[12]. A permanent magnet which can slide into a solenoid leads to a very simple linear actuator. The force that can be exerted by such an actuator is sufficient as long as the range of motion is not too large and the control is basically simple while the stiffness of the actuator is low. The mechanical efficiency of these actuators are good (no reduction gear and the friction of the magnet in the solenoid is reduced with the centering effect of the system). Note also that the leg inputs we have presented for performing simple motions can be easily produced with on-board simple electronic hardware and battery power source.

We have build a first version of a magnetically driven linear actuator (figure 3). The plunger is constituted of three iron mini-magnets coated with Teflon connected by two aluminium cylinders. This plunger slides into a cylinder fixed to the base with two coils along the main axis. The current in each coil is controlled independently so that the resulting magnetic fields of the coils create two forces enabling the motion of the plunger or its station keeping.

Each extremity of the plunger is topped by a miniature head so that the motion is limited. The mass of the actuator is about 20 grams, the diameter is about 6 mm, the total length about 55 mm and the stroke is about 5 mm. The first tests have shown that the actuator force is in the range 0.3-0.5 N for a peak current of 2A and is therefore sufficient. However to improve the design we will like to decrease the actuator length and be able to use a lower peak current. Hence we are currently designing a second version of linear actuator with the following changes:

- we plan to use either Neodymium Iron Boron (NdFeB) or Samarium-Cobalt (SmCo) magnets instead of iron magnet. Their energy product

Figure 3. Principle and first version of the magnetic linear actuator

is about 20 to 60 times higher than the low-cost iron magnet.

- to increase the magnetic field produced by the coils they will be disposed on a 0.2 mm thick film of FPC (ferrite polymer composite) with permeability 9.
- we will investigate the use of only one coil, the plunger being held by a spring. This will enable to simplify the control scheme, reduce the power consumption and actuator length. The lower force provided by a lone coil will be compensated by the higher force provided by the FPC/SmCo coil/magnet.

The initial and potential versions of MIPS are presented in figure 4. To in-

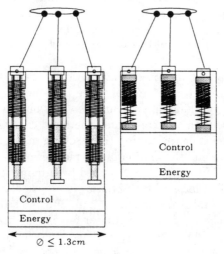

Figure 4. Two possible versions of MIPS: the first one uses the early version of the linear actuator while the second will use the improved version of the actuator

crease the autonomy of the end-effector we intend to have a modular energy cell which can be attached to the mechanism and will provide the necessary power. Similarly a control cell could also be attached to the end-effector: predefined motion of the mechanism will be stored in this cell and they should be triggered by an external input sequence (via ultrasound waves for example). This modularity will enable to change the level of autonomy of the robot according to the task to be performed (from a completely autonomous robot with on-board control and power to a teleoperated robot with external power and control).

4. Position sensing

The linear motion of the actuators should be measured for the close-loop control of the robot. Currently we are investigating three possible methods:

- optical measurement: a tilted mirror will be fixed to the bottom magnet and will deflect the light of a laser diode toward a light sensor array: the position of the spot on the array is a function of the height of the mirror

- LVDT measurement: two auxiliary coils will be fixed at the bottom of the actuator. One of them will be connected to an AC source and the tension in the second coil will be a function of the position of the magnet

- Hall effect measurement: a fixed Hall effect sensor will be put at the bottom of actuator and the current in this sensor will enable the measurement of the distance between the sensor and the bottom magnet.

5. Design of the passive joints

The passive joints are the revolute joint at A_i and the universal joint at B_i. A miniature revolute joint is not difficult to manufacture. Thus remains the problem of constructing the joint at B_i. Three possible solutions are possible: elastic joints constituted of miniature flexible coupler which can be obtained by drilling appropriate holes in a metal block[13], or more classical miniature joints. In an early design each leg has been made of a needle with a sphere as end point. This sphere will be disposed in a hole of the mobile platform which will then closed by a small ring whose hole diameter will be slightly less than the diameter of the sphere.

6. Conclusion

We have presented the preliminary result of the design of a micro in-parallel magnetically actuated 3 d.o.f robot. Our main task has been to test the linear actuator. Our early version exhibit interesting properties that we hope to improve in a second version by using more sophisticated materials.

References

[1] Arai T., Stoughton R., and Jaya Y.M. Micro hand module using parallel link mechanism. In *Japan/USA Symp. on Flexible Automation*, pages 163–168, San Francisco, July, 13-15, 1993.

[2] Fukuda T. and others . Characteristics of optical actuator-servomechanisms using bimorph optical piezo-electric actuator. In *IEEE Int. Conf. on Robotics and Automation*, volume 2, pages 618–623, Atlanta, May, 2-6, 1993.

[3] Grundfest W.S., J.W. Burdick, and Slatkin A.B. Robotic endoscopy, August, 16, 1994. United States Patent n° 5,337,732, Cedars-Sinai Medical Center.

[4] Guo S. and others . Development of the micro pump using icpf actuator. In *IEEE Int. Conf. on Robotics and Automation*, pages 266–271, Albuquerque, April, 21-28, 1997.

[5] Hegao C. and others . A new force-controlled micro robotic worktable driven by piezoelectrical elements. In *IMACS/SICE Int. Symp. on Robotics, Mechatronics, and Manufacturing Systems*, pages 637–644, Kobe, September, 16-20, 1992.

[6] Hollis R.L., Allan A.P., and Salcudean S. Six degree of freedom magnetically levitated variable compliance fine motion wrist. In *4th Int. Symp. of Robotics Research*, pages 65–73, Santa Cruz, , 1987.

[7] Khatib O. and Bowling A. Optimization of the inertial and acceleration characterics of manipulators. In *IEEE Int. Conf. on Robotics and Automation*, pages 2883–2889, Minneapolis, April, 24-26, 1996.

[8] Lee K-M. and Arjunan S. A three-degrees-of freedom micromotion in-parallel actuated manipulator. *IEEE Trans. on Robotics and Automation*, 7(5):634–641, October 1991.

[9] Lu A., Grant D., and Hayward V. Design and comparison of high strain shape memory alloy actuators. In *IEEE Int. Conf. on Robotics and Automation*, pages 260–265, Albuquerque, April, 21-28, 1997.

[10] Merlet J-P. Direct kinematics and assembly modes of parallel manipulators. *Int. J. of Robotics Research*, 11(2):150–162, April 1992.

[11] Merlet J-P. and Gosselin C. Nouvelle architecture pour un manipulateur parallèle à 6 degrés de liberté. *Mechanism and Machine Theory*, 26(1):77–90, 1991.

[12] Nakamura Y., Kimura Y., and Arora G. Optimal use of non-linear electromagnetic force for micro motion wrist. In *IEEE Int. Conf. on Robotics and Automation*, pages 1040–1045, Sacramento, April, 11-14, 1991.

[13] Pernette E. and Clavel R. Parallel robot and microrobotics. In *6th ISRAM*, pages 535–542, Montpellier, May, 28-30, 1996.

[14] Pierrot F., Dauchez P., and Fournier A. Fast parallel robots. *Journal of Robotic Systems*, 8(6):829–840, December 1991.

[15] Shahinpoor M. Microelectro-mechanics of ionic polymeric gels as artificial muscles for robotic applications. In *IEEE Int. Conf. on Robotics and Automation*, pages 380–385, Atlanta, May, 2-6, 1993.

[16] Sturges R.H. and Laowattana S. A flexible, tendon-controlled device for endoscopy. In *IEEE Int. Conf. on Robotics and Automation*, Sacramento, April, 11-14, 1991.

[17] Treat M.R. and Trimmer W.S. Self-propelled endoscope using pressure driven linear actuators, January, 21, 1997. United States Patent n° 5,595,565, Columbia University.

[18] Wendlandt J.M. and Sastry S.S. Design and control of a simplified Stewart platform for endoscopy. In *33nd Conf. on Decision and Control*, pages 357–362, Lake Buena Vista, December, 14-16, 1994.

Chapter 10

Robots in Surgery

The increasing research activity in the field of robots in surgery is a natural consequence of the great and diverse possibilities that robotics provides for improving the efficiency in surgical procedures. The papers of this session show the diversity of activities in this promising sector.

Surgical training is an important issue, mainly in the domain of minimally invasive surgery. The work presented by Hayward et al. describes a haptic device, the Freedom-7. After an analysis of different families of small instruments, their similarities and differences, a 7 degree of freedom haptic device was designed. Besides the device structure, its physical implementation considering positioning, orienting, transmission and sensing are described, as well as the effective results in what refers to accessibility, dynamic range and frequency response.

Mitshuishi et al. describe in their paper the analysis of microsurgery operation to define the specifications of a tele-microsurgery system. This research has been addressed to develop a master slave prototype aimed to be applied to microblood vessels connection. After a description of the manual operation process the requirements of the device are defined (motions and sensing). Then, the configuration of the designed system is presented and its experimentation with two master and two slave manipulators. The study of how to increase dexterity having multiple master-slave manipulators and the possibility to understand the operator's intention so as to predict actuation constitutes the rest of the work. The system is also evaluated from the experimental results of suturing a blood vessel with a diameter less than 1mm.

The paper of Nakamura Y., Onuma, Kawakami, and Nakamura T., describes an active forceps designed for endoscopic surgery. The use of shape memory alloy pipes allow the development of bending but stiff devices to facilitate the access of instruments into the operating area. The paper describes the SMA active forceps structure, its functionality and performance. Finally the authors show the master-slave surgical robot developed to teleoperate the active forceps.

Davies describes the experiments carried out using the concept of a robot with a force control handle moved by the surgeon, implemented with the ACROBOT for knee surgery applications. This concept comes from the study of the advantages and drawbacks of active and passive robots, looking for the best synergy between humans and robots. With this concept the robot has been experimented for replacement knee surgery, programming the robot to allow free motion within the given area but restricting its intrusion to forbidden zones, thus reducing risk situations and avoiding dangerous errors. Bilateral sensing also enables the surgeon to feel what he is doing.

Transfontaine, Bidaud and Dario study the behavior of Shape Memory Alloy, SMA, for its use in microrobotic applications. They experiment with two control algorithms based on position and temperature control respectively, and apply the microactuator to a microgripper and a steerable endoscope. Both open and closed control are evaluated as well as the effects of temperature changes with regard to the dynamic performance

Freedom-7: A High Fidelity Seven Axis Haptic Device With Application To Surgical Training

V. Hayward, P. Gregorio, O. Astley, S. Greenish, M. Doyon
Dept. of Electrical Engineering and Center for Intelligent Machines
McGill University, Montréal, Canada
http://www.cim.mcgill.ca/~hayward/Home.html

L. Lessard, J. McDougall
Dept. of Mechanical Engineering
McGill University, Montréal, Canada
http://www.mecheng.mcgill.ca/~zen/Home.html

I. Sinclair, S. Boelen, X. Chen, J.-G. Demers, J. Poulin,
I. Benguigui, N. Almey, B. Makuc, X. Zhang
MPB Technologies Inc., Pointe Claire, Canada
http://www.total.net/~mpbt

Abstract: A seven axis haptic device, called the Freedom-7, is described in relation to its application to surgical training. The generality of its concept makes it also relevant to most other haptic applications. The design rationale is driven by a long list of requirements since such a device is meant to interact with the human hand: uniform response, balanced inertial properties, static balancing, low inertia, high frequency response, high resolution, low friction, arbitrary reorientation, and low visual intrusion. Some basic performance figures are also reported.

1. Introduction

It is suggested that the future of surgical simulation for training [1] will follow a path similar to that of flight simulation for aviation training [2], which has now become an industry justifying significant research in a plurality of domains.

We describe an electromechanical device capable of supporting the simulation of tasks carried out with a variety of surgical instruments including, knives, forceps, scissors, and micro-scissors. In these categories, individual instruments differ from each other by their business-end and by the interface they present to the human hand. What most of these instruments have in common are seven degrees of kinematic freedom, with the exception of the knife, which has only six. The device incorporates a mechanical interface which enables the interchange of handles, for example to support these categories of instruments and to provide the force feedback needed to simulate the interaction of the instrument with a tissue. The Freedom-7 is described in relation to surgical simulation, but its general purpose design makes it relevant to most haptic applications: multi-dimensional data set exploitation, teleoperation, computer aided design and animation, and so-on. The present paper is a follow-up to [3]

Figure 1. Five ways humans manipulate small handles.

where much of the motivation is discussed. It is argued that there is considerable motivation for the development of devices designed to support the haptic emulation of tasks involving small tools: pens, styli, brushes, screwdrivers, cutter knives, lancets, scapels, and so-on, see Figure 1. In all these cases, a person is gripping a small handle which is free at one end, while the other marks the location of the task where the tool interacts with an object. The handle proximal end must be free of attachments for realism of the simulation given the great variety of hand configuration and re-configurations. Surgical instruments are no exception to this.

In the present paper we focus on the design of the Freedom-7 as it relates to instruments selected from the basic dissection set illustrated in Figure 2. Some basic performance figures are also reported and discussed.

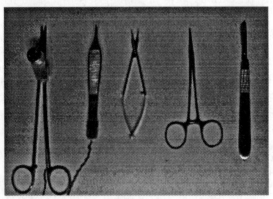

Figure 2. Five families of instruments: scissors (please ignore the attachment), tissue forceps, micro-scissors, hemostatic forceps, and knife.

2. The Basic Dissection Kit

Referring to Figure 2, from left to right we find scissors, a tissue forceps, micro-scissors, a hemostatic forceps, and a knife. Actual sets contains families of instruments in these category [4]. For example, scissors may come in various lengths and sturdiness; forceps can be curved, have teeth, etc. The selected subset nevertheless represents most encountered structural characteristics.

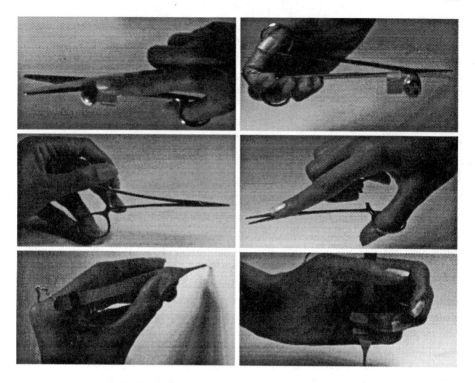

Figure 3. Various ways of holding scissors and forceps.

2.1. What These Instruments Have in Common

For purposes of simulation, it should be possible to position and orient these instruments in the greatest achievable work volume. Forces and torques should also be reflected with a sufficient amount of fidelity. They all locate the task at one extremity, but the way they are held may vary greatly. Scissors, for example, may be held in drastically different poses. Forceps and knives are also subject to a diversity of holding positions. See Figure 3.

2.2. How They Differ

The knife category requires six degrees of freedom, but all the others involve a closing action which requires a seventh freedom with force feedback. Instruments differ by the interface they present to the hand. They also greatly differ from the view point of their own internal dynamics. For example, tissue forceps and micro-scissors can be represented by spring dynamics. Scissors have instead dissipative dynamics due to friction occurring at the hinge and between blades, introducing high frequency textural sensations. Hemostatic forceps have complicated potential and dissipative dynamics when the ratchet is engaged or disengaged. With scissors and hemostatic forceps, internal dynamics can create high level of forces.

The approach was to design the device to deliver three forces and three torques with respect to the ground, plus one differential force available at the tip. The device includes a quick attachment interface to accommodate actual instruments. For all closing instruments, one branch is connected to a flange where forces and torques are available, while the other is attached to a sliding rod to provide for differential force feedback. With this approach, the haptic device is only responsible for producing forces and torques resulting from the interaction of an instrument with a tissue which are combined with the natural dynamics of each instrument. In the case of knives, the differential flange is constrained so the device becomes a six axis haptic device with improved performance due to actuator redundancy [5].

3. Design Rationale

A haptic device is a two-way transducer to address the human hand motor and sensorial capabilities. By analogy with any transducer designed to stimulate human senses, it is believed that there is more to be gained from wide dynamic range and high resolution than from the sheer amplitude of the transduced signals (force, position and derivatives) in terms of realism and intelligibility.

Throughout the history of master arm designs [6, 7, 8], direct drive manipulators [9, 10], and now haptic devices [11], static balancing is viewed as a major design goal. One motivation is to fully allocate actuator torques to the generation of acceleration. The maximization of acceleration capability was found to be a major factor of performance for haptic devices [12]. Static balancing satisfies one additional crucial requirement for the Freedom-7. Due the diversity of the hand positions that must be accommodated, the device must be easily re-oriented and must preserve its entire set of dynamic properties under re-orientation. Because haptic devices should be general purpose transducers, a uniform response as well as balanced inertial properties (uniform principal inertia terms and minimized coupling terms) must also be included in the list of requirements. The need for a high frequency response imposes severe constraints on the structural design and the elimination of transmission backlash. Absence of friction is needed for output dynamics range. Large work-space and minimum intrusion in the operator's visual and manipulation space must be achieved. We set out to meet these requirements for a total of seven degrees of freedom, while minimizing the complexity of the construction and maintenance.

Many six degrees of freedom parallel linkages were examined, but none were found to meet a reasonable subset of the above requirements (lack of angular workspace, no balancing, heavily coupled inertial tensor, too many joints, bulk and intrusion). Purely serial designs could satisfy even fewer of these requirements. Consequently, a hybrid design was developed. It consists of a distal orienting-plus-sliding stage with a parallel structure supported by a three degree-of-freedom positioning stage including a four-bar mechanism, a structure commonly found in manipulator design. In both cases, the merits of each structure have been best taken advantage of to optimize the system as a whole. The position stage is directly driven, but the orienting stage has remotized and grounded actuation, leading to a third transmission subsystem. Each of them are now discussed in turn.

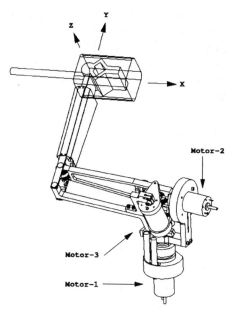

Figure 4. Position stage.

3.1. Positioning Stage

A line drawing view of the positioning stage connected to the distal stage enclosed in a protective cover (discussed in the next section) is shown on Figure 4. It is built around a four-bar mechanism which achieves static and dynamic balancing as well as minimization of inertia. The concept is inspired by the design of turntable tone-arms, actually subject to similar constraints. Consider

Figure 5.

Figure 5: m refers to the mass of a cantilevered structure set at distance d from a pivot point. Parameters m and d are given by geometrical and structural requirements. Find M and D such that the system is balanced under gravity and such that inertia is minimized. Static balancing yields $md = MD$. Find M such that $I = MD^2 + md^2 = md^2(m/M + 1)md$ is minimum. This expression is minimum when M is maximum which means that the best design has the heaviest counter-weight. The mass is assumed to be concentrated at the motors, so we will use them as counter-weights. The first requirement is to place the motors so that the center of mass of the system coincides with its support center and the second is to keep this property invariant under any motion of the device.

This led to what is shown on Figure 4. Motor-1 is fixed to the ground to produce a force at the tip principally in the Y direction (the sweep of joints is limited to $\pm 30^\circ$). Motor-2 drives four-bar regional structure, producing a force principally in the Z direction. It is placed to balance exactly the gravity contribution of the four bar structure around the axis of motor-1. Motor-3 actuates the four-bar, producing a force principally in the X direction. The center of mass of a four-bar mechanism travels on a circle. If the circle vanishes to point, once balancing is achieved for one position, it is achieved for all. The center of mass of the four-bar is placed by design on the axis of motor-2, symmetrically to the center of mass of motor-2 with respect to axis of motor-1. Thus, static balance is invariant with position.

The offsets of motor-2 and motor-3 are similar, so the inertia experienced at the tip is also similar in the X, Y, and Z directions. The design thus follows the "tone-arm rule" for all directions. All joints have a yoke design and are assembled throughout with pre-loaded bearings to guarantee absence of backlash even under conditions of high speed oscillatory motions. All parts are symmetrical so the Freedom-7 can be assembled in "righty" or in "lefty" configuration.

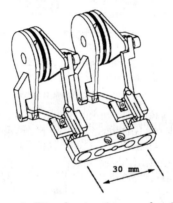

Figure 6. Distal orienting mechanism.

3.2. Orienting Stage

The mechanism provides four degrees of freedom, yet is relatively simple. It has a total of 15 parts which averages to only 4 per degree of freedom. An important consideration is made here: when mechanisms are scaled down, the angular work-space remain invariant. There are two limits to down-scaling in the present case: manufacturing and preservation of the sliding motion. A characteristic length (average length of moment arms) of about 30 mm was found to be best. At that small scale, the mechanism can be made quite light (less than 50 g) for good structural properties even with conventional materials.

A line drawing view of the orientation stage is shown on Figure 6. It comprises two five bar linkages driven by twin pulleys and connected to the driven member by twinned spherical joints. The twin pulleys are supported by bearings (not shown) which allows them to swivel. The driven member clamps to an output rod (not shown) which is constrained to four degrees of freedom by a gimbal. Its last joint is a cylindrical pair. When the sliding motion is constrained this mechanism has three degrees of orientation freedom. In all cases, it preserves high kinematic conditioning throughout its work-space which is limited to 90° of pitch, 100° of yaw combined with a roll motion of 120° in the present application. When not constrained, the differential sliding motion occurs when the two five bars deform identically. As an added bonus, a handle of standard size made of lightweight material will keep the orientation stage statically balanced. Moreover, because the positioning stage applies acceleration at the center of mass of the distal assembly, dynamic cross coupling terms are also minimized.

3.3. Transmission

In [3], the relative merits of transmission techniques for haptic interfaces have been reviewed, including: linkages, flexible elements (cables, steel belts, or polymeric tendons), shafts plus gears, or fluid lines connected to bellows. It was concluded that since transmissions work by exposing structural elements to stress, the best results are obtained when the largest amount of material is exposed to the most uniform stress, which led us to consider linkages or tendons. For reasons of bulk and complexity, the use of linkages should be limited to low numbers of degrees of freedom. In the case at hand, linkages provide for transmission of the positioning stage. In the orientation stage, linkages transmit motion from the four driven pulleys to the differential output interface. The motors of the orientations are all grounded and motion is transmitted to the orientation stage via high modulus polymeric tendons. The principal disadvantages of polymeric tendons over cables or steel belts is creep under permanent loading, and higher dissipation when routed around idlers. On the other hand, idlers diameters can be made small without causing fatigue. A tensioning technique has been devised for keeping the tendons under

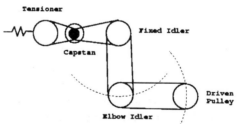

Figure 7. Transmission structure.

low stress when not transmitting torque. This reduces fatigue and long term creep. Its principle is shown on Figure 7. The routing is integrated within the position stage so that the tendon path lengths remains constant to a very small approximation (0.1 % of their length) under any motion of the end effector. Sensing is combined with the tensioning pulley so that even if slip accidentally occurs at the capstan, the device remains calibrated. Because the actuators are grounded, the tendon path length is constant, and idlers and driven pulleys all have the same diameter, all kinematic and dynamic couplings between the two stages are eliminated.

3.4. Sensors and Electronics

All sensors are non-contact hall-effect angle transducers which guarantee high resolution and low noise. Signal conditioning as well as motor current drivers are presently linear electronics. They are also co-located which benefits closed loop control. As a bonus, no electric signal need to be transmitted beyond motor-3 which simplifies assembly, reduces noise and promotes reliability.

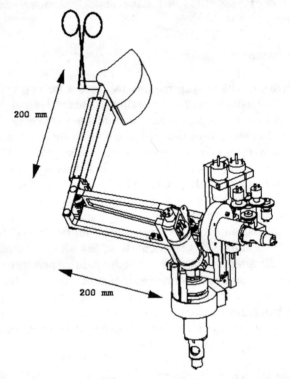

Figure 8. Freedom-7 shown without its holding stand.

4. Results

Figure 8 shows the complete integrated device without its table-top holding stand. The stand includes a clamping gimbal which permit the device to be positioned at any height and oriented in any pose with respect to the user. The figure also shows also the four motor-sensor-tensioner assembly which is tightly packaged and placed not to interfere with the device's workspace.

The results are those of a research prototype in operation at McGill University.

A subset of the guidelines outlined in [13] is being followed:

Work-space: The device is wrist partitioned. Its workspace is limited by mechanical stops to its most dextrous region. The link lengths (200 mm each) and the mechanical stops have been optimized so that the translation workspace contains an ellipsoidal volume of axis lengths $130 \times 160 \times 180$ mm. The angular workspace is 90° of pitch, 100° of yaw combined with a roll motion of 120°. It's origin is not changed by more than ±15° throughout. These figures target the manipulation of tools with full wrist motions and elbow resting. The seventh differential output range is 10 mm.

Intrusion: The device has the general appearance of an "elbow manipulator". The secondary link has a box-beam structure (200 mm long, 30 × 50 mm in section). The distal stage is enclosed in a protective case with section 50 × 50 mm and length 80 mm. The case follows the angular motions of the output handle and may be used as a handle.

Output and Input Dynamic Range, Inertia: The mechanical noise of the device was measured under conditions of low velocity. A calibrated load-cell (Transducer Techniques model MDB-2.5/conditioner 308 calibrated at ±10.0N) was set on the table of a milling machine and the device slowly back driven. A level of about 0.06 N of friction was measured in translation and $8 \, 10^{-3}$ Nm in angular motion. The short term peak force and torque are 5 N and 0.6 Nm. This corresponds to about one part in 1000 of dynamic range in translation and one in 100 in rotation. The position sensing resolution was also measured by driving the device back by known distances and was found to be better and 0.02 mm. The inertia was measured. Known torques were applied and the resulting accelerations measured (Analog Device ADXL05 evaluation board). It was estimated to be 133 g in the Y direction and 84 g in the other two. The angular inertia at the handle is of the order of $0.1 \, gm^2$.

To develop an intuitive idea of why that level of inertia is targeted, suppose that during a simulation an operator moves an instrument between two locations 10 cm apart in a one second motion. A symmetrical parabolic time optimal trajectory will require 0.5 second to accelerate and 0.5 second to stop, which yields an acceleration of $.4 \, ms^2$. The spurious inertial forces introduced by the device are then .04 N, a small number indeed, similar to the friction level of the device.

Frequency Response: The frequency response was measured under two loading conditions: under isometric condition with the load cell, and under isotonic condition with the accelerometer (using a DSP Technology SIGLAB Model 20-22 signal analyzer).

The response in X reveals a typical two-mass resonance/antiresonance at 25 Hz which is attributed to a weakness in the secondary link. The response in Y has a smooth roll off after 30 Hz. This was traced to lack of rigidity in the motor-1 connection. The angular response shows excellent results given that it has remotized actuation. The isotonic acceleration response has a ±3dB response up to 50 Hz, and was shown to transmit significant energy beyond 200 Hz. The isometric response is more resonant and rolls off faster, as it is to be expected, due to the elasticity of the tendons. It is nevertheless considered excellent for a device of that size. Since sensing is co-located, it is possible to contemplate open loop frequency response shaping to extend the bandwidth without compromising the closed loop performance. Such a technique is however not as easily applicable to the metallic linkages of the positioning stage

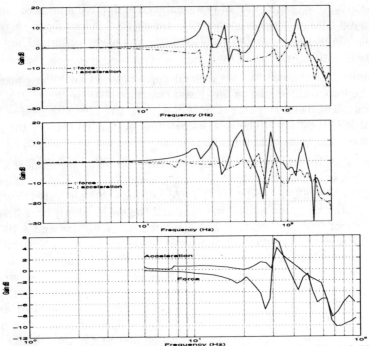

Figure 9. Linear acceleration and force response along the X and Y directions (top two plots). Angular acceleration and torque response around the Z direction (bottom plot).

due to their sharp resonances, until they are replaced by better materials.

5. Present and Future Construction

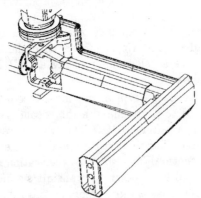

Figure 10. Advanced composite construction under development.

The Freedom-7 is presently made of aluminum for prototyping, however, this material has many limitations in this application. Research is under way to explore the use of advanced composite materials for the fabrication of haptic devices. Some preliminary results are available [14]. Studies are conducted on finite elements models in conjunction with the fabrication of prototype links for experimentation. These studies indicate that is possible to simultaneously raise the natural frequency and decrease the weight of these links. It is also possible to tailor other properties such as structural damping. With conventional construction methods, these requirements oppose each other. A view of the Freedom-7 composite construction is shown on Figure 10.

6. Kinematics, Control and Computer Interface

The coordinate transformations for position, velocity, and force admit closed forms for all seven degrees of freedom in the forward and inverse directions. When only the six first freedoms of the Freedom-7 are needed, actuators torques are calculated without having to resort to posing the problem as overconstrained: the output differential force is simply set to equal zero. Their implementation indicate that closed loop control in Cartesian coordinates can be achieved at a rate greater than a KHz on a Pentium class computer.

A computer control interface has been written to provide access to the device in Cartesian coordinates. The interface includes also sensor calibration functions and safety monitors that make sure that the device remains inside its thermal operational envelop.

Finally, the device has been included the library of devices supported by Armlib a package developed at the University of North Carolina [15] which facilitates the use of haptic devices in graphical applications.

7. Acknowledgments

Initial funding for this research was provided by the project "Haptic Devices for Teleoperation and Virtual Environments" (HMI-6) supported by IRIS (Phase 2), the Institute for Robotics and Intelligent Systems part of Canada's National Centers of Excellence program (NCE), and an operating grant "High Performance Robotic Devices" from NSERC, the National Science and Engineering Council of Canada. Additional funding for this research is provided by an NSERC University/Industry Technology Partnership Grant, McGill/MPBT "A Balanced Haptic Device for Computer/User Interaction".

References

[1] Satava, R., M. 1996. Advanced simulation technologies for surgical education. *Bull. A. Coll. Surg.*, 81(7):77–81.

[2] Kuppersmith, R. B., Johnston, R., Jones, S. B., Herman A. J. 1996. Virtual reality surgical simulation and otolaryngology, *Arch. Otolaryngol. Head Neck Surg.*, 122:1297–1297.

[3] Hayward, V. 1995 (August, Pittsburgh). Toward a seven axis haptic interface. *IROS'95, Int. Workshop on Intelligent Robots and Systems.* IEEE Press, Vol. 2, pp. 133–139.

[4] Brooks Tigue, S. M. 1994. *Instrumentation for the Operating Room: A Photographic Manual.* Fourth edition. Toronto:Mosby.

[5] Kurtz, R., Hayward, V. 1992. Multi-goal optimization of a parallel mechanism with actuator redundancy. *IEEE Transactions on Robotics and Automation.* Vol. RA-8, No. 5. pp. 633–651.

[6] Goertz, R. C. 1952. Fundmentals of General Purpose Remote Manipulators. *Nucleonics*, 10(11), pp. 36–42.

[7] Vertut, J. 1976. Advance of the new MA 23 force reflecting manipulator system. *Proc. 2nd International Symposium on the Theory and Practice of Robot and Manipulators*, CISM-IFToMM, pp. 307–322.

[8] Bejczy, A. K., Salisbury, K. 1980. Kinesthetic coupling between operator and remote manipulator. Proc. *International Computer Technology Conference, ASME*, San Francisco, pp. 197–211.

[9] Asada, H., Yousef-Toumi, K. 1987. *Direct Drive Robots: Theory and Practice* MIT Press.

[10] Takase, K. 1984. Design of torque controlled manipulators composed of direct and low reduction ration drive joints. In *Robotics Research: The First International Symposium*, M. Brady and R. Paul (eds.). MIT Press, pp. 655–675.

[11] Massie, T. H., Salisbury, J. K. 1994. The Phantom haptic interface: a device for probing virtual objects. *Proc. ASME Winter Annual Meeting, Symposium on Haptic Interfaces for Virtual Environment and Teleoperator Systems.*

[12] Hayward, V., Choksi, J. Lanvin, G. Ramstein, C. 1994. Design and multi-objective optimization of a linkage for a haptic interface. In *Advances in Robot Kinematics.* J. Lenarcic and B. Ravani (Eds.). Kluver Academic. pp. 352-359.

[13] Hayward, V. Astley, O.R. 1996. Performance measures for haptic interfaces. In *Robotics Research: The 7th International Symposium*, Giralt, G., Hirzinger, G., (Eds.), Springer Verlag. pp. 195-207.

[14] McDougall, J., Lessard, L. B., Hayward, V. 1997, Applications of advanced materials to robotic design: The Freedom-7 haptic hand controller, to appear in the *Proc. of 11th Int. Conf. on Composite Materials*, Gold Coast, Australia, July 1997.

[15] Mark, W.R., Randolph, S.C., Finch, M., Van Verth, J.M., Taylor, R.M. 1996. Adding force feedback in graphics systems: issues and solutions, *Computer Graphics Proceedings, Annual Conference Series*, pp. 447-452.

Tele-micro-surgery: analysis and tele-micro-blood-vessel suturing experiment

Mamoru MITSUISHI*, Hiroyoshi WATANABE**, Hiroyuki KUBOTA*, Yasuhiro IIZUKA* and Hiroyuki HASHIZUME**

*Department of Engineering Synthesis
Faculty of Engineering, The University of Tokyo
Hongo 7-3-1, Bunkyo-ku, Tokyo 113, Japan
e-mail: mamoru@nml.t.u-tokyo.ac.jp
http://www.nml.t.u-tokyo.ac.jp/~mamoru

** Department of Orthopaedic Surgery
Okayama University Medical School
Shikata-cho 2-5-1, Okayama 700, Japan

Abstract: This paper describes the analysis of actual micro-surgery operation to determine the specifications of a tele-micro-surgery system. The implemented system for the connection of micro blood vessels consists of multiple micro co-located operation point slave manipulators, macro rotational-force-feedback-free master manipulators and a vision system which is comprised of a fixed viewpoint microscope and a movable monitor system. The parameters of the vision system can be controlled in accordance with the system's understanding of the operator's intention by monitoring the posture change of the operator. The experimental results showed the effectiveness of the developed system. In the experiment, a blood vessel of a rat with a diameter less than 1 mm was successfully sutured using needle with a curvature radius of 2 mm.

1. Introduction

The authors' group has developed a master-slave type tele-micro-surgery system[1]. The main purpose of the system is the connection of micro-blood-vessels. The operation is necessary in case of amputation accident for hands and legs. In particular, the system aims at the micro-blood-vessel connection which is currently executed by a medical doctor's hands under the microscope, placing an extreme burden in stress and time on the doctor.

The features of the system are as follows: (i) The axes of all rotational degrees of freedom intersect at the tip of the slave manipulators. (ii) Moments are not fed back to master manipulators because no moments generate if the tip of the slave manipulators are fixed. (iii) The axes of the rotational degrees of freedom intersect at the focal point of the microscope. (iv) Visual information acquisition and display system which consists of microscope and monitor can be controlled using the detected information of the operator's posture change. (v) Multi-axis force information obtained at the slave manipulators are converted into the auditory information and presented augmentedly to the operator.

In this paper, analyzed results of actual micro-surgery operation is presented. The mechanism of the slave manipulators were revised to cover the all necessary area by checking the working area of the previous system. The experimental results of the performance concerning the visual information control such as microcope viewing direction and magnification ratio control and displaying direction control are also discussed.

2. Related work

It is discussed that precision and safety are the most important attributes of a surgical system [2]. Concerning tele-medicine and tele-surgery, [3] have developed a mobile telepresence system for laparoscopic surgery using paired CCDs and 5 DOF manipulators. They have already executed animal surgery. [4] have developed a telemedicine system using ISDN. In [5] a multimedia tele-surgery system was developed using optical fiber network for an intravascular surgery. Another example of a tele-surgery is the automatic and real-time tracking system of a laparoscopic instruments using visual information [6]. A teleoperation system considering the scale effect is reported in [7]. The authors have already developed a master-slave system for tele-surgery [1] and the system was operated using the Internet and an artificial satellite. In the system discussed in this paper compared with the systems mentioned above, the following items are the main features: (a) The structure and the rated value of the system were determined considering the analysis of actual micro-surgery. (b) Intention of the operator was understood by unconscious posture change of the operator. The visual information presentation and display system was controlled according to the inferred intention of the operator.

3. Analysis of a micro-surgery operation

3.1. Analysis of micro-blood vessel connection motion

Among the advancements of medical technologies, the ability to restore seriously injured organs has progressed remarkably. With the development of micro-surgery technology, it becomes possible to connect hands, legs and other tissues even if they are amputated, for example, by an accident. The following sequence is necessary for the connection of hands, for instance: (1) bone connection, (2) muscle connection, (3) nerve connection, (4) artery connection, (5) vein connection, and (6) skin connection. The operations from (3) to (5) require the connection of micro-blood vessels and nerves. They are the most difficult kinds of work in micro-surgery operations. In particular, connection of blood vessels is difficult because the blood vessels have to be connected without twisting so that blood flow is not encumbered, while the connection of nerves is easier. It is possible to connect small diameter blood vessels of up to 0.4 or 0.5

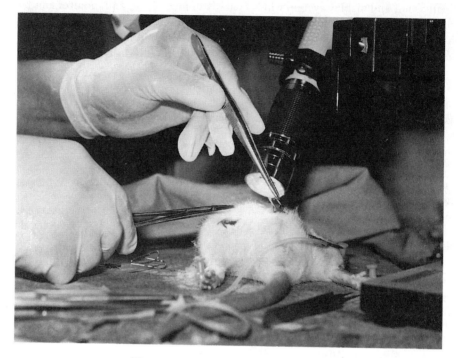

Fig.1 Overview of micro-surgery

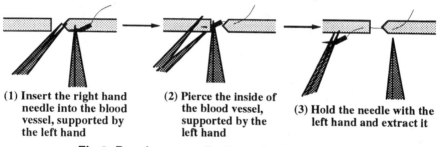

(1) Insert the right hand needle into the blood vessel, supported by the left hand

(2) Pierce the inside of the blood vessel, supported by the left hand

(3) Hold the needle with the left hand and extract it

Fig.2 Passing a needle through a blood vessel

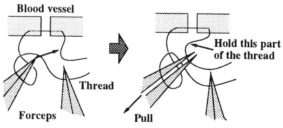

Fig.3 Tying a thread

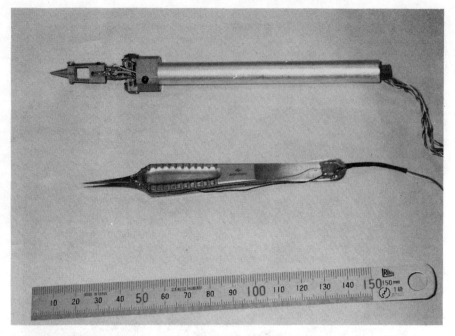

Fig.4 Multi-axis force measurement apparatus (upper) and force sensing forceps (lower)

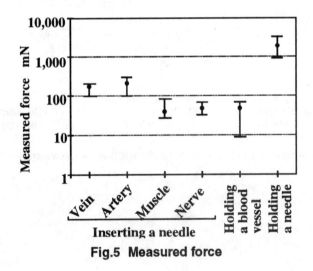

Fig.5 Measured force

mm by human skill. The operation is performed generally by two doctors. One is a main operator and the other is an assistant. The operation is performed under a stereo-type optical microscope. Fig.1 shows the actual micro-surgical operation.

Connecting blood vessel is accomplished by the following sequence: (i) fixture blood vessels with a microvascular clamp, (ii) wash, (iii) set the ends to be connected, (iv) align the edges of the blood vessels, (v) suture, (vi) unclamp the fixture, (vii) arrest the bleeding using a sponge, and finally, (viii) confirm blood flow. The operations in Fig.2 and Fig.3 are executed using a needle holder and forceps, respectively. The features of these operations are as follows: (a) The operation is three dimensional work using left and right hands cooperatively. (b) One hand is used to guide the needle and the other is used to hold the wall of the blood vessel. (c) The right hand performs the translational motion with left hand's assistance while passing a needle through the blood vessel. (d) After passing the needle through the blood vessel, left hand receives the needle by rotating the wrist. (e) A needle holder which can hold the needle without finger force and forceps which do not have the holding capability are used properly depending on the situation. (f) While passing a needle through the blood vessel, the operator has to be careful not to injure the opposite wall of the blood vessel. (g) The suturing operation is repeated to connect several points.

3.2. Force measurement during micro-surgery operation

Force applied at the tip of the forceps was measured through the operation. In the experiment, (1) the holding force of the needle, (2) the holding force of the blood vessel, and (3) the force required to insert the needle through the blood vessel wall were measured. To measure the force while a needle is being inserted in the blood vessel, a special apparatus for measuring multi-axis force was developed(Fig.4). Parallel plate structures are combined in series. The holding function of the needle is equipped at the tip side of the forcep so as not to affect the force measurement. Furthermore, a pair of forceps with strain gauges was used to measure the force during the operations.

In the experiment, the forces in the suturing of an abdominal artery and vein of a rat were measured. The diameter of the blood vessel was approximately 1mm. Force for inserting a needle, holding force of a needle while inserting the needle and holding force of a blood vessel were measured. Fig.5 shows the summary of measured force range.

From Fig.5 it is clear that the force ranges measured for the holding force of the needle, the insertion force of the needle and the holding force of a blood vessel are completely different. The required resolution of the force sensor equipped on the slave manipulator should be sub-mN to know the contact state between the needle and the blood vessel. However, it is impossible to apply a large force if the resolution of the force is high. On the contrary, the holding force of the needle must be more than 1,000 mN to ensure gripping even if excessive force is applied to the needle. Therefore, if the rated value of the force sensor equipped on the slave manipulator is in the range from 500 mN to 1,000 mN, it is possible to hold the needle and to measure the subtle force to know the contact state.

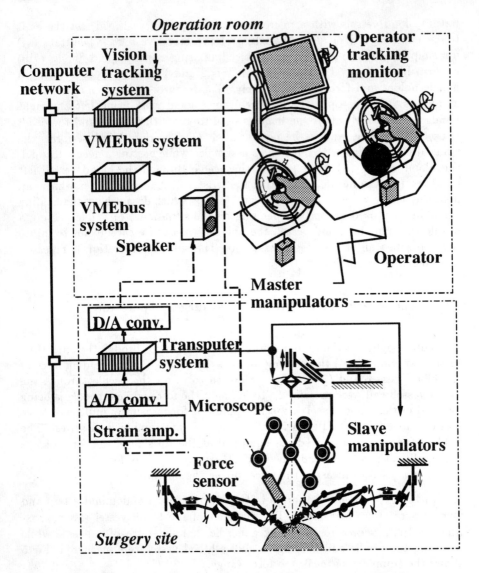

Fig.6 Overview of the developed tele-micro-surgery system

4. Overview of the tele-micro-surgery system

Fig.6 shows an overview of the system. The input data to the left and right master manipulators is sent to the two slave manipulators via the real-time controller and computer network. The force information from the two 3-axis force sensors is used not only for bilateral control but also for auditory information generation to display the contact state between the tweezers and the object[1]. Visual information from the CCD camera at the top of the monitor is used to control the viewing direction and the magnification ratio of the mi-

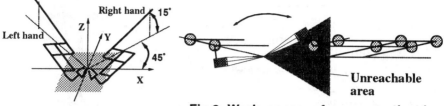

Fig.7 Arrangement of slave manipulators

Fig.8 Work space of a conventional slave manipulator

croscope as well as the posture of the visual information display system. The visual information from the microscope is sent directly to the movable monitor for the operator.

5. Dexterity enhancement with multiple co-located operation point master-slave manipulators

5.1. Assingment and the structure of slave-manipulators

Multiple slave manipulators are necessary to accomplish the suturing operation because the needle and thread have to be handed over during the operation. Furthermore, one hand is used to fix the blood vessel. There are several ways to realize the multiple slave-manipulator system: (a) multiple slave manipulators, (b) a couple of slave manipulators operated by multiple surgeons, and (c) left and right manipulators. In the system developed in this paper, method (c) was adopted. Structure of each slave manipulator is discussed in [1].

Limited space has to be shared by the microscope and left and right slave manipulators. Because the space occupied by each manipulator is reduced by adopting a link mechanism, slave manipulators can move independently from the location of the microscope. However, the following functions are required for the cooperation of multiple manipulators: (1) Manipulators should have a wide rotational motion range. (2) Motion range of each slave manipulator should be appropriately overlapped. (3) Handover between the multiple manipulators should be easily performed. (4) The motion of each manipulator should avoid its singular points. (5) The motion of slave manipulators should not disturb the visual field of the microscope. In the implemented system, the left and right manipulators were arranged as shown in Fig.7.

In the conventional system, unreachable area was overlapped with the required motion area (Fig.8). Therefore, the mechanism was revised as shown in Fig.9 and Fig.10 not to occupy the same space (Fig.11).

5.2. Realization of high-precision and wide working range

Rough positioners for the three translational directions were installed to cover the wide working range (Fig.12). Three-directional precise translational positioning was realized using hydraulic actuators driven by oil pressure. These

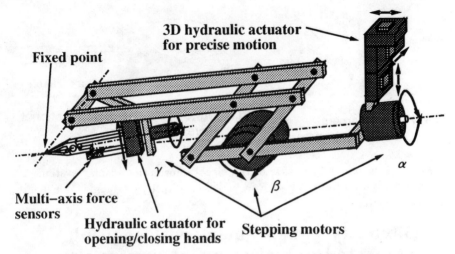

Fig.9 Structure of a slave manipulator

Fig.10 Overview of a slave manipulator

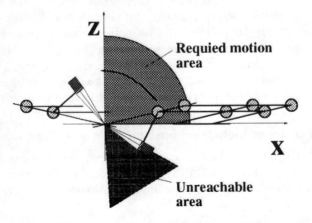

Fig.11 Work space of a revised slave manipulator

Fig.12 Slave manipulators with rough and fine positioners

Fig.13 Overview of master manipulators and operator tracking monitor

actuators allow the location of the power source at a position remote from the manipulator. In the implemented system the stroke of the fine positioner is approximately 10 mm for each direction. Cooperation of rough and precise positioners are necessary to prevent inconvenience of operation caused by reaching the travel limit of the precise mechanism during surgical operation.

5.3. Rotational-force-free-master manipulators

Rotational-force-free-master manipulators for left and right hands were implemented(Fig.13). If it is assumed in the system that the tip of the tool is fixed, only translational force will occur and moments will not generated. Therefore, only the translational force need be fed back to the force feedback type master manipulator. The rotational and translational degrees of freedom are realized by the combination of rotation rings and orthogonally combined linear guides, respectively. In particular, all axes of the rotational degrees of freedom intersect at the same point where the human operator applies the force and displacement through the motion of the human fingers. It is possible to keep the balance at the arbitrary attitude of the joystick even if the operator is not holding it, by adopting a mechanism such as the combination of rings mentioned above. The mechanism allows the distance between the point where force is applied by the operator and the 3-axis force sensor to be kept small, to reduce the interference to the 3-axis force sensor which is caused by the applied moment. The tweezers are controlled by two fingers with force feedback.

6. Operator's intention understanding by detecting operator's posture change

6.1. Basic concept for system implementation

Because the operator's left and right hands are occupied for the master manipulators, it is appropriate to be able to control the visual information acquisition and display system according to the unconscious posture change of the operator. Therefore, in this paper the posture change of the operator is detected as a visual information and the information is used to control the micro-world visual information acquisition and display system.

6.2. Micro-world visual information acquisition and display system

Concerning the microscope system, the axes of all rotational degrees of freedom intersect at the focal point of the microscope to be able to observe an object from an arbitrary direction. The parallel link mechanism was adopted in the system. Translational degrees of freedom were realized by an orthogonal combination of linear guides. Concerning the visual information display system, the display has the motion degrees of freedom around yaw- and pitch-axis to become its surface perpendicular to the operator's viewing direction(Fig.13). In the implemented system, 8.6 inch thin crystal liquid display was used. A small CCD camera was attached at the upper part of the monitor.

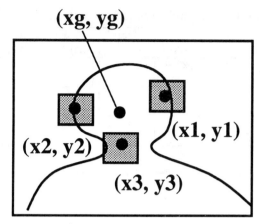

Fig.14 Tracking of an operator's face

6.3. Viewing direction and zoom control

To detect the posture change of the operator, several positions of the operator's face were tracked. The information is used to control the viewing direction and the magnification ratio of the microscope. At the vision tracking board used in this paper, a scene is presented as 512 × 512 pixels and 4-bit depth. A correlation operation is performed for local template information with the size 8 × 8 (or 16 × 16) to obtain a translation vector with the highest correlation. More than 10 templates can be tracked at the video frame rate by specifying low calculation templates.

(a) Posture change detection

Several templates are fixed to the face of a human operator for tracking. If the correlation value was less than the threshold value, the information from its template was omitted. A low pass filter with a frequency of approximately 10Hz was used to detect the operator's motion and to stabilize the motion of the microscope. The command to the velocity controller is determined as proportional to the value $\bar{x}$ and $\bar{y}$ in the following equations:

$$\bar{x} = \frac{1}{n}\sum_{i=1}^{n}(x_i - x_{i_0}) \tag{1}$$

$$\bar{y} = \frac{1}{n}\sum_{i=1}^{n}(y_i - y_{i_0}) \tag{2}$$

where n is the number of templates, and x_{i_0} and y_{i_0} represent the initial coordinates of a template.

(b) Zoom index detection

The distance of the operator can be detected by the size change of the operator's face on the monitor. More concretely, several parts of the face are tracked, as illustrated in Fig.14 and the sum of the squares of the distances

Table 1 Execution time for "thread in hole"

Operation conditions	1st trial (s)	2nd trial (s)	3rd trial (s)	4th trial (s)	Average (s)
Microscope: fixed Monitor: fixed	27	21	18	22	22
Microscope: fixed Monitor: movable	17	12	24	13	17
Microscope: movable Monitor: movable	16	18	12	9	14

from the gravity point (x_g, y_g) was selected as the "zoom index" as shown in the following equations:

$$x_g = \frac{1}{n}\sum_{i=1}^{n} x_i, \qquad (3)$$

$$y_g = \frac{1}{n}\sum_{i=1}^{n} y_i \qquad (4)$$

and

$$(\text{Zoom index}) = \frac{1}{n}\sum_{i=1}^{n}\{(x_i - x_g)^2 + (y_i - y_g)^2\} \qquad (5)$$

The maximum and minimum limits are set to control the zoom in and out. Furthermore, an insensitive area is implemented to prevent the unnecessary motion. The zoom index is effective even if the operator inclines the neck.

6.4. Performance of operator's intention understanding system

6.4.1. Tracking experiments of an operator

In the experiment, three templates with 8 × 8 pixels were attached to the operator's face and tracking was performed according to the information obtained from these templates. The time delay was approximately 0.13 second. However, the inertia of the visual information acquisition system including microscope is larger than the monitor and rapid acceleration will cause the vibration of the system and reduce the operator comfort. Therefore, an upper limit velocity was set for the motion of the monitor. Moreover, a low pass filter at approximately 10 Hz cutoff was applied to remove unexpected vibration. If the face moved with an angular velocity of $\pi/4/s$, the observed delay was approximately 0.55 second. Even if the observed time delay was half a second, the operator did not feel discomfort if the motion of the monitor was synchronized with that of the microscope.

6.4.2. Zooming in/out control experiments

The distance from the face to the monitor was measured by the size of the face on the vision image. Three templates were set at the boundary of the operator's face. In the experiment (i) the sum of the square of the distance and (ii) the size of triangle were used to calculate the zoom index. The zoom

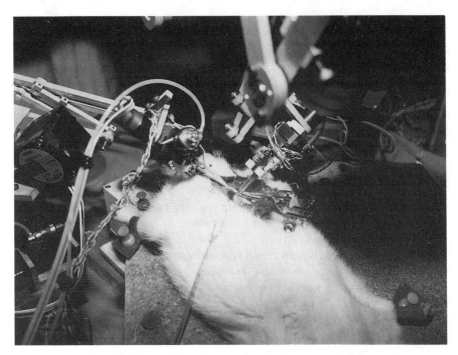

Fig.15 Micro-blood-vessel suturing experiment

indices obtained using both methods were good when an operator is changing the face's distance in a normal way. However, method (i) was much more robust when the operator made an action such as inclination and rotation of the neck. Therefore, method (i) was adopted in the implemented system.

6.4.3. Evaluation of the system for executing a "thread in hole" task

In the experiment, a thread with a diameter 80 μm was passed through the eye (800 μm × 400 μm) of a needle. The experiments were performed under three different conditions: (i) the microscope and the monitor were fixed, (ii) the microscope was fixed and the monitor was moved, and (iii) the microscope and the monitor could both be moved. Time to execute the task was measured. The initial postures of the master and slave manipulators were varied in each experiment. Table 1 shows the results. If the microscope is fixed, it was difficult to know the relative position and trial and error was necessary to thread the needle. On the contrary, if the microscope can be moved, it was easy to recognize the relative position. The motion range required was more than 60 degrees. Continuous motion of the microscope was effective for the recognition of the relative position. If the monitor is fixed, the discomfort of the operation increased when the view angle around the vertical axis became large.

7. Micro-blood-vessel suturing experiment

In the micro-blood-vessel suturing experiment an artery from a rat abdomen was successfully sutured using the implemented system as shown in Fig.15. The diameter of the artery was less than 1 mm and the curvature radius of the needle was approximately 2 mm. Cooperation of rough and precise motion of slave manipulators was necessary. Operator's intention understanding and multimodal information transmission, in particular, force-to-auditory information transformation were found to be effective.

8. Conclusions

This paper described the analysis of micro-surgical operation, in particular, connection of blood vessels and implementation of a tele-micro-surgery system. The implemented system consists of left and right micro co-located operation point slave manipulators, macro rotational-force-feedback-free master manipulators and a vision system whose motion is determined depending on an understanding of the intention of the operator. Experimental results showed the effectiveness of the proposed system. A blood vessel with a diameter less than 1 mm was successfully sutured using a needle.

References

[1] Mitsuishi,M., et al., "A Tele-micro-surgery System with Co-located View and Operation Points and a Rotational-force-feedback-free Master Manipulator," *MRCAS'95*, pp.111–118, Baltimore, U.S.A., 1995.

[2] *Proc. 1st Intern. Symp. Medical Robotics and Computer Assisted Surgery (MR-CAS)*, Pittsburgh, PA, Sep. 22–24, 1994.

[3] Green,P.S., et al., "Mobile Telepresence Surgery," *MRCAS'95*, pp.97–103, Baltimore, 1995.

[4] Rovetta,A., et al., "Robotics and Telerobotics Applied to a Prostatic Biopsy on a Human Patient," *MRCAS'95*, pp.104–110, Baltimore, 1995.

[5] Arai,F., et al., "Multimedia Tele-surgery Using High Speed Optical Fiber Network and Its Application to Intravascular Neurosurgery," *Proc. IEEE Intern. Conf. on Robotics and Automation*, pp.878–883, Minneapolis, 1996.

[6] Wei,G.-Q., Arbter,K. and Hirzinger,G., "Automatic Tracking of Laparoscopic Instruments by Color Coding," *CVRMed-MRCAS'97*, pp.357–366, Grenoble, France, 1997.

[7] Salcudean,S.E., et al., "Performance Measurement in Scaled Teleoperation for Microsurgery," *CVRMed-MRCAS'97*, pp.789–798, Grenoble, France, 1997.

Active Forceps for Endoscopic Surgery

Yoshihiko Nakamura, Kensuke Onuma
Hiroo Kawakami, and Tsutomu Nakamura
Department of Mechano-Informatics
University of Tokyo
7-3-1 Hongo, Bunkyo-ku, Tokyo 113, JAPAN
nakamura@mech.t.u-tokyo.ac.jp

Abstract: Endoscopic surgery has recently been adopted in various surgical operations. Forceps stuck through trocars into the abdominal or chest cavity, dissects and grasps internal organs. Due to their fixed shapes, the current forceps limit the skill of surgeons. We have worked on the development of active forceps adoping shape-memory-alloy pipes and coaxicially assembling them. In this paper, we show several models that we prototyped to find out the functional conditions of SMA pipes and their material. A 3DOF active forceps is also introduced, which posesses 6DOF inside the body being manipulated by the AESOP manipulator. A master-slave robot system is developed using it.

1. Introduction

Laparoscopic surgery has shown a rapid and wide expansion as a minimary invasive therapy. Straight shaped equipments, however, limit the surgeons' maneuvers in particular behind complicated spots behind organs. In order to unload the surgeons' load due to the above, we have continued our research since 1993 focussing on the development of actively bending yet stiff forceps employing Ti-Ni shape memory alloy pipes [1][2]. In this paper, we show several models that we prototyped to find out the functional conditions of SMA pipes and their material. A 3DOF active forceps is also introduced, which posesses 6DOF inside the body being manipulated by the AESOP manipulator. A master-slave robot system is developed using it.

2. The Active Forceps: Prototype D

2.1. Structure

The current prototype consists of a co-axially assembled SMA-pipe pair and a drain pipe as seen in **Fig.1**. The pipes are prepared as follows:

PIPE1 Ti-Ni-Cu alloy is adopted since it has an advantage of low yield stress at the Martensite phase. The A_f and M_f temperatures (T_{Af}, T_{Mf}) are designed such that $T_{opL} < T_{Mf} < T_{Af} < T_{opH}$, where T_{opL} and T_{opH} are the lower and higher bound of the operational temperature. A circular shape is memorized by heat treatment.

PIPE2 Ti-Ni alloy is used with the M_s temperature (T_{Ms}) lower than T_{opL}. Since a straight shape is memorized, this pipe maintains straight and stiff at the range of the operational temperature.

PIPE3 This pipe is made of tefron and used for returning path of water circulation.

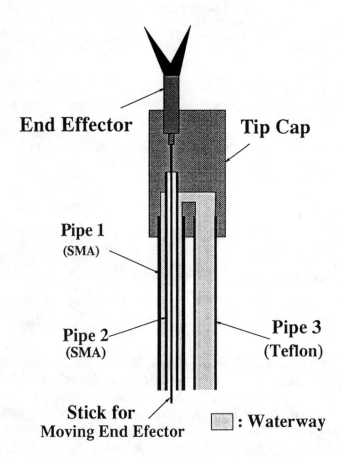

Fig. 1. Structure of SMA Active Forceps

The shape of the forceps changes according to the temperature of water circulation as follows: At the higher temperature, the returning force of PIPE1 to recover the circular shape and the elastic force of PIPE2 to keep the straight shape find an equilibrium point at a circular shape that has a larger radius than that of the PIPE1's memory. On the contrary, at the lower temperature, PIPE1 behaves like a plastic material with rather low yield stress. The elastic force of PIPE2 dominates and deternines the shape of the forceps almost straight.

The temperature of water circulation is controlled by mixing prepared hot and cold water.

2.2. Performance

In this section, the performance of the prototype introduced in the previous section is to be described.

Fig.2 shows the photos of the stretched and bent shapes. The minimum bent angle of the stretched shape was 18.0 degree, while the maximum bent angle of the bent shape was 63.6 degrees.

The response from the minimum bent angle to the maximun one took 9.6 sec, among which 3.9 sec was used for sending water from the pump to the forceps. Figure 3 shows the graph of time-response. The pumping pressure was rather low due to the limited performance of pump and the low-pressure connection of water tubes. Although it is anticipated that the improvement of pumping system will significantly lower the response time, to achieve a few seconds response time there needs a fundamental development for the new water circulation system.

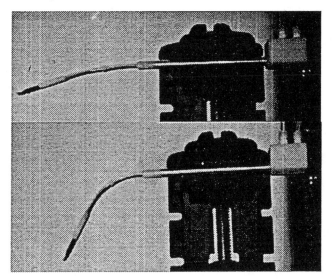

Fig. 2. Prototype of Active Forceps in Stretching and Bending State

3. Other Prototypes

3.1. Prototype A

The prototype A was developed in 1994 using three SMA pipes as shown in Fig.4. The dimensions were chosen as Table ??. Figure 5 is a photo of the prototype A. Since the outer diameter was fairly large, the maximun strain limited the curvature radius.

3.2. Prototype B

In order to get shrap bending, we chose to use pipes with smaller diameters. The stiffness was not sacrificed since the length of the SMA pipes were now one third of the prototype A (30 cm). The prototype used double pipes, while the prototype A used triple pipe structure.

474

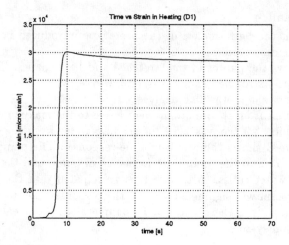

Fig. 3. Step Responce of Prototype

Pipe1(SMA2)	$\phi 7.3 \sim \phi 7.0$ [mm] ($t=0.15$ [mm])
Pipe2(SMA1)	$\phi 5.2 \sim \phi 4.2$ [mm] ($t=0.5$ [mm])
Pipe3(SMA2)	$\phi 2.3 \sim \phi 2.0$ [mm] ($t=0.15$ [mm])
Length	298 [mm]

Table 1. Dimensions of SMA pipes: external and internal diameters, and thickness

The structure is shown in Fig.6. The dimensions of several different designs are in Table 2.

Figure 7 show the results of bending experiments.

The large bending angle was attained by adopting diffrent material for the outer pipe, which contains a little Cu in Ti-Ni alloy. This combination provided lower yield stress.

The prototype D in the previous section adopted the same materials. The effect of having a drain tefron pipe was not significant as seen in Fig. 8.

4. Master-Slave Surgical Robot

We have developed a master-slave surgical robot system incorporating the active forceps developed in the previous section. Figures 9 and 10 show the master device whose 3D location and orientation is mesured by a Polhemus sensor. The handle is made of Pilling's handle.

The developed slave robot is shown in Fig.11

The forceps has three degrees of freedom, of which one is of SMA pipes and the other two are motor-driven as seen in Fig.12.

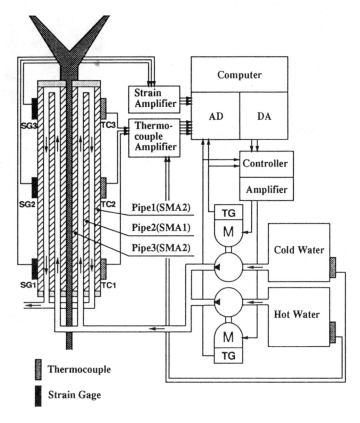

Fig. 4. System configuration of SMA active forceps

5. Conclusion

We have developed an actively bending yet stiff forceps for laparoscopic surgery. Use of Ti-Ni-Cu and Ti-Ni alloy was useful to make the effective bending angle large (from 18.0 to 63.6 degrees). The response time of current prototype was 9.6 sec. The improvement of pumping system will shorten the time. For the significant improvement of response time, however, we will need a fundamental new development of water circulation system. A master-slave surgical robot system is proposed and developed using a 3DOF active forceps.

References

[1] Y. Nakamura, A. Matsui, T.Saito and K. Yoshimoto: "Shape-Memory-Alloy Active Forceps for Laparoscopic Surgery," Proceedings of the IEEE International Conference on Robotics and Automation, pp. 2320–2327, vol, 3, 1995.

[2] k. Ohnuma, Y. Nakamura: "Research of Active Forceps for Laparoscopic Surgery –Development of Prototype with End Effector–", Proceedings of Conference of the Japan Society of Computer Aided Surgery, pp. 75–76, 1996.(in Japanese)

[3] K. Ikuta, M. Tsukamoto, S. Hirose, "Shape Memory Alloy Servo Actuator System with Electric Resistance Feedback and Application for Active Endoscope",

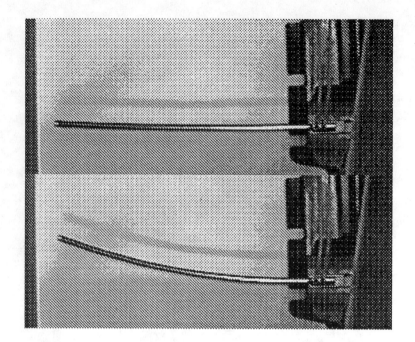

Fig. 5. The active forceps in stretching and bending states

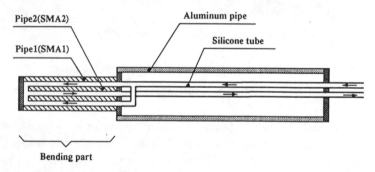

Fig. 6. Structure of SMA active forceps B

Proceedings of IEEE International Conference on Robotics and Automation, 1988, pp.427-430.
bitemikuta4 K. Ikuta, M. Nokata, S. Aritomi, "Hyper-Redundant Active Endoscope for Minimum Invasive Surgery", Proceedings of the First International Symposium on Medical Robotics and Computer Assisted Surgery, Pittsburgh, PA, September, 1994, pp.230-237.

[4] K. Ikuta, M. Nokata, S. Aritomi, "Biomedical Micro Robots Driven by Miniature Cybernetic Actuator", IEEE International Workshop on Micro Electro Mechanical System(MEMS-94), 1994, pp.263-268.

[5] T. Fukuda, S. Guo, K. Kosuge, F. Arai, M. Negoro, and K. Nakabayashi, "Micro Active Catheter System with Multi Degrees of Freedom", Proceedings of the IEEE International Conference on Robotics and Automation, San Diego, CA,

Table 2. Sizes and memorizing angle of SMA pipes using at prototype B3–6 (D_1:external diameter, D_2:internal diameter, t:thickness, θ:memorizing bending angle)

		Pipe1	Pipe2	Length[mm]
B3	$D_1[mm]$	3.825	2.28	100
	$D_2[mm]$	3.33	1.49	
	$t[mm]$	0.25	0.40	
	$\theta[°]$	47	0	
B4	$D_1[mm]$	3.825	2.48	100
	$D_2[mm]$	3.33	1.69	
	$t[mm]$	0.25	0.40	
	$\theta[°]$	47	0	
B5	$D_1[mm]$	3.825	2.69	100
	$D_2[mm]$	3.33	1.88	
	$t[mm]$	0.25	0.40	
	$\theta[°]$	47	0	
B6	$D_1[mm]$	3.825	2.69	100
	$D_2[mm]$	3.33	1.88	
	$t[mm]$	0.25	0.40	
	$\theta[°]$	101	0	

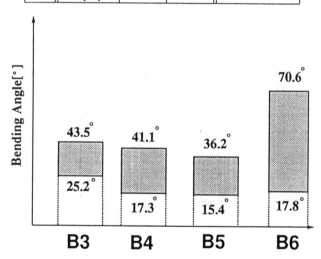

Fig. 7. Maximum bending angle and minimum bending angle

May, 1994, pp.2290-2295.

[6] P. Dautzenberg, B. Neisius, H. Fischer, R. Trapp, G. Bue, "Robotic Manipulator for Endoscopic Handling of Surgical Effectors and Cameras", Proceedings of the First International Symposium on Medical Robotics and Computer Assisted Surgery, Pittsburgh, PA, September, 1994, pp.238-244.

[7] P. Dario, R. Valleggi, M. C. Montesi, F. Selsedo, and M. Bergamasco, "Flexi-

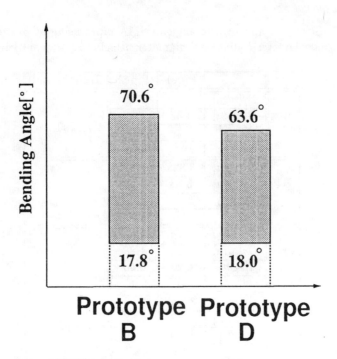

Fig. 8. Maximum bending angle and minimum bending angle of B6 and D1

Fig. 9. Polhemus sensor

ble Active Structure Incorporating Embedded SMA Actuators", IFToMM-jc International Symposium on Theory of Machines and Mechanisms, Proceedings, Nagoya, Japan, September, 1992, pp.851-855.

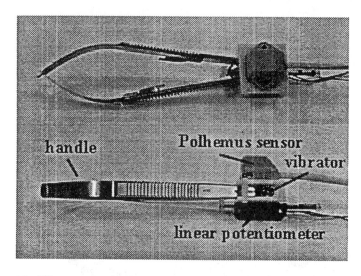

Fig. 10. The master device : Pilling's handle and a Polhemus sensor

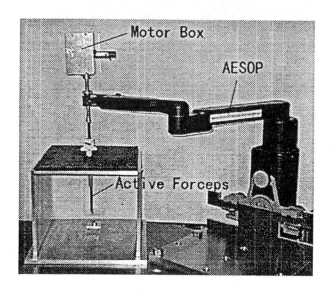

Fig. 11. The slave manipulator: an AESOP holding a 3DOF active forceps

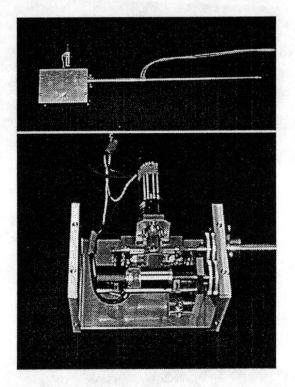

Fig. 12. 3DOF active forceps: 1DOF SMA pipes and motor-driven 2DOF

SYNERGISTIC ROBOTS IN SURGERY -SURGEONS AND ROBOTS WORKING CO-OPERATIVELY.

BRIAN DAVIES

Department of Mechanical Engineering,

Imperial College, London, SW7 2BX

b.davies@ic.ac.uk

Abstract: This paper describes experiments using a new concept in robot surgery; that of a robot with a force control handle moved by the Surgeon. The surgeon back-drives the robot under servo assistance, whilst feeling the force from a rotating cutter during surgery, for example, in total knee replacement. Thus, when machining the knee bones to take a prosthetic metal replacement, the surgeon can use his inherent sensing to slow down, or take a lighter cut, when cuting hard bone. The robot however, can be provided with regions of force constraint so that, say, a flat plane can be cut accurately into the bone to allow the prosthetic implant to be subsequently fitted. At the same time the robot can also be programmed to not allow adjacent regions to be entered, thus preventing damage to features such as ligaments. An experimental system using this active constraint robot (or ACROBOT) for knee surgery is described.

This concept is one of 'synergy' between the best capabilities of the robot and those of the surgeon. The robot is able to develop accurate geometry in 3D space, and constrain motions to a safe region, whilst the surgeon can use his innate sensing and judgement to adapt motions and tasks within the pre-defined limits. This ability to adapt procedures in the light of the changing circumstances, makes a synergistic robot concept ideally suited to soft tissue which can move unpredictably when pushed or cut.

1. Introduction

The Imperial College Group has previously produced an active robot for surgery, namely the PROBOT, which has been used for prostate removal[1]. Clinically applied in 1991, this was the first time in the World that an active robot had been used to remove tissue from human patients. The term 'active' robot means that the robot is fully powered and is programmed to move autonomously throughout the procedure whilst under the supervision of the surgeon. This is in contrast to unpowered manipulators, which are often used as tracking devices to hold tools, and locate them relative to organs whose position is also represented in a computer data base. Another alternative is powered robots which are used passively. That is, tools are positioned next to the patient at an appropriate location by the robot. The robot is then locked in position and power removed, so that subsequent actions depend on the surgeon using the robot as a locating fixture. Both unpowered manipulators and passive powered robots are considered safer

Imperial College suggest that whilst these active robots can have advantages in simple operations, there are circumstances in more complex procedures when the surgeon wishes to have more direct control, and be able to adapt his actions depending on what he senses.

For these procedures the concept has recently been evolved of a 'synergy' between the best capabilities of the robot and those of the surgeon. The idea is that tools are placed on the end of the robot, adjacent to a handle mounted on force sensors. The handle can then be gripped by the surgeon to act as a servo input to the robot. Whilst the Imperial College approach has been to develop an Active Constraint Robot (called the ACROBOT) for this task, a group at Grenoble, France, have developed a passive three axis manipulator, called PADYC, which uses a series of clutches which control the motions[2]. Each joint has two clutches, each of which are motor driven with an encoder to act as a free wheel. Powering one motor will allow rotations of the joint in one direction, whilst powering the other will allow motions in the opposite direction. Thus, by computer control of the sequence of motors, each joint can permit or restrict motion. Since the clutches act to unlock the braking action and no power to the joint is provided by the motor, it has been claimed that this type of device is intrinsically safer than ACROBOT. However, PADYC still relies on real time computer control for its correct functioning. Also clutch based systems have been said to be inherently less precise than active systems.

2. The Acrobot

The ACROBOT concept was first demonstrated on a prototype 'Scara' style arm with two axes under force control, whilst a third, (base), axis was under position control[3]. This arm demonstrated precision in being able to exactly follow complex inputs such as in handwriting, as well as more conventional machine-tool types of motion. Subsequently a special purpose 4 axis robot has been built to explore the synergy concept further[4]. Fig 1 shows a diagram of the robot motions.

This 'synergistic' approach has been developed in a demonstration project for knee surgery, in which the end bones of the knee are machined to take a prosthetic implant as a total knee replacement. Fig 2 shows a photograph of the ACROBOT. A rotary cutter is held at the end of the robot and a force control handle is positioned around its motor. By holding the handle, the surgeon can back-drive the mechanism under force control. The mechanism has been chosen to have admittances which are roughly similar, so that no motion will be preferred by the mechanism. This is achieved by the use of pitch and yaw motions and a linear motion, which all intersect their axes at a common point. A fourth, vertical base motion, is also provided. By the use of servo-assistance the surgeon can impose his motions on the robot within a pre-programmed region, without having to "fight-against" the robot impedance.

The robot can be programmed to allow free motion within a central 3-D region, whilst other regions can be programmed to prevent entry by the use of high-gain position control. An intermediate zone between these two regions uses a gradually increased damping so that, as the prohibited region is met, no motion is permitted. This is achieved by the use of controlled damping so that vectors orthogonal to the boundary and pointing into the prohibited region have very high damping, whilst motions tangential to the boundary, (or into the central region), have relatively low damping. This enables the robot to be used to prevent motion into regions where damage may

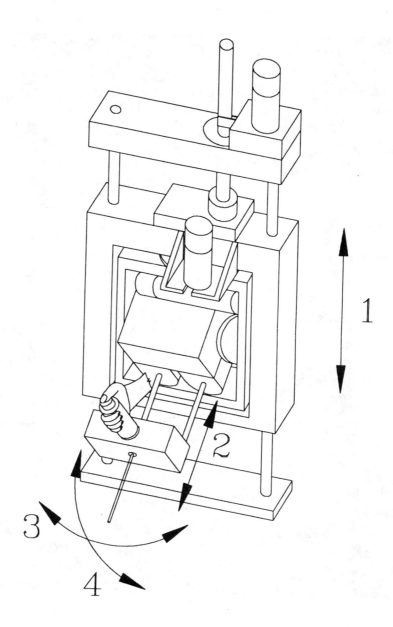

Fig 1 Diagram of the robot motions.

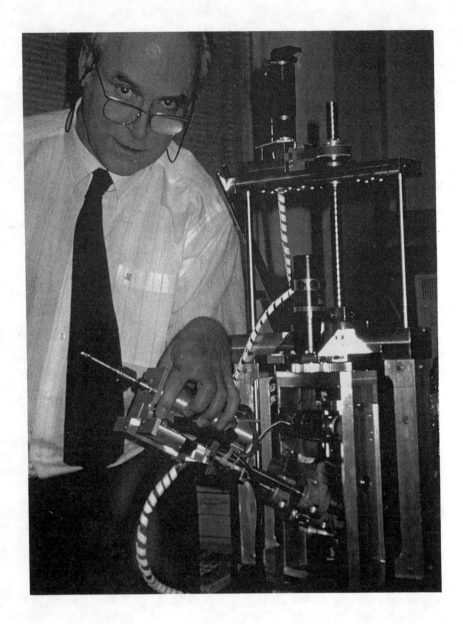

Fig 2 Photograph of ACROBOT System, showing Author with finger on "Dead-Man" Switch.

result, such as inadvertent cutting of the ligaments. The region of constraint can also form a plane, inclined in 3D to the robot axes, so that a precise plan can be accurately generated at a desired angle. In a further "trajectory" mode, motions such as drilling a hole in a precise trajectory can be generated.

The synergistic concept has some similarity to the use of telemanipulators for hazardous environments (such as the nuclear industry) or for "telepresence" surgery. In the latter case, a 'slave' manipulator is proposed at some distance from the 'master' input system which is usually a kinematic equivalent to the slave. By this means the surgeon can operate remotely over a land-line, or even a satellite link, with the 'slave' in remote regions (such as in 3rd world areas) or unsafe areas (such as on the battle field). In some procedures, a scaling function between master and slave can give micro-motions and high precision, for example for use in eye surgery.

In contrast to telemanipulators, the benefit of the synergy devices derives from having the force controlled handle held by the surgeon at the tip of the robot. For sensitive 'feel' of tool actions, the robot requires a low (and equal) impedance between the surgeon and the tool tip in all directions, as well as a relatively high constraint force, a high bandwidth and a smooth servo-assistance for back-driving the robot. All of these are different requirements from, say, the use of a force controlled joystick "master" acting as part of a nuclear telemanipulator. Because the surgeon has direct control of the cutting tools, the ACROBOT is also suitable for soft tissue work, in which distortions and motions of the tissue during the procedure can require the surgeon to track the tissue in real time. The prototype robot has been shown to have a typical accuracy of 0.5 mm over a 100 mm cube region when generating a series of 3-D planes. A repeatability of 0.3 mm is also achieved.

3. Conventional Total Knee Replacement.

Orthopædic surgery to replace damaged knee joints is a complex process requiring accurate alignment of sufaces of prosthetic implant components and their accurate positioning within the knee in order to regain normal leg functioning. Typically, for total knee replacement (i.e. replacement of all mating surfaces within the knee), two prosthesic components are used. One is mounted on the proximal tibia, the other on the distal femur. A third component is also used to replace the patella. In total, these parts may require that 6 flat planes be cut from the bone ends, along with two circular holes and two locating slots. Figure 3 shows a sketch of a pair of typical prosthetic components. The relationships between these planes, holes and slots, along with their relationship to the geometry of the knee, is crucial in obtaining good mating of the prosthesis with bone and an appropriate alignment of the femur and tibia. These are necessary to provide a long life from the prosthesis with good, pain-free, leg motions for the patient [5].

Conventionally, the manual procedure is performed using a sequence of templates mounted on the bones [5] to cut the flat planes required for the tibia (labelled '6' on figure 3) and femur (1 to 5 on figure 3), and to drill positioning holes (A,B on figure 3) and slots (C on figure 3). The sequential application of each of the many templates required, may result in a cumulative error that compromises the requirement for accuracy in the registration of the parts to each other and to the patient's leg. This can

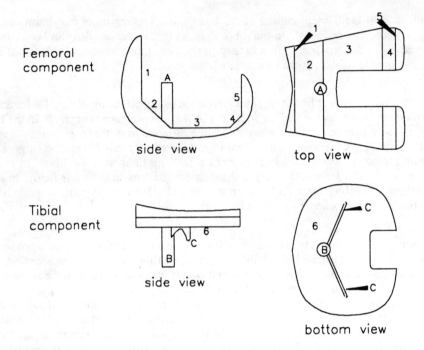

Fig. 3. Diagram of knee prosthesis (anterior at left of each sketch).

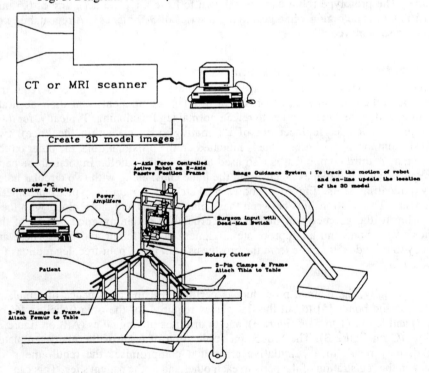

Fig. 4. Diagram of the total robotic system

lead to poor fitting of the joint and poor gait in the patient's subsequent motion. The use of such templates also requires that the surgeon become familiar with a large number of mechanical parts which must be used in an exact sequence.

Discussions with surgeons have indicated that the planes should be cut to within 1 mm of the required template positions for optimum prosthesis fit and bone regrowth. An additional difficulty lies in the use of an oscillating saw blade to cut the flat planes. The flexible blade can bounce off or dig-into the knee, in spite of the use of jigs, leading to further inaccuracies. For this reason a rotary cutter has been utilised which is stiffer and more precise than the saw. It is hoped to overcome these saw and jig problems by the use of robots and computers to assist in the knee surgery. Robots, once programmed, can make accurate cuts with the high geometric precision required to obtain good registration between the bone and prosthesis surfaces. The use of robotic techniques may also allow new, improved, complex designs for mounting the prosthesis to be produced, that would be difficult to cut into the bones using template based manual techniques.

4. The Knee Surgery Robot

Computer Tomography (CT) images of the knee are taken pre-operatively and loaded into the computer system. These images can be used to form three-dimensional models that can be manipulated and viewed from various directions, allowing the surgeon to select a particular size of prosthesis to match the bone dimensions. By overlaying computer models of the selected prosthesis onto the bones in the CT images, the surgeon can preview the relative positions of the prostheses and bones, and can define safe areas within which the robot can work. The prosthesis models can be used to compute the relative cutting planes required to obtain a good fit of the prosthesis on the bone. The model can be rotated and translated to the appropriate positions relative to the bone as defined by the surgeon. Thus the surgeon is able to pre-operatively plan the procedure, trying out a variety of strategies to ensure all is correct before proceeding with the actual operation.

In addition, for registration between the pre-operative computer model of the procedure and the patient's bones at the time of the operation, fiducial markers are used to locate fixed points on the bones. These can be identified in the CT image during the modelling procedure. Then, by touching onto them with the robot at the beginning of surgery, the robot's frame of reference is matched with that of the model. (Fig. 4). The use of fiducial markers is regarded as the "gold-standard" for this type of CAS work in orthopaedics. However such markers are regarded as invasive and so an alternative of anatomical markers is being attempted. This uses a number of points which are registered by the robot touching a region on the exposed bone surface. These points are built into a surface which is then matched with the pre-operative 3D model of the knee to perform registration of the model to the robot and to the knee location. In order to proceed with the surgery, the patient's femur and tibia are fixed to the table by a light frame which clamps the bone, which is also referenced to the robot. Thus the bones are treated as a fixed structure whihc can be machined by a pre-programmed robot. The bones are separately monitored for motion, in which case the procedure is suspended.

To access all parts of the knee, and cut at the necessary angles, requires a large amount of flexibility and a large amount of movement. It is necessary to access planes on both the anterior and posterior sides of the femur, hence many degrees of freedom would be required in the robot linkages. This can prove a problem since: (i) large movements of the robot will result in relatively large positional inaccuracies, partly from the scaling up of any rotational inaccuracies in the position sensors attached to the motors, and partly because of physical distortions of long robot linkages when forces are applied to them; and (ii) controlling many axes simultaneously in real-time uses a complex structure which is unnecessary for this simple task. It was therefore decided that the robot would have a restricted amount of movement - just enough to access some of the resection planes at any one time, and that this movement would be confined to four axes, as shown in figure 1.

The robot is mounted on a passive carrier which can be manually adjusted to position and then locked so that the robot is approximately in the correct position for each sequence of cuts required. For example, it could be re-positioned for cuts in each of the three regions: anterior femur, posterior femur, and tibia. Having obtained an approximate position, the robot's end-effector can be touched onto the fiducial markers, located in the CT planning stage, to regenerate accurate registration before each plane is cut. The author beleives that this type of concept is best suited to the safety critical aurgery task. The use of the locked, passive gross positioning robot permits the use of a small precise active robot whose motions and forces are just enought o perform the task safely [6].

5. Conclusions

The surgeon's experience with the ACROBOT to date, suggests that the 'synergy' concept is a good compromise between a fully autonomous active robot for surgery and a passive manipulator arm. The 'synergy' robot has the benefit that the surgeon feels in direct control of the tool. He can sense how hard he is pushing, feel any vibration, hear if the cutter motor is working too hard and smell if the cutter is burning the bone. These sensory perceptions are difficult to provide in a robot. Thus in addition to the psychological benefits of the surgeon "feeling in control", there is also a clear advantage from the surgeon sensing what is happening and being able to use his experience and judgement to change the procedure appropriately.

The robot is able to use its position sensors to show the surgeon on the computer display where the cutter is, relative to the imaged bones, in the same way that an unpowered manipulator arm or camera based tracking system can. However, in addition the robot is able to provide physical constraints which ensure no motion is made into prohibited unsafe areas, whilst ensuring that a required physical geometry is accurately achieved.

The 0.5 mm accuracy (0.3 mm repeatability) of the ACROBOT in knee surgery laboratory trials, shows that the total system can achieve the overall 1mm accuracy required for cutting the knee bones. This overall accuracy is also dependent on the CT scan imaging resolution, the 3D modelling and the registration accuracies. Further laboratory tests of the ACROBOT as part of the total knee replacement surgery system are continuing. These should result in a system for clinical trials in about one year. Since the cadaver studies are being conducted to a realistic level of detail and knowledge of the needs of the operating room environment, it is not envisaged that the clinical trials will result in any difficulty. However, with a complex computer controlled system, it is inevitable that details of protocols and techniques will require development within the clinical setting. The system should also be applicable to a range of other orthopaedic procedures. The ability for the surgeon to programme an adaptable region of constraint makes the concepts of synergy and active constraint also applicable to soft tissue surgery, where the deformation of tissue as it is cut and pressed makes it necessary to dynamically adapt the constraint region as the surgery progresses.

6. Acknowledgements

The authors wish to acknowledge the financial assistance of the UK Department of Health and the 'LINK' Medical Implants Fund in support of this project. The collaboration of Mr. Justin Cobb, Consultant Surgeon at Middlesex Hospital, London, is also gratefully acknowledged. Optimised Control Ltd. and of Harmonic Drives Ltd.are also thanked for their assistance.

7. References

1. Davies, B.L.,Harris, S.J., Arambula-Cosio, F., Mei, Q., Hibberd, R.D., : *"The Probot - An Active Robot for Prostate Resection".* J. Eng. in Medicine, Proc. H. of IMechE. Vol 210, M.E.P. ltd., June 97.
2. Troccaz, J., Delnondedieu Y. *"Semi-Active Guiding Systems in Surgery. A Two-DOF Prototype of the Passive Arm with Dynamic Constraints (PADyC)"* Mechatronics Vol 6, No. 4 - pp 399-421, 1996
3. Davies,B.L., Ho,S.C., Hibberd, R.D. *"The Use of Force Control in Robot Assisted Knee Surgery",* MRCAS 1994 - pp258-262
4. Harris, S.J., Fan, K.L., Hibberd, R.D., Davies, B.L.,.: *" Experiences with Robotic Systems for Knee Surgery".* Proc. 3rd Int Conf. Medical Robotics and Computer Assisted Surgery, Grenoble, France. pp757-766,March 1997
5. Howmedica *"Kinemax modular total knee system - Instrument Surgical Technique"* 1995
6. Davies, B.L. *"Safety of Medical Robots",* Safety Critical Systems, Chapman Hall Press, Part 4, chapter 15, pp 193-201, 1993

Control Experiments on two SMA based micro-actuators

N. Troisfontaine, Ph. Bidaud and P. Dario*
Laboratoire de Robotique de Paris
10-12 Av. de l'Europe -78140 Vélizy - France
* ARTS Lab. - MiTech Lab. Pisa Italy
e-mail : troisfontaine,bidaud@robot.uvsq.fr

Abstract: In this paper, two different control technics for SMA actuators are proposed : a position control and a temperature one. The position controller is based on a two stage (P and PI) structure. The temperature controller uses a PID with a lag compensator in the feedback loop. Both of these simple technics have been designed for integration in micro-actuators. They have been experimented and have shown good performances with regard to perturbations in environmental conditions.

1. Introduction

In recent years, Shape Memory Alloy (SMA) actuation has been considered in numerous proposals for micro-actuators. Its capacities in both tensile stress (up to 250 MPa) and motion range (up to 6% of its length) make it very attractive for micro-robotic applications. Basically, SMA actuators exploit a phase transformation in the alloy, called martensite transformation, by heating and/or applying external stress. At high temperature, an SMA wire is basically elastic. At low temperature, in the martensite phase, it exhibits an elasto-plastic behavior. Restoring the initial length of an SMA actuator after heating requires an external bias force. This force is usually produced by the use of either an antagonist SMA actuator or a pre-stressed spring.
Cyclic transitions between the austenite phase and the martensite phase are highly non linear with large hysteresis. Repeated heating/cooling process and/or applied stress variations induce major and minor loops and lead to great difficulties for controlling the actuator position. Moreover, the SMA actuator behavior is influenced by thermal exchanges. Experiments have clearly shown that open loop control is not suitable for robotic applications [1] [2].
Only a limited number of approches for controlling SMA actuator can be consider positively when taking into account integration constraints for microsystems (i.e number of sensors and controller hardware have to be minimized) [2].
In this paper, position and temperature feedback control schemes are proposed and experimented. Their good performances and their remarkable simplicity open perspectives for integration in micro-actuator structures. They have been designed for two SMA based micro-actuators. An inter-phalangeal actuator for

dexterous micro-grippers [1] and a steerable endoscope [3].

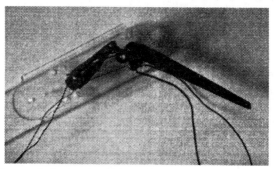

Photo 1 : Inter-phalangeal actuator for dexterous micro grippers

In micro-grippers, joint positions have to be accurately controlled in order to perform precise manipulation tasks. This need justifies the integration of a position sensor. In this context, a minimal controller may use a single position feedback loop.

Photo 2 : Steerable endoscope tip

The steerable endoscope is for applications in computer-assisted arthroscopy (photo 2). The endoscope tip trajectory is remotely controlled through a joystick and visual video feedback. However, in order to preserve the system controllability, to avoid thermal drift, and to maintain the desired bending, the actuator temperature has to be controlled. Thus, only a temperature sensor is integrated in the actuator and temperature measurement is used to obtain the feedback loop.

Both SMA actuators have been implemented by using Ni-Ti(50%/50%) wires. The bias force is produced, in the first one by an elastic element, and in the second one, by a second wire mounted in an antagonist mode.

2. Open loop experiments
2.1. Principle

In [1], a mathematical model has been proposed and experimentally validated to describe the thermo-mechanical behavior of an SMA wire. This model expresses the strain of the wire ϵ as a function of the two parameters governing the martensite transformation : the temperature T and the applied stress σ. Since ϵ relies directly on the displacement x produced by the SMA actuator, the system can be considered as having two inputs variables (T,σ) and one output (x). The applied stress depends on the force produced by the bias element and the external force applied to the SMA actuator. The temperature is generated by Joule effect with an electrical current controlled by a transistor (see fig. 2.1). The electrical resistance in the wire depends upon the transformation state. For this reason, it is preferable to use electrical power P_{elec} rather than electrical voltage (u) [2]. The open loop control scheme takes the form described on the figure 2.1. Considering the hight non-linearity of the actuator, a simple transistor has been prefered to a PWM driver to minimize the complexity of the electronic circuit.

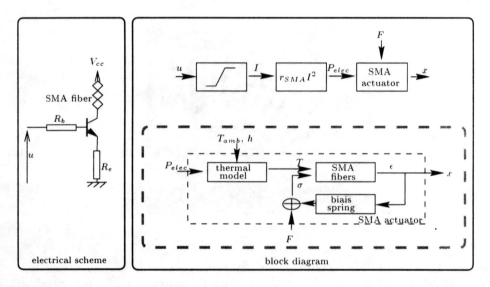

Figure 1. Block diagram of the SMA actuator

The velocity of the martensite transformation is closed to the sound velocity in materials, so that its dynamics will be neglected. Thus, the system bandwidth is limited mainly by the thermal bandwidth. Heating and cooling delays depend on the material geometry, alloy transition temperatures and on external conditions, such as ambient temperature T_{amb} and convection mode h, which both vary with respect to time.

In the next section, the sensitivity of the thermo-mechanical system to thermal and stress disturbances is analyzed. This is achieved by exploiting a static thermo-mechanical model. Considering the complexity of the thermal model with time varying parameters, an experimental approch has been prefered for analizing the actuator dynamics.

2.2. Thermo-mechanical behavior sensibility analysis

The strain inducing the output motion in SMA actuators depends upon thermo-mechanical conditions. In order to control the actuation process, the SMA behavior has to be analyzed.

In [1], we proposed an analytical thermo-mechanical model for SMA actuators. We have experimentally verified for SMA wires, that their strain ϵ can be described by :

$$\epsilon = (1-R)\frac{\sigma}{D_A} + R(\frac{\sigma}{D_M} + \alpha\sigma + \epsilon_r) \qquad (1)$$

where the reference value to compute ϵ is taken in austenite phase without any external stress. $R(T,\sigma)$ represents the martensite fraction :

$$R(T,\sigma) = \frac{R_1}{1 + \exp K(T - T_m - c\sigma)} + R_2 \qquad (2)$$

K and T_m depend on the transformation way. R_1 and R_2 rely on the material history. For more details on this model and its identification, see [1]. The response to variations in thermo-mechanical conditions can be estimate from the model sensitivity to both parameters temperature T ansd stres. From the relationship :

$$\dot{\epsilon}(T,\sigma) = \frac{\partial \epsilon}{\partial T}\dot{T} + \frac{\partial \epsilon}{\partial \sigma}\dot{\sigma} \qquad (3)$$

with :

$\frac{\partial \epsilon}{\partial T} = (\epsilon_{M_0} - \epsilon_{A_0})\frac{-R_1 K exp[K(T_0-T_m-c\sigma_0)]}{(1+exp[K(T_0-T_m-c\sigma_0)])^2} \leq 0$

$\frac{\partial \epsilon}{\partial \sigma} = (1-R_0)\frac{1}{D_A}$
$+R_0(\frac{1}{D_M} + \alpha) + (\epsilon_{M_0} - \epsilon_{A_0})\frac{R_1 c exp[K(T_0-T_m-c\sigma_0)]}{(1+exp[K(T_0-T_m-c\sigma_0)])^2} \geq 0$

We can observe that temperature and strain variations vary in opposite direction, while stress and strain variations vary in same direction. This is true even in discontinuities caused by changes in transformation directions. It is important to notice that both partial derivatives are strongly dependent on the operative point, on the transformation way (i.e foward and reverse tranformations) as well as on the material history (throught R_1 and R_2).

In position control of the SMA actuators, only the temperature can be considered as control variable. Stress variations in position control has to be compensated through an adequate temperature variation. An important feature, which is not reflected by the model, is the natural damping. The material experimentally exhibits a behavior that we call natural damping in the thermo-mechanical loading variations. This "damping" comes from internal friction in the phase transformation process and is dissipated in thermal energy. When both phases are present simultaneously, the internal damping effect can induce oscillations in control. This is due to coupling between temperature corrections coming from the position controller and internal temperature variations.

2.3. Environmental conditions sensibility analysis

The SMA actuator dynamics rely mainly on the ambient temperature and convection mode. Open-loop experiments over a large set of different environmental conditions is a direct way to access to sensitivity analysis.

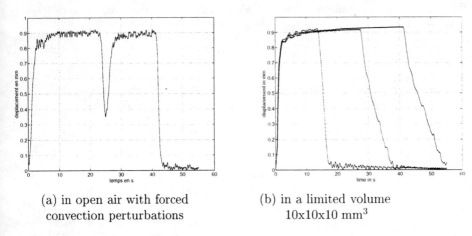

(a) in open air with forced convection perturbations

(b) in a limited volume 10x10x10 mm^3

Figure 2. Open loop experiments

In figure 2-a, a very large perturbation is observed when switching the convection mode in a short time. On figure 2-b, experiments are performed in a limited ambient volume. The drift on output position is induced by a drift on the surrounding temperature caused by heat propagation. This ambient temperature rising also modifies the falling response time, and it can make the actuator uncontrollable after a limited number of cycles. It is also clear that the falling response time increases when the volume of ambiant air decreases. The two main conclusions for this analysis are :
1. open-loop control is not suitable for SMA actuators,
2. thermal exchanges have to be controlled.

3. Temperature feedback control
3.1. Controller design
Thermal exchanges can be controlled by a temperature feedback loop.
Since the SMA wire strain is a function of temperature, applied stress and its transformation history, temperature feedback is not suitable to control the output position of the SMA actuators. Nevertheless, it can be sufficient in case of teleoperated systems such as the steering end-effector.
The temperature measurement of the wire is not strictly necessary. A thermocouple with a 50μm diameter has been glued on the wire with a thermal conductive paste. The measured temperature is in fact the average temperature of the interface between the wire and the surrounding air. Moreover,

by considering the heat propagation phenomena, the measure reflects the average temperature of the wire with a time delay in which the response time of the thermocouple can be neglected. To compensate this time delay, a lag compensator is integrated in the feedback loop (fig. 3).

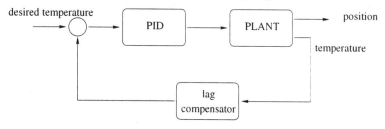

Figure 3. Temperature feedback control structure

3.2. Experimental results

To evaluate the performances of the proposed controller, experiments were performed on wires with $100\mu m$ in diameter which activation temperature is $70°C$ in the 10x10x10 mm^3 volume.

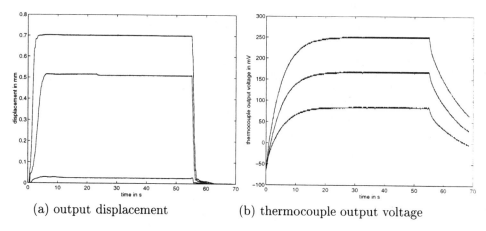

(a) output displacement (b) thermocouple output voltage

Figure 4. Temperature feedback control for three different orders.

In figure 4, it can be noticed that the output position signal is established before the temperature signal because of heat propagation phenomena.
In comparison to open loop experiments, the system response time is reduced in by a ratio of 40. This allows to maintain the system in a position controlled over a long period of time.
The major drawback of temperature feedback control is that very small perturbations on environmental conditions can produce large variations in the output position. Figure 5 shows that a temperature perturbations, not perceptible by the thermocouple, modifies the strain (which is not compensated) in the SMA actuator because of its hysteresis.

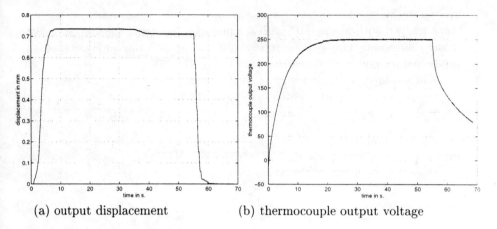

(a) output displacement (b) thermocouple output voltage

Figure 5. Experiments on temperature feedback control with small environment perturbations

4. Position feedback loop
4.1. Controller design

In [1], we proposed a feedback position control based on a switching structure. This solution is generaly well adapted to non linear systems but it leads to oscillations on the output position. These oscillations are intrinsic to the controller structure. Hayward & al. [2, 4] exploited a variable structure with three gain levels. They studied the onset of limited cycles which can be avoided with small gains. As said before, they are also due to damping in the material. This last point forces to select a smallest gain around the desired position, especially if it corresponds to a transformation state with both phases. Consequently, it leads to a significant steady state error.

Here, we propose to apply a feedback position loop with a two-stage controller by defining a boundary layer around the desired position as described in figure 6.

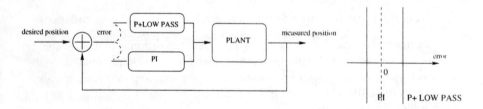

Figure 6. Position control schem

In this controller:
- when the position error is large, a high proportional gain is applied with a low pass filter. This reduces the velocity when the position error is closed the boundary layer.
- In the layer, a PI controller is applied to minimize the steady state error with a smallest proportional gain.

Gorbet & al. have studied the $\mathcal{L}_2$ stability of shape memory alloy actuated position control systems with PI controller [5]. Their stability analysis concludes that gain optimization is difficult. The stability of the system depends upon the thermo-mechanical and the thermal models used. Consequently, controller parameters were set experimentally.

4.2. Experimental results

Experiments were performed in same conditions as for temperature control. The two-stage variable controller provides very good position accuracy. We can see that the control voltage u is adapted to environmental conditions and decreases slowly when the surrounding temperature increases.

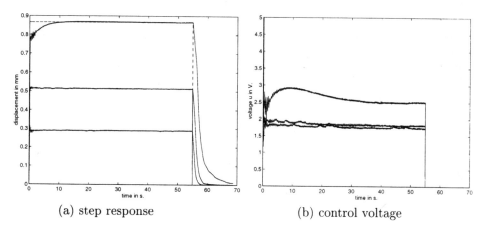

(a) step response (b) control voltage

Figure 7. Experiments with position feedback loop

When we applied a full range motion sinus command (8-a), the maximum frequency value is low (0.1Hz) but performances can be improved by using only the half range motion. In this case, the maximum frequency value is approximatily the double (8-b).

Moreover, this controller has been tested for different values of the ambient temperature and has demonstrated good reproductibility.

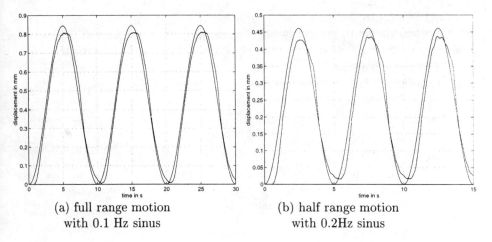

(a) full range motion
with 0.1 Hz sinus

(b) half range motion
with 0.2Hz sinus

Figure 8. tracking sinus trajectory

4.3. Comparisons in dynamic performances between the two control technics

It is clear that temperature feedback does not allow to control the displacement of the SMA actuator. But, if we consider the experimental results represented in figure 9-a made in the same conditions for both control technics to obtain at least the same output position (in dotted line temperature feedback control), we can observe that the falling response time is smallest in case of temperature feedback control. This is due to a drift on wire temperature in case of position feedback control (fig. 9-b). As a matter of fact, to maintain the desired position, the control voltage u variations produce minor thermal cycles.

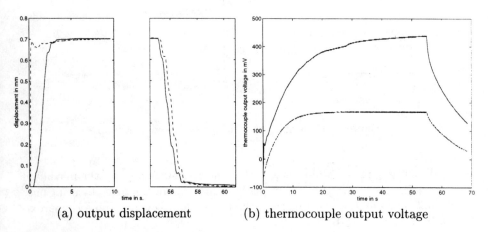

(a) output displacement

(b) thermocouple output voltage

Figure 9. Comparison of the performance between both controllers for the same output position

Notice that temperature feedback loop can be integrated with the position feedback loop to improve the dynamic performances of SMA actuator. Results will be presented in a future communication.

5. Conclusions

We have proposed here two different modes for SMA micr-actuator control depending on the concerned robotic application. Experimental results have shown good performances of both technics with regard to external perturbations. In both case, experimental tests have proved that thermal exchanges limit dynamic performances and they have to be considered during the design of such actuators. Dynamic performances can be improved by (1) adapting activation temperatures of the alloy to the working conditions and its geometry, (2) not exploiting the full range of the maximum available motion produced by the wire, (3) choosing body material with good thermal properties, and (4) controlling forced convection phenomena inside the volume of the micro-system.

References

[1] N. Troisfontaine, P. Bidaud, and G. Morel, "A new inter-phalangeal actuator for dexterous micro-grippers," in *International confernce on Robotics and Automation, Albuquerque USA*, IEEE, 1997.

[2] D. Grant and V. Hayward, "Design of shape memory alloy actuator with high strain and variable structure control," in *IEEE International Conference on Robotic and Automation, Nagoya*, pp. 2305–2312, 1995.

[3] P. Dario, C. Paggetti, N. Troisfontaine, E. Papa, T. Ciucci, M. Carrozza, and M. Marcacci, "A miniature steerable end-effector for application in an integrated system for computer-assisted arthroscopy," in *International confernce on Robotics and Automation, Albuquerque USA*, IEEE, 1997.

[4] D. Grant and V. Hayward, "Controller for a high strain shape menory alloy actuator:quentching of limit cycles," in *IEEE International Conference on Robotic and Automation, Albuquerque*, pp. 254–259, 1997.

[5] R. Gorbet and D. Wang, "General stability criteria for a shape memory alloy position control system," in *IEEE International Conference on Robotic and Automation, Nagoya*, pp. 2313–2319., 1995.

Chapter 11

Actuation Control

Actuation control requires not only to design a control strategy oriented to execute some planed actuation based on the tasks objectives; it has to consider the actuators, transmissions and joints behavior. In this chapter three different problems are addressed.

Lahdhiri and ElMaraghy design an optimal nonlinear position tracking controller for an experimental two-link-flexible joint robot. In articulated arms, in which the flexibility of the joints can not be neglected, the problem of position tracking control can not be solved by simple linear approximate models. The work presented starts with the dynamic modeling of the two link flexible joint robot and from this model a linearizing state transformation is derived to take advantage of much simpler linear design methodologies. The controller designed has been simulated and tested on an experimental robot manipulator system showing good tracking performance in the presence of noise.

The work of Prisco, Madonna and Bergamasco deals with joint position control of an articulated tendon driven finger. The use of cable transmissions allows to locate the actuators on the robot base, thus avoiding weight and volume constraints. A testbed prototype finger has been built to study its behavior and the dynamic model formulation is derived. First, the model is theoretically studied, and afterwards experimented with the finger prototype.

Modeling for the control of an hydraulic manipulator joint is the work presented by Bilodeau and Papadopoulos. Hydraulic actuators control implies a more complex control than electrical driven actuators. The control signal does not act proportionally to the final actuation due to hydraulic control devices (valves, distributors, ...) that present non liniarities, hysteresis, etc. The modeling of the hydraulic manipulator used in the experiments includes hysteresis, valve tip and fluid dynamics. While some parameters are taken from the manufacturer, other (servovalve parameters, shaft stiffness, volumetric displacement and internal leakage) have been obtained from experiments. Their obtention had required to build a specific apparatus for their measurements. The validation of the identified parameters is carried out comparing the simulated results with those obtained from experimentation.

Optimal Nonlinear Position Tracking Control of a Two-Link Flexible-Joint Robot Manipulator

Tarek Lahdhiri
Intelligent Manufacturing Systems Center
University of Windsor
Windsor, Ontario, Canada N9B 3P4
tl007@ims.uwindsor.ca

Hoda A. ElMaraghy
Faculty of Engineering, Dean
University of Windsor
Windsor, Ontario, Canada N9B 3P4
hae@ims.uwindsor.ca

Abstract: This paper presents the design of an optimal nonlinear position tracking controller for an experimental two-link flexible joint robot manipulator. The controller is designed based on the concept of exact feedback linearization and LQG/LTR control design techniques. The proposed control approach reduces the number of required measurement sensors and takes into account the effects of measurement noises. Simulation and preliminary experimental results demonstrate the potential benefits of the proposed control approach in reaching the desired system performance with minimum control effort and equipments.

1. Introduction

In many robot manipulator systems, the joint flexibility factor can not be neglected because of the flexibility in the drive systems. Thus, poorly damped oscillations may take place whenever the joint resonant frequencies are excited. This problem is due to the fact that these resonant frequencies are usually located within the control bandwidth [1]. Also, joint flexibility may degrade the controller performance and in some cases it may cause unstable behavior. Therefore, the joint flexibility has to be considered in the control design in order to achieve the desired system performance.

The problem of position tracking control of flexible joint robot manipulators has been discussed by several researchers and several solutions have been reported in the literature. The first class of solutions uses linear approximate models of the robot manipulator system and existing linear controllers range from proportional derivative (PD) to optimal multivariable and adaptive controllers, e.g. [2] and [3]. The problem with this control approach is that the validity of linear approximate models ceases whenever the state of the system exhibit large deviations from the operating point around which the system is linearized. The second class of solutions uses reduced order nonlinear models, derived using singular perturbation techniques, and the controller design is based on Lyapunov method or inverse dynamics techniques, e.g. [4] and [5]. The singular perturbation approach is based on the assumption that the flexible motion dynamics are very fast when compared to the rigid body dynamics. However, such an assumption may be violated specially for high speed applications and therefore, this method can be considered limited to weakly elastic joint robot systems [6]. The third class of solutions uses detailed nonlinear models and the control laws are derived using the concept of feedback linearization or inverse dynamics techniques, e.g. [7] and [8]. The main challenge in using such control methods is the need for measurements of the full state vector which is required for the computation of the inverse input transformation. Some of these control designs assume that all states are available. However, this assumption may not be realistic and an observer is needed in general to provide an estimate of the state. Existing observer designs range from sliding mode to nonlinear adaptive observers. However, the design of nonlinear observers is very complex, in general, and defeats the whole purpose of using the concept of feedback linearization, since the control problem becomes a nonlinear design problem. In addition to the previously discussed limitations, most controller designs assume noise free measurements. This is not generally the case and any realistic controller design should take into consideration the measurement noise.

The objective of this paper is to present an optimal nonlinear position tracking controller for an experimental two-link robot manipulator. The proposed control approach takes into account the joint flexibility and the measurement noise in order to guarantee the achievement of the desired tracking performance. The controller is designed using the concept of feedback linearization (FL) [9] and linear quadratic Gaussian/Loop transfer recovery techniques (LQG/LTR) [10]. Also, a new method for computing the state estimate is presented. This method takes advantage of the linear structure of the transformed system in order to get around the need for designing nonlinear observers. In addition, this estimation method has the advantage of reducing the number of measurement sensors: It requires sensors that provide measurements of the joints angular positions and velocities only.

This paper is organized as follows: Section 2 presents the system model. Section 3 outlines the derivation of the linearizing state transformation and presents the new approach for computing the nonlinear state estimate. Section 4 contains the controller design. Section 5 discusses the simulation results. Section 6 contains some conclusions and future work.

2. System Model

Consider the two-link flexible-joint robot manipulator system shown in Figure 1. The equations of motion of a two-degree of freedom flexible joint robot manipulator, in the absence of contact forces, are given by [1]:

$$D(q)\ddot{q} + C(q,\dot{q}) + B_l \dot{q} - K(q_m - q) = 0 \qquad (1)$$

$$I_m \ddot{q}_m + B_m \dot{q}_m + K(q_m - q) = \tau \qquad (2)$$

where q is the link angular vector, q_m is the motor angular position vector, $D(.)$ is the manipulator inertia matrix, $C(.)$ is the Coriolis and centrifugal force vector, K is the joint stiffness matrix, I_m is the rotor inertia matrix, B_m is the motor viscous friction matrix, B_l is the joints viscous friction matrix, and τ is the control input torque.

The system dynamics, based on (1) and (2), can be put in a state space form as:

$$\dot{X} = f(X) + \sum_{i=1}^{m} g_i u_i \quad , \quad m = 2 \qquad (3)$$

where the state vector X and the control input u_1 and u_2 are given by

$$X = \begin{bmatrix} x_1 & \cdots & x_8 \end{bmatrix}^T = \begin{bmatrix} q_1 & q_2 & q_{m_1} & q_{m_2} & \dot{q}_1 & \dot{q}_2 & \dot{q}_{m_1} & \dot{q}_{m_2} \end{bmatrix}^T \quad , \quad u_i = \tau_i \quad , \quad i = 1,2 \qquad (4)$$

To take advantage of well developed linear design methodologies, the nonlinear model (3) is transformed into an equivalent linear dynamical system using the concept of feedback linearization.

3. Derivation of the Linearizing State Transformation

In the design of feedback control laws for nonlinear systems, the concept of FL has emerged as a promising technique. Using this concept, a nonlinear system can be transformed, using a suitable state transformation and an appropriate control law, to an equivalent linear system for which well developed linear design techniques can be applied. Necessary and sufficient conditions for the solvability of the FL problem for systems of the form of (3) are stated by the following Theorem.

Theorem 1 [9]

A nonlinear system of the form of (3), with the assumption $rank\{[g_1 \cdots g_m]\} = m$, is feedback linearizable in an open and dense set W_x of $\mathbf{R}^n$, where n is the dimension of the system, if and only if the following conditions are satisfied.
(1) The distributions G_i $i=0...n-1$ have constant dimension in W_x, with G_i defined as

$$G_i = span\left\{ ad_f^k g_j \quad , \quad 0 \le k \le i \quad , \quad 1 \le j \le m \right\} \qquad (5)$$

where $ad_f^k g$ denote the Lie brackets of f and g.
(2) For each $i=0...n-2$, the distribution G_i is involutive in W_x.
(3) The distribution G_{n-1} has dimension n in W_x. □

To apply the concept of feedback linearization to the problem of position tracking of flexible-joint robot manipulators, the nonlinear system model (3) must satisfy the conditions stated by Theorem 1. In the present case, the dimension of the system is equal to 8; i.e. $n=8$. Computing the Lie brackets vectors $ad_f^k g_j$, $k=0...7$, $j=1,2$, and their different Lie brackets, using Lie algebra and differential geometry, it is shown that G_0, G_1, and G_2 are involutive distributions of rank 2, 4, and 6, respectively. Also, appropriate calculations show that the distributions G_i $i=3...7$ satisfy $G_i = \mathbf{R}^8$. These results prove that the nonlinear system model (3) satisfies all three conditions of Theorem 1 and therefore, the robot model (3) is feedback linearizable in the open and dense set $W_x = \mathbf{R}^8$.

Using Frobenius Theorem and Leibnitz Formula [9], it can be shown that the linearizing state transformation is of the form

$$T = \begin{bmatrix} T_1 & \cdots & T_8 \end{bmatrix}^T = \begin{bmatrix} \lambda_1 & L_f^1 \lambda_1 & L_f^2 \lambda_1 & L_f^3 \lambda_1 & \lambda_2 & L_f^1 \lambda_2 & L_f^2 \lambda_2 & L_f^3 \lambda_2 \end{bmatrix}^T \qquad (6)$$

where the scalar functions $\lambda_1(X)$ and $\lambda_2(X)$ are solutions of the differential equations:

$$\frac{\partial \lambda_l}{\partial x_m} = 0 \quad , \quad 1 \le l \le 2 \quad , \quad 3 \le m \le 8 \tag{7}$$

$$d_3 \frac{\partial \lambda_l}{\partial x_1} - (d_3 + d_2 \cos x_2) \frac{\partial \lambda_l}{\partial x_2} \ne 0 \quad , \quad 1 \le l \le 2 \tag{8}$$

$$(d_3 + d_2 \cos x_2) \frac{\partial \lambda_l}{\partial x_1} - (d_1 + 2d_2 \cos x_2) \frac{\partial \lambda_l}{\partial x_2} \ne 0 \quad , \quad 1 \le l \le 2 \tag{9}$$

Choosing $\lambda_1(X) = x_1$ and $\lambda_2(X) = x_2$, it can be easily verified that equations (7)-(9) hold, which yields the following linearizing state transformation

$$T(X) = [\, q_1 \quad \dot{q}_1 \quad \ddot{q}_1 \quad \dddot{q}_1 \quad q_2 \quad \dot{q}_2 \quad \ddot{q}_2 \quad \dddot{q}_2 \,]^T \tag{10}$$

Note that T, given in (10), is not the unique solution of equations (9)-(11). However, the chosen solution has the advantage that the new state space coordinates consists of the joint angles, angular velocities, angular accelerations, and angular jerks, which are physical system variables that can be measured or computed and yet reducing the complexity of the controller design. The resulting linearizing control law is given

$$v = \alpha(X) + \beta(X)\, u \tag{11}$$

where the vector field $\alpha(X)$ and the matrix fields $\beta(X)$ are given by

$$\alpha(X) = \begin{bmatrix} L_f^4 \lambda_1 \\ L_f^4 \lambda_2 \end{bmatrix} \quad , \quad \beta(X) = \begin{bmatrix} L_{g_1} L_f^3 \lambda_1 & L_{g_2} L_f^3 \lambda_1 \\ L_{g_1} L_f^3 \lambda_2 & L_{g_2} L_f^3 \lambda_2 \end{bmatrix} \tag{12}$$

Using the transformation (10), it can be shown that the transformed system, in the new state space coordinates, is a linear system whose dynamics are given by

$$\dot{Z} = A\, Z + B\, v \tag{13}$$

where v satisfies (11)-(12) and the matrices A and B are given by

$$A = \begin{bmatrix} A_1 & 0_{4\times 4} \\ 0_{4\times 4} & A_1 \end{bmatrix}, \; A_1 = \begin{bmatrix} 0 & 1 & 0 & 0 \\ 0 & 0 & 1 & 0 \\ 0 & 0 & 0 & 1 \\ 0 & 0 & 0 & 0 \end{bmatrix}, \; B = [B_1 \; B_2] = \begin{bmatrix} 0 & 0 & 0 & 1 & 0 & 0 & 0 & 0 \\ 0 & 0 & 0 & 0 & 0 & 0 & 0 & 1 \end{bmatrix}^T \tag{14}$$

As mentioned previously, the new state vector Z consists of the joint angles q, angular velocities $\dot{q}$, angular accelerations $\ddot{q}$, and angular jerks $\dddot{q}$. In general, sensors that provide measurements of the joint angle and angular velocity are available in most robot manipulator systems. However, sensors that provide accurate measurements of the accelerations are not usually available because of their high cost. In addition, the jerks can not be measured or computed with high accuracy. In order to reduce the cost of the proposed nonlinear controller, compared with the cost of existing linear and nonlinear controllers, it is assumed that only measurements of the joint angles and angular velocities are available. Therefore, the outputs of the original nonlinear system (3) and the linear transformed system (14) are given by:

$$Y = \begin{bmatrix} I_{2\times 2} & 0_{2\times 2} & 0_{2\times 2} & 0_{2\times 2} \\ 0_{2\times 2} & 0_{2\times 2} & I_{2\times 2} & 0_{2\times 2} \end{bmatrix} X + \eta_s \triangleq C X + \eta_s = C Z + \eta_s \quad (15)$$

where η_s represents the measurement noise which is assumed to be a Gaussian process with zero mean and covariance R_s.

From equations (11)-(12), it can be seen that knowledge of the control input v and measurements of the nonlinear state vector X are required to compute the actual control input u of the nonlinear system (3). However, measurements of the full state vector X are partially available. Therefore, the control input u can only be computed using an estimate of the state process X. That is

$$u = [\beta(\hat{X})]^{-1} (v - \alpha(\hat{X})) \quad (16)$$

One method of computing the estimate of X, $\hat{X}$, is to design a nonlinear stochastic observer for the nonlinear system (3). This approach requires solving Kushner [20] equation for the conditional probability density function (CPDF) of the state X based on the measurements vector Y, $P_x(X/Y)$. The problem with this approach is that an analytical solution does not exist, in general, for such equations. In addition, even if the analytical expression of the CPDF is known, the maximum likelihood estimate of X can only be approximated. Nevertheless, the problem of computing the nonlinear control input u can be solved as follows. Since the transformed system (14)-(15) is linear and the measurement noise η_s is Gaussian, then a linear stochastic observer can be designed to generate an estimate of Z defined as $\hat{Z} = E\{Z/Y\}$. Since T is guaranteed to be a diffeomorphism in the set W_x, then T^{-1} exists and it is also a diffeomorphism in the image set W_z, defined as $W_z = T(W_x)$. Given the statistics of state process Z, the statistics of the state process X can be computed using the analytical expression of T^{-1}. In such case, the control input u can be computed using (16) with $\hat{X} = E\{T^{-1}(Z)\}$. To reduce the complexity of the computation of $\hat{X}$ one may approximate the expression $\hat{X} = E\{T^{-1}(Z)\}$ by

$$\hat{X} = T^{-1}(\hat{Z}) \quad (17)$$

In order to verify the validity of this approximation, define the error e as

$$e = X - T^{-1}(\hat{Z}) = T^{-1}(Z) - T^{-1}(\hat{Z}) . \quad (18)$$

After some manipulations, it can be shown that the mean of the error e is given by

$$E\{e\} \approx \left[\text{trace}\left(\text{cov}(Z) \left[\frac{\partial^2 T_1^{-1}}{\partial Z^2} \right]_{Z=\hat{Z}} \right) \quad \ldots \quad \text{trace}\left(\text{cov}(Z) \left[\frac{\partial^2 T_8^{-1}}{\partial Z^2} \right]_{Z=\hat{Z}} \right) \right]^T \quad (19)$$

Since T^{-1} is diffeomorphic then the magnitude of the term $\frac{\partial T_3^{-1}}{\partial Z^2}|Z=\hat{Z}$ is guaranteed to be bounded for stable operations. Therefore, the approximation (17) is valid if a good estimate of Z is provided. In what follows, the estimate $\hat{X}$ will be computed using equation (17). The controller design is presented next.

4. Controller Design

The controller is designed into three steps. First, a linear quadratic Gaussian regulator (LQG) is designed for the linear system (14) based on a linear quadratic optimization, where the performance measure is given by

$$J = E\left\{\int_0^\infty \left[(Z-Z_{ref})^T Q_c (Z-Z_{ref}) + v^T R_c v\right] dt\right\} \quad , \quad Q_c \geq 0 \quad , \quad R_c > 0 \tag{20}$$

where R_c and Q_c are the control weighing matrices and Z_{ref} is the desired tracking trajectory. The control law that minimizes the performance measure (20) is given by

$$v = -K_c (Z - Z_{ref}) \tag{21}$$

where K_c is the controller gain. The choice of Q_c and R_c is based on the following criteria: (a) good tracking performance, (b) minimum control effort, and (c) the cross-over frequency of the LQR U-node loop return ratio, $L_1(s)$ defined as

$$L_1(s) = -K_c [sI - A]^{-1} B \quad , \tag{22}$$

is less than the frequency bandwidth of the system, w_b. Constraint (c) is necessary for noise rejection and to guarantees robustness towards unmodeled dynamics and model uncertainties, which become active for frequencies above w_b.

The second step consists of designing an observer to estimate the transformed state vector Z. From linear control theory [10], it is known that the desired performance, achieved by the LQR design, are no longer guaranteed when using a state estimator in cascade with a state feedback controller. In order to maintain the performance achieved by the LQR, the LTR techniques are used to design the observer. In such case, the control law (21) is modified as

$$v = -K_c (\hat{Z} - Z_{ref}) \tag{23}$$

where $\hat{Z}$ is the output of the state estimator whose dynamics are given by

$$\frac{d}{dt}\hat{Z} = (A - K_f C)\hat{Z} + Bv + K_f Y \tag{24}$$

where K_f is the estimator gain. The LTR procedure is based on designing the estimator with the following set of weighing matrices:

$$Q_f(q) = q_0 I + q^2 B V B^T \quad , \quad V > 0 \quad q, q_0 \in \mathbb{R}$$
$$R_f(q) = R_s \tag{25}$$

where the matrix B is defined in (14). The design parameter q has to be chosen so that the singular values of the LTR U-node return ratio L_2, defined as

$$L_2 = -K_c[sI - A + BK_c + K_fC]^{-1} K_fC[sI - A]^{-1} B , \qquad (26)$$

are close to those of L_1. In this way, the performance achieved by the LQR is recovered while using an observer in cascade with a the LQG feedback controller.

Given the control input v generated by the linear controller, the analytical expression of T^{-1}, and the estimate $\hat{Z}$, the actual nonlinear control input u of the original nonlinear system (3) is computed using

$$u = [\beta(\hat{X})]^{-1} [v - \alpha(\hat{X})]_{\hat{X} = T^{-1}(\hat{Z})} . \qquad (27)$$

Note that this nonlinear control input u is guaranteed to be a bounded stabilizing control input for the nonlinear system (3) as long as the linear transformed system is stable. This is due to the fact that the original nonlinear system and the transformed system are dynamically equivalent. Also, it is important to note that in the present case, the tracking performance achieved by the linear transformed system is the same as the performance achieved by the nonlinear compensated system. The block diagram of the overall compensated system is shown in Figure 2.

5. Simulation and Experimental Results

The proposed controller was simulated and implemented on the experimental two-link flexible joint robot manipulator system of Figure 1. A sampling rate of 1 Khz was used for the simulation as well as the experiment. The system parameters are given in Table 1 [1]. These parameters were obtained using I-DEAS solid modeler and Sine Sweep identification techniques. The desired characteristic frequency is 35.0 rad/s and the desired damping ratio is 1.0. Two tests were conducted.

5.1. Test 1: Noise-free Measurements

In this first test, the measured variables are the links and the motors angular positions and velocities. All measurements are assumed to be noise free. The links angular acceleration and jerks are computed using the analytical expression of the coordinate transformation. Figures 2 and 3 show the simulated angular position and velocity errors. These figures demonstrate the good tracking performance provided by the FL controller. Figure 4 and 5 show the experimental angular position and the angular velocity tracking errors for both joints, respectively. As can be seen from these figures, the maximum position error is less than 0.018 rad while the maximum velocity error is less than 0.2 rad/sec. However, the steady-state position error is not equal to zero. This imperfection is mainly due to the effects of measurement noises which were not taken into account in this first experiment.

5.1. Test 2: Noisy Measurements

In this second test, the controller was designed taking into consideration the measurement noise and assuming that only measurements of the links angular positions and velocities are available. The simulations results are shown in figures 6-11. Figures 6 and 7 show plots of the measured, estimated, and actual angular positions and velocities of the two joints. These figures indicate that excellent tracking performance is achieved with the proposed controller. The system reaches the desired trajectory within 1.5 sec without any overshoot, despite the presence of measurement noise. Figures 6 and 7 show plots of the actual and the estimate of the motor angular positions and velocities. These figures demonstrate the high tracking performance of the new state estimator presented in Sections 3 and 4. Thus, with the proposed control approach, the use of sensors that measure the motor dynamics can be omitted without affecting the performance of the FL-based controller. This controller is being implemented and the results will be available in the near future.

6. Conclusions

In this paper, an optimal nonlinear position tracking controller is designed for an experimental robot manipulator, using the concept of FL and LQG/LTR techniques. The proposed control approach has the advantage of achieving excellent tracking performance in presence of measurement noise. The new approach for computing the nonlinear state estimate was shown to be very efficient in reducing the cost and the complexity of the controller design. Future work will consider the problem of position/force control of flexible joint robot manipulators.

Acknowledgement

The authors wish to acknowledge the support provided by the Institute for Robotic and Intelligent Systems (IRIS) and the Natural Sciences and Engineering Research Council of Canada (NSERC) through research grants to Dr. Hoda A. ELMaraghy. The contribution made by Dr. Atef T. Massoud, particularly to the experimental results, are appreciated.

References

[1] Masoud A. T. M., " Motion and Force Control of Flexible Joint Robot Manipulators", Ph. D. Dissertation, McMaster University, Hamilton, Canada, 1994.
[2] Lin S. H., et al., "Control of multi-link robot manipulator with compliant joints," Winter Annual Meeting of the ASME, San Francisco, DSC-vol 14, pp.299-307, 1989.
[3] Tomei P., et al., " An approach to the adaptive control of robots elastic at joints," Proc. IEEE Conf. on Robotics and Automation, San Francisco, pp.552-558, 1986.
[4] Ficolo A., et al., " A singular perturbed approach to the dynamic control of elastic robots," Proc. 21st Ann. Allerton Conf. on Comm., Cont., and Computer, University of Illinois at Urbana Champaign, October 1983.
[5] Mrad F. T., and Ahmad S., "Adaptive control of flexible joint robot using position and velocity feedback," Int. Journal of Control, Vol. 55, No. 5, pp.1255-1277, 1992.
[6] Zaki A. S. A., Modeling and Control of a 3D Manipulator with Flexible Links, Ph.D. Dissertation, University of Western Ontario, London, Ontario, Canada, 1994.
[7] Jankowski K. P., and ElMaraghy, H. A.," Inverse dynamics and feedforward controllers for high precision position/force tracking of flexible joint robots," Robotica, Vol. 12, pages 227-241, 1994.
[8] Spong M. W., Khorasani K., and Kokotovic P. V., " An Integral Manifold Approach to the feedback control of flexible joints robots," IEEE Trans. on Robotics and Automation, Vol. 3, No. 4, pp. 291-300, 1988.
[9] Isidori A., "Nonlinear Control Systems", Springer-Verlag, London, England, 1995.
[10] Maciejowski J. M., " Multivariable Feedback Design," Addison Wesley, England 1996.
[11] Jazwinski A. H. " Stochastic Process and Filtering Theory," Academic Press, NY, 1970.

Parameters	I-DEAS	Sine Sweep
l_1 (m)	0.400	
l_2 (m)	0.350	
d_1 (Kg.m^2)	2.110	2.087
d_2 (Kg.m^2)	0.223	0.216
d_3 (Kg.m^2)	0.085	0.084
b_1 (N.m.s/rad)		2.041
b_2 (N.m.s/rad)		0.242
b_{m1} (N.m.s/rad)		1.254
b_{m2} (N.m.s/rad)		0.119
k_1 (N.m/rad)	198.49	125.56
k_2 (N.m/rad)	51.11	31.27
I_{m1} (Kg.m^2)	0.1226	0.1224
I_{m2} (Kg.m^2)	0.017	0.0168

Table 1. Experimental robot parameters from design and Sine Sweep identification

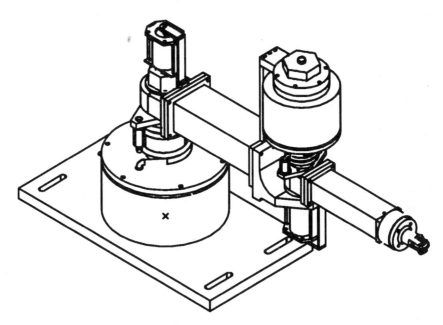

Figure 1. Experimental robot manipulator system

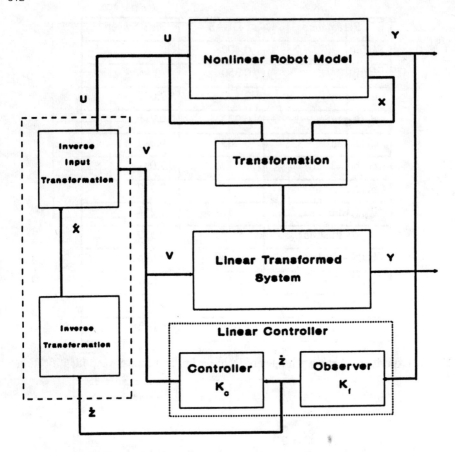

Figure 2. Block diagram of the overall compensated system

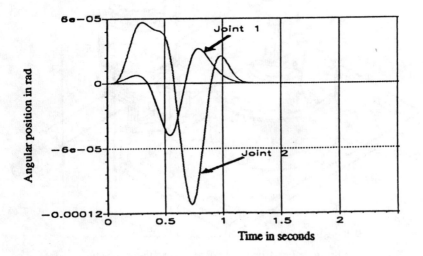

Figure 3. Experiment 1: Angular position error (simulation)

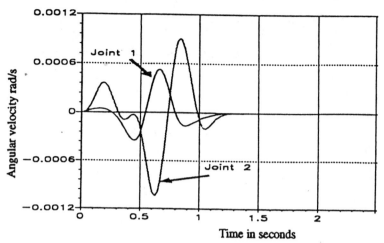

Figure 4. Experiment 1: Angular velocity error (simulation)

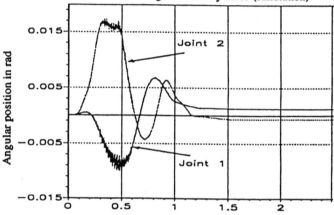

Figure 5. Experiment 1: Angular position error (experimental)

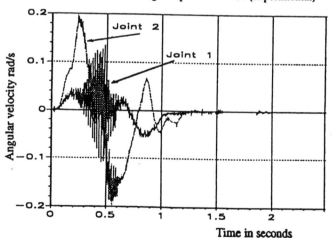

Figure 6. Experiment 1: Angular velocity error (experimental)

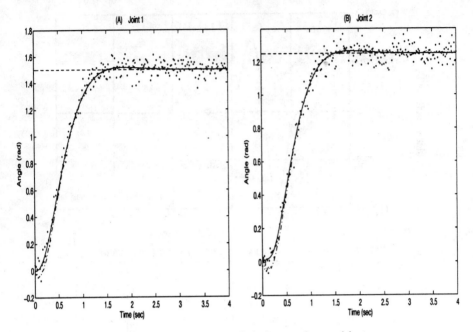

Figure 7. Experiment 2: Joint's angular positions
--- Desired, —— Actual, -.-. Estimate, ... Measured

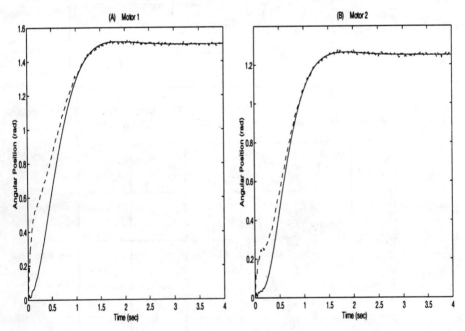

Figure 8: Experiment 2: Motor's angular positions
—— Actual, ----- Estimate

Motion Control of Tendon Driven Robotic Fingers Actuated with DC Torque Motors: Analisys and Experiments

G. M. Prisco, D. Madonna, M. Bergamasco
PERCRO
Simultaneous Presence, Telepresence and Virtual Presence
Scuola Superiore S. Anna, PISA, Italy
gmprisco@sssup.it

Abstract: This paper deals with the joint position control of a robotic finger designed to test the force feedback performance of an hand controller. The robotic finger is actuated with 3 DC motors located on the base link, which drive the joints by means of bidirectional cable transmissions. The focus of the work is on the analisys of the influence of the elastic cable transmission on the robot dynamics and control. Experimental data are presented to characterize the system behaviour and verify the performance of different control laws. Guidelines for the design of robotic joints actuatated with cables and DC motors are deduced.

1. Introduction

Shortcomings in the actuation system hinder many interesting applications of industrial and service robotics. It is therefore strategic to make a clever use of the actuators to drive a robot. Many studies on kinematics optimization applied to serial and parallel robots have been guided by this general criteria of saving on limited actuators performance.

The design of robots with tendon transmission is another way to go to overcome the limitations of state-of-the-art actuator technologies. In fact tendon transmissions allow to locate the actuators away from the joint axis; for instance it is possible to drive a robot placing all the actuators on the base link, i.e. on ground. In this way the constraints on the weight and volume of the actuators are greatly relaxed and, at the same time, the moving mass of the robot can be reduced and its payload in turn results increased.

Tendon transmission though introduces new phenomena which have to be understood for the design and control of such systems. First of all, the presence of elastic tendons connecting the motor inertia to the link inertia determines a resonant mode at a frequency usually lower than the one associated to the structural elasticity of the links. Secondly, tendon transmissions in multi-DOF robots with serial kinematics cause a phenomenon, known as coupling between joint torques and motor torques, not present at all in robots with actuators on

Figure 1. *Photo of the testbed robotic finger with the Hand Force Feedback system mounted aside*

the joints. The torque at a joint is not determined exclusivly by the actuator directly connected to the joint, but also by the actuators connected to the tendons passing the joint to reach the successive links of the robot.

A correct treatment of these two issues, elasticity and coupling, is of key importance in the understanding of many relevant properties of such robots, such as its mechanical bandwidth, the disturbances introduced by the control of a joint on the other joints, the impact of friction in the motors on the joint motion.

In this paper we analise of the Testbed Robotic Finger, a tendon driven 3 DOF serial robot actuated with DC motors built in the PERCRO lab [1] (see Figure 1). Experimental work has beed done to assess its performance. At the same time we have to outline a methodology for its analisys which remains valid for a whole class of tendon driven, electrically actuated robots.

Our idea is that only relating quantitatively the final performance to the mechanical characteristics of the robot, it is possible to determine which are really the limit performances achievable with this class of robots. At the same time, gathering indications for their mechanical design and their control development, it is possible to push even further such limits.

1.1. The TRF - Testbed Robot Finger

The TRF - Testbed Robotic Finger, analized in the present paper, has been designed on purpose to test the performances of an anthropomorphic hand controller for teleoperation and virtual environment applications, the Hand Force Feedback system described in [1]. The HFF is composed of four finger exoskeletons which the user can wear as a glove; each HFF finger is controlled to exert desired normal forces on the three phalanxes of the user finger so that the HFF system altogether is able to apply forces on the user hand which

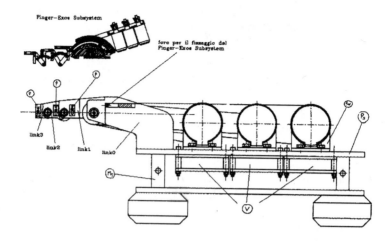

Figure 2. *Drawing of the testbed robotic finger*

approximate those occurring during manipulative procedures.

The problem of the quantitative evaluation of the haptic feedback realizable with the HFF ([1] [5]) originated the idea of building a robotic finger which could "wear" the HFF fingers just like a human finger does. In this way the actual forces fed-back by the HFF could be monitored while the robotic finger is executing a set of test movements. The interesting part of such approach is that the performance of the hand controller can be measured in any working condition and in a repeatable way, avoiding all the complicacies associated to having a user involved in the measurement. It is possible to investigate the dependency of performances of the HFF from the human finger posture, the non-linear phenomena taking place during large motions, the frequency dependency of the dynamic backdrivability and of the force control responce at each phalanx; all the measurements can be repeated to compare the effects of different controllers of the HFF.

Figure 1 shows the operating conditions of the testbed robotic finger. One of the HFF finger exoskeletons is secured to the base plate of the TRF and connected to the links of the TRF. Both isometric and isotonic tests are executed. For instance, during the basic isotonic tests the finger exoskeleton is commanded to exert constant forces at each phalanx, while the TRF is controlled to impose sinusoidal motions to the finger exoskeleton joints. The objective of this test is to record the steady state force error as a function of the frequency of motion of the TRF.

The TRF is required to track sinusoidal waves of 45° of amplitude up to 8 Hz and to have a positioning bandwidth larger that 10 Hz for small movements. In fact the voluntary human movement have frequency contents up to 5 Hz [6] and only large joint motions permit to elicit non-linear phenomenon in the HFF functioning. Such performances are very demanding for the actuation system especially in terms of maximum and continuos torque required to the motor.

A solution however has been found with the employment of a tendon trans-

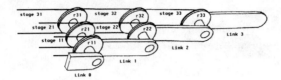

Figure 3. *Scheme of transmissions with nomenclature*

mission system of the type, called 2N, which makes use of a pair of opposed tendons to actuate each joint with three DC motors located on the fixed base. This choice has allowed to find actuators meeting the requirements of peak and continuous torque without having to be worried about constraints on their weight or volume (see Figure 2).

2. Dynamic model formulation

In Figure 3 the structure of the TRF transmissions is depicted together with the nomenclature adopted for its description. The transmission connecting motor i with link i is indicated with index i. A second index is used to indicate the stages in which each transmission is divided: the stage $(i,1)$ is composed of the pair of opposed cables connecting the motor pulley $(r_{i,0})$ with the pulley $r_{1,1}$ pivoted on the axis of joint 1. The cables belonging to the transmissions 2 and 3 are routed over idle pulleys $r_{3,1}$, $r_{3,2}$ and $r_{2,1}$ to reach the respective joints. The idle pulleys, which are mounted on ball bearings, have the task of guiding the cables so, that independently of the finger posture, the cable tensions have always constant lever arm with respect to the joint axis. More general arrangement of tendons, for instance as those present in the human finger, do not have this property which, as we will see, ensures the linearity of the coupling effect induced by the cables.

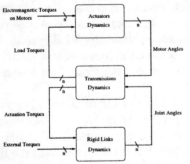

Block diagram of the dynamic model

For our analisys, the robot can be divided in three interacting subsystems: the actuators, i.e. the three DC motors, the transmission system, composed of the cables and the idle pulleys, and the robot links. Figure 2 shows the interaction between these three subsystems.

2.1. Actuators

The actuators are current controlled DC motors. The input command variable is the electromagnetic torque $\tau_{em,i}$ applied to the rotor. This is the case when a power amplifier is used to control the current in the motor windings. The dynamic model for the actuators takes as input also the load torque $\tau_{load,i}$ exerted by the transmission on the rotor and offers as output the motor angular position $q_{i,m}$. Indicating with $J_{i,m}$ and $b_{i,m}$, for $i = 1, 2, 3$, respectively the motor inertia and the viscous coefficient of the motor the actuator equations are:

$$J_{1,m}\ddot{q}_{1,m} + b_{1,m}\dot{q}_{1,m} = \tau_{em,1} - \tau_{load,1}$$
$$J_{2,m}\ddot{q}_{2,m} + b_{2,m}\dot{q}_{2,m} = \tau_{em,2} - \tau_{load,2}$$
$$J_{3,m}\ddot{q}_{3,m} + b_{3,m}\dot{q}_{3,m} = \tau_{em,3} - \tau_{load,3}$$

2.2. Links

Our analisys is based on the hypothesis that the robot links are ideally rigid. The classical lagrange equations for serial manipulators can be written for the rigid links:

$$\mathbf{M(q)\ddot{q} + h(q,\dot{q}) + g)q) = \tau_{act} - \tau_{ext}} \quad (1)$$

The torques acting on the joints are divided in two different contributions. The external torques τ_{ext} are the effect of forces and torques applied on the manipulator by the environment. The actuation torques τ_{act} are the effect at the joints of the tendon tensions. In robots mounting the motors on the joint axes, such actuation torques are simply the torques exerted by the motors.

2.3. Transmission

The transmission system interconnects the actuators with the links. In our formulation of the dynamics equation, it takes as input the motor and joint angles $q_{i,m}$ and $q_{i,i}$ and furnishes the load torques and the actuation torques.

The cables are modelled as a linear spring and a linear dashpot in parallel. This is an improvement with respect to considering the tendons ideally rigid; it allows to take into account at least the linear elastic effects and predict the frequencies and damping of the elastic resonant modes. The physical behaviour of cables, especially those composed of many strands, is much more complex including non linear and hysteretic behaviours and delays in tension propagation.

Indicating with $t_{i,j}$ the sum of the tensions in the pair of opposed tendons of stage i, j, the idle pulleys dynamics is described by the equations:

$$J_{2,1}\ddot{q}_{2,1} = r_{2,1}(t_{2,1} - t_{2,2})$$
$$J_{3,1}\ddot{q}_{3,1} = r_{3,1}(t_{3,1} - t_{3,2})$$
$$J_{3,2}\ddot{q}_{3,2} = r_{3,2}(t_{3,2} - t_{3,3})$$

The actuation torques τ_{act} are computed from the stage tensions with the relations:

$$\tau_{act,1} = r_{1,1}t_{1,1} + r_{2,1}t_{2,2} + r_{3,1}t_{3,2}$$

$$\tau_{act,2} = r_{2,2}\, t_{2,2} + r_{3,2} t_{3,3}$$
$$\tau_{act,3} = r_{3,3}\, t_{3,3}$$

These equations can be interpreted as equations of static balance at the joints. They are derived in the general case of a n DOF robot in [2]. They show the nature of the tendon induced coupling: the actuation torque at joint i depends linearly by the tensions in all the stages passing link i.

The transmission dynamic model is completed by the equations relating the stage tensions $t_{i,j}$ to the angular rotation of the joints, of the idle pulleys and of the motor pulleys. The tensions in the first stage of each transmission is function of the difference of rotation of the motor pulley and the pulley $r_{i,1}$. Letting $k_{i,j}$ and $b_{i,j}$ be the total spring and viscous coefficients of stage (i,j), n_i the reduction ratio of the gearbox mounted on motor i we have:

$$t_{1,1} = \tau_{load,1}\frac{n_1}{r_{1,m}} = k_{1,1}\left(\frac{r_{1,0}}{n_1} q_{1,m} - r_{1,1}\, q_{1,1}\right) + b_{1,1}\left(\frac{r_{1,0}}{n_1} \dot{q}_{1,m} - r_{1,1}\, \dot{q}_{1,1}\right)$$

$$t_{2,1} = \tau_{load,2}\frac{n_2}{r_{2,m}} = k_{2,1}\left(\frac{r_{2,0}}{n_2} q_{2,0} - r_{2,1}\, q_{2,1}\right) + b_{2,1}\left(\frac{r_{2,m}}{n_2} \dot{q}_{2,m} - r_{2,1}\, \dot{q}_{2,1}\right)$$

$$t_{3,1} = \tau_{load,3}\frac{n_3}{r_{3,m}} = k_{3,1}\left(\frac{r_{3,0}}{n_3} q_{3,0} - r_{3,1}\, q_{3,1}\right) + b_{3,1}\left(\frac{r_{3,m}}{n_3} \dot{q}_{3,m} - r_{3,1}\, \dot{q}_{3,1}\right)$$

the tensions in the stages $(2,2)$, $(3,2)$ and $(3,3)$, which are stages solidal with link 2 and 3, are instead a function not only of the rotation of the two pulleys between which each stage is stretched, but are also dependent on the rotation of the joint of the link which carries the transmission stage. For instance, when the idle pulleys and the joints 2 and 3 do not rotate, the tension in the stages $(2,2)$ and $(3,2)$ increases/decreases if joint one is rotated. The equations describing this are:

$$t_{2,2} = k_{2,2}\left[r_{2,1}(q_{2,1} - q_{1,1}) - r_{2,2}\, q_{2,2}\right] + b_{2,2}\left[r_{2,1}(\dot{q}_{2,1} - \dot{q}_{1,1}) - r_{2,2}\, \dot{q}_{2,2}\right]$$
$$t_{3,2} = k_{3,2}\left[r_{3,1}(q_{3,1} - q_{1,1}) - r_{3,2}\, q_{3,2}\right] + b_{3,2}\left[r_{3,1}(\dot{q}_{3,1} - \dot{q}_{1,1}) - r_{3,2}\, \dot{q}_{3,2}\right]$$
$$t_{3,3} = k_{3,3}\left[r_{3,2}(q_{3,2} - q_{2,2}) - r_{3,3}\, q_{3,3}\right] + b_{3,3}\left[r_{3,2}(\dot{q}_{3,2} - \dot{q}_{2,2}) - r_{3,3}\, \dot{q}_{3,3}\right]$$

The model presented has the advantage of being physically based and modular so that it is possible to add to it other features. For instance friction torques can be added to the equation of motion of the motors, of the idle pulleys and of the links.

2.4. Model semplifications

The dynamic model for the transmission can be usefully simplified in some cases.

The first case is when the tension in the different stages of the same transmission can be approximated as constant i.e. $t_{2,1} \simeq t_{2,2}$ and $t_{3,1} \simeq t_{3,2} \simeq t_{3,3}$. This happens if the friction torque on the idle pulleys is negligible and the torques due to the idle pulleys inertias are small. It is then possible to define t_1, t_2 and t_3 respectively the (constant) differential tension of transmission 1, 2 and 3; the total elasticities of each transmission k_1, k_2 and k_3 can be estimated as the serial connection of the elasticities of each stage. The transmission dynamic equations become:

$$t_1 = k_1 \left(\frac{r_{1,0}}{n_1} q_{1,m} - r_{1,1} q_{1,1} \right)$$

$$t_2 = k_2 \left(\frac{r_{2,0}}{n_2} q_{2,m} - r_{2,1} q_{1,1} - r_{2,2} q_{2,2} \right)$$

$$t_2 = k_3 \left(\frac{r_{3,0}}{n_3} q_{3,m} - r_{3,1} q_{1,1} - r_{3,2} q_{2,2} - r_{3,3} q_{3,3} \right) \quad (2)$$

and:

$$\begin{pmatrix} \tau_{act,1} \\ \tau_{act,2} \\ \tau_{act,3} \end{pmatrix} = \begin{pmatrix} r_{1,1} & r_{2,1} & r_{3,1} \\ 0 & r_{2,2} & r_{3,2} \\ 0 & 0 & r_{3,3} \end{pmatrix} \begin{pmatrix} t_1 \\ t_2 \\ t_3 \end{pmatrix} \quad (3)$$

In this case the relationship between motor load torques τ_{load} and joint actuation torques τ_{act} is:

$$\tau_{act} = \mathbf{S}^T \tau_{load} \quad (4)$$

with:

$$\mathbf{S} = \begin{pmatrix} \frac{n_1 r_{1.1}}{r_{1.0}} & 0 & 0 \\ \frac{n_2 r_{2.1}}{r_{2.0}} & \frac{n_2 r_{2.2}}{r_{2.0}} & 0 \\ \frac{n_3 r_{3.1}}{r_{3.0}} & \frac{n_3 r_{3.2}}{r_{3.0}} & \frac{n_3 r_{3.3}}{r_{3.0}} \end{pmatrix} \quad (5)$$

$\mathbf{S}$ is the constant structure matrix which is used in most cases to describe the coupling effect of tendon transmissions.

A further simplification of the model is possible if the tendon elastic constants k_1, k_2, k_3 go to infinity. In this case Equations 2 become:

$$\mathbf{q}_m = \mathbf{S} \, \mathbf{q} \quad (6)$$

2.5. Single joint model

A physical model for a single joint which takes into account the transmission elasticity, but of course neglects the coupling, is represented in Figure 4. The mass on the left represents the rotor and gearbox inertia, the mass on the right the inertia at the joint when the other joints are blocked; the spring and damper stand for the transmission.

Defining:

$$N = n_i \cdot \frac{r_{i.i}}{r_{i.0}} \quad J_m = J_{i,m} \quad J_2 = \frac{J_{link}}{N^2} \quad (7)$$

$$k_t = \frac{2 k_i r_{i.0}^2}{n_i^2} \quad b_t = \frac{2 b_i r_{i.0}^2}{n_i^2} \quad (8)$$

and

$$\theta_m = q_{i,m} \quad \theta_2 = q_{i,i} \cdot n_i \cdot \frac{r_{i.i}}{r_{i.0}} \quad (9)$$

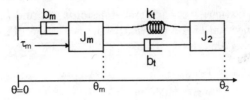

Figure 4. *Simplified model of a joint with elastic transmission*

the transfer function between the electromagnetic torque τ_m and the motor angle θ_m is:

$$\frac{\theta_m}{\tau_m}(s) = \frac{J_2 s^2 + b_t s + k_t}{s\left[J_m J_2 s^3 + (b_m J_2 + J_m b_t + J_2 b_t)s^2 + (J_m k_t + J_2 k_t + b_m b_t)s + b_m k_t\right]} \quad (10)$$

3. Analisys of the TRF Transmissions Dynamics

The values for all the parameters in the presented dynamic model of the TRF have been estimated or measured and are reported in Appendix.

The TRF model results linear if the friction effects are neglected and the link dynamic equations are linearized; under these hypothesis the poles of the systems can be computed and furnish an estimation of the resonant modes of the transmission. In the following, the link dynamic equation are considered linearized at zero joint velocities and joint angular position equal to $\pi/2$ for all the joints.

The poles associated to the slow motor dynamics are respectively at 0.63, 0.59 and 1.02 Hz; the pairs of complex poles associated to the principal oscillation mode of each transmission are respectively at 92, 150 and 216 Hz; three more pairs of high frequencies complex poles are forecast by the model at frequencies 1205, 1462 and 2820 Hz. These are the oscillating modes associated to the idle pulleys; in our case they are clearly neglegible.

For the estimation of the three principal elastic modes, the model simplified as indicated in Section 2.4 yields quite accurate estimations of the three principal elastic modes (see Table 1 Model #2), just a little bit too high because the inertias of the idle pulleys are neglected. If the inertia of the idle pulleys are associated to the link inertias, the estimations are on the other side a little lower (see Table 1 Model #3).

The estimations of the system poles using the single joint model presented in 2.5 are affected by a considerable error (see Table 1 Model #4). In particular they result lower, showing that the transmission coupling stiffens the robot.

4. Preliminary experiments

The presence of gearboxes on the motor shafts required preliminary experiments to characterize the behaviour of the motor-gearbox subsystem. These experiments have been carried out detaching the first stage of the tendon transmissions from the pulley $r_{i,0}$ mounted on the output shaft of the gearboxes.

	Model #1	Model #2	Model #3	Model #4
Motor poles	0.63,0.59,1.02	0.7,0.59,1.03	0.64,0.58,0.98	0.62,0.54,0.64
Elastic poles	92,150, 216	92,151,227	92,143,206	78,117,206

Table 1. *Estimation of poles from the models: #1 complete model; #2 simplified model without idle pulleys inertias; #3 simplified model with idle pulleys inertias; #4 single joint models*

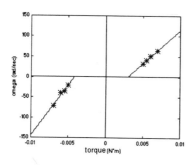
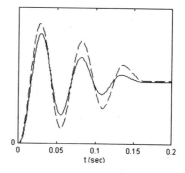

Figure 5. Estimation of bm and Jm from experimental data

With respect to the behaviour of the motor with no gear-box, the static torques resulted increased and the presence of position dependent friction and temperature dependent viscosity was noticed. These observations convinced us not to proceed with friction identification and compensation. Instead just a rough estimate of the average viscous and static friction has been made. This allowed to test with simulations the robustness of different controllers structure against friction.

In Figure 5 it is reported the average angular motor shaft velocity reached in correspondence of different constant values of electromagnetic torques by motor #3. The viscous component of friction is estimated as $5.5 \cdot 10^{-5}$ Nm/s and the coulomb friction as $3.5 \cdot 10^{-3}$ Nm.

Two other paramenters required estimation from experimental data. The motor torque constant, whose nominal value was imprecise, as been measured as in [10]. The additional inertia on the motor shaft due to the presence of the gearbox is not declared by the constructor and has been estimated from the closed loop step response of the motor-gearbox system with constant position error feedback gain (see Figure 5).

5. Controller design

The TRF is sensorized with motor angular position sensors under the hypothesis that joint position control can be implemented using the estimates for the joint angular positions provided by:

$$\mathbf{q} = \mathbf{S}^{-1} \mathbf{q_m} \tag{11}$$

This equation, which is the inverse of 6, is exactly valid only if the transmissions are in equilibrium and no external forces are applied to the TRF. The TRF

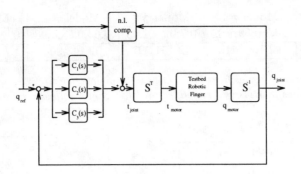

Figure 6. *Structure of the TRF joint controller*

transmissions have been designed with an overall stiffness which guarantees that, in every forecast operating condition, the errors committed using 11 to estimate the joint angles are more than one order of magnitude smaller than the desired accuracy for joint measurements, estimated as $3.5 arcmin$ [1].

Under the hypothesis of constant tension in the different stages of each transmission, which leads to the TRF model simplification presented in Section 2.4, the relationship between motor load torques τ_{load} and joint actuation torques τ_{act} is given by the static, linear equation $\tau_{act} = \mathbf{S}^T \tau_{load}$ and the controller can easily compensate for it.

The controller structure which has been adopted is shown in Figure 6. Under the aforementioned hypothesis, the matrices $\mathbf{S}^T$ and $\mathbf{S}^{-1}$ introduced in the control law get rid of the transmission coupling effects. Then a classical approach for joint control can be followed: a compensation of nonlinear effects, such as gravity and rigid link dynamics plus independent linear joint controllers. In the presented experiments, gravity compensation has been implemented using the parameters reported in Appendix.

The digital control implementation has a sampling frequency of 1 KHz and introduces one cycle of delay for control law computing. The control design has been carried out in continuous time on the single joint model 10, adding to the plant a 1.5 ms finite delay to take into account the presence of the sampling and hold and of the computing delay. The controllers were tested with simulations on the complete TRF model presented in Section 2.

A critical point in the design of controllers using model 10 is that an overestimation in the model of the damping of the elastic poles (which is associated to the value of b_t) can lead to the synthesis of controllers which result unstable when tested on the TRF.

For instance, the PD gains in Table 2 are stable controllers for plant if a 0.1 damping factor for the elastic poles is assumed; nonetheless, applied to the TRF, they cause sustained oscillations as shown in Figure 7.

It is worth noticing that the frequency of these oscillations is 90 Hz in good accordance to the prediction for the elastic poles frequency of the first transmission given in Table 1. Moreover, the presence of the oscillation on all 3 joint signals shows that the decoupling realized with relation 11 doesn't work in this condition.

Link	k_p	k_d	B_{cloop}	Margins
#1	0.2	$2.2 \cdot 10^{-3}$	30 Hz	47°, 10 dB
#2	$3 \cdot 10^{-2}$	$3.3 \cdot 10^{-4}$	37 Hz	47°, 10 dB
#3	$3 \cdot 10^{-3}$	$3.3 \cdot 10^{-5}$	38 Hz	47°, 8 dB

Table 2. *Unstable PD gains*

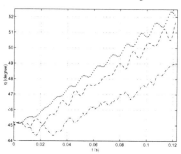

Figure 7. *Unstable behaviour of PD controllers.* $q1 \cdots$, $q2 - -$, $q3 - \cdot -$

As can be seen with Bode and Nyquist diagrams, the instability of the elastic poles is a consequence of the finite delay of 1.5 ms, which increases the phase delay at the frequency of the elastic poles and of the resonance peak introduced by the elastic poles in the open loop system transfer function.

Keeping an eye on these two issues and reducing the bandwidth required for the first joint, PD controllers (see Table 3) which are actually stable on the TRF have been synthesized.

From experiments (see Figure 8), the performance of these controllers in tracking slow sinusoidal reference signals has proved to be very sensitive to friction. In order to gain more insight in the problem, the static and Coulomb friction simulation model proposed in [8] has been included in the TRF model in order to represent the effects of the friction in the motors and gear-boxes noticed in Section 4. The simulation results (see 8), which are quite similar to real data, confirm that the tracking problem of the PD controller is due to the static and Coulomb friction in the motor and gearbox.

Comparing with simulations the performance of sinusoidal waves tracking of different types of compensators we have found that compensators of the

Link	k_p	k_d	B_{cloop}	Margins
#1	$2 \cdot 10^{-2}$	$6.7 \cdot 10^{-4}$	10 Hz	55°, 20 dB
#2	$1 \cdot 10^{-2}$	$1.7 \cdot 10^{-4}$	20 Hz	48°, 14 dB
#3	$2 \cdot 10^{-3}$	$2.2 \cdot 10^{-5}$	28 Hz	41°, 10 dB

Table 3. *Stable PD gains*

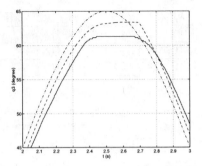

Figure 8. *Tracking performances of q3 of PD at 0.5 Hz. Reference $\cdots$, simulated $--$, real $-\cdot-$*

Link	k	$z_1 = z_2$ rad/s	B_{cloop}	Margins
#1	$15 \cdot 10^{-2}$	15.08	10 Hz	53°, -19 dB
#2	$12 \cdot 10^{-2}$	30.78	20 Hz	42.7°, -14 dB
#3	$4.3 \cdot 10^{-2}$	40.8	28 Hz	42°, -14.2 dB

Table 4. *Paramenters of the compensators*

form:
$$C(s) = \frac{(s - z_1)(s - z_2)}{s} \qquad (12)$$

which have high gain at low frequencies and zero steady state error to ramp inputs, are particularly robust to friction. Compensators of this type have been sinthetized for each joint (see Table 4). The real tracking performance of slow and fast sinusoids of joint 3, reported is Figure 9, confirms the simulation results.

When simultaneous fast motions of the three links are commanded, the disturbances due to dynamic forces increase the tracking error, as can be seen in 10. The TRF model presented in Section 2, including the nonlinear terms of Equation 1 evaluated with the parameters in Appendix yields outputs which are very similar to the experimental measurements (see for example Figure 10); this confirms the possibility of improving the tracking performance with a feedforward compensation of nonlinear rigid links dynamics.

6. Design guidelines

The experience gained in the design of controllers for the TRF, shows that it is not difficult to obtain the desired closed loop performance for joint position tracking with standard control tecniques, provided that the crossover frequency calculated from the single joint model 10 is matched to the desired closed loop bandwidth and that the frequency of the elastic poles is sufficiently higher.

Revising the methodology followed for the TRF design, we have noticed that first of all the mechanical design of the links and of the routing of cables has been carried out. As second step the motors have been chosen to meet

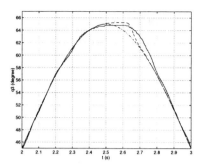
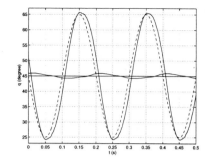

Figure 9. *Tracking performances of q3 of compensation network at 0.5 Hz and 5 Hz. Reference* $\cdots$, *simulated* $--$, *real* $-\cdot-$

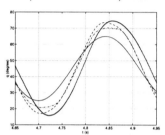
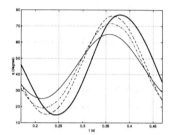

Figure 10. *Tracking performances of compensation network at 3.5 Hz; real (left) and simulated (right); q1* $\cdots$, *q2* $--$, *q3* $-\cdot-$

the specification of peak torque and the gear ratio has been determined as the one giving maximum joint acceleration. The maximum temperature reached by the motors during worstcase functioning has been checked. In these choices the final dynamic characteristics of the joint were not explicitly considered.

In the following, we show a design methodology to choose the motor, the gear-box reduction ratio and the cable stiffness, which allows the designer to take into account the desired final dynamical performance and at the same time to consider different design alternatives.

The single joint model represented by Equation 10 has a Bode diagram which qualitatively looks like that in figure 11.

In the approximation that $b_t \simeq 0$ and $b_m \simeq 0$, the crossover frequency and the elastic poles frequency are computed from 10 as:

$$\omega_{el}^2 = \frac{k_t}{J_2} \cdot \left(1 + \frac{J_2}{J_m}\right)$$

$$\omega_{cr}^2 = \frac{1}{2} \left[\frac{1+k_t}{J_m} + \frac{k_t}{J_2} - \sqrt{\left(\frac{1+k_t}{J_m} + \frac{k_t}{J_2}\right)^2 - \frac{4k_t}{J_m J_2}} \right]$$

(13)

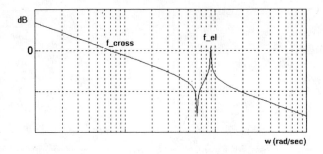

Figure 11. *Magnitude Bode diagram of simplified model*

A practical approximation for ω_{cr} when k_t goes to infinity, i.e. $\omega_{el} \gg \omega_{cr}$ is:

$$\omega_{cr}^2 = \frac{1}{J_m + J_2} = \frac{N^2}{J_l + J_m \cdot N^2} \tag{14}$$

The peak joint acceleration, which is a specification coming directly from the application, is related to the peak torque requested to the motor by:

$$\dot{\omega}_{l,peak} = \frac{\tau_{m,peak}}{N(J_2 + J_m)} = \frac{\tau_{m,peak} N}{J_{link} + J_m N^2} \tag{15}$$

The motor inertia, the cable stiffness, the motor peak torque and the gear box reduction ratio are related to ω_{cr}, ω_{el} and $\dot{\omega}_{l,peak}$ by the equations:

$$\frac{J_m}{k} = \frac{2\, r_{i,i}^2}{J_l \omega_{cr}^2 \omega_{el}^2} \tag{16}$$

$$N^2 = \frac{\omega_{cr}^2 J_{link}}{1 - \omega_{cr}^2 J_m} \tag{17}$$

$$\tau_{m,peak} = \frac{\dot{\omega}_{l,peak} \left(J_{link} + J_m N^2 \right)}{N} \tag{18}$$

The procedure of analytic design starts choosing desired values for ω_{cr}, ω_{el} and $\dot{\omega}_{l,peak}$. The designer, using equation 16, chooses the values for J_m and k. The Equations 17 and 18, given the chosen value of J_m, provide values for N and $\tau_{m,peak}$. Of course in the end it is necessary to check if a motor with the calculated J_m and $\tau_{m,peak}$ really exists and is compatible with the application from the point of view of thermal dissipation.

The design can be eased drawing a chart that helps in finding a motor compatible with the design specifications. We will present this graphical method with an example: the re-design of the motor, gearbox and cable of the first joint of the TRF which currently has the following characteristics:

$$f_{cr} = 35 \text{ Hz} \quad f_{el} = 71 \text{ Hz} \quad \dot{\omega}_{l,peak} = 1984 \text{rad}/s^2 \tag{19}$$

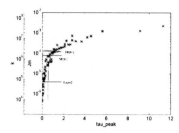 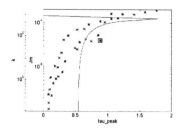

Figure 12. *Chart to design joint 3 actuation/transmission system*

As noticed in Section 5, the present value of f_{el} limits the achievable position control bandwidth, let's then set the following new specifications:

$$f_{cr} = 13.5 \text{ Hz} \quad f_{el} = 270 \text{ Hz} \quad \dot{w}_{l,peak} = 1984 \text{rad}/s^2 \qquad (20)$$

On the design chart 12, a set of existing DC torque motors are represented as points in a Cartesian plane so that on the x and y axis respectively the peak motor torque and its rotor inertia can be read. On such plane the curve represented by equation 18 is drawn: each point of the curve is associated to a value of N (these values can be read on the horizontal lines in 12). Since k and J_m are proportional to each other as specified by equation 16, a second y-axis can be drawn to let the designer immediately read the values of k associated to the values of J_m according to 16.

On the chart, each point of the curve represents a choice for the parameters k, j_m τ_{peak} and N, whose values can be read directly on the chart, which satisfies Equations 16 17 18. The points on the right side of the curve represent motors which satisfy the specifications 20. In our case the data reported in Figure 12 are taken from the Vernitron range of DC brushed torque motors [13].

In Figure 12 a possible design choice is evidenced: it corresponds to the model the 1937V -147 of catalog [13], the value of N is 2.37, the value of k is $8.35 \cdot 10^5$ N/m, corresponding to a cable of 1.6 millimiter of diameter. The proposed design has the additional advantage of eliminating the need for a gearbox diminishing the friction in the system.

7. Conclusions

The elasticity and coupling introduced by cable transmission have been theoretically analized; this has led to the design of joint position controller for the Testbed Robotic Finger whose performance has been verified experimentally. The work outlines a methodology for the dynamical analisys and control which we is applicable to a whole class of tendon driven, electrically actuated robots. It is shown how the choice of motors, gearboxes and cable stiffness can be based on the desired joint tracking performances.

The presented work is a first step towards the improvement of the performance achievable with serial manipulators with actuators on the base link and bidirectional cable transmission.

8. Acknowlegemets

The authors wish to thank Fabio Salsedo, Carlo Alberto Avizzano and Lucio De Paolis which have been involved respectively in the TRF mechanical design, control analysis and control board developments.

9. Appendix

The TRF dynamic model presented in section 2 has been simulated using the values for the parameters reported in the following. The rigid link dynamics has been evaluated with the parameters:

Link	Length	Mass	Center of mass	Baricentral Inertia (I_{zz})
#1	0.046	0.147	{0.0145 , -0.0040}	$49.4 \cdot 10^{-6}$
#2	0.024	0.073	{0.0175 , -0.0075}	$10.0 \cdot 10^{-6}$
#3	n.d.	0.044	{0.0060 , 0.0005}	$3.3 \cdot 10^{-6}$

The pulleys radii are:

$$r_{1,0} = 0.030 \quad r_{1,1} = 0.017$$
$$r_{2,0} = 0.015 \quad r_{2,1} = 0.017 \quad r_{2,2} = 0.013$$
$$r_{3,0} = 0.030 \quad r_{3,1} = 0.017 \quad r_{3,2} = 0.013 \quad r_{3,3} = 0.0085$$

The motor inertias, identified as described in section 4, are

$$J_{1,m} = 6.5422 \cdot 10^{-6}$$
$$J_{2,m} = 6.5285 \cdot 10^{-6} \quad J_{2,1} = 3.1 \cdot 10^{-6}$$
$$J_{3,m} = 7.9344 \cdot 10^{-6} \quad J_{3,1} = 3.1 \cdot 10^{-6} \quad J_{3,2} = 1.3 \cdot 10^{-6}$$

The reduction ratios of the gear-boxes are:

$$n_1 = 14 \quad n_2 = 3.71 \quad n_3 = 3.71$$

The lenght of the stages of the transmissions are:

$$l_{1,1} = 0.168$$
$$l_{2,1} = 0.248 \quad l_{2,2} = 0.046$$
$$l_{3,1} = 0.330 \quad l_{3,2} = 0.046 \quad l_{3,3} = 0.025$$

The tendons have diameter $\phi = 0.0003$ m and are composed of several strands. The Jung modulus j of such type of cable has been measured and resulted 6010^7 N/m^2. The elasticities associated to each transmission stage have been determined with the classical formula:

$$k_{i,j} = \frac{j\pi \left(\frac{\phi}{2}\right)^2}{l_{i,j}}$$

The motors used are three *MINIMOTOR 3557 K 024 CR* equipped with optical encoders providing 2000 ticks per rotor revolution.

References

[1] Bergamasco M 1995 Force Replication to the Human Operator: the Development of Arm and Hand Exoskeletons as Haptic Interfaces. *ISRR*, Munich

[2] Prisco G M, Bergamasco M 1997 Dynamic Modelling of a Class of Tendon Driven Manipulators. *ICAR*, Monterey

[3] Jacobsen S C, Ko H, Iversen E K, Davis C C 1989 Antagonistic Control of a Tendon Driven Manipulator. *ICRA*, Scottsdale

[4] Hayward V, Cruz-Hernandez J M 1995 Parameter Sensitivity Analisys for Design and Control of Tendon Transmissions. *ISER*, Stanford

[5] Hayward V, Astley O 1995 Performance measures for haptic interfaces *ISRR*, Munich

[6] Burdea G 1996 Force and Touch Feedback for Virtual Reality *Wiley*, New York

[7] Townsend W T 1988 The Effect of Transmission Design On The Performance of Force-controlled Manipulators. PHD thesis, MIT

[8] Haessig D A, Frieland B 1991 On the modelling and simulation of friction. *J. of dynamic systems, measurements and control* September

[9] Armstrong-Helouvry B, Dupont P, Canada de Wit C 1994 A survey of models, analysis tools and compensation methods for the control of machines with friction. *Automatica* V. 30 n.7

[10] Corke P 1996 In situ measurement of DC motors electrical constant, *Robotica* V. 14

[11] Townsend W, Salisbury J K 1989 Mechanical Bandwidth as a Guideline to High-Performance Manipulator Design. *ICRA*,

[12] Morrell J B, Salisbury J K 1996 Performance Measurement for Robotic Actuators. *ASME Dynamic Sysems and Control Division*,

[13] Vernitron Corp. 1996 Brush Type DC Torque Motors *New York*

Experiments on a High Performance Hydraulic Manipulator Joint: Modelling for Control

Glen Bilodeau and Evangelos Papadopoulos
Department of Mechanical Engineering & Centre for Intelligent Machines
McGill University
Montreal, Canada
gbilod@cim.mcgill.ca, egpapado@cim.mcgill.ca

Abstract: Modelling and experimental identification of a hydraulic servoactuator system is presented. The development of the model is important for further understanding the system and for developing a robust force controller. System parameters are identified using the elbow joint of the SARCOS slave experimental hydraulic manipulator. Experimental work is central to achieving the modelling objectives. Physical parameters are identified using specially designed experiments and apparatus which isolate various subsystems of the joint. Several modelling assumptions are justified by experimental observations. The model is validated by comparing simulation and experimental results. Correlation between model and actual system response proved to be very good. Hence, the developed model predicts well system dynamics behavior and will prove useful in the development of a robust force controller.

1. Introduction

Teleoperated robotic systems can improve both the safety and efficiency of manipulation tasks in hazardous environments. Some applications include live-line maintenance, firefighting, hazardous waste management and underwater operations. These tasks are characterized by the need for applying large forces on an environment that may be stationary or moving. Actuator, link, and sensor dynamics of the manipulator may also be important and influence the overall system performance. Of particular interest are manipulators with hydraulic actuators due to their high output force to mass ratio, to their fire inertance and to the availability of hydraulic power in mobile systems.

To achieve accurate force control, one needs to have precise control of joint torque. Hydraulic actuators introduce additional complexities to force control of manipulators. Unlike electrically actuated manipulators, actuator torque output is not proportional to motor current input. In hydraulic actuators, current input modulates valve orifice area. In addition, actuator effects may include hysteresis, stiction and other valve-related nonlinearities which further complicate their dynamics. In order to develop a robust and effective controller and to ensure controller performance, an accurate model of the actuator is required.

Prior work in modelling and control of hydraulic actuators deals mostly with the common spool valve for which orifice areas are generally linear with respect to the valve position. On the other hand, the servovalve used in this work is of the jet-pipe/suspension type which is more complex. In these valves, there is no contact between moving surfaces. Also, they have a small moving mass and therefore can be very fast resulting in high bandwidth. For the jet-pipe servovalve, a model was proposed and studied in [1], and [2]. In the present work, the suspension type design is studied.

A number of previous studies have dealt with position and force control of hydraulic actuators. A linearized model was used for position control of a spool valve and rotary actuator system, [3]. A model was employed in a feedforward simulation filter, an alternative to the inverse dynamics method, for control of a hydraulically actuated flexible manipulator, [4]. Additional research emphasized temperature variations, friction and limit cycling, [5]. In force control applications, limited work has been done. Use of a model of a hydraulic system to evaluate the hybrid position/force control scheme, inherently not model-based, was demonstrated by [6], [7], and [8]. Explicit force control of hydraulic actuators was treated by [1], and [9]. A position-based impedance control law was applied to a hydraulic manipulator, [10]. Although the focus is in control, modelling is essential to understand the system to be controlled.

In this research work, the objective is to develop an accurate model of a hydraulic actuator joint, to experimentally identify associated parameters, and to validate the derived model experimentally. The final result should be useful in designing and implementing an effective force controller. Section 2 describes the experimental manipulator. Section 3 discusses the physical effects within the system and their modelling. Section 4 describes the experimental parameter identification procedures and additional apparatus, and Section 5 compares experimental results with simulation results validating the model. Finally, conclusions and future work are given in Section 6. Table 1 details the notation used throughout this paper.

Table 1. Nomenclature.

Variable	Definition	Variable	Definition
i, i_{hvs}	current before and after hysteresis.	V_{p1}, V_{c1}	volume in line of port 1 and in chamber 1.
m_v, b_v, k_v	servovalve suspension arm mass, damping, and stiffness.	V_{p2}, V_{c2}	volume in line of port 2 and in chamber 1.
x_v	valve tip displacement.	C_d	discharge coefficient.
F_{ff}	flow force at valve tip.	D_v	rotary actuator volumetric displacement.
B	servovalve motor torque constant.	R_v	leakage coefficient of rotary actuator.
ρ, μ, β	density, viscosity and bulk modulus of oil.	J_v, J_l	vane and load rotary inertia.
A_s, l_s, d_s	cross-sectional area, length and diameter of supply line.	b_{vn}, b_l, b_s	vane, load and shaft damping.
A_r, l_r, d_r	cross-sectional area, length and diameter of return line.	k_s	shaft angular stiffness.
P_s, P_t	pump pressure and tank pressure.	ω_{vn}, θ_{vn}	vane angular velocity and angular position.
Q_{sv}, P_{sv}	flow through supply line, supply pressure before servovalve.	ω_l, θ_l	load angular velocity and angular position.
P_{sv2}	pressure at valve tip.	τ_{ext}, τ_{coul}	external torque and torque due to friction.
Q_{rl}, P_{rl}	flow through return line, return pressure after servovalve.	W_l	weight of load
P_{p1}, P_{p2}	chamber pressure, port 1 and port 2.	C_A	accumulator capacitance

2. Description of the Experimental Setup

The high performance SARCOS hydraulic manipulator is used for the experimental determination and validation of the model parameters. The SARCOS manipulator has ten degrees of freedom, seven in the arm and three in the hands. Overall, the hardware support consists of a 486 PC, a digital signal processor (DSP), an I/O card and Advanced Joint Controller (AJC) cards. For modelling, the elbow joint of the manipulator is used. Sensors available at this joint include an optical encoder angular position sensor, a rotary variable differential transformer (RVDT) for analog position measurement, and a strain gauge full-bridge joint torque sensor. The input current is also measurable.

3. System Characteristics and Modelling

To obtain an accurate model of the hydraulic joint, a description of the physical effects within each subsystem is required. These effects are mainly due to servovalve dynamics, fluid dynamics, and vane and load dynamics. A schematic of the joint is shown in Figure 1. A bond graph of the system was obtained in our previous work, [11]. Each subsystem and the modelling assumptions are discussed next.

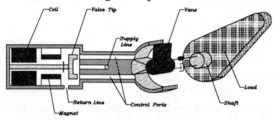

Figure 1. Schematic of hydraulic joint.

3.1. Servovalve Dynamics

The servovalve used in this work is a single-stage, suspension-type valve. As can be seen in Figure 1, fluid impinges the valve tip while current in the coil modifies the magnetic field generated by the magnet which modulates valve tip motion directing the flow from the supply line to one of the control port. Hysteresis and valve tip dynamics and orifice geometry must be addressed to model the servovalve accurately.

3.1.1. Hysteresis

An important phenomenon in the servovalve is hysteresis. Several researchers have observed and characterized hysteresis in the jet-pipe servovalve [1], [2], and [12]. Physically, the hysteresis occurs between the input current and valve tip position. For simplification, in this research, the physical hysteresis and the valve tip dynamics are taken as decoupled. Essentially, the hysteresis is modelled as being between the input current and some virtual output current, i_{hys}, which in turn modulates the valve tip position. Overall, the effect is a hysteresis relation between the input current and the output valve tip position.

To analytically represent this phenomenon, a model based on the Jiles-Atherton theory for magnetization of ferromagnetic material is used [13]. The model is a nonlinear first order differential equation which accounts for major and minor loops. The only requirement for this model is the knowledge of the reversal point, i.e., the point at which the slope of the input current changes sign. Thus, the virtual output current is related to the input current as

$$i_{hys} = \frac{\Lambda \mu_o i \left(i_s L(i, i_{hys}) - i \right)}{k\delta - \mu_o \alpha \left(i_s L(i, i_{hys}) - i \right)} \quad (1)$$

where L, the Langevin function, and δ are given by

$$L(i, i_{hys}) = \coth\left(\mu_o (i + \alpha i_{hys}) \right) - \frac{1}{\mu_o (i + \alpha i_{hys})} \quad (2)$$

$$\delta = \text{sign}(i)$$

The scaling factor, Λ, which is less than unity for minor loop generation, depends on the switching point and the major loop which saturates at i_s. Parameters, μ_o, α and k affect the inclination and width of the hysteresis. A hysteresis curve generated by this model is shown in Figure 2 for a decreasing amplitude sinusoid input.

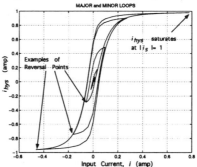

Figure 2. Model generated hysteresis.

3.1.2. Valve Tip Dynamics

The moving part of the servovalve may be taken as a cantilever beam with an end mass, see Figure 1. Thus the dynamics of the valve tip may be approximated by a second-order lumped parameter system with mass, m_v, damping, b_v, and stiffness, k_v. In reality, these parameters would be nonlinear since the cantilever has a distributed mass and is submerged in oil. Forces acting on this cantilever include the input force due to current and flow forces acting at the valve tip. As fluid passes through an orifice, flow forces develop due to fluid acceleration. Reaction forces result which tend to close an opening valve, [14]. These forces are difficult to model, and for the suspension type valve, no model based on the physics is available. Thus, the dynamics of the valve tip may be expressed as

$$\dot{x}_v = v_v$$
$$\dot{v}_v = \frac{1}{m_v} \left(Bi_{hys} + F_f - b_v v_v - k_v x_v \right) \quad (3)$$

where F_f are fluid-induced forces. Note that the input to the system is essentially a force, Bi_{hys}, where i_{hys} lumps the hysteresis part of the model. In all, the hysteresis and valve tip dynamics combine to describe the behavior of the servovalve. Further to the dynamics of the servovalve, geometric modelling of the valve tip orifices is essential for the fluid dynamics subsystem, which is discussed next.

3.2. Fluid Dynamics

Fluid flow through lines, orifices and the rotary actuator are modelled including turbulent flow, leakages, and line losses. In addition, fluid inertance as well as fluid capacitance due to fluid compressibility are taken into account, [14].

Flow through orifices is taken as turbulent, thus, the square root law relating the pressure drop across the orifice and the flow through the orifice is used, [15]

$$Q = C_d A_{orifice}(x_v) \sqrt{\frac{2}{\rho}(P_{hi} - P_{lo})}$$

(4)

$$= g_{orifice}(x_v, P_{hi}, P_{lo})$$

This relation contributes to the nonlinearity of the joint model. In addition, the orifice area $A_{orifice}$ is also nonlinear. In the case of the suspension valve design used in the experiments, the orifice areas are eye or slit shaped. Furthermore, as opposed to the usual matched and symmetrical orifice configuration of spool valves, the present servovalve design was found to be symmetric but unmatched, which adds further to the complexity of the model.

The model accounts for servovalve leakage as a result of the clearance between valve tip and receiver. Two stages of pressure drops are present. First, as the supply flow impinges on the valve tip, the pressure drops from P_{sv} to P_{sv2}, see Figure 3.

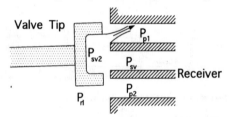

Figure 3. Valve tip and receiver with pressure notation.

Second, P_{sv2} drops to one of the port pressures as the flow is diverted to the actuator. Meanwhile, with each drop, leakage to low pressure is also evident. The two dependent variables, P_{sv2} and P_{rl} are cumbersome to solve for assuming the square root law. Thus, linear resistances were assumed. They may be found through compatibility equations giving the following

$$P_{sv2} = w_1(P_{sv}, P_{p1}, P_{p2}, x_v, Q_{rl})$$ (5)

$$P_{rl} = w_2(P_{sv}, P_{p1}, P_{p2}, x_v, Q_{rl})$$ (6)

In the actuator, leakage between chambers as a result of the gap between the vane and the vane housing was also accounted for. This leakage was considered as a linear resistance denoted as R_v. In the end, the dynamic equations for flow from the servovalve to the actuator take on the form

$$\dot{Q}_{sv} = \frac{A_s}{\rho l_s}\left(P_s - \frac{128\mu l_s}{\pi d_s^4}Q_{sv} - P_{sv}\right)$$ (7)

$$\dot{P}_{sv} = \frac{1}{C_A}(Q_{sv} - g_{lk}(x_v, P_{sv}, P_{rl}) - g_v(x_v, P_{sv}, P_{sv2}))$$ (8)

$$\dot{P}_{p1} = \frac{\beta}{(V_{c1} + V_{p1})}\left(g_1(x_v, P_{sv2}, P_{p1}) - g_4(x_v, P_{p1}, P_{rl}) - D_v\omega_{vn} - R_v(P_{p1} - P_{p2})\right) \quad (9)$$

$$\dot{P}_{p2} = \frac{\beta}{(V_{c2} + V_{p2})}\left(g_2(x_v, P_{sv2}, P_{p2}) - g_3(x_v, P_{p2}, P_{rl}) + D_v\omega_{vn} + R_v(P_{p1} - P_{p2})\right) \quad (10)$$

$$\dot{Q}_{rl} = \frac{A_r}{\rho l_r}\left(P_{rl} - \frac{128\mu l_r}{\pi d_r^4}Q_{rl} - P_r\right) \quad (11)$$

3.3. Vane and Load Dynamics

The vane is modelled as a second-order mechanical rotational system. For an ideal hydraulic rotary actuator the input torque is related to the load pressure as

$$\tau = D_v P_{load} \quad (12)$$

This relation allows identification of the volumetric displacement, D_v. This parameter also relates the flow through the actuator with its angular velocity

$$Q_{load} = D_v \omega_{vn} \quad (13)$$

Continuing, the load is connected to the vane via a shaft which is modelled as a spring and damper. Also modelled is viscous friction and Coulomb friction due to the contact of the seals with the housing. Thus the mechanical equations for the vane and load are

$$\dot{\omega}_{vn} = \frac{1}{J_v}\left(D_v(P_{p1} - P_{p2}) - \tau_{coul}(\omega_{vn}) - b_{vn}\omega_{vn} - k_s(\theta_{vn} - \theta_l) - b_s(\omega_{vn} - \omega_l)\right) \quad (14)$$

$$\dot{\omega}_l = \frac{1}{J_l}\left(k_s(\theta_{vn} - \theta_l) + b_s(\omega_{vn} - \omega_l) - \tau_{coul}(\omega_l) - b_l\omega_l - W_l\sin(\theta_l) + \tau_{ext}\right) \quad (15)$$

4. Parameter Identification

Model parameters were obtained from various sources and methods. Oil properties like bulk modulus, β, viscosity, μ, and density, ρ, were taken from manufacturer tables and plots. However, other parameters needed to be obtained through specially devised experiments. These parameters include actuator volumetric displacement, servovalve dynamics parameters, actuator leakage, and shaft stiffness. In the following, additional apparatus built and the experimental procedures devised for the purpose of identifying these parameters are discussed.

4.1. Experimental Apparatus

Two pieces of apparatus were designed and built to obtain additional measurements and to isolate subsystems, see Figure 4. First, the joint brace allows identification of

(a)   (b)

Figure 4. Additional apparatus: (a) Joint brace; (b) Manifold.

shaft stiffness, servovalve dynamics and actuator leakage. Second, the manifold equipped with pressure transducers allows pressure measurements of the supply line, control ports and the return line. Appropriate ports may be blocked in order to deviate the return flow to a graduated cylinder for volume measurements. In addition, the manifold may be installed at other joints with similar servovalve/robot interface. The usefulness of the equipment is three-fold: (1) identification, (2) validation of the model, and (3) validation of a controller.

4.2. Experimental Identification

Several experiments were performed to identify key parameters of each subsystem. First, a discussion of the identification of servovalve dynamics is given followed by shaft stiffness identification and finally identification of actuator volumetric displacement and leakage.

4.2.1. Servovalve Parameters

The principal concerns for modelling the servovalve are the geometry of the valve tip and receiver orifices, and the dynamic characteristics of the valve tip. The geometric information includes the size and layout of the orifices at the valve tip and the receiver. These were obtained by direct measurement of the valve tip and receiver. An approximate valve tip range of motion was also obtained from these measurements.

The dynamic characteristics of the valve tip were obtained by isolating the servovalve from the actuator by immobilizing the load with the brace. The valve tip dynamics were assumed to be second order. Neglecting flow forces for the identification process, the valve tip equation of motion may be expressed as

$$\ddot{x}_v + 2\zeta_P \omega_{nP} \dot{x}_v + \omega_{nP}^2 x_v = \frac{B}{m_v} i_{hys} \qquad (16)$$

Since the valve tip position is not measurable, it is expressed in terms of the load pressure which is accessible. This is written as

$$x_v = f(P_{load}) = K_v P_{load} \qquad (17)$$

Note that this is essentially the static load characteristic of the servovalve. To simplify the model further, it was assumed that this characteristic is linear, see second part of Equation (17). A curve obtained by a simulation and its linear approximation is illustrated in Figure 5. Substituting Equation (17) into Equation (16), the second order dynamics may be expressed in terms of the load pressure

$$\ddot{P}_{load} + 2\zeta_P \omega_{nP} \dot{P}_{load} + \omega_{nP}^2 P_{load} = \frac{B}{m_v K_v} i_{hys} = K_{dc} \omega_{nP}^2 i_{hys} \qquad (18)$$

As a consequence of the highly nonlinear nature of this system, the damping ratio, natural frequency and the DC gain, K_{dc}, will depend on the input current. Figure 6a shows several experimentally obtained Bode plots of the transfer function for currents of increasing amplitude. As it can be seen, the DC gain between load pressure and input current decreases with increasing amplitude current. It may also be observed that the dynamic characteristics are similar. The curves are close to those of a second order critically damped system. Figure 6b shows one of the experimentally obtained Bode plots fitted with one which corresponds to a second order system with appropriate natural frequency and critical damping. For simplicity, it is assumed that the damping ratio and the natural frequency are constant. Any nonlinearities in the servovalve, such as discharge coefficients, are lumped into the DC gain, K_{dc}. A reasonable relationship between the DC gain and the valve tip position was obtained for best correspondence with experimental data.

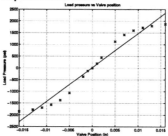

Figure 5. Static load characteristic of servovalve and its linear approximation.

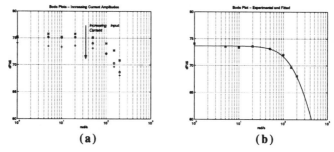

Figure 6. Bode Plots: (a) Increasing amplitude input current; (b) Experimental plot (*) with fitted plot.

4.2.2. Shaft Stiffness

The shaft stiffness was determined by measuring angular position and torque while the load was braced. Plotting torque versus angular position, an approximate straight line results whose slope is approximately the angular joint stiffness. This plot is shown in Figure 7a. The joint shaft stiffness was found to be 8.9×10^4 lb·in/rad. Now, to verify this value, assuming a solid shaft, the stiffness was computed as

$$k_s = GJ/l_s = 11.06 \times 10^4 \text{ lb} \cdot \text{in/rad} \qquad (19)$$

Since the shaft is, in fact, nonsolid, this discrepancy is expected. In all, the experimentally determined stiffness closely matches that obtained theoretically.

4.2.3. Actuator Volumetric Displacement and Internal Leakage

An important property of the actuator is its volumetric displacement, D_v. This parameter relates torque and load pressure as well as load flow and angular velocity as depicted in Equations (12) and (13). To identify D_v, measurements of torque and load pressure are required. Actuator nonlinearities are assumed to insignificantly affect the identification of D_v, so that the torque/pressure relationship may be used. Thus, with the manifold installed and the elbow free to rotate, a sinusoidal current was sent in open-loop resulting in an oscillation of the arm. The torque versus load pressure is plotted in Figure 7b. Here, the slope of the inclined line segments is the sought after volumetric displacement of the rotary actuator. Now, in order to identify the actuator leakage, the brace was installed, and one of the control ports was diverted to a graduated cylinder for rate measurements for a series of constant input currents. In effect, the flow to the graduated cylinder is that through the vane clearance. The actuator leakage was identified according to

$$Q_{leak} = R_v(P_{p1} - P_{p2}) \qquad (20)$$

Under the experimental conditions, P_{p2} is close to atmospheric pressure.

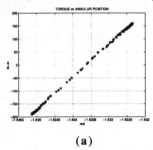 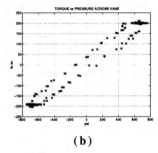

(a) (b)

Figure 7. Determination of shaft and actuator parameters: (a) Shaft stiffness; (b) Volumetric displacement.

4.2.4. Load and Friction Parameters

The load parameters such as mass and damping were obtained by a least squares estimation. The load mass, was verified under static conditions also. Viscous friction and Coulomb friction were both accounted for. In the case of viscous friction, it was lumped into the damping term of the load. They were also identified by a least-squares fit. In order to obtain more certainty in the Coulomb friction model, the torque versus load pressure curve was used, shown in Figure 7b. As the pressure measurements are made before the vane actuator and the torque measurements include friction, the difference between the two is due to the stick-slip phenomenon. In short, the horizontal portions of the curve are due to Coulomb friction.

On the whole, through these experiments, several parameters were identified with good accuracy since each parameter was obtained by isolating the subsystem of interest. Those parameters that were not estimated with good certainty, including the clearance between valve tip and receiver, were tuned until satisfactory correlation between simulation and experiments were obtained. In the next section, the model is evaluated by comparing its response to those of the actual system.

5. Validation

Having developed and identified the model and its parameters, a comparison of simulation and experimental results is performed to test how well the model predicts system behavior. The *s-function* approach in Matlab with Gear integration method was used. Experiments were performed in open loop mode at an operating supply pressure of 3000 psi. Experiments were done to validate servovalve dynamics, and the overall joint model in statics and dynamics cases. Before the execution of each experiment, an exponentially decaying sinusoidal current was commanded in order to begin each experiment with a load pressure close to zero, resulting in a valve tip position practically at the null. The decaying sinusoid also removed any memory due to hysteresis. Refer to Figure 2 for an example.

5.1. Braced Joint Experiments

As it was done for identifying the servovalve dynamic characteristics, the joint was blocked in order to validate the model of the servovalve dynamics. In effect, the valve tip dynamics are isolated from the load dynamics. In simulation, the vane and load positions and their derivatives were constrained to be zero. Initial conditions for simulation were set to match those of the experiments.

For sinusoidal currents of amplitudes 0.2 A and 0.3 A and frequency of 1.0 rad/sec, the supply pressure and the two chamber pressures are shown in Figure 8 for simulation and experiment. The two responses match well. An interesting feature is that one of the chamber pressures is not symmetric. It is suspected to be related to the unmatched characteristic of the servovalve. However, it was observed only for currents of 0.3 A and higher. In the model, K_v was made to be asymmetric.

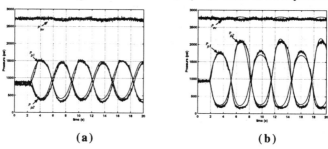

Figure 8. Supply and chamber pressures, braced joint tests: (a) Current amplitude 0.2 A; (b) Current amplitude 0.3 A.

5.2. Unbraced Joint - Statics Case

Static tests allowed the verification of the model in the case for which the load remains stationary, where all the state rates are practically zero. The joint was unbraced and a constant input current was commanded. Measurements were taken after steady state was achieved. Thus, for two different constant input currents of 0.05 A and 0.1 A the chamber pressures and supply pressure before the servovalve are shown in Figure 9a and 9b. As it may be seen, the model captures well the static behavior of the system. Figure 9c illustrates that the load position is also well modelled in the statics case. Next, the dynamic behavior of the model is discussed.

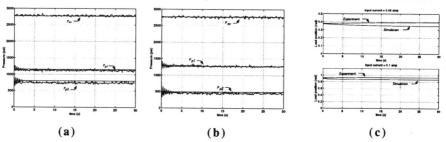

Figure 9. Static Response: (a) Pressures for $i = 0.05$ A; (b) Pressures for $i = 0.1$ A, (c) Load position - (top) 0.05 A; (bottom) 0.1 A.

5.3. Unbraced Joint - Dynamics Case

In this set of experiments, the joint is free to rotate according to an input current of

$$i = 0.1\sin(0.25t) \text{ A} \quad (21)$$

The pressure response was plotted and compared to the simulated response, as shown in Figure 10a. As depicted, the simulation curves match well the experimental curves for supply and chamber pressures. Of importance in control is the load pressure, P_{load}, which is plotted in Figure 10b. Again, simulation and experimental plots correspond well. The response of the load position is illustrated in Figure 10c and is quite close to the experimental load position. As the arm approaches the highest parts of its trajectory, it can be seen that the stick-slip friction model is satisfactory. On the whole, the model for the load dynamics is also good.

Some differences between simulation and experiments are present due to unmodelled effects and to the lumped parameter approach in modelling. These factors include temperature effects on the oil parameters as well as the reduction of oil bulk modulus due to air entrainment. The above results indicate that these effects are not significant for the purpose of control and therefore no further modelling of these effects is required.

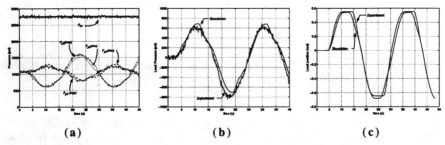

(a)   (b)   (c)

Figure 10. Dynamic response for simulation and experiment: (a) Supply and chamber pressures; (b) Load pressure; (c) Load position.

6. Conclusions

The contributions of this research work is both analytical and experimental. An accurate model of a hydraulic joint of a manipulator has been developed. The associated parameters were identified through a series of specially designed experiments and equipment. In turn, this lead to a model that accounts for the major effects of an electrohydraulic actuator such as hysteresis, flow through orifices, and line losses. In addition, the model is able to characterize the servovalve dynamics well. The developed model represents well the behavior of the real system and can be extended to other joints of the SARCOS slave manipulator as well as the master in such a way as to obtain a complete model of the hydraulics of the SARCOS manipulator. It is expected that this model will be useful in designing a robust force controller in order to reduce control effort, to mask unwanted nonlinear behavior and to improve control performance.

Acknowledgements

The support of this work by the Fonds pour la Formation de Chercheurs et l'Aide à la Recherche (FCAR), Québec, and by the Natural Sciences and Engineering Council of Canada (NSERC) is gratefully acknowledged.

References

[1] Boulet B, Daneshmend L, Hayward V, Nemri C 1992 System Identification and Modelling of a High Performance Hydraulic Actuator. In: Chatila R, Hirzinger G(eds) 1992 *Exp Robotics II - The Second Int Symp.* Springer-Verlag, New York.

[2] McLain T W, Iversen E K, Davis C C, Jacobsen S C 1989 Development, Simulation, and Validation of a Highly Nonlinear Hydraulic Servosystem Model. *Proc of the 1989 Am Control Conf.* Pittsburgh PA, pp 385-391.

[3] Heintze J, Schothorst G v, v d Weiden A J J, Teerhuis P C 1993 Modeling and Control of an Industrial Hydraulic Rotary Vane Actuator. *Proc of the 32nd Conf on Decis and Control.* San Antonio TX, pp 1913-1918.

[4] Kwon D S, Babcock S M, Burks B L, Kress R L 1995 Tracking Control of the Hydraulically Actuated Flexible Manipulator. *Proc of the 1995 Int Conf on Robotics and Autom, ICRA'95.* Nagoya Japan, pp 2200-2205.

[5] Love L, Kress R, Jansen J 1997 Modeling and Control of a Hydraulically Actuated Flexible-Prismatic Link Robot. *Proc of the 1997 IEEE Conf on Robotics and Autom, ICRA'97.* Albuquerque NM, pp 669-675.

[6] Bluethmann B, Ananthakrishnan S, Scheerer J, Faddis T N, Greenway R B 1995 Experiments in Dexterous Hybrid Force and Position Control of a Master/Slave Electrohydraulic Manipulator. *Proc of the 1995 Int Conf on Intell Robots and Syst, IROS'95.* Pittsburgh PA, pp 27-32.

[7] Dunnigan M W, Lane D M, Clegg A C, Edwards I 1996 Hybrid position/force control of a hydraulic underwater manipulator. *IEE Proc Control Theory & Appl.* 143:145-151.

[8] Unruh S, Farris T, Greenway B, Hibbard W 1994 A Hybrid Position/Force and Positional Accuracy Controller for a Hydraulic Manipulator. *SPIE Telemanipulator and Telepresence Technol.* 2351:207-213.

[9] Laval L, M'Sirdi N K, Cadiou J-C 1996 H_∞-Force Control of a Hydraulic Servo-Actuator with Environmental Uncertainties. *Proc of the 1996 IEEE Int Conf on Robotics and Autom, ICRA'96.* Minneapolis, MN, pp 1566-1571.

[10] Heinrichs B, Sepehri N, Thornton-Trump, A B 1996 Position-Based Impedance Control of an Industrial Hydraulic Manipulator. *Proc of the 1996 IEEE Int Conf on Robotics and Autom, ICRA'96.* Minneapolis, MN, pp 284-290.

[11] Bilodeau G, Papadopoulos E 1997 Development of a Hydraulic Manipulator Servoactuator Model: Simulation and Experimental Validation *Proc of the 1997 IEEE Int Conf on Robotics and Autom, ICRA'97.* Albuquerque, NM, pp 1547-1552.

[12] Henri P D, Hollerbach J M 1994 An Analytical and Experimental Investigation of a Jet-Pipe Controlled Electropneumatic Actuator. *Proc of the 1994 IEEE Int Conf on Robotics and Autom, ICRA'94.* San Diego, CA, pp 300-306.

[13] Carpenter K H 1991 A Differential Equation Approach to Minor Loops in the Jiles-Atherton Hysteresis Model. *IEEE Trans on Magnetics.* 27:4404-4406.

[14] Blackburn J F, Reethof G, Shearer J L (eds) 1960 *Fluid Power Control.* MIT Press, Cambridge.

[15] Merritt H E 1967 *Hydraulic Control Systems.* John Wiley and Sons Inc., New York.

Chapter 12

Sensor-Based Control

Real-time sensor-based control is essential for extending robot capabilities to operate in unstructured environments. The papers in this session focus on different problems; calibration, stabilization and on applications of sensor based control to underwater systems.

Hosoda and Asada propose an adaptive visual servoing method that avoids the tedious process of camera calibration. The approach combines an on-line estimator of the robot/image instantaneous relationship with a feedforward/feedback controller to achieve robust and smooth tracking. The estimator does not need to know a priori the kinematics structure nor the parameters of the camera-manipulator system.

Visual servoing is also addressed by Tsakiris, Kapellos, Samson, Rives, and Borrelly in the context of non-holonomic mobile manipulator position control. The visual data are used to implement a continuous time-varying state feedback for the stabilization of the system.

The paper by Kapellos, Simon, Granier, and Rigaud focuses on the connection between low-level control and high-level programming in the implementation and validation of autonomous missions for underwater robot systems. The paper addresses this issue in the context of distributed computer controllers using the ORCCAD programming environment.

The paper by Amat, Aranda, and Villà is also concerned with underwater systems. They propose a compact underwater teleoperated camera system. Connected by a flexible umbilical to a fixed base, the camera is actuated by water jets and controlled by visual feedback. This device is of use for exploration in recondite places or as an appendix of another underwater vehicle.

Adaptive Visual Servoing for Various Kinds of Robot Systems

Koh Hosoda and Minoru Asada

Dept. of Adaptive Machine Systems, Osaka University
Yamadaoka 2-1, Suita 565, Japan
hosoda@mech.eng.osaka-u.ac.jp

Abstract: This paper propose an adaptive visual servoing method consisting of an on-line estimator of the robot/image Jacobian matrix and a feedback/feedforward controller for uncalibrated camera-manipulator systems. The estimator does not need *a priori* knowledge on the kinematic structure nor on parameters of the camera-manipulator system. The controller consists of feedforward and feedback terms to make the image features converge to the desired trajectories using the estimated results. Some experimental results are given to show the validity of the proposed method.

1. Introduction

Visual information plays an important role for a robot to accomplish given tasks in an unknown/dynamic environment. Many vision researchers have been adopting deliberative approaches to the problem of reconstructing 3-D scene structure from visual information, which are not only very time consuming but also brittle to noise, therefore it seems hard to apply these methods to real robot tasks. Recently, there have been several studies on visual servoing, using visual information in the dynamic feedback loop to increase robustness of the closed loop system [1]. For vision-based robots, image features on the image planes are primitive descriptions of the environments. In this sense, feature-based visual servoing control is the most fundamental one for the vision-based robots in which image features are controlled to converge to the desired ones, and therefore has been focused by a number of researchers [2-13].

In most of the previous work on visual servoing, they assumed that the system structure and parameters were known [2-7], or that the parameters could be identified in an off-line process [14]. Such a controller, however, is not robust for disturbances and changes of the parameters. To overcome this problem, some on-line parameter identification schemes have been proposed [8-13]. Weiss et al. [8] assumed that the system could be modeled by linear SISO (Single Input Single Output) equations, and applied an MRAC (Model Reference Adaptive Control) controller. In [9], the structure and parameters of a camera-manipulator system were assumed to be known, and an ARMAX (auto-regressive with external inputs) model was used to estimate disturbances and positions of the target points. Papanikolopoulos et al. [10, 11] modeled the system using an ARMAX model and estimated the coefficients of the model.

They also estimated the depth related parameters in [12]. Yoshimi and Allen [13] used a special camera-manipulator setup to realize an uncalibrated camera system. Thus, in these approaches, there were restrictions and assumptions on the system that the camera-manipulator system was described as SISO equations [8], that *a priori* knowledge on the system structure was required [9, 12, 13], or that the depth was assumed to be constant [10, 11].

On the other hand, most of the previous work have paid their attentions only to the feedback servoing. They sensed positions of targets and made feedback inputs by subtracting the sensed positions from the desired ones. Using their controllers, the manipulator does not move until the error is observed, which can be considered to be *reactive*. To increase the ability of trajectory tracking, there have been several researches on *feedforward*, in which the dynamic motion of the target is predicted [15–17], but no one has mentioned to feedforward control to predict the motion of the robot itself to the best of our knowledge. If the estimator can obtain a kinematic model of the robot system appropriately, the servoing controller can feedforward the obtained kinematics to realize smooth trajectory tracking motion along the desired trajectories designed to accomplish a certain task. For example, a trajectory generator for obstacle avoidance is proposed by authors utilizing the adaptive visual servoing method [18]. In this paper, we propose an adaptive visual servoing method consisting of an on-line estimator and a feedback/feedforward controller for uncalibrated camera-manipulator systems. It has the following features:

1. The estimator does not need *a priori* knowledge on the system parameters nor on the kinematic structure of the system. That is, we need not to devote ourselves to tedious calibration processes, or to separate the unknown parameters from the system equations, which depends on the detailed knowledge on the kinematic structure of the system.

2. There is no restriction on a camera-manipulator system: the number of cameras, kinds of images features, structure of the system (camera-in-manipulator or camera-and-manipulator), the numbers of inputs and outputs (SISO or MIMO). The proposed method is applicable to any kinds of systems.

3. The aim of the estimator is not to obtain the true parameters but to ensure asymptotical convergence of the image features to the desired values under the proposed controller. Therefore, the estimated parameters do not necessarily converge to the true values. In [8–12], they tried to estimate the true parameters, and therefore they need their restrictions and assumptions.

4. The proposed controller can realize smooth tracking motions along the desired trajectories because not only the feedback terms but also feedforward terms are utilized based on the estimated results.

This paper is organized as follows. First, we propose an estimator for an image Jacobian that represents the relation between the image features

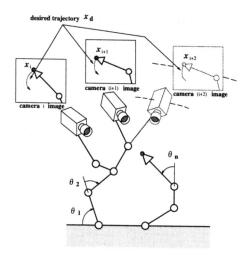

Figure 1. Camera-manipulator system

and the variables that describe the state of the system. Then, a visual feedforward/feedback servoing controller is proposed based on the estimated Jacobian. Finally, experimental results show the validity of the proposed method.

2. Adaptive visual servoing

2.1. Estimation of the relation between image features and system describing variables

A camera-manipulator system consists of manipulators and cameras, as shown in Figure 1. From cameras, one can observe quantities of image features such as position, line length, contour length, and/or area of certain image patterns. The task of the system is to make the quantities of image features converge to the given desired values. The image features are assumed to be on the tip/body of the manipulators or fixed to the ground. That is, a set of measurable variables (which we call system describing variables in the rest of this paper) can describe the state of the camera-manipulator system. Changes of the image features caused by an independently moving object with unknown velocity are not dealt with here.

Let $\boldsymbol{\theta} \in \Re^n$ and $\boldsymbol{x} \in \Re^m$ denote the vectors of the system describing variables and the image features obtained from visual sensors, respectively. A relation between $\boldsymbol{\theta}$ and $\boldsymbol{x}$ is

$$\boldsymbol{x} = \boldsymbol{x}(\boldsymbol{\theta}), \qquad (1)$$

because we assume that the system describing variables can describe the state of the system. Differentiating eq.(1), we obtain a velocity relation,

$$\dot{\boldsymbol{x}} = \boldsymbol{J}(\boldsymbol{\theta})\dot{\boldsymbol{\theta}}, \qquad (2)$$

where $\boldsymbol{J}(\boldsymbol{\theta}) = \partial \boldsymbol{x}/\partial \boldsymbol{\theta}^T \in \Re^{m \times n}$ is a Jacobian matrix of time-derivatives of the quantities of image features with respect to those of system describing variables.

This Jacobian matrix depends on the kinematic structure of the system, the internal camera parameters such as focal length, aspect ratio, distortion coefficients, and the kinematic parameters such as the length of links and the relative position and orientation of cameras with respect to the tip of the manipulator.

Assuming that movement of the camera-manipulator system is slow enough to consider the Jacobian matrix J to be constant during the sampling time, we obtain

$$x(k+1) = x(k) + J(k)u(k), \qquad (3)$$

as a discrete model of the system, where $J(k)$ and $u(k)(=\dot{\theta}\Delta T)$ denote the constant Jacobian matrix and a control input vector in k-th step during sampling rate ΔT, respectively. From eq.(3), i-th row vector of the matrix J, j_i^T, satisfies

$$\{j_i(k+1)^T - j_i(k)^T\}u(k+1) = \{x(k+2) - x(k+1) - J(k)u(k+1)\}_i. \qquad (4)$$

Among an infinite number of solutions of eq.(4), we pick up one to make the norm of weighted time-derivatives of $\widehat{j}_i$ as small as possible by the iteration,

$$\widehat{j}_i(k+1) - \widehat{j}_i(k) = \frac{\{x(k+1) - x(k) - \widehat{J}(k)u(k)\}_i}{u(k)^T W_i(k) u(k)} W_i(k) u(k). \qquad (5)$$

Theoretically, the right-hand side of eq.(5) does not tend to infinity when $\|u\|$ tends to 0, because $|\{x(k+1) - x(k) - \widehat{J}(k)u(k)\}_i|$ also tends to 0 at the same or faster speed. In real situations, however, the right-hand side is prone to be unstable because of disturbances. To increase the stability of estimation (5), particularly when $\|u\|$ tends to 0, the estimating law is modified as

$$\widehat{j}_i(k+1) - \widehat{j}_i(k) = \frac{\{x(k+1) - x(k) - \widehat{J}(k)u(k)\}_i}{\rho_i + u(k)^T W_i(k) u(k)} W_i(k) u(k), \qquad (6)$$

where ρ_i is an appropriate positive constant that makes the iteration (6) stable. When $\|u\|$ tends to 0, the denominator tends to ρ_i and the stability is ensured even if the numerator does not tend to 0 because of disturbances. The positive constant ρ_i is determined so small that ρ_i can be neglected with respect to $\|u\|$ when $\|u\|$ is large. Note that when ρ is in the range $0 < \rho \le 1$ and the matrix W_i is a covariance matrix, the proposed estimator coincides with the least-mean-square method[19].

The proposed estimator is intended not to obtain the true Jacobian matrix/parameters, but to estimate a matrix that satisfies eq.(3). This is the main difference from [9], [10], and [11], in which they tried to estimate the true parameters. To estimate the true parameters, one have to make restrictions and assumptions on the camera-manipulator system. The proposed estimator, however, is not intended to estimate the true parameters, but to make the closed loop system consisting of this estimator and a controller stable. Therefore, there is neither restrictions nor assumptions on the camera-manipulator system.

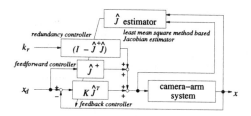

Figure 2. Block diagram of the proposed method

2.2. Feedforward/feedback visual controller

In this section, a feedforward/feedback visual controller is proposed, based on the estimated Jacobian matrix $\hat{\boldsymbol{J}}$. The aim of the controller is to ensure convergence of the image feature vector $\boldsymbol{x}(k)$ to the desired vector $\boldsymbol{x}_d(k)$.

From eq.(3), we can derive a feedforward/feedback controller,

$$\begin{aligned} \boldsymbol{u}(k) &= \hat{\boldsymbol{J}}(k)^+\{\boldsymbol{x}_d(k+1) - \boldsymbol{x}_d(k)\} \\ &\quad + \{\boldsymbol{I}_n - \hat{\boldsymbol{J}}(k)^+\hat{\boldsymbol{J}}(k)\}\boldsymbol{k}_r \\ &\quad + \boldsymbol{K}\hat{\boldsymbol{J}}(k)^T\{\boldsymbol{x}_d(k+1) - \boldsymbol{x}(k)\}, \end{aligned} \qquad (7)$$

where $\hat{\boldsymbol{J}}(k)^+$, $\boldsymbol{I}_n$, and $\boldsymbol{K}$ denote a pseudo-inverse matrix of $\hat{\boldsymbol{J}}(k)$, an $n \times n$ identity matrix, and a positive-definite gain matrix, respectively. Let $\boldsymbol{k}_r$ be an arbitrary vector.

The first and second terms on the right-hand side are feedforward terms. The second term on the right-hand side denotes the redundancy of the camera-manipulator system. The third term on the right-hand side is a feedback term that ensures stability of the closed loop system. Note that one can use $\hat{\boldsymbol{J}}(k)^+$ instead of $\hat{\boldsymbol{J}}(k)^T$ to ensure the closed loop stability [3].

We propose an adaptive visual servoing method consisting of the proposed estimator and controller shown in Figure 2.

3. Experiments

To show the validity of the proposed method, some experimental results are given in this section. First step response results of two kinds of camera-manipulator systems are given to show how the estimator and the feedback terms of the controller can realize *reactive* tasks well and how the proposed method can be applied to various kinds of system structures. Then, a result of trajectory tracking is given to show how the feedforward terms work.

3.1. Experimental equipment

In Figure 3, a camera-manipulator system used for the experiments is shown. The video signals from two cameras (UN401, ELMO) are sent to an image processing board MV200 (DataCube, image size: 512[pixel] × 480[pixel]) and compressed into the half along the horizontal axis (256[pixel] × 480[pixel]). Two images are pasted onto one image (512[pixel] × 480[pixel], see Figure 5 for example), which is sent to a tracking module equipped with a high-speed

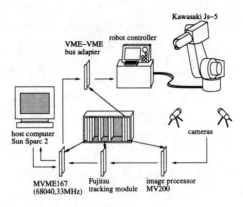

Figure 3. Experimental equipment

correlation processor utilizing a SAD (Sum of Absolute Difference) measure by Fujitsu [20]. Before starting an experiment, we specify target images to be tracked by the module. During the experiment, the module tracks the target images, and it feeds coordinates of the images in the image plane to the main control board MVME167 (CPU:68040, 33MHz, motorola). The board calculates a desired posture of the manipulator by the proposed method and sends it to the manipulator controller through a VME-VME bus adapter. We use a 6 d. o. f. manipulator Js-5 (Kawasaki Heavy Industry Co.) as a 3 d. o. f. manipulator, maintaining fixed desired orientation of the tip of the manipulator. Therefore, the system describing variable vector $\boldsymbol{\theta}$ is a tip position vector of the manipulator in these experiments. Using this experimental equipment and writing programs using C language on VxWorks (Wind River), the sampling ratio is 30[Hz].

3.2. Step responses of two kinds of systems

To show that the estimator and the feedback terms of the controller can realize *reactive* tasks well and that the estimator and controller can be applied to different system structures, step responses of two kinds of systems are given in this subsection. At $t = 0$, the desired image feature coordinates which are *taught by showing* are fed to the controller. In these experiments, we show step responses, and therefore the feedforward terms in eq.(7) equal zeros.

The positive constants $\rho_i = 0.8$ $(i = 1, \cdots, 4)$ are selected as small as possible in the trial and error manner. We set the weighting matrices $\boldsymbol{W}_i(k) = \boldsymbol{I}_3 (i = 1, \cdots, 4)$. The feedback gain matrix $\boldsymbol{K}$[m/pixel] in eq.(7) is also selected in the trial and error manner,

$$\boldsymbol{K} = \text{diag} \begin{bmatrix} 1.5 \times 10^{-4} & 1.5 \times 10^{-4} \\ 1.5 \times 10^{-4} & 1.5 \times 10^{-4} \end{bmatrix}.$$

An initial Jacobian matrix, which is needed at the beginning of the control phase, can be given arbitrarily as far as its rank is full. In the experiment, therefore, we give the initial Jacobian matrix as

$$\hat{J}(0) = \begin{bmatrix} 0.1 & 0 & 0 \\ 0 & 0.1 & 0 \\ 0 & 0 & 0.1 \\ 0 & 0 & 0.1 \end{bmatrix}.$$

First, we fix the two cameras on the ground. The reference images are windows of a pattern (a cross) which is fixed at the tip of the manipulator (see Figure 4).

In Figure 5, we can find the initial posture of the manipulator, the initial positions of the reference images, and the desired positions of the reference images. Responses of two cases, (a) applying the proposed controller with the proposed estimator, and (b) applying the proposed controller without on-line estimation, are shown in Figure 6. Because the initial Jacobian matrix is given arbitrary, the controller cannot eliminate error without on-line estimation. On the other hand, using the proposed method, the manipulator can be controlled to make the image features converge to the desired ones.

Second, in order to show that it can be applied to various kinds of system structures without *a priori* knowledge on the systems, we apply the proposed method to a different system in which two cameras are mounted on the tip of the manipulator (see Figure 7). The reference images are the windows of a pattern (a cross) fixed on the ground. The positive constants ρ_i, the gain matrix K and the weighting matrices W_i are the same as the previous case. That is, the estimator and the controller are all the same as in the previous experiment. The reference images, their initial positions, and the desired positions are shown in Figure 8, and the results are shown in Figure 9.

From these experimental results, we can conclude that the proposed on-line estimator and the feedback term of the proposed controller are effective to realize a reactive task, and that the proposed method is applicable to different kinds of systems.

3.3. Trajectory tracking task

In this subsection, a result on trajectory tracking task is given to show the validity of the proposed purposive visual control.

We fix two cameras on the ground. The reference images are windows of a pattern (a cross) fixed at the tip of the manipulator (see Figure 4). The desired trajectories must be realizable. Therefore, the desired trajectories are *taught by showing* to satisfy this constraint (Figure 10). The squares in the figure indicate the location along the desired trajectory every 0.2[s]. The target moves along each trajectory in 12[s] in each image.

The initial Jacobian matrix is the same one given in the previous experiments. The tracked image feature vector is $x \in \Re^4$, and the controlled tip position vector of the manipulator is $\theta \in \Re^3$, therefore the second feedforward term on the right-hand side of eq.(7) equals zero. Because our experimental system has a time-delay problem, we cannot stabilize the closed loop system with 100% feedforward terms in eq.(7). Therefore we apply 30% feedforward terms in the following experiment.

One of experimental results is given in Figure 11, where desired trajectories and realized ones are indicated. From this figure, we can see how the proposed method can track the desired trajectories better than the controller without the feedforward terms.

4. Conclusion and Discussion

In this paper, we have proposed an adaptive visual servoing method consisting of an on-line estimator and a feedback/feedforward controller for uncalibrated camera-manipulator systems. We have proposed an estimator for the Jacobian matrix that describes the relation between the image features and the system describing variables. Then, a feedforward/feedback controller has been proposed making use of the estimated relation. Finally experimental results are given to show that the proposed method is validity to various kinds of robot systems.

We have to mention to the redundancy of the system. If the camera-manipulator system is redundant to accomplish given tasks, the redundancy can be utilized to realize other sub-tasks such as obstacle avoidance. In the proposed controller, the redundancy is denoted as the second term on the right-hand side of eq.(7), but we have not mentioned how to utilize the redundancy in this paper. *A study on redundancy in the uncalibrated system* is one of our future major work.

One alternative to deal with such redundancy is to introduce another servoing method, and to build a hybrid servoing controller. The authors have shown some theoretical and experimental results: (1) hybrid adaptive visual servoing/force servoing control [21], and (2) adaptive visual servoing control for legged robots [22]. When we build a fast/robust robot system, such kinds of hybrid servoing controllers would be powerful and essential.

References

[1] P. I. Corke. Visual control of robot manipulators – a review. In *Visual Servoing*, pages 1–31. World Scientific, 1993.

[2] W. Jang and Z. Bien. Feature-based visual servoing of an eye-in-hand robot with improved tracking performance. In *Proc. of IEEE Int. Conf. on Robotics and Automation*, pages 2254–2260, 1991.

[3] K. Hashimoto, T. Kimoto, T. Ebine, and H. Kimura. Manipulator control with image-based visual servo. In *Proc. of IEEE Int. Conf. on Robotics and Automation*, pages 2267–2272, 1991.

[4] N. Maru, H. Kase, et al. Manipulator control by visual servoing with the stereo vision. In *Proc. of the 1993 IEEE/RSJ Int. Conf. on Intelligent Robots and Systems*, pages 1865–1870, 1993.

[5] P. Allen, A. Timcenko, B. Yoshimi, and P. Michelman. Automated tracking and grasping of a moving object with a robotic hand-eye system. *IEEE Trans. on Robotics and Automation*, RA-9(2):152–165, 1993.

[6] A. Castano and S. Hutchinson. Visual compliance: Task-derected visual servo control. *IEEE Trans. on Robotics and Automation*, 10(3):334–342, 1994.

[7] G. D. Hager, W.-C. Cang, and A. S. Morse. Robot feedback control based on stereo vision: Towards calibration-free hand-eye coordination. In *Proc. of IEEE Int. Conf. on Robotics and Automation*, pages 2850–2856, 1994.

[8] L. E. Weiss, A. C. Sanderson, and C. P. Neuman. Dynamic sensor-based control of robots with visual feedback. *IEEE J. of Robotics and Automation*, RA-3(5):404–417, 1987.

[9] J. T. Feddema and C. S. G. Lee. Adaptive image feature prediction and control for visual tracking with a hand-eye coordinated camera. *IEEE Trans. on System, Man, and Cybernetics*, 20(5):1172–1183, 1990.

[10] N. P. Papanikolopoulos and P. K. Khosla. Adaptive robotic visual tracking: Theory and experiments. *IEEE Trans. on Automatic Control*, 38(3):429–445, 1993.

[11] B. Nelson, N. P. Papanikolopoulos, and P. K. Khosla. Visual servoing for robotic assembly. In *Visual Servoing*, pages 139–164. World Scientific, 1993.

[12] N. P. Papanikolopoulos, B. Nelson, and P. K. Khosla. Six degree-of-freedom hand/eye visual tracking with uncertain parameters. In *Proc. of IEEE Int. Conf. on Robotics and Automation*, pages 174–179, 1994.

[13] B. H. Yoshimi and P. K. Allen. Alignment using an uncalibrated camera system. *IEEE Trans. on Robotics and Automation*, 11(4):516–521, 1995.

[14] R. Y. Tsai and R. K. Lenz. A new technique for fully autonomous and efficient 3d robotics hand/eye calibration. *IEEE Trans. on Robotics and Automation*, 5(3):345–358, 1989.

[15] C. Brown. Gaze controls with interactions and delays. *IEEE Trans. on System, Man, and Cybernetics*, 20(1):518–527, 1990.

[16] E. D. Dickmanns, B. Mysliwetz, and T. Christians. An integreted spatio-temporal approach to automatic visual guidance of autonomous vehicles. *IEEE Trans. on System, Man, and Cybernetics*, 20(6):1273–1284, 1990.

[17] W. J. Wilson. Visual servo control of robots using kalman filter estimates of robot pose relative to work-pieces. In *Visual Servoing*, pages 71–104. World Scientific, 1993.

[18] K. Hosoda, K. Sakamoto, and M. Asada. Trajectory generation for obstacle avoidance of uncalibrated stereo visual servoing without 3d reconstruction. In *Proc. of the 1995 IEEE/RSJ Int. Conf. on Intelligent Robots and Systems*, pages 29–34, 1995.

[19] P. Eykhoff. *System Identification*, chapter 7. John Wiley & Sons Ltd., 1974.

[20] M. Inaba, T. Kamata, and H. Inoue. Rope handling by mobile hand-eye robots. In *Proc. of Int. Conf. on Advanced Robotics*, pages 121–126, 1993.

[21] Koh Hosoda, Katsuji Igarashi, and Minoru Asada. Adaptive hybrid visual servoing/force control in unknown environment. In *Proc. of the 1996 IEEE/RSJ Int. Conf. on Intelligent Robots and Systems*, pages 1097–1103, 1996.

[22] K. Hosoda, M. Kamado, and M. Asada. Vision-based servoing control for legged robots. In *Proc. of IEEE Int. Conf. on Robotics and Automation*, pages 3154–3159, 1997.

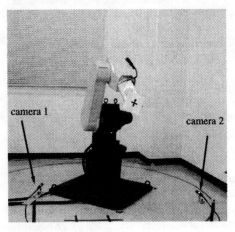

Figure 4. Eye and arm system used for experiments

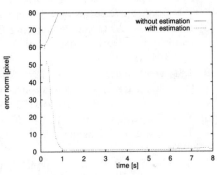

Figure 6. Experimental result 1 (step response, eye and arm system, with and without estimation)

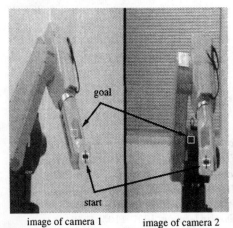

Figure 5. Initial and final position of prememorized image pattern(initial posture, eye and arm system)

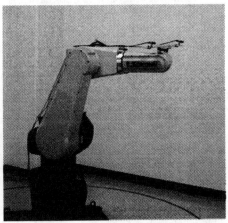

Figure 7. Eye on arm system used for experiments

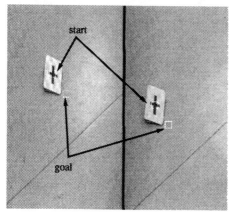

image of camera 1   image of camera 2

Figure 8. Initial and final position of prememorized image pattern(initial posture, eye on arm system)

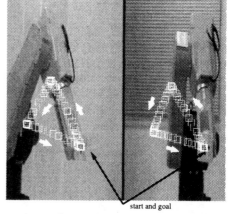

image of camera 1   image of camera 2

Figure 10. Given desired trajectories for tracking control

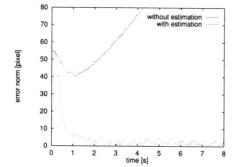

Figure 9. Experimental result 2 (step response, eye on arm system, with and without estimation)

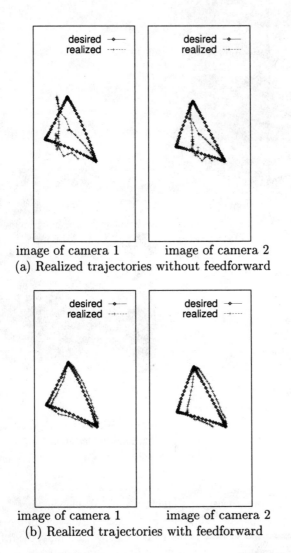

Figure 11. Realized trajectories with and without feedforward terms

Underwater hidrojet explorer camera controlled by vision

Josep Amat
Institut de Robòtica Industrial (UPC-CSIC)
Barcelona, Catalonia
e-mail: amat@esaii.upc.es

Joan Aranda, Ricard Villà
Dept. Automatic Control and Computer Engineering (UPC)
Barcelona, Catalonia
e-mail: aranda@esaii.upc.es

Abstract

Access to restricted spaces underwater requires a small unencumbered camera. A low cost solution has been envisaged. It consists of an hermetic capsule enclosing the camera and lights, joined to the host by an umbilical. Along the umbilical go six water carrying ducts. Three of them end at backwards pointing nozzles located at the capsule body. The jets of water flowing from the nozzles are controlled and their forces allow some restricted positioning of the camera. Another three ducts end at nozzles located some distance up the umbilical, and allow a greater manoeuvrability.

1. Introduction

Underwater work has been enhanced by the use of ROV's. They allow the realization of many tasks very difficult to be carried out directly. The growth of informatics and electronics in recent years [1], has allowed virtual reality to be incorporated to the bag of tools available for teleoperation.

Nonetheless there remain at least three limitations. First, the smallness of the forces allowable for manipulation due both to the mechanical limitations of the arms, and to problems in stabilizing the vehicle in the working zone. Secondly, the difficulties for the vehicle in accessing the working zone, at times compounded by the presence of the umbilical. Thirdly, the question of a good imaging of the working zone.

It is precisely for this question of getting the best point of view of the working zone, that we have designed a multiarticulated sensorial appendix, with a TV camera as the terminal element, adapted for underwater work. So as to make this appendix telepositionable and of the least possible dimensions, its mobility has been accomplished by means of water jets, fed from the host.

2. Architecture of the sensor appendix.

This sensorial appendix is composed of a base which connects it to the host, its body, totally flexible, and the terminal element which is an hermetic container for the TV camera, ligths, and the ducts for propulsion, (Fig. 1). The flexible body acts as a polyarticulated backbone and as an umbilical transporting the video and sensors signals, electrical power and comprises also the three ducts for the propulsive water.

Mobility of the terminal element, the sensors capsule, is provided by three water jets at 135° off the z-axis and in planes 120° apart, as seen in fig.2.

Figure 1 The hidrojet explorer

Modulation of the strength of each of the three water jets provides the means of orienting the resultant force and also of varying its strength. To obtain an static equilibrium the resultant of the three jet forces $F_a+F_b+F_c=F_r$ must equal the reaction of the umbilical

To get a resultant force $\vec{F_r}$ not axial, a differential modification of the three jet rates is done. This force produces an acceleration a of the capsule, and now the dynamical equilibrium of the capsule is given by

$$\vec{F_a}+\vec{F_b}+\vec{F_c}+\vec{F_t}+m\vec{a}-k\vec{v}=0$$

being m the mass of the capsule, v its velocity and k the friction coefficient of the capsule against the medium.

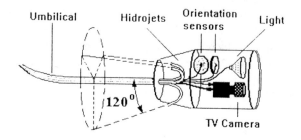

Figure 2 Sensor appendix layout

As the viscosity of the medium produces also a resisting force along all the umbilical, a deviation of the resulting force F_r produces also a displacement of the capsule as shown in fig. 3a, and a curving of the umbilical as shown in fig. 3b.

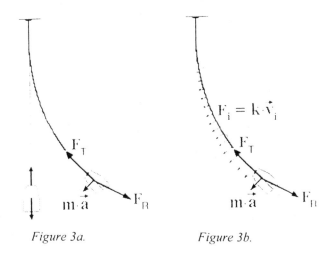

Figure 3a.   *Figure 3b.*

So the architecture of the sensor appendix is equivalent to a polyarticulated chain in which the angle between each element Δl_i and Δl_{i+1} is given by the direction of the propulsive vector at each instant, $\vec{F}_p(t)$ and the dynamics of the capsule and umbilical.

To study the accessible space of this appendix a simulation was done with an approximate model. In it the capsule moves in the plane of $\vec{F}_t$ and the instantaneous z-axis of the capsule. Velocity is proportional to $F_t \sin\theta$ being θ the angle between F_t and the z-axis. The time of acceleration to the equilibrium velocity, a few seconds in reality, is taken as zero.

The angle between consecutive elements Δl_i and Δl_{i+1} is $\beta_i(k)=c\beta_i(k-1)$

In this way it can be seen that the volume accessible by the appendix, by a judicious modulation of the jet rates, q_a, q_b, q_c has the shape shown in fig. 4. If the flow rate of the three jets, a, b and c, which is usually $q_a=q_b=q_c$ is unbalanced during some time t, the capsule describes curved arcs, as shown in fig. 5a. If after time t, the vector V_r is not zeroed, but aligned with the z-axis, with $q_a=q_b=q_c$, the capsule coasts to an equilibrium position with the umbilical in a straight line, as shown in fig. 5b.

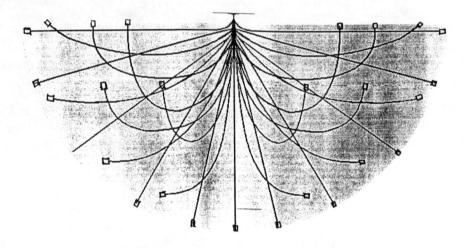

Figure 4. Space accessible modifying Fr(t)

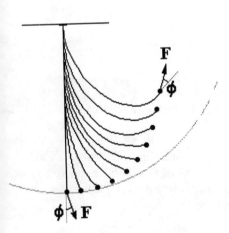

Figure 5a. Arcs described by the capsule when Fr is rotated.

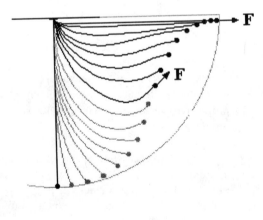

Figur 5b. Coasting trajectories

3. Control of the polyarticulated sensing appendix.

The sensing capsule is balanced in mass/volume to make its density as similar to the medium as possible. It has 6 degrees of freedom. The capsule position control in an weightless environment, with propulsion by liquid jets, has been studied and used in space applications [2],[3]. In underwater applications and with a feeder umbilical, this system presents some mobility restrictions, due specially to the length of the umbilical. When this length is fixed so as to have the capsule just at the desired distance of the working zone, its local mobility is restricted to an spherical surface with radius the length of the umbilical. In this case there are only 2 degrees of freedom, and with unidirectional jets, the minimum number necessary to move along this surface is 3.

So as to miniaturize the capsule and minimize its cost, only three jets are used. This implies that the capsule can't be oriented in stable equilibrium. Stable equilibrium means that the tension of the umbilical equals the resulting force of the jets, which must be along the z-axis and grater than zero, $F_a = F_b = F_c > 0$.

On the other hand, the volume of space accessible (i.e. not only the surface of the sphere) can be reached, and with a given orientation, following some trajectory. But in this case equilibrium is unstable, as it means $F_a = F_b = F_c = 0$ and this equality can (and will) be broken by any perturbation.

3.1. 2D position control

There are cases when it is sufficient to keep the capsule position in front of a given point, with no need of keeping its orientation. In these cases, the control of the capsule is done (fig. 6) by playing out a length of umbilical l=R and controlling F_a, F_b and F_c in function of the observed position error.

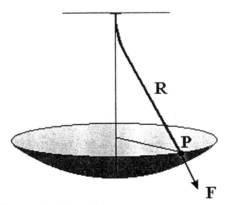

Figure 6. 2D stabilization in front of point P

The position error can be measured manually or automatically. Manual control using the images obtained by the camera is adequate for exploring and locating uses, when an autonomous control strategy would defeat the low cost objective. Once the object is located and the camera is in place, keeping it in place manually would be expensive in man-hours. An automatic solution is in order.

This can be done by the automatic analysis of the visualized image and tracking of some relevant object in the scene, [4], [5]; or it can be done via position sensors in the capsule.

In the experimental system developed, the two methods have been simultaneously used. The tracking system has not mean an overcost, as the host in which this appendix has been mounted is a ROV, GARBI [6],[7],[8], with a real time tracking system on board. As position sensor, a bidimensional angular sensor has been used. The clinometers used, one each in the X- and Y-axis of the capsule, move a 270° potentiometer through a 3:2 reduction. This allows pitch and yaw of up to $\pm 200°$ This angular limitation has been introduced so as to reduce the possibility of self-strangulating loops in the umbilical, which would cut the feeding of the jets, with the result of a capsule out of control. The station keeping of the capsule in front of a point can be done using the known angles of pitch and yaw, as the torsion of the umbilical in a limited environment is minimal. The low resolution of those measures give a great uncertainty zone, but it is of great help as an incremental measure to help in stabilizing the capsule by visual tracking, fig. 7.

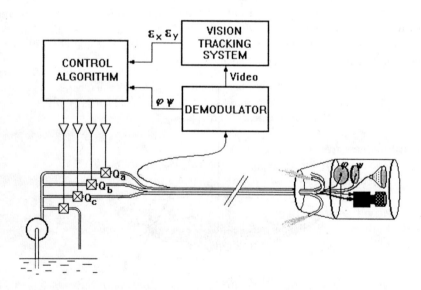

Figure 7a. *Control loop hardware structure*

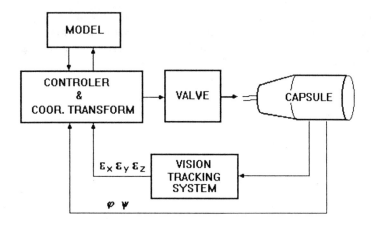

Figure 7b. Control loop logical structure

The control mission is essentially one of transforming the errors (ε_x and ε_y are angular errors in the x and y directions, ε_z is the relative variation of distance along the z-axis) into the forces along the three jet axes, and to adjust the parameters of the control action to compensate the variations due to the differences between the valves.

3.2. 3D position control

The elasticity of the umbilical allows flexion but practically no axial torsion. This means that the capsule movements have 5 degrees of freedom. With the 3 water jets it is theoretically possible to reach a point (x,y,z) in the spherical volume and with a given orientation. This, though, needs calculation and tracking of some trajectory. And when in place, the equilibrium is unstable, as said before. To be able to control position and orientation in 3D space (not only in the spherical surface), that is, with 5 degrees of freedom, another 3 water jets are necessary.

Two distributions are possible:
    -the 6 jets on the capsule
    -3 jets on the capsule, as before, and the other three on the umbilical.

As the second solution allows the minimizing of the sensor capsule, and most important, allows a smaller and more flexible umbilical in the last section, it has been chosen for the experimental prototype.

With the geometry of fig. 8, jets J_a, J_b and J_c are antagonistic to jets J_d, J_e and J_f which are situated at an preadjustable distance up the umbilical. This distance can be modified to adapt it to the dimensional characteristics of different working environments.

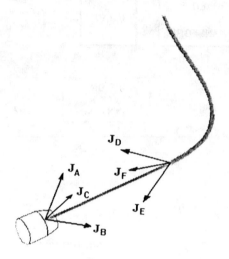

Figure 8. Capsule control with 5 degrees of freedom and 6 jets

It can be shown by simulation that with 6 jets and 5 degrees of freedom, the capsule orientation (2 freedoms) can be used to keep the point of view X,Y,Z (fig. 9)

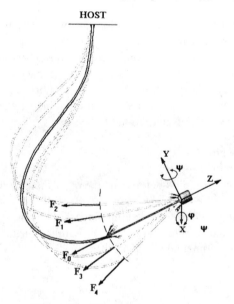

Figure 9. 3D sensor positioning.

4. The vision tracking system

A tracking system based on computer vision has been implemented to servocontrol the position of the appendix relative to the target. The vision system is based on a PC fitted with a special purpose image processing card. The computer on the host ROV gets the image through the cable associate to the camera in the appendix.

From the camera image the special purpose board implemented performs, at video rate, a polar transformation of some regions of interest previously selected by the system itself (or over the whole image). This kind of image transformation has been used by a lot of authors [9][10].

This polar transformation has been optimized in order to permit a low cost hardware implementation [11]. The result is a 8 radii polar codification $r(\theta)$ ($\theta=0:7$). It is used in fast-recognition of local features in selection and tracking algorithms. Thus, the PC processing unit only has to read this information and apply the recognition algorithm. (Fig. 10).

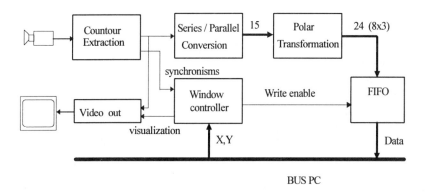

Figure 10. Special purpose image processor architecture

For tracking initialization the user has to indicate the area (or areas) of the image where the object to be tracked is included (*searching window*). From this area the system selects those singular points in the image that will be tracked. This selection is based on the reliability that these singular points show in front of the recognition algorithm [12]. The total selection time is never longer than 100 *ms*.

To track each of these singular points the following algorithm is used. For every new image, the program searches for the singular point in an environment centered on its last position (*tracking window*). The search is based on the evaluation of an error function which compares the current image region analyzed with that corresponding to the singular point acquired during the initialization. The comparison is performed over the transformed image by means of the following error function:

$$E(i,j) = \sum_{\theta=0}^{7}\left[r(\theta) - m_{ij}(\theta)\right]^2 \qquad \forall (i,j) \in \textit{tracking window}$$

where $r(\theta)$ is the polar codification of the tracked singular point and $m_{ij}(\theta)$ is the polar codification corresponding of each pixel (i,j) in the tracking window.

The point (i,j) where $E(i,j)$ gets its minimum value is taken to be the current position of the tracked singular point if this value does not surpass a given threshold. Else, a new image is acquired and $E(i,j)$ is evaluated again. If after a few frames that singular point is not localized, a new searching window is open near the last predicted position to select some new singular points (*reinitialization*).

Three of these tracked points can be taken, defining a plane. Their relative positions over the time determine the relative pan and tilt angles of the appendix and also the relative distance to the selected plane [6]. More than one set of three points can be tracked in order to increase the reliability of the deviation measurements. The overall computation of this tracking process takes less than 20 *ms*, thus new data can be fed to the control loop at video rate.

5. Results

The positionable sensor appendix studied has been tested with an experimental prototype with 3 jets. Mobility has resulted very similar to the one obtained by simulation.

The pump is a centrifugal one, half an HP of power, giving a flow of 100*l/min* and a mean propulsive force of about 2+2+2 *kg*. Proportional valves for the flow control have the handicap of their hysteresis and dead zone, greatly hindering the control at low speeds. To solve this problem servovalves would be indicated, but their high price defeat the objectives of low cost. Instead, a fourth proportional valve with its flow connected directly to the return, has been used (fig. 7), to move the working point of the valves well out of the dead zone. This means working with higher control inputs, not only for position changes but also for low correcting flows.

This appendix can be very useful as a complement for a ROV, or directly as a tool for inspecting very restricted spaces, where conventional ROV's can't go.

6. References

[1] Becjzy, A.K. 1995. "Virtual reality in telerobotics". 7th ICAR. Sant Feliu de Guixols. Catalonia.
[2] Ullman, M.A. 1993. "Experiments in autonomous navigation and control of multimanipulator, free-flying space robots". PhD thesis. Stanford University.
[3] Russakow, J., Rock, S.M. and Khatib, O. 1995. "An operational space formulation for a free-flying, multi-arm space robot". ISER'95.
[4] Moravec, H.P. 1977. "Towards automatic visual obstacle avoidance", Proc. Int. Joint Conf. on Artificial Intelligence, p. 584. Cambridge, MA.
[5] Rosenthaler, L., Heitger, F., Kübler, O. and Heydt, R. 1992. "Detection of general edges and keypoints", ECCV'92. pp. 78-86.
[6] Amat, J., las Heras, M., Villà, R. 1994. "Vision based underwater robot stabilization". 2nd IARP. Monterey. California.
[7] Amat, J., Codina, J., Cufí, X., Puigmal, J. 1995. "Vision based control of underwater vehicle for long duration observations". 7th ICAR. Sant Feliu de Guixols. Catalonia.
[8] Amat, J., Batlle, J., Casals, A., Forest, J. 1996. "GARBÍ: A low cost ROV, constraints and solutions". 6th IAPR. Toulon-La Seyne.
[9] Amat, J. and Casals, A. 1989. "Real time tracking of targets from a mobile robot". Intelligent Autonomous Systems 2, vol. 1, Amsterdam.
[10] Friedland, N.S. and Rosenfeld, A. 1992. "Compact object recognition using energy-function-based optimization", IEEE Trans. on Pattern Anal. and Mach. Int., vol. 14, n. 7, pp. 770-777.
[11] Aranda, J. 1996. "Image Processing card for real time target tracking". DCIS'96. Sitges. Catalonia.
[12] Aranda, J. 1996. "Selection and identification of image singular regions to track them" (in Catalan). 1st Workshop in Control, Robotics and Perception. UPC. Barcelona.

Experiments in Real–time Vision–based Point Stabilization of a Nonholonomic Mobile Manipulator

D.P. Tsakiris, K. Kapellos, C. Samson, P. Rives and J.-J. Borrelly
Project Icare, INRIA, U.R.-Sophia Antipolis,
2004, Route des Lucioles, BP 93,
06902 Sophia Antipolis Cedex, France
{first_name}.{last_name}@sophia.inria.fr

Abstract:

The stabilization to a desired pose of a nonholonomic mobile robot carrying a manipulator arm, based on sensory data provided by a camera mounted on the end-effector of the arm, is considered. Instances of this problem occur in practice during docking or parallel parking maneuvers of such vehicles. The visual data obtained as the camera tracks a target of known geometry are used to implement a continuous time–varying state feedback for the stabilization of the mobile robot. We focus on the case where the camera moves parallel to the plane supporting the mobile robot. The experimental evaluation of the proposed techniques uses a mobile manipulator prototype developed in our laboratory and dedicated multiprocessor real–time image processing and control systems. The real–time programming aspects of the experiments are handled in the context of the ORCCAD control architecture development environment.

1. Introduction

We consider here the experimental evaluation of certain recently–developed techniques for the stabilization to a desired pose of a mobile manipulator composed of a nonholonomic wheeled unicycle–type vehicle on which a holonomic manipulator arm is mounted (Tsakiris, Samson and Rives [18], [19]).

The presence of nonholonomic constraints makes this task non–trivial. These constraints arise from the rolling–without–slipping of the mobile platform's wheels on the plane supporting the system and constrain its instantaneous motion, whose component lateral to the heading direction has to be zero.

The kinematics of such nonholonomic mobile robots can be modeled as drift–free controllable nonlinear systems with fewer controls than states, with the controls entering linearly in the state equations. One of the approaches developed to solve the stabilization problem for such systems is the use of time–varying state feedback, i.e. control laws that depend explicitly, not only on the state, but also on time, usually in a periodic way. Samson [14] introduced them in the context of the unicycle's point stabilization and raised the issue of the rate of convergence to the desired equilibrium. M'Closkey and Murray [6], [7], Pomet and Samson [11] and Morin and Samson [8] derived continuous non-Lipschitz periodic time–varying exponentially stabilizing controls, which make the closed–loop system homogeneous of degree zero.

The implementation of such a closed-loop control scheme requires the state of the mobile robot to be estimated at every time instant. In our work, we use visual data from a camera which is mounted at the tip of the manipulator arm. The extra degrees–of–freedom associated with the arm, make it possible to position the end–effector, and thus the camera, independently from the mobile platform using a visual servoing approach. In this way, the camera is able to track a target of interest, enabling the localization of the system with respect to this target, while the nonholonomic mobile platform performs the maneuvers necessary to its own positioning.

For static manipulator arms, the visual–servoing approach provides a way of accomplishing such vision–based stabilization tasks by introducing directly visual feedback in the system's control loop (Espiau, Chaumette, Rives [2], Hashimoto [5], Hager and Hutchinson [4]). To use this approach for mobile manipulators, we extend it to the case of mobile robotic systems with nonholonomic motion constraints. Previous work in this area by Pissard–Gibollet and Rives [9] shows how to position a camera mounted on a nonholonomic mobile robot in front of a given target, without, however, explicitly controlling the final position and orientation of the mobile robot.

The time–varying stabilizing control scheme presented in [18] uses the unicycle's absolute position and orientation, which are reconstructed from the visual data. Alternatively, a scheme is proposed in [19] where the visual data enter directly in the control loop, without the intermediate step of state reconstruction. This gives rise to a visual–servoing scheme for stabilizing the nonholonomic mobile manipulator to a desired configuration, the first such scheme, to our knowledge, to appear. Details are presented in section 2.

The software implementation of the control and sensing schemes for our experiments and the treatment of the corresponding real–time aspects need to be done in a coherent way. To do that, we use the ORCCAD control architecture design environment (Simon, Espiau, Castillo and Kapellos [16], [3]), as explained in section 3.

Our experiments were carried out using a mobile manipulator prototype developed at INRIA. Initial experiments using odometry and the state reconstruction–based method of [18] agreed with the results of the corresponding simulation studies. Experiments in progress demonstrate the validity of the visual servoing method of [19], focusing on its comparison with the previous technique and on improving its robustness in the presence of noise, delays and parameter uncertainties. Details on the experiments are given in section 4.

2. Vision–based Point Stabilization

Let (x, y) be the position of the midpoint M of the wheel axis and θ be the orientation of the mobile robot with respect to an inertial coordinate system coinciding with the desired equilibrium configuration in point O. We consider

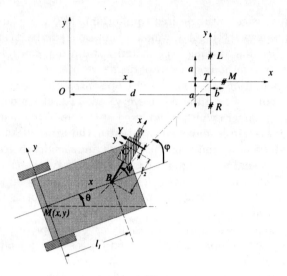

Figure 1. Unicycle with Camera

the unicycle kinematic model with heading speed v and angular velocity ω:

$$\dot{x} = v\cos\theta, \quad \dot{y} = v\sin\theta, \quad \dot{\theta} = \omega. \tag{1}$$

Let ψ be the angle of the manipulator arm with respect to the body of the mobile robot and let ω_ψ be the corresponding angular velocity. Then

$$\dot{X} = B_3(X)\,(v,\,\omega,\,\omega_\psi)^\top, \tag{2}$$

where $X \stackrel{\text{def}}{=} (x, y, \theta, \psi)^\top$ is the system state and $B_3(X)$ can be easily specified from equation 1.

Consider also a fixed target containing three easily identifiable feature points arranged in the configuration of fig. 1, which we suppose to be at a distance d from the point O. The distances a and b (fig. 1) are assumed to be known. The coordinates of the three feature points with respect to the camera are $(x_p^{\{C\}}, y_p^{\{C\}})$, $p \in \{l, m, r\}$ and their corresponding coordinates on the (1-dimensional) image plane are Y_p, $p \in \{l, m, r\}$.

From the system kinematics and the perspective projection camera model, we get the mapping $Y = \Phi(X)$ between the system state X and the sensory data $Y \stackrel{\text{def}}{=} (Y_l, Y_m, Y_r, \psi)^\top$. The corresponding Jacobian is $J(X) \stackrel{\text{def}}{=} \frac{\partial \Phi}{\partial X}(X) = B_2(X)\,B_1(X)$, where the matrices $B_1(X)$ and $B_2(X)$ are given in [19]. The matrix $B_2(X)$ corresponds to the interaction matrix of [2], [9]. The dimensions l_1 and l_2 of the mobile robot are shown in fig. 1 and f is the focal length of the camera.

The problem that we consider is to stabilize the unicycle to the desired configuration $X_\star = (x_\star, y_\star, \theta_\star, \psi_\star)^\top = 0$. The corresponding sensory data $Y_\star = (Y_{l\star}, Y_{m\star}, Y_{r\star}, \psi_\star)^\top = \Phi(0)$ can be directly measured by putting the system in the desired configuration or can be easily specified, provided d is also known, along with the target geometry a and b.

An exponentially stabilizing control is considered for the unicycle, while a control that keeps the targets foveated is considered for the camera.

The following state transformation brings equations 1 in the so-called *chained form* [11], [8]:

$$(x_1, x_2, x_3)^\top = \Psi(X) \stackrel{\text{def}}{=} (x, y, \tan\theta)^\top. \quad (3)$$

The unicycle control, that can be used if the state is known or reconstructed, is given by:

$$v(t, X) = u_1(t, \Psi(X)) / \cos\theta, \quad \omega(t, X) = \cos^2\theta\, u_2(t, \Psi(X)) \quad (4)$$

where u_1 and u_2 are the time–varying state–feedback controls, developed by Morin and Samson [8] for the 3–dimensional 2–input chained–form system and which are given in terms of the chained–form coordinates of equation 3 by:

$$\begin{aligned} u_1(t, x_1, x_2, x_3) &= k_1\left[\rho_3(x_2, x_3) + \alpha(-x_1 \sin wt + |x_1 \sin wt|)\right]\sin wt, \\ u_2(t, x_1, x_2, x_3) &= -\frac{k_3}{\rho_3(x_2, x_3)}\left[|u_1|x_3 + k_2 u_1 \frac{x_2}{\rho_2(x_2)}\right], \end{aligned} \quad (5)$$

where $\rho_2(x_2) \stackrel{\text{def}}{=} |x_2|^{\frac{1}{3}}$, $\rho_3(x_2, x_3) \stackrel{\text{def}}{=} \left(|x_2|^2 + |x_3|^3\right)^{\frac{1}{6}}$, w is the frequency of the time–varying controls and α, k_1, k_2, k_3 are positive gains. The exponential convergence to zero of the closed–loop system can be demonstrated using the homogeneous norm $\rho(x_1, x_2, x_3) \stackrel{\text{def}}{=} \left(|x_1|^6 + |x_2|^2 + |x_3|^3\right)^{\frac{1}{6}}$.

The arm control ω_ψ is chosen to keep the targets foveated by regulating the angular deviation of the line–of–sight of the camera from the targets to zero, while the mobile robot moves. It is specified so that Y_m is made to decrease exponentially to $Y_{m\star}$, i.e. by making the closed–loop system for Y_m behave as $\dot{Y}_m = -k_4(Y_m - Y_{m\star})$, where k_4 is a positive gain. This gives

$$\omega_\psi(t, X, Y) = -\frac{k_4}{\mathcal{J}_{2,3}}(Y_m - Y_{m\star}) - \left(\frac{\mathcal{J}_{2,1}}{\mathcal{J}_{2,3}}v + \frac{\mathcal{J}_{2,2}}{\mathcal{J}_{2,3}}\omega\right), \quad (6)$$

where $\mathcal{J}_{2,i}$ is the $(2,i)$–entry of the 4×3 matrix $\mathcal{J}(X) \stackrel{\text{def}}{=} B_2(X)B_1(X)B_3(X)$. In particular, $\mathcal{J}_{2,3} = -f - \left(\frac{Y_m^2}{f} + \frac{l_2 f}{x_m^{\{C\}}}\right)$. The first term of equation 6 makes the arm track the targets, while the term in parenthesis pre–compensates for the motion of the mobile robot.

A useful simplification of this law is obtained by ignoring the pre–compensation term and by approximating the element $\mathcal{J}_{2,3}$ by its first term, giving finally

$$\omega_\psi(t, Y) = k_4(Y_m - Y_{m\star}) / f. \quad (7)$$

This does not depend explicitly on our system's models, whose parameters may be imperfectly known, and will suffice to keep the targets in the field–of–view of the camera.

The control $\mathcal{U}$ for the full system is then

$$\mathcal{U}(t, X, Y) = \bigl(v(t, X),\ \omega(t, X),\ \omega_\psi(t, X, Y)\bigr)^\top. \quad (8)$$

A procedure for the estimation of the relative planar configuration of the target and the camera using three target feature points has been considered [18], similar to the one presented in Sugihara [17]. From the projections of the target features on the image plane, it is possible to estimate the angles with which these points are seen from the optical center C. Assuming that we know the target geometry, the relative configuration of the target with respect to the camera can be computed and the state (x, y, θ) can be estimated using standard geometry.

However, since we are interested in positioning the mobile robot to the desired configuration $X_\star = 0$, while starting relatively close to it, we could attempt to do so without reconstructing explicitly its state. Since $Y = \Phi(X)$, the state X can be approximated up to first order by

$$\hat{X}(Y) = (J_\star)^{-1}(Y - Y_\star), \qquad (9)$$

where $J_\star \stackrel{\text{def}}{=} \frac{\partial \Phi}{\partial X}(0) = B_2(0)\, B_1(0)$.

The proposed control law for the mobile manipulator can thus be expressed as a function of only the sensory data:

$$\mathcal{U} = \mathcal{U}(t, Y). \qquad (10)$$

3. Real–time Programming

Robotic applications are real time dynamical systems involving the combination of hardware, system software and dedicated algorithms. The successful design and implementation of an application strongly depends on the methodology used to organize these elements. The availability of CAD tools is a great help to the end-user to specify and to ensure the quality of the produced programs in terms of safety and performance. ORCCAD is a control architecture design environment (Simon, Espiau, Castillo and Kapellos [16], [3]) that we are currently developing to satisfy such requirements.

Systems where continuous and discrete event aspects interact are known as hybrid [1]. Robotic applications belong to this class of systems. Given that an established unified computational framework is not available to describe and analyze all the different aspects of hybrid systems, the ORCCAD system proposes a development methodology and a set of tools addressing coherently problems arising in all these aspects. Each step of the design of the application may involve the use of different classes of models: their interfacing is an important part of this procedure. This system makes some parts of this procedure transparent to the user by selecting and merging two computational models for the continuous and discrete-event parts of the application : the data-flow model for the specification of the former and the computation model defined by the semantics of the synchronous languages used for the specification of the latter. Their merging is obtained by the definition of two key entities: the *Robot-task* (RT), representing an elementary robotic action, where control aspects are predominant, encapsulated in a logical behavior expressed in terms of input/output events constituting its interface with the environment and the other entities of the ORCCAD system; the *Robot-procedure* (RP), the basic element of which are the RTs, and where only behavioral aspects are considered. These aspects are specified, verified and implemented using the ESTEREL synchronous language whose semantics are expressed in terms of transition systems (automata). The above-mentioned events are characterized as pre-conditions, indicating that the RT can begin, exceptions, indicating critical situations inhibiting the execution of the RT and post-conditions, indicating the normal termination of the RT.

According to the proposed programming methodology, the first step towards the implementation of our experiment consists in identifying all the elementary actions needed to perform it: these correspond to the RTs. The basic idea is to associate a control law to a sub-objective of the experiment and to identify the set of events related to its execution. Subsequently, we specify and validate the identified RTs and we compose them in the form of a RP. After its validation, the whole specification is translated into an executable program to be loaded on the target architecture.

Robot-task Specification and Validation : Three sub-objectives, corresponding to three RTs, are identified for our application : the initialization, the stabilization of the mobile manipulator and the ending.

The INITIALIZATION RT drives the base and the arm in the working area and sets them in a desired given configuration. Two events are observed: i) a type-3 exception *joint_lim* monitoring the joint limits of the arm and ii) a post-condition *pos_reached* indicating that the pre-defined fixed configuration of the base and the arm have been reached.

The POINTSTABILIZATION RT starts after checking that the connection with the vision system has been established. During its execution the vision-based stabilization control law described in section 2 is executed and possible system errors, such as the camera losing sight of the target, are monitored. The control input $\mathcal{U}$ is given by equation 10. From the user point of view, the RT is specified by interconnecting a set of modules implementing algorithmic and behavioral functionalities. The internal structure of this RT is designed through the Graphical User Interface of ORCCAD and depicted in Figure 2a, where the following modules appear:
− READENC: reads the mobile manipulator state from wheel and arm joint encoders (wheel and joint position and velocity),
− VISION: reads the results of the image processing $Y = (Y_l, Y_m, Y_r)^\top$,
− POSVISION: interprets the results of the image processing in order to compute $\hat{X}(Y)$ given by equation 9 ,
− CONTROL: computes $\mathcal{U}(t, Y)$ given by equation 10.

The logical behavior of the RT is specified by the VIS_PREC, POST and RTA modules as follows: i) the *pre-conditions* of the RT are that the connection with the vision machine is established and a valid information is obtained (*signal_ok* event), ii) a type-2 *exception* is the loss of the video signal (*signal_lost* event), iii) type-3 *exceptions*, like the arm reaching its joint limits and hardware failures, lead to an emergency stop of the vehicle (*joint_lim, robot_fail*, etc. events), iv) the *post-condition* of the RT is that the mobile robot reached its

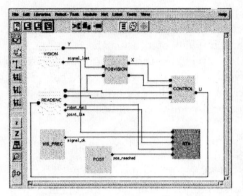 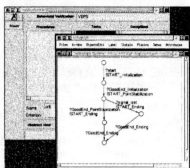

Figure 2. a)POINTSTABILIZATION RT b) VBPS automaton observation

final position (*pos_reached* event).

The specification is completed by assigning a sampling period of 40 msec (video rate) to all modules. In addition, non-blocking message passing mechanisms are selected for inter-module communication.

The ENDING RT drives the arm to its home position in a similar way as for the INITIALIZATION RT.

Robot-procedure Synthesis : The VBPS RP is designed to specify the evolution of the application as the composition of the previously defined RTs. It simply states that the application will start after user confirmation, which corresponds to the pre-condition *start*; its nominal execution consists of the sequencing of the three RTs: INITIALIZATION, POINTSTABILIZATION and ENDING:

```
RP VBPS
[
pre-condition:          start
<main_program>
                        INITIALIZATION();
                        POINTSTABILIZATION();
                        ENDING()
exception_type2:        (signal_lost),
                        ENDING();
exception_type3:        (robot_fail), (sensor_fail), (joint_lim), ...
]
```

This specification of the RP is programmed in ESTEREL. Its validation can be done in the ORCCAD environment. First, the safety property (any fatal exception is appropriately handled by emission of a specific signal leaving the system in a safe situation) and the liveliness property (the RP always performs its goal in a nominal execution) are automatically proven.

Then, the conformity of the RP behavior with mission constraints can be verified, by observing the global automaton with respect to criteria restricted only to events *relevant* to these constraints and indicated by the user in terms of RTs and exceptions. Figure 2b illustrates this process, for the case when the user indicates that he is interested to see the relation between the RTs and the type-2 exception *signal_lost*. The resulting automaton, after observation, is shown: at the occurrence of a *signal_lost* event, the ENDING action is required to start.

Real-Time Robot Control Software : ORCCAD provides tools for the specification and formal verification of robotic applications as well as the automatic generation of the corresponding real-time code [10], integrated within a set of dedicated graphical user interfaces, several aspects of which are illustrated below. The whole robot control software is obtained by the translation of the RP specification into C++ code, independent of the real-time target environment (operating system, computational hardware), which is subsequently integrated with its run-time libraries. It is made of three parts:
• a set of real-time tasks, usually periodic and synchronized according to consumer/producer protocols, dedicated to the computation of the continuous aspects of the system (mainly the control laws);
• the discrete event automaton directly generated in C by the ESTEREL compiler;
• the interface between the two previous parts. Its role is dual: it transforms significant changes occurring in the continuous aspects of the system into the discrete events that are fed to the automaton. And vice versa, it links each output event produced by the automaton with appropriate requests made to the real-time task layer.

Robot Control Software Generation : From a graphical and textual high-level specification of the robotic application, the ORCCAD environment automatically produces the corresponding real-time program to down-load and execute on the target architecture. The real-time operating system used is VXWORKS 5.2. The generated code runs on a single processor at video rate.

4. Experimental Results

The Robotic System : Our test-bed is a unicycle-type mobile robot carrying a 6 d.o.f. manipulator arm with a CCD camera and equipped with a belt of eight ultrasonic sounders (see figure 3) ([12], [13]). The on-board computer architecture is built around a VME backplane. The robot controller is based on a Motorola MVME 162 with IP modules ensuring an actuator velocity servo-loop at a rate of 1ms.

Vision Hardware and Software : To overcome difficulties in image processing, induced by strong real-time constraints and the processing of large amounts of data, we have developed a vision machine [12] characterized by its modularity and its real-time capabilities. Its architecture is based on VLSI chips for low level processing and on DSP processors for more elaborate processing. To facilitate program development, the vision machine has been implemented on an independent VME rack outside of the mobile robot, keeping however the possibility to transfer the vision boards to the on-board rack. The software used on the vision machine implements the concept of active window [12]. An active window is attached to a particular region of interest in the image and its task is to extract a desired feature in this region. At each active window we associate a Kalman filter able to perform the tracking of the feature along the image sequence. Several active windows can be defined, at the same time, in the image and different types of processing can be done in each window. In our application, the visual data (Y_l, Y_m, Y_r) are extracted at video rate from 3 active windows, which track the targets as the camera moves (see figure 4).

Figure 3 : Mobile Manipulator Figure 4 : Active windows

Experiments : In the experimental results presented below, we use the control law 10 with the unicycle controls 4, the arm control 7 and the state approximation 9 by sensory data. The following parameters corresponding to the models of section 2 are used: $l_1 = 0.51\ m$, $l_2 = 0.11\ m$, $d = 2.95\ m$, $f = 1\ m$. The following gains are used for the above control laws: $w = 0.1$, $k_1 = 0.25$, $k_2 = 2$, $k_3 = 100$, $\alpha = 10$, $k_4 = 12$.

The visual data (Y_l, Y_m, Y_r) are acquired and pre–processed by the vision system, which transmits them to the robot controller for use in implementing the above control laws. This process introduces a delay δ to the system, which, in our case, equals to 3 sampling periods (120 msec). Thus, the visual data used to control the system at time t correspond to its state at time $t - \delta$. This degrades the performance of the control and cannot be corrected by e.g. increasing the gains of the control laws. To compensate for these delays, we use the arm joint and robot wheel encoders to calculate corrections to be added to the controls for the arm and for the wheel angles. In the case of e.g. the arm, this compensation takes the form of the term $-k_4[\psi(t) - \psi(t - \delta)]$, which is added to the control law 7 and where $\psi(t)$ is the arm encoder reading at time t. A comparison of the step responses of the arm, in the case where delay compensation is used and where no such compensation is used, shows that the long oscillations that the uncompensated arm does before settling down can be significantly reduced.

Initial experiments used the raw visual data to calculate the state and the controls. The resulting (x, y)-trajectory is plotted in figure 5. In the following figures, the dotted lines represent data obtained by odometry, while solid ones are data obtained by vision. As it is evident from this figure, implementation of this scheme leads to significant small oscillations and jerks during the motion of the system.

Subsequent experiments used Kalman filtering of each of the state variables (x, y, θ), to make the corresponding trajectories smoother and compensation of the vision-induced delays was introduced. No filtering was used on the visual data themselves. The resulting (x, y)-trajectory, the evolution of the state (x, y, θ), of the unfiltered visual data Y_m, as well as the corresponding controls v, ω are shown in figures 6-10 when the system starts at $(x, y, \theta) = (-0.03, -0.23, 0.02)$. Each period of the time-varying controls corresponds to 1570 samples (data on the state of the system are recorded every 40 msec). The system motion is noticeably ameliorated.

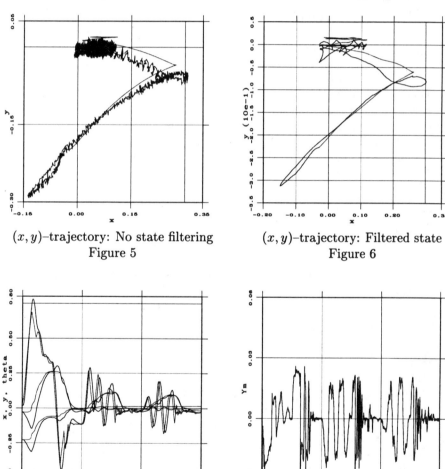

(x, y)-trajectory: No state filtering
Figure 5

(x, y)-trajectory: Filtered state
Figure 6

x, y, θ : Filtered state
Figure 7

Y_m : Visual data (Unfiltered)
Figure 8

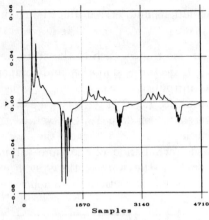 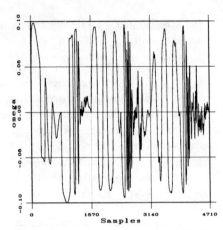

v :Time–varying heading speedω : Time–varying angular velocity
Figure 9Figure 10

The results presented here are preliminary, demonstrating the feasibility of the proposed method, leaving however several issues to be further studied, such as the difference between the (x, y)-trajectories obtained by vision and odometry in figure 6, possibly due to calibration and state approximation errors, and the excitation of small oscillations near the desired final configuration (c.f. figures 7–10), possibly due to robustness problems of the controls in the presence of model parameter errors.

5. Conclusions

The feasibility of a novel class of vision-based point stabilization techniques for nonholonomic mobile manipulators was demonstrated in this work. Their robustness with respect to noisy data, delays and modeling errors is currently under study. We consider the planar case, where a one-d.o.f. arm carrying a camera moves parallel to the plane supporting the mobile base. The case of an arm with more d.o.f. and a more general visual setting is also currently under investigation.

References

[1] R. Alur, C. Courcoubetis, N. Halbwachs, T. Henzinger, P. Ho, X. Nicollin, A. Olivero, J. Sifakis and S. Yovine, " The Algorithmic Analysis of Hybrid Systems", *Theoretical Computer Science*, Vol. 137, 1995.

[2] B. Espiau, F. Chaumette and P. Rives, "A New Approach to Visual Servoing in Robotics", *IEEE Trans. on Robotics and Automation* **8**, 313-326, 1992.

[3] B. Espiau, K. Kapellos and M. Jourdan, "Verification in Robotics: Why and How?", *Robotics Research*, The 7th Intl. Symposium, Eds. G. Giralt and G. Hirzinger, Springer Verlag, 1995.

[4] G.D. Hager and S. Hutchinson, Eds., "Vision–based Control of Robotic Manipulators", Special section of *IEEE Trans. Robotics and Automation* **12**, 649-774, 1996.

[5] K. Hashimoto, Ed., *Visual Servoing*, World Scientific, 1993.

[6] R.T. M'Closkey and R.M. Murray, "Nonholonomic Systems and Exponential Convergence: Some Analysis Tools", *IEEE Conf. on Decision and Control*, San Antonio, Texas, 1993.

[7] R.T. M'Closkey and R.M. Murray, "Exponential Stabilization of Driftless Nonlinear Control Systems Using Homogeneous Feedback", preprint, Caltech, 1995.

[8] P. Morin and C. Samson, "Application of Backstepping Techniques to the Time–Varying Exponential Stabilization of Chained Form Systems", INRIA Research Report No. 2792, Sophia–Antipolis, 1996.

[9] R. Pissard–Gibollet and P. Rives, "Applying Visual Servoing Techniques to Control a Mobile Hand–Eye System", *IEEE Intl. Conf. on Robotics and Automation*, 1995.

[10] R. Pissard-Gibollet, K. Kapellos, P. Rives and J.J. Borrelly, "Real-Time Programming of Mobile Robot Actions Using Advanced Control Techniques", *Experimental Robotics IV*, The 4th International Symposium, Eds. O. Khatib and J.K. Salisbury, pp. 565-574, Stanford, USA, June 30-July 2, 1995.

[11] J.-B. Pomet and C. Samson, "Time–Varying Exponential Stabilization of Nonholonomic Systems in Power Form", INRIA Research Report No. 2126, Sophia–Antipolis, 1993.

[12] P. Rives, J.J. Borrelly, J. Gallice and P. Martinet, "A Versatile Parallel Architecture for Vision Based Applications", *Workshop on Computer Architecture for Machine Perception*, New Orleans, 1993.

[13] P. Rives, R. Pissard-Gibollet and K. Kapellos, "Development of a Reactive Mobile Robot Using Real–Time Vision", Third International Symposium on Experimental Robotics, Kyoto, Japan, October 28-30, 1993.

[14] C. Samson, "Velocity and Torque Feedback Control of a Nonholonomic Cart", in *Advanced Robot Control*, Ed. C. Canudas de Wit, Lecture Notes in Control and Information Sciences, No. 162, Springer-Verlag, 1990.

[15] C. Samson, "Time–varying Feedback Stabilization of Car–like Wheeled Mobile Robots", *The International Journal of Robotics Research* **12**, 55-64, 1993.

[16] D. Simon, B. Espiau, E. Castillo and K. Kapellos, "Computer-aided design of generic robot controller handling reactivity and real-time control issues", *IEEE Trans. on Control Systems and Technology* **1**, 1-24, 1993.

[17] K. Sugihara, "Some Location Problems for Robot Navigation Using a Single Camera", *Computer Vision, Graphics and Image Processing* **42**, 112-129, 1988.

[18] D.P. Tsakiris, C. Samson and P. Rives, "Vision–based Time–varying Mobile Robot Control", Final European Robotics Network (ERNET) Workshop, Darmstadt, Germany, September 9–10, 1996. Published in *Advances in Robotics: The ERNET Perspective*, Eds. C. Bonivento, C. Melchiorri and H. Tolle, pp. 163-172, World Scientific Publishing Co., 1996.

[19] D.P. Tsakiris, C. Samson and P. Rives, "Vision–based Time–varying Stabilization of a Mobile Manipulator", Proceedings of the Fourth International Conference on Control, Automation, Robotics and Vision *(ICARCV'96)*, pp. 2212-2216, Westin Stamford, Singapore, December 4-6, 1996 (Available on WWW in URL: http://www.inria.fr/icare/personnel/tsakiris).

Distributed Control of a Free-floating Underwater Manipulation System

K. Kapellos   D. Simon
INRIA, BP 93, 06902 Sophia-Antipolis, France
email: <kapellos, dsimon>@sophia.inria.fr

S. Granier   V. Rigaud
ISIA, 06904 Sophia-Antipolis   IFREMER, BP 330, 83500 La Seyne/Mer
email: <sgranier, rigaud>@toulon.ifremer.fr

Abstract: Robotic applications are real-time dynamical systems which intimately combine different components ranging from high-level decision making and discrete-event controllers to low-level feedback loops. Tightly coupling the two last components and considering them in a formal framework permitted, in a centralized approach, both crucial properties to be proved and efficient implementation. We examine this coupling in the case of an architecture where discrete-event and low-level controllers are spatially distributed and we propose, in the framework of the ORCCAD methodology, different methods for their implementation. Their impact to the verification process is analyzed. The experimental evaluation of the proposed techniques uses the IFREMER free-floating underwater manipulation system VORTEX-PA10. The real-time programming aspects of the experiment are handled in the framework of the ORCCAD programming environment targeting the PIRAT real-time controller.

1. Introduction

In applications like autonomous vehicles, aircrafts and robots, stand alone computerized controllers are being integrated to form cooperating subsystems which can together provide improved functionality, reliability, and reduced costs. Modern and future machinery will therefore, and to some extend already does, include *embedded distributed real-time computer control systems*. In this context, considering a given hardware architecture, programming of an application must deal with algorithms distribution over the different locations and implies exchange of real-time data between subsystems. The present contribution gives insight on the programming of some aspects of such a distributed robotic applications.

The work reported here is motivated by the particular objective of specifying, validating and implementing missions on the Vortex-Pa10 underwater manipulation system developed at Ifremer, where the control computers are distributed on a network. In [8] we specified and validated by formal methods and realistic simulations an underwater structure inspection mission considering the whole system. Distinct control laws were designed for each subsystem which were coordinated by a centralized discrete-event controller which rhythmed the logical evolution of the mission. It has been shown that a reasonable stabilization of the vehicle was achievable using acoustic or visual sensor-based control despite the arm motions. Besides, the correctness of the logical behavior of the mission has been formally verified taking advantage of the centralized implementation of the discrete-event controller.

This work is carried out in the framework of the development of the ORCCAD programming environment [9], which allows the specification, the validation and the implementation of robotic reactive applications. It clearly separates the specification of the discrete-event controller which rhythms the logical evolution of the application from the data handling used to implement the control laws on the target architecture. This is achieved by designing an application as a hierarchical and structured composition of *Robot-Tasks* (RTs), which harmoniously integrate discrete and continuous time aspects, in *Robot-Procedures* (RPs) which mainly handle logical behaviors. Using the synchronous approach as the semantics of composition for RTs and RPs pave the way to apply formal verification methods (and especially 'model-checking' approaches) to prove the mission correctness from a logical and temporal point of view [4].

This approach has been already used in various robotic applications [6]: even in the case of multi-processor systems the discrete-event controller was always centralized (i.e. compiled as a single automaton) thus allowing for its formal analysis. However, this centralized approach is not always timely efficient or may be unsafe or impossible to implement on a naturally distributed system as failures in the communication links between the controllers may leave some subsystems totally out of control. It is expected that distributing the logical control code over the distributed hardware may improve subsystems survivability.

As an extension of this previous work we now experiment parts of an underwater mission in a pool [8]. In this paper our interest is twofold and combines theoretical considerations and experimental validations:

- propose a methodology to distribute the discrete-event controller of the mission. Two solutions are envisioned:
  1) separately design the controllers of the subsystems (i.e independently of its context) and the global controller, and compile them into several object programs located on the different nodes. In that case, we obtain asynchronously connected controllers for which global formal verification is impossible and validation must be carried out using simulation.
  2) globally design the mission controller, compile the mission specification into a single object program, and then distribute it on the nodes. Thus, the specification of the logical evolution of the mission and its formal analysis can be globally carried out while its implementation will be automatically distributed ensuring that the verified properties hold in the resulting distributed program. This methodology is based on recent theoretical results in the field of reactive systems theory [3] and is currently integrated in our control architecture.
- experimentally validate the simulated control laws. Simulations taking into

account most of the temporal features of the controlled system enlightened limits of the hardware and suggested improvements e.g. adaptive triggering of acoustic sensors. Preliminary experiments will be used to assess these simulation results and will be used to better calibrate the simulation model. We consider several phases of the mission: preparing the vehicle for cruising, reaching the working area, and finally performing motions of the arm while the vehicle is actively stabilized.

2. The Discrete Event part of the Hybrid Controller

2.1. ORCCAD overview

ORCCAD ([9]) is a development environment for specification, validation by formal methods and by simulation, and implementation of robotic applications. Its conception is articulated around two entities which formally characterize a robotic action, the *Robot-Task* and the *Robot-Procedure*.

The *Robot-Task* (RT) models basic robotic actions where control aspects are predominants. Typical examples are the trajectory tracking of an arm manipulator, the positioning at a desired pose of an underwater vehicle, ... It characterizes in a structured way continuous time closed loop control laws, along with their temporal features related to implementation and the management of associated events. A data-flow well identified mode of execution implements the algorithmical (control law) part of the RT. The considered events may be pre-conditions, post-conditions and exceptions which are themselves classified in type 1 (weak), type 2 (strong) and type 3 (fatal) exceptions. The reception of these events rhythms the evolution of the action according to a pre-defined scheme: the satisfaction of the pre-conditions leads to the activation of the control law. During its execution, if a specified exception occurs, it is handled according to its type; the reception of the post-conditions implies the ending of the action. A reactive process is used to implement this discrete-event part of the RT. Therefore, in a RT intimately cooperate a data-flow scheme and a discrete-event controller; section 2.2 details this connection. RT design mainly falls in the field of automatic control design using block-diagrams and sampled time while calls to the underlying operating system and the encoding of the reactive program are automatically generated.

The *Robot-Procedure* (RP) entity models a robotic action of variable complexity. For example, the cruising phase of the inspection mission schedules trajectory tracking navigation followed by automatic wall following associated to starting and stopping conditions. It specifies in a hierarchical and structured way a logical and temporal *composition* of RTs and RPs in order to achieve an objective in a context dependent and reliable way, providing pre-defined corrective actions in the case of unsuccessful execution of RTs. The composition concerns only the logical aspects of the RT and it is systematized thanks to a well defined interface. Therefore, the RP can be considered as a discrete-event controller which rhythms the sequencing of RTs (through their local discrete-event controllers) following a user-defined scheme (section 2.3 details this aspect). Thus the end-user is able to design a full robotic application using a library of pre-defined validated high-level actions without worrying with

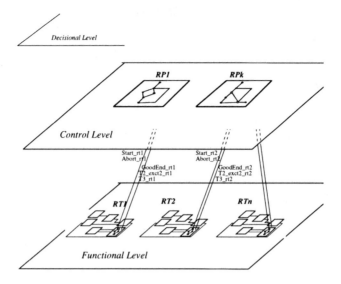

Figure 1. Control architecture

low level programming and implementation tricks.

Using robust control-laws and tuning the gains and parameters with a simulation tool like SIMPARC ([1]) ensures the stability of the physical system during RTs execution with specified performance. On the other hand the logical behavior of the RT is verified to satisfy critical properties of liveness and safety. The RP ensures a robust control of the physical system seen as a collection of RTs. Here, simulations are used to validate the transition phases and formal verification to prove the correctness of the logical and temporal behavior by checking critical properties and conformity with application's requirements. These well defined structures allows to systematize and thus automatize formal verification on the expected controller behavior.

The resulting control architecture is organized in three levels (Figure 1): in the *functional* one reside the RTs executing the low level control laws; thanks to the event driven interface they are sequenced by the RPs which are elements of the *control* level. Finally a *decisional* level should be ideally added on the top of the architecture to provide automatic or manual replanning.

2.2. Robot-Task Discrete-Event Controller

The evolution of a RT execution is characterized by three main phases, briefly speaking initialization, control law application and controlled ending. The activation of a RT starts after an external request, follows its local behavior rhythmed by the events monitored during the RT execution and signals its end, thus providing an *external abstract view* allowing for its manipulation. The role of the RT discrete-event controller is to model and implement this evolution.

We illustrate in Figure 2) the discrete-event RT controller in the form of an automaton where transitions are labeled by events : input one are prefixed by '?' and the output by '!'. The first phase copes with initialization aspects and starts after reception of the synchronization pre-condition event *Start_ rt*. By

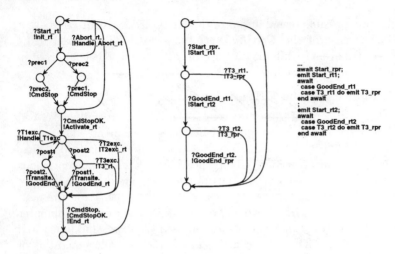

Figure 2. Robot-task and Robot-procedure local behaviors

the emission of the *Init_rt* output signal the discrete-event controller asks the services of a concurrently running process to prepare the RT execution : the necessary threads implementing the data-flow part of the RT are spawned and the used algorithms are initialized (communication devices, filters, memories, ...). Sequently, the inter-connected modules network is periodically executed, without sending commands to the actuators while the previous RT ensures the robot command until all the pre-conditions are satisfied. A request is then send to the previous RT discrete-event controller and after acknowledgment (*CmdStopOK*) the *Activate_rt* output signal asks from the current RT to start sending commands to the actuators thus entering in its second phase. During this phase the control law is applied to the robot and the presence of the specified exceptions and post-conditions are observed and handled. The post-conditions satisfaction implies the logical end of the RT, signaled by the *GoodEnd_rt* event and drives the RT to its third phase of execution, named transition phase. During this phase the commands are always send to the robot during the initialization of the following RT after which the ending code of the algorithms is executed and the threads are deactivated. We note that *Abort_rt* requests and type 2 or type 3 exceptions are handled at each phase of the execution of the RT signaled by the *T2_exception_rt* and *T3_rt* events. For clarity purposes in figure 2 a limited number of these transitions are designed.

2.3. Robot-Procedure Discrete-Event Controller

The RP discrete-event controller is aimed both to coordinate RTs and RPs in order to achieve a predefined behavior and to offer, as for the RT, the adequate event based interface which constitutes its abstract view. Consider for example a very simple RP, named *rpr*, sequencing two RTs, *rt1* and *rt2*. The behavior of the corresponding discrete-event controller is illustrated by the automaton figure 2 : upon the reception of the *Start_rpr* synchronization pre-condition signal the first RT is asked to be launched (*Start_rt1* event) and sequently the controller waits possible terminations of the RT issued by its discrete-event controller. In the case of the *GoodEnd_rt1* the second RT is asked to be launched (*Start_rt2* event) while in the case of *T3_rt1* the RP is aborted in order to apply a security procedure.

2.4. Specification using the ESTEREL language

The activity of the RT and RP discrete-event controllers is naturally described in terms of responses to stimuli originated by the environment. The use of the *reactive* model to formalize their behavior is therefore a natural choice. Following this model one considers communicating systems that continuous interact with their environment. When activated with an *input event* a reactive system *reacts* by producing *output events* and returns waiting for the next activation. Moreover this choice is justified by the efforts of the computer science community to provide mathematically sound formalisms and integrated environments for the specification, verification and efficient code generation of such systems.

The ESTEREL language [2] is a member of the synchronous languages family based on a rigorous mathematical semantics, especially designed to program reactive systems where control aspects are predominant. Due to an adequate set of primitives an ESTEREL program is very close to the system's behavior specification. The RP discrete-event controller given as example in section 2.3 is simply specified by the ESTEREL program and is easy to understand without particular knowledge of the language. For verification and simulation/implementation purposes the ESTEREL code is compiled into automata which are translated into several target languages such as C or Ada.

In ORCCAD system, ESTEREL is used both to encode RT and RP discrete-event controllers. From the user point of view, the ORCCAD programming environment automatically generates RTs discrete event controllers through a dedicated window while currently the RP programming must be left to the end-user. However, an ORCCAD compliant task-level language targeting ESTEREL is currently under development [5].

3. Centralized vs. Distributed Control

In the ORCCAD methodology for the design of the robot controller the specification of the RTs and RPs is followed by: i) their verification, aiming to prove that the specification is conform to the end-user's requirements and ii) their implementation w.r.t. the specification. For the discrete-event part of the robot controller a crucial question is the choice of the semantics of the composition of the specified discrete-event controllers in order to obtain the global one which rhythms the logical evolution of the whole application. Let us take the particular example of the *Start_rt* synchronization event which is an output one for the RP discrete-event controller and input for the RT; how this synchronization signal is exchanged with respect to the reaction of the two controllers? Two solutions can be adopted: the *centralized* one where the global discrete-event controller is executed as a single process and the synchronization signals are synchronously broadcast within each controller, and the *distributed*

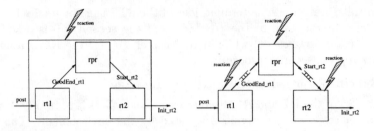

Figure 3. a) Synchronous and b) asynchronous compositions

one where the discrete-event controllers run on different processes (possibly on different processors) and the synchronization signals are asynchronously transmitted. These two solutions and their impact to the verification process are now detailed.

3.1. Centralized approach

The centralized approach consists in *synchronously* [2] compose the specified discrete-event controllers. This composition paradigm is illustrated figure 3a.

In reaction to some external event an atomic reaction occurs which propagates activation instantaneously to all discrete-event controllers at all levels of the hierarchy. Let us suppose the presence of the postcondition external event of the RT *rt1*. The reaction of the global controller to this event consists in activating the RT controller which broadcasts the *GoodEnd_rt1* signal. During the same reaction the RP controller receives this synchronization event and emits the *Start_rt2* event to the controller of the second RT. Thus *rt2* enters in its initialization phase emitting to the environment the *Init_rt2* output event which asks, as indicated in section 2.2, the preparation of the RT execution. We note here that the events *post, GoodEnd_rt1, Start_rt2, Init_rt2* are received and emitted at the same atomic reaction. Given also that ESTEREL compiler completely disappears the instantaneous communications from the resulting code time efficiency is ensured.

A second major advantage of the centralized approach is related to the verification aspects. It is known that a property verified by each of two discrete-event controllers separately is not necessarily verified when their interaction is considered. In the centralized approach a particular property is verified over the global behavior of the system taking into account even the particular details of each RT local behavior.

3.2. Distributed approach

Robotic or more generally mechanical systems are often composed by spatially distributed actuators and sensors in order to improve modularity, functionality and performance. Therefore, a centralized approach is not always feasible nor desirable in particular from the dependability point of view. Firstly, external events handling, in particular those concerned with fatal exceptions must be fast processed at the location they have been produced. In addition, for safety reasons, a minimum logical control must be provided at each robotic sub-system in order to give a certainly restricted but crucial decisional autonomy facing failures such as communication interrupts.

Therefore, face to a distributed system a distributed approach for the implementation of the discrete-event controller of the system must be envisioned. Several methods are available to built such controllers :

- separately consider each RT discrete-event controller and the RP one, compile them into several object programs located on the different nodes and make them communicate through asynchronous channels (first-in/first-out files for example). Let us revisit our pedagogic example: in Figure 3b the three reactive discrete-event controllers communicate through two channels. In this configuration *rt1* discrete-event controller as reaction to the presence of the *post* event sends through the network the *GoodEnd_ rt1* event to the *rpr* discrete-event controller and ends its reaction; the *rpr* reacts emitting the *Start_ rt2* event to *rt2* and ends its reaction; finally the *rt2* reacts and emits the *Init_ rt2* event.
  This approach has a major drawback. Even if each reactive part can be analyzed and verified individually their composition through non-deterministic communications inhibit any verification on the global system. A second approach tends to remediate this drawback.
- synchronously compose all the RTs and RPs discrete-event controllers, compile them into a single reactive program and then distribute it on the nodes in such a way that the parallel execution of these codes implements the initial program and each node has only to perform its own computations. Given a synchronous program (in oc format) and distribution directives locating each signal to a node, the *ocrep* tool [3] automatically distributes the centralized program to the corresponding nodes. The major advantage of this approach lies on the global verification capability as properties verified on the centralized program remains valid on its distributed implementation. Nevertheless, this approach increases the load on the communication links since additional dummy signals are exchanged in order to correctly synchronize each program.

4. Experimental results

Let us now focus to the case study we used to evaluate some of the proposed methods. This example is an extension of the work done in the Union project [7] to assess the design and verification methods of ORCCAD with an underwater test-bed application. The mission takes place in a pool, using the Vortex vehicle fitted with a manipulator.

4.1. Experiment setup

Vortex is a Remotely Operated Vehicle (ROV) designed by Ifremer as a test-bed for control laws and control architectures. It is equipped with a set of six screw propellers and with a traditional sensing set such as compass, inclinometers, depth-meter and gyrometers allowing to measure most of its internal state. A video camera is used for target tracking tasks and a belt with eight ultrasonic sounders allows to perform positioning and wall following tasks. Vortex is also equipped with an electric powered Mitsubischi Pa10 arm with 7 degrees of freedom to perform manipulation tasks (see Figure 5).

The vehicle control algorithms are computed on an external VME backplane running Pirat, a C++ library of objects dedicated to underwater vehicles control. At a higher level, i.e. control laws and mission management, the reactive synchronous ESTEREL language is used to design Robot-Tasks and Procedures, consistently with the ORCCAD approach. The control algorithms for the arm are run on a second VME backplane. As the two controllers only have a low bandwidth communication capability, control laws for Vortex and the Pa10 arm run independently and only short synchronization messages are exchanged on the communication link. This distributed architecture (Figure 4) is used to test the ORCCAD approach given the following informal end-user's requirements.

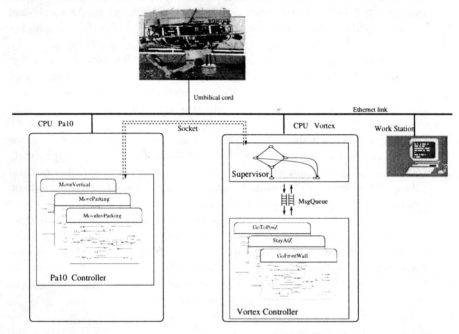

Figure 4. Experimental plant

Starting from the initial position, Vortex is set in station keeping mode at a pre-defined depth, using the on-board flux-gate for heading control. After completion of all necessary initialization[1] (e.g. the heading and depth set points are reached), the arm is moved backward in a safety position for navigation. Then the vehicle locks itself at a predefined distance of the front wall using acoustic sensors servoing. Once the vehicle is locked in the chosen corner the arm is moved forward and backward to check the platform stability under inertial and hydrostatic disturbances. The vehicle is then driven back to its

---

[1] It must be pointed out that the first initialization phase is also mission dependent e.g. choosing the initial heading set-point pointing towards the East direction will further drive the vehicle in the South-East corner of the pool.

initial position under the crank.

At any time the detection of a water leak or of an hardware failure must lead to a mission abortion leaving the system in a safe situation, i.e. setting an alarm and emergency surfacing with the arm locked in folded position. Other exceptions more specific to a given system or control algorithms are also defined inside the subtasks involved in the mission. Some actions are automatically triggered e.g. the first arm motion is triggered by an observer checking that the vehicle has reached its working depth. Others, e.g. the event starting the mission, are given through the keyboard thus sketching an embryonic teleoperation master station to be further developed.

This mission scenario naturally lead to split the mission in five main procedures: INITCRUISECONFIG consists in preparing the vehicle for cruising with the GoToPosZ RT. REACHWORKINGAREA is used to navigate in the pool (using the GoFrontWall vehicle's RT) until the vehicle reaches the inspection place. DOINSPECTION is used to coordinate actions of the platform and of the arm to simulate the inspection of an underwater structure with the arm tip: it respectively uses the StayFrontWall and MoveJS RTs. GOHOME is in charge of driving the vehicle to its homing position and preparing it to be pulled out. EMERGENCY is always active to handle permanent recovery behaviors like triggering a fast ascent in case of water leak. Eleven RTs are used in this mission even if some of them are very simple like the BRAKE used to lock the arm during vehicle's motions.

Each of the reactive programs is generated automatically by the ORCCAD programming environment in the following way. The discrete-event controllers of the set of the RTs concerning the same physical resource are composed synchronously forming as many reactive program as the number of the controlled physical resources. RPs are synchronously composed in a separate program downloaded in the vehicle backplane. Asynchronous communications are also generated and established following a description of the target architecture. In our particular example (Figure 4) sockets are created for the communication between the RP and the PA10 reactive programs running on different backplanes while message queues are established between the RP and the Vortex programs which run on the same VME board. Instanciating the right communication mechanisms is automatically done from the software mapping specification.

4.2. Experiment results

Figure 5 on the right presents experimental results carried out while running the aforementioned mission scenario. The top plot shows the front acoustic sensors signals which are used in particular to stabilize the vehicle in front of the pool's wall while the arm is in motion. The bottom plot shows the pitch angle of the vehicle which is, due to the low accuracy of time profiling between the two backplanes, the best way to detect the moments during which the arm is in motion.

The necessary communicating devices (sockets or message queues according to the respective controllers localization) are automatically instantiated at boot-time. Even if the two backplanes are linked through a non real-time Ethernet, signals between the different controllers are exchanged according to the mission specification e.g. observers checking the state of the vehicle are able to trigger actions of the arm and conversely. Signals sent through the keyboard are also correctly handled.

Some properties of the mission program have been checked before launching: for example, for safety reasons we want that the arm be motionless and locked when the vehicle is cruising. Thus, an abstract view of the mission automaton is built by reducing the global automaton to the only relevant signals. Figure 5 on the left shows this reduced automaton where it can be easily check that this property is verified whatever is the state of the mission: we can see that the activation of motions of the vehicle always immediately follows the activation of the BRAKE RT.

Besides the logical correctness of the reactive controllers behaviors the success of such an experiment relies on the efficiency of the used control laws. Although the gains which have been used for the experiment are very close to those found by simulation using Simparc [8] for most basic actions, the stabilization front of a wall using acoustic sensors was disappointing as the acoustic sensors loose a lot of data or provide absurd measures as soon as the pitch velocity is not negligible, may be due to multiple path propagation in the pool's corners. The gains of this control law had to be set to rather low values thus reducing the robustness of the stabilization w.r.t. the tether stiffness leading to non-repetitive results. It is expected that after careenage the new set of sensors and actuators will allow to improve the vehicle control efficiency.

4.3. Further improvements

The work presented here is a preliminary phase in the design of an underwater *distributed* teleoperated manipulation system. Further improvements will include the following:

- In this experiment we have only actually test one way of distributing the control code using asynchronous communication between the controllers. The *ocrep* tool for automatic distribution of synchronous programs will be tested and the additional communication cost will be evaluated.
- The code repartition has been done on a Robot-Task basis rather than on a Procedural one. Thus, some signals which are only relevant for the vehicle control are still exchanged between the two backplanes while the granularity of the actions described at the mission level should be made larger. The way of functionally distribute a robotic application will be further explored.
- Signals sent through the keyboard only sketches a master teleoperation station where an additional reactive program will be in charge of the management of the station (e.g. for managing the displays on a context dependent basis) to provide assistance to the human operator.

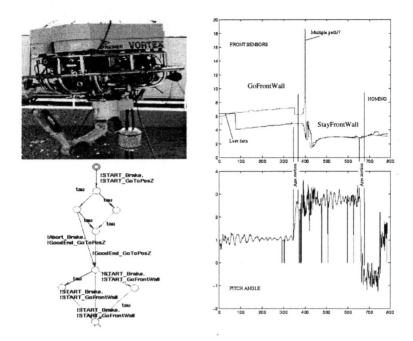

Figure 5. Some experimental results

References

[1] C. Astraudo, J.J. Borrelly, "Simulation of Multiprocessor Robot Controllers", *Proc. IEEE Int. Conf. on Robotics and Automation*, Nice, May 1992

[2] G. Berry, G. Gonthier: " The Synchronous Programming Language ESTEREL: Design, Semantics, Implementation", *Science Of Computer Programming*, Vol 19 no 2, pp 87-152, 1992.

[3] P. Caspi, A. Girault, and D. Pilaud. Distributing reactive systems. In *7th Int. Conf. on Parallel and Distributed Computing Systems*, Las Vegas, Oct. 1994.

[4] B. Espiau, K. Kapellos and M. Jourdan : *Verification in Robotics: Why and How?*, Robotics Research, the seventh International Symposium, octobre 1995, Giralt and Hirzinger editor, Springer Verlag.

[5] B. Espiau, K. Kapellos, E. Coste-Manière and N. Turro: "Formal Mission Specification in an Open Architecture", *Isram'96*, Montpellier, May 1996.

[6] K. Kapellos, S. Abdou, M. Jourdan and B. Espiau: "Specification, Verification and Implementation of Tasks and Missions for an Autonomous Vehicle", *4th Int. Symp. on Experimental Robotics*, Stanford, 1995.

[7] V. Rigaud e.a. (Union team): "Union: Underwater intelligent operation and navigation", to appear in *IEEE Robotics and Automation Magazine*, Special Issue on Robotics & Automation in Europe.

[8] D. Simon, K. Kapellos and B. Espiau, "Control laws, Tasks and Procedures with ORCCAD: Application to the Control of an Underwater Arm", *preprints of 6th IARP workshop on Underwater Robotics*, Toulon, March 1996.

[9] D. Simon, B. Espiau, E. Castillo and K. Kapellos: "Computer-aided Design of a Generic Robot Controller Handling Reactivity and Real-time Control Issues", *IEEE Trans. on Control Systems Technology*, vol 1, no 4, Decembre 1993.

Chapter 13

Cooperative Multirobots

Multirobot cooperation clearly provides possibilities to advance in robotizing complex tasks. The concept of distributing the task among different small and simple robots as an alternative to big and expensive robots is widely used. The cooperation of a set of such robots requires to define a collective behavior that is frequently inspired on biological systems.

Martinoli, Franci and Matthey conduct bio-inspired collective experiments using the family of small robots Khepera. The experiment consists on gathering and clustering randomly distributed passive seeds. The collective behavior based on each robot individual adaptive behavior is discussed. To operate in an adequate way the robots need to exchange information, to have enough autonomy and to define a control strategy. Both radio and IR communication is used for global and local data exchange. If the seeds are endowed with an IR sensor, they can be detected by the robots using simple presence sensors. To define the robots behavior the authors believe that the integration of learning methods can contribute to self-program robots specifically for a given task.

Jung, Cheng and Zelinski experiment on cleaning tasks, using a wide range of cooperating robots schemes, from emergent cooperation with no communication to explicit cooperation and communication. After discussing three control approaches: purely reactive, planner-based and behavior-based, they select the last one due to its potential in real-time response and cognitive processing in a uniform manner. The experimentation required the implementation of a proportional whisker sensor, sensor data fusion and to implement basic robot behaviors able to improve its learning performance from observation.

A self-reconfigurable robot can adapt its shape and function to tasks. A set of identical reconfigurable modules can then cooperate more effectively since they are able to change their individual and collective organization. Kotay and Rus describe the Inchworm robot that can adopt different structures: legged-robot, snake shape, or a manipulator. The four-links robot can navigate in 3D space, can perform transitions among different surfaces, and also lift, place or pull objects. Up to now a unique unit has been experimented but new modules are being constructed.

The problem of a human controlling multiple robots is studied by Beltrán, Arai, Nakamura, Kakita and Oka. After evaluating interface systems to perform gestures and teleperation they propose to control four robots using a unique joystick. To make this task possible they introduce the concept of semi-autonomy, so as to combine the human expertise with the robot capability to be programmed to actuate in front of some given situations. The experimentation is shown in two examples. First, pushing an object with two robots and second, assembling with four such units grouped two by two.

Towards a Reliable Set-Up for Bio-Inspired Collective Experiments with Real Robots

A. Martinoli, E. Franzi, and O. Matthey
Microcomputing Laboratory, Swiss Federal Institute of Technology
IN-F Ecublens, CH-1015 Lausanne
E-mail: martinoli@di.epfl.ch, franzi@di.epfl.ch, matthey@di.epfl.ch

Abstract: This paper describes a set of tools developed at our laboratory that provide a reliable set-up for conducting bio-inspired experiments with real robots. We focus on the hardware tools needed to monitor team performances as well as those to achieve collective adaptive behaviours. We propose concrete solutions to some of the main problems in collective robotics. The four main results we derive are: *first*, the hardware modularity of the miniature robot Khepera [1] allows us to build a flexible set-up; *second*, the energy autonomy problem is solved in a reliable way for experimenting with real robots during several hours; *third*, the communication architecture among teammates and/or with a supervisor unit is designed to prevent bandwidth bottlenecks with bigger robot teams; *fourth*, the use of programmable active pucks (also called "seeds" below) extends the set of possible bio-inspired experiments without increasing the sensorial complexity of the robots. A simple bio-inspired collective experiment, the gathering and clustering of randomly distributed passive seeds, is presented as an example as well as a test-bed for the extended autonomy tool. The results are compared with those reported in [2, 3].

1. Introduction

In the last few years, we have observed a growing collaboration between biologists and engineers [4, 5]. Robots running bio-inspired controllers allow biologists to better understand living organisms, while engineers manage to solve problems which are hard to tackle using classical control methods. Unfortunately, the difficulty to build adequate and reliable set-ups for experiments with real robots prompts many researchers in autonomous robotics to carry out investigations with simulated robots in simulated environments. This is especially true in collective autonomous robotics, where the autonomy of the robots depends mainly on their on-board computational power and their energy supply. In many bio-inspired single-robot experiments, the robot is connected to a workstation through a cable, which supplies the required energy and supports intensive computing, such as learning algorithms [6]. With many robots using cables becomes impossible: they would become entangled. Two further robot features, not necessarily required in single-robot experiments but essential in collective robotics, are the capability to explicitly communicate with and distinguish the other teammates from the rest of the environment. Providing the robots with these capabilities in a noisy real environment is not a trivial task,

a) b)

Figure 1. a) Three Kheperas equipped with different combinations of modules on the energy supply board (from left to right): gripper and IrDA modules, IrDA and radio modules, and gripper and radio modules. The active seeds complete the set-up picture. b) A closer look at the Khepera with IrDA and radio modules.

particularly with miniature robots. This paper addresses these problems and describes the solutions developed and currently tested at our laboratory.

In the following paragraph we describe what we mean by bio-inspired collective robotics, and our motivation for developing this particular set-up (see fig. 1a).

1.1. Collective Behaviour Synthesis and Analysis in Bio-Inspired Experiments

Bio-inspired collective robotics favours decentralised solutions, i.e. solutions where coordination is *not* taken over by a special unit using private information sources, or concentrating and redistributing most of the information gathered by the individual robots. Inspired by the so-called collective intelligence demonstrated by social insects [7], bio-inspired collective robotics studies the robot-robot and robot-environment interactions leading to robust, goal-oriented, and perhaps emergent group behaviours.

One way to generate robust collective behaviours is to apply bio-inspired adaptive algorithms at the team level. We believe that the integration of learning methods can contribute strongly to design a team of self-programming robots in view of predefined task. In the last few years reinforcement learning and genetic algorithms have been used to produce adaptive behaviour in the context of single-robot applications [8, 6]. In multiple-robots applications, where fitness is measured at team level, robots are faced with the *credit assignment problem*, which means the problem of deciding to what extent their own behaviour has contributed to the team's overall score [9]. Two ways for bypassing this problem have been proposed. *First*, by integrating *global communication* among teammates [10]. However, this is not a completely decentralised solution and does not match the above definition of bio-inspired robotics. Furthermore, depending on the environmental conditions, global communication is not always possible and tends to bottleneck with great team sizes. *Second*, by measuring each robot individual performance instead of team performance [11]. A main drawback of this approach is to force collective behaviour to be the sum of identical individual behaviours, which is not necessarily the optimal strategy for every boundary condition of the shared mission.

We can achieve *real team solutions*, whose form depends strongly on task boundary conditions (such as the number of robots involved in the experiment or their functionalities), only at the price of dealing with the credit assignment problem. Attempts in this direction in simulated environments have been recently published [9, 12]. In order to implement the learning process according to this approach on a team of real robots, enough energy autonomy and a reliable communication with the workstation (learning supervisor) is required. Via radio link, the team performances are computed on the supervisor unit and the adapted parameters are sent back to the robots.

Let us consider now behavioural analysis. Currently, autonomous mobile robotics is dominated by the experimental approach. Very few researchers have performed quantitative measurements of robot performances. This is also true for collective robotics. However, recently, the research community in this field has focused on this problem [3, 13, 14]. In [3, 14] the environmental key parameters were collected by filming or by observing the team behaviour. It would be of great interest to automatically collect robot key parameters, parallel to the environment evolution, i.e. to quantitatively correlate team strategies with team performances. A first attempt in this direction has been conducted in [13]. The authors demonstrated that a quantitative measurement of the interference rate can be used as tool for evaluating the multi-robots controller.

2. Experimental Set-Up

2.1. The Robots

Khepera is a miniature mobile robot developed to perform "desktop" experiments [1]. Its distinguishing characteristic is its small size (55 mm in diameter). Other basic features are: important processing power (32 bits processor at 16 MHz), energetic autonomy of almost half an hour, precise odometry, and light and proximity sensors. The wheels are controlled by two DC motors with an incremental encoder, and can rotate in both directions. The simple geometrical shape and the motor layout allow Khepera to negotiate any kind of obstacle or corner. Modularity is another characteristic of Khepera. Each robot can be extended with several modules: a gripper module which can grasp and carry objects with a maximum diameter of 50 mm and a weight of 20 grammes, a radio module, an IR local communication module, a vision module, a KPS (Khepera absolute Position System) module, and other general purpose or custom modules. Thanks to its size, Khepera is a convenient platform for both single- and multi-robot experiments: 20 Kheperas can easily work on a 2 m^2 surface, which is equivalent to a workspace of 10×20 m for robots with a 50 cm diameter.

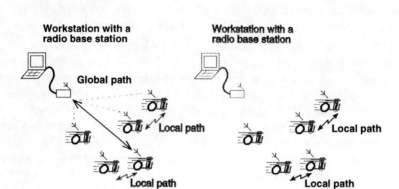

Figure 2. a) Global communication path. b) Local communication path.

2.2. The Communication Tools

In collective robotics, communication is crucial for coordinating behaviour among robots. Furthermore, a communication link between the workstation and teammates enables supervision of the robots. In bio-inspired robotics, team behaviour is obtained with a completely decentralised control, which limits communication to neighbouring teammates. These considerations have facilitated the definition of a hierarchical communication strategy which optimises robot-to-robot (local path) and workstation-to-robot (global path) communication (see fig. 2). The multi-microcontroller architecture of Khepera has a software layer which supports any kind of communication turrets. Specific communication implementation is taken in account by an on-board dedicated microcontroller connected to Khepera by its standard local network.

According to the multi-microcontroller architecture of Khepera, the size of the environment, the possible number of robots involved in the experiments and the technological constraints, the choice of the global and local communication paths have led to radio and infrared physical implementations (see fig. 1b). Notice that, due to the completely different range of the these two physical links (see further paragraphs), implementing the local path using radio communication would imply that the robots are aware of their absolute position (see [10] as example). This is not coherent with our definition of bio-inspired robotics and would further reduce the available bandwidth for local and global communications.

2.2.1. The Radio Turret

The radio communications are managed by a low speed star topology network. Two possible modes are available: standard (the communication master is the base unit) and robot-based (the communication master is a Khepera). In the *standard mode*, which is currently used for the bio-inspired collective experiments, all the transactions are started and controlled by the radio base unit which can address either a single Khepera (selective standard mode) or all the

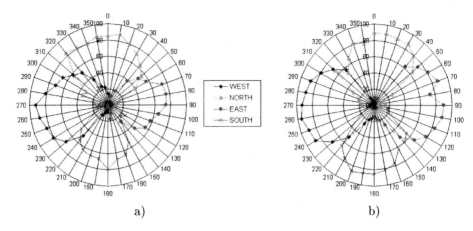

Figure 3. a) Reception covered area of the IrDA turret. b) Emission covered area of the IrDA turret. Data registered with low emission power and high sensitivity level (level 3, see fig. 4).

Kheperas used in the experiment (broadcasting standard mode). In this mode only messages which use the standard ASCII protocol of Khepera (see [1]) can be sent. In the *robot-based mode*, no protocol is a priori defined. Each Khepera can start a transaction with another Khepera (selective robot-based mode) or with all other Kheperas in the experiment (broadcasting robot-based mode). The supervising communication algorithm is then running on the transmitting teammate. The radio turret is composed of a 418-MHz FM radio module and a microcontroller at 8 MHz, which establishes the interface with the local network. A 32000 bits/s half-duplex protocol, similar to HDLC (High-Level Data Link Control), is used to ensure reliable communications. The real user throughput at hand is about 4800 bits/s which can slightly decrease if the environment is too polluted (e.g. strong electro-magnetic field generated by poorly shielded computers). Of course, the real user throughput represents the available bandwidth of the radio channel and it has to be divided by number of robots to have the full global bandwidth by robot. The area covered by the radio network is about 100 m^2.

2.2.2. The IrDA Turret

The infrared communication is based on the standard IrDA (Infrared Data Association) physical layer. The implementation allows selective point-to-point communication. The local emitting-receiving area of each turret is divided in four regions (south, north, east, and west). The IrDA turret is composed of a microcontroller and four IrDA devices, whose placement creates an omnidirectional local path (similar to the solution proposed in [15], see fig. 3).

The physical specifications of IrDA devices allow for a very high bandwidth, typically above 100000 bits/s. However, the real throughput is determined by the hardware implementation of the whole turret. Currently, the baudrate is 20833 bits/s. Notice that the real throughput is not decreased

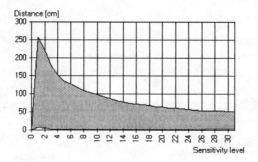

Figure 4. Covered range as a function of the sensitivity level of the IrDA device. The grey zone represents the useful range for reliable communication.

by the number of robots engaged because the channel is shared by only two Kheperas at a time. According to the IrDA chip specifications, the emission power and reception sensitivity can be set to obtain the desired range. The area covered can be modified from $0.03\,\mathrm{m}^2$ to $7\,\mathrm{m}^2$ (see fig. 4).

A greater area can be covered using higher level of sensitivity. However, due to the IR activity of Khepera's proximity sensors, the communication is only reliable from a distance of 15 cm. If the required range lies between a few centimetres and the maximal obtainable distance (about 250 cm), a spatial filter must to be added to the Khepera's infrared receiver, to suppress the influence of the proximity sensors.

2.3. The Extended Autonomy Tool

As mentioned in the introduction, our idea is to apply adaptive bio-inspired algorithms to self-programming in a team of robots with a shared mission. Due to the many iterations currently required by these algorithms to converge (hours or days), the robots should have a long autonomy. That is why we have developed a supply floor board and have modified the Khepera basic module to get energy from it.

2.3.1. General Description of the Device

To achieve extended autonomy, we have considered two possible solutions, using a special floor board as an interface between an external supply source and the robots. In the first option, energy is transmitted by electrical contact. An original contact and floor board layout (see next paragraph) allows each robot to take advantage of the external supply source, regardless its position. In the second option energy is transmitted by induction [16]. The idea is to equip Khepera with secondary windings of a multiple transformer whose primary windings are placed on the special floor board. The latter option has two advantages: no additional friction is added between the robots and the board (the odometry is not influenced) and there are no contact rebounding problems. However, the induction option is significantly more expensive and difficult to miniaturise; obtaining a high energy transmitting ratio is not trivial. We are currently investigating both options, but focus on the first, more advanced solution.

The extended autonomy tool is composed of a common supply generator, a special floor board, and a modified Khepera basic module. When the robot is moving on the supply floor board, it is able to keep its own batteries charged and, if necessary, to recharge them during the experiment, without stopping to work. This suppresses pauses and the need of special behaviours for recharging (with the batteries and charger currently used, about 45 minutes are necessary for 30 minutes of work); the power supply level is constant during the whole experiment which can last hours or days. Furthermore, thanks to its on-board batteries, the robot preserves its traditional autonomy which can be useful for short experiments or demonstrations. Past experience has revealed a further reason to preserve the batteries on the robot: the 100% efficiency in energy transmission can be assured only if all the robot poles are permanently in contact with the floor board surface. Short transient fluctuations can be filtered with condensers but if dust has been accumulating, power failure could occur at any moment. We have conducted tests without batteries during one week with a robot moving on the arena in a pseudo-random way: it stopped on the average every half hour. To restart the robot, it was sufficient to slightly push it.

2.3.2. Layout Optimisation of the Floor Board and Robot Contacts

The floor board should fulfill the following requirements: modularity (the size of the useful surface can be chosen by the experimenter), work surface as flat as possible to avoid contact rebounding, and simple layout pattern. The robot contacts should fulfill the following requirements: matched with the floor board for high energy transmission efficiency, rotation symmetric placement of poles as well as low friction (the odometry of the robot should be not influenced), minimal rebounding, and possible miniaturisation for matching with the active seeds size (see next section).

Copper has been chosen both for the robot poles and floor board surface. This offers good contact (resistance smaller than 0.3 Ohm) and low friction. The isolated gaps between the conducting bands are made of unconnected conducting surfaces, achieving minimal discontinuity between bands.

How many poles are needed and how should they be placed in order to fulfill the mentioned requirements? How should the floor board upper surface be designed? We solved this optimisation problem with an atypical procedure, using first simple geometric considerations, then selected the optimal solutions with a genetic algorithm and finally we demonstrated their validity mathematically. The width of floor board band and the placement of the robots' poles were encoded in a genome string. The fitness function was the number of "external powered" positions in a set of samples generated with discrete translations and rotations of the robot. Some noise was added to simulate the mechanical vibrations of the contacts. The obtained set of solutions assures the 100% efficiency with 4 poles if transient mechanical phenomena (rebounding and dust on the pole-board contact surface) are neglected.

2.3.3. Test Results

We have tested the performances of the extended autonomy tool with an obstacle avoidance algorithm (mean power consumption of about 1.5W) and with the clustering algorithm presented in [3] (mean power consumption of about 1.7W). The batteries were recharged at a very low rate (C/10) during the experiment. The tests consist of three phases: working on the floor board without external supply until the batteries are discharged, working on the floor board for a hour with external supply starting with charged batteries, and replaying phase 1 immediately after phase 2, without recharging the batteries. The tests were conducted several times with several robots running the two different algorithms and the discharging time of phase 1 and 3 were compared. The results show that the power consumption during phase 2 was always smaller than 5% of the whole batteries power. Managing to charge the batteries faster should lead to open end autonomy.

2.4. The Active Seeds

Khepera's 8 proximity sensors allow obstacle detection at a maximal distance of 6 cm in nearly all directions, depending on the obstacle size and material [17]. This basic configuration has demonstrated its reliability in simple tasks such as obstacle avoidance. However, the limited number of sensors as well as their variable sensivity generate vague information about the form and the size of the encountered objects. Object-gathering experiments with real Kheperas showed the difficulty to distinguish seeds, the objects to gather, from obstacles, i.e., arena walls and other teammates (see [3]). A possible solution to improve the discriminating capabilities of Khepera, without increasing the complexity of its hardware, was to develop active seeds easily recognised by the proximity sensors.

2.4.1. General Description of the Device

The "active" seed (see fig. 5a) has a diameter of 30 mm, is equipped with 5 IREDs whose overlapped spatial lobes cover all the required 360 degrees. It is able to synchronously respond to the IR pulses of the Khepera's proximity sensors. By emitting at a given rate, for instance one pulse for each two received, the seed is seen by the robot as a "blinking" object. The IREDs are controlled by a low-power RISC microcontroller which allows great flexibility. Currently, three basic operation modes are implemented: answer once each 2, each 3 and each 4 pulses received. A LED on the seed top signals to the experimenter its activity and selected mode. These modes and their temporal validity can be completely and permanently reprogrammed in a few seconds by connecting a dedicated programmer to the seed. Four other interesting features distinguish our active seed. *First*, their robustness: the sensitivity is adapted to the ambient lighting conditions. *Second*, their complementary features for recognition: each type of seed is provided by another software conditioned "internal" resistance, which Khepera is able to measure with the gripper resistivity sensor. *Third*, their energy autonomy: the primary supply source is represented by 2 replaceable 225 mA/h lithium coin cells. Moreover, the seed's lower face is equipped with 4 contacts, whose placement is identical to those on Khepera. If

 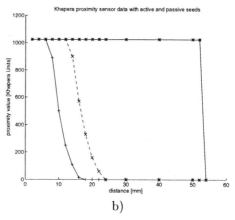

a) b)

Figure 5. a) Front and bottom view of an active seed. b) Seed detection range (values of Khepera's proximity sensors): "*-*" represents the emitting seed signal, "+-+" the idle active seed, and "x-x" the passive reference seed (small wood cylinder).

used together with the extended autonomy tool, the seeds can take advantage of their secondary supply source (the floor board) bypassing the on-board batteries. Without the extended autonomy tool, the batteries last about 3 days of continuous operation, depending on the seed's activity and the lighting conditions. *Four*, their weight and mechanical stability: an active seed weighs only 17 grammes, can be lifted without problem by Khepera's gripper and does not impede the movements of the robot. The seed centre of mass is very low. As a consequence, even after inaccurate dropping operations the seed body tends to keep its upright position.

2.4.2. Test Results

We have conducted a series of tests to evaluate the spatial range of the seed prototype. We should be able to assure a 5 cm detection under various lighting conditions. Fig. 5b shows the useful detection range of a seed measured with a Khepera.

3. Seed Clustering: The Bio-Inspired Test Experiment

A biologically inspired experiment concerned with the clustering and gathering of scattered seeds was presented in [2] and replaid in [3]. We can summarise the resulting robot behaviour with the following simple rules. The robot moves on the arena looking for seeds. When its sensors are activated by an object, the robot begins the discriminating procedure. Two cases can occur: if the robot is in front of a large obstacle (a wall, another robot or an array of seeds), the object is considered as an obstacle and the robot avoids it. In the second case, the object is identified as a seed. If the robot is not already carrying a seed, grasps the seed with its gripper; if the robot is carrying a seed, it drops the seed close to the one it has found; then, in both cases, it turns

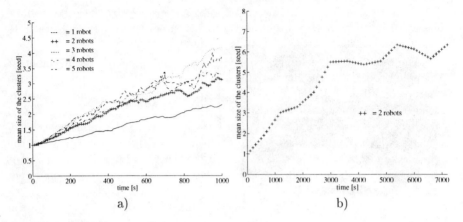

Figure 6. a) Team performances in the clustering bio-inspired experiment in in [3] (average over 5 replications) and b) with the help of the extended autonomy tool for 2 robots (average over 3 replications).

Figure 7. The arena at the beginning (a) and at the end of the experiment (b).

about 180 degrees and begins searching again.

In [2], due to the difficulty for the recognition algorithm to distinguish between a seed and another robot, a robot often dropped a seed in front of a fellow, and the latter grasped the seed (seed exchange) or a seed was dropped in an isolated position in the middle of the arena or beside the wall. For the same reason, the robots tried to grasp each other and often became entangled for a few seconds. The experiment was conducted with a group of 1 to 5 Kheperas equipped with the gripper module and 20 scattered seeds in an arena of 80 × 80 cm. The measured team performance was the average size of the cluster created in about 30 minutes. Due to the low reliability of the distinguish algorithm, the performance evolution showed a saturation phase after about 10 minutes.

In [3] we carried out the same experiments with a group of 1 to 10 Khepera, different arena sizes, different number of seeds, and with a more reliable distinguishing algorithm. Furthermore, in order to improve the discrimination, each robot was equipped with an IR reflecting band: the size of the robots indicated by the proximity sensors was therefore increased and, at a distance of 4-5 cm a robot was already recognised as an obstacle by the other robots. The team performance evolution did not saturated, even after more then 16 minutes (see fig. 6a).

The question arises spontaneously: what would happen after 16 minutes? A simple probabilistic model developed in [3] suggests that, because of the geometry conditioned probabilities of building or destroying a cluster, the size of the clusters should continuously increase until all the seeds are gathered together. The model forecast does not take in account the interference phenomena, which play a crucial role in experiments with real robots. With the help of the developed extended autonomy tool, we have performed an experiment with two robots, using the same algorithm as in [3]. The experiment lasted 2 hours and was replicated 3 times. As we can see in fig. 6b, the cluster size does not increase very much after 80 minutes. Fig. 7 shows a typical scattering of the seeds at the beginning and at the end of the experiment.

4. Conclusion

We presented a set-up for bio-inspired experiments in collective robotics. The four main results we derive are: *first*, the hardware modularity of Khepera allows us to build a flexible set-up; *second*, the energetic autonomy problem is solved in a reliable way for experimenting with real robots during several hours; the seed clustering experiment has demonstrated the efficiency of the developed tool; *third*, the communication architecture among teammates and/or with a supervisor unit is designed to prevent bandwidth bottlenecks with bigger robot teams; *fourth*, the use of programmable active seeds extends the set of possible bio-inspired experiments without increasing the sensorial complexity of the robots. We hope that the design ideas described in this paper will help other researchers to develop their own set-up.

Acknowledgements

We would like to thank André Guignard for the important work in the design of Khepera, Masakazu Yamamoto for programming the new clustering algorithm, Georges Vaucher and our students Rémy Blank and Gilbert Bouzeid for helping us in the tool development, Jean-Bernard Billeter for the reviewing of this paper, Prof. Jean-Daniel Nicoud, Francesco Mondada, Luca Gambardella, and Cristina Versino for helpful discussions on autonomous mobile robotics. Alcherio Martinoli has been partially supported by the Swiss National Research Foundation.

References

[1] F. Mondada, E. Franzi, and P. Ienne. Mobile robot miniaturization: A tool for investigation in control algorithms. In *Proceedings of the Third International Symposium on Experimental Robotics ISER-93*, pages 501–513, Kyoto, Japan, 1993.

[2] A. Martinoli and F. Mondada. Collective and cooperative group behaviours: Biologically inspired experiments in robotics. In O. Khatib and J. K. Salisbury,

editors, *Proceedings of the Fourth International Symposium on Experimental Robotics ISER-95*, pages 3–10, Stanford, U.S.A., June 1995. Springer Verlag.

[3] A. Martinoli, M. Yamamoto, and F. Mondada. On the modelling of bio-inspired collective experiments with real robots. In *Proceedings of the Fourth European Conference on Artificial Life ECAL-97*, Brighton, UK, July 1997. http://www.cogs.susx.ac.uk/ecal97/present.html.

[4] J. C. Deneubourg, P. S. Clip, and S. S. Camazine. Ants, buses and robots self-organization of transportation systems. In P. Gaussier and J-D. Nicoud, editors, *Proceedings of the conference From Perception to Action*, pages 12–23. IEEE Press, Los Alamitos, CA, 1994.

[5] N. Franceschini, J.-M. Pichon, and C. Blanes. Real time visuomotor control: From flies to robots. In *Proceedings of the Fifth International Conference on Advanced Robotics*, pages 91–95, Pisa, June 1991.

[6] D. Floreano and Mondada F. Evolution of homing navigation in a real mobile robot. *IEEE Transactions on Systems, Man and Cybernetics*, 26:396–407, June 1996.

[7] E. G. Bonabeau and Theraulaz G. *Intelligence Collective*. Hermès, Paris, France, 1994.

[8] Millan J. del R. Reinforcement learning of goal-directed obstacle-avoiding reaction strategies in an autonomous mobile robot. *Robotics and Autonomous Systems*, 15:275–299, 1995.

[9] C. Versino and L. M. Gambardella. Ibots learn genuine team solutions. In M. Van Someren and G. Widmer, editors, *Proceedings European Machine Learning ECML-97*, pages 298–311, Kyoto, Japan, 1997. Springer Verlag. Lecture Notes in Artificial Intelligence.

[10] L.E. Parker. The effect of action recognition and robot awareness in cooperative robotic teams. In *Proceedings of IEEE International Conference on Intelligent Robots and Systems IROS-95*, volume 1, pages 212–219, Pittsburgh, PA, August 1995. Springer Verlag.

[11] M. J. Matarić. Learning in multi-robot systems. In G. Weiss and S. Sen, editors, *Adaptation and Learning in Multi-Agent Systems*, volume 1042, pages 152–163. Springer Verlag, Lecture Notes in Artificial Intelligence, 1996.

[12] A. Murciano and J. del R. Millán. Learning signaling behaviors and specialization in cooperative agents. *Adaptive Behavior*, 5(1):5–28, 1997.

[13] D. Goldberg and M. J. Matarić. Interference as tool for designing and evaluating multi-robot controllers. CS-96-186, Brandeis University Computer Science Technical Report, 1996.

[14] M. Maris and R. te Boekhorst. Exploiting physical constraints: Heap formation through behavioral error in a group of robots. In *Proceedings of IEEE/RSJ International Conference on Intelligent Robots and Systems IROS-96*, volume 3, pages 1655–1660, Osaka, Japan, November 1996.

[15] T. C. Lüth, J. Hellqvist, and T. Längle. Distributing real-time control tasks among multi-agent robot systems. In *Proceedings of the World Automation Congress WAC-96*, volume 3, pages 477–482, Montpellier, France, May 1996.

[16] M. Jufer, L. Cardoletti, P. Germano, B. Arnet, M. Perrottet, and N. Macabrey. Induction contactless energy transmission system for an electric vehicle. In *Proceedings of International Conference on Electrical Machines ICEM-96*, volume II, pages 343–347, Vigo, Septembre 1996.

[17] K-Team SA. Khepera user manual. Version 4.06, 1995.

Experiments in Realising Cooperation between Autonomous Mobile Robots

David Jung, Gordon Cheng and Alexander Zelinsky
Robotic Systems Laboratory, Department of Systems Engineering
Research School of Information Science and Engineering
Australian National University
Canberra, ACT 0200, Australia
http://syseng.anu.edu.au/rsl

Abstract: This paper describes experiments in cooperation using autonomous mobile robots to perform a cleaning task. The robots have heterogeneous capabilities and the task is designed so that cooperation is required. Each experiment increases the sophistication of the cooperation scheme to assess the effect on task performance. The experiments range from using emergent cooperation with no communication to explicit cooperation and communication. We also propose an action selection mechanism that can also be used for distributed planning of joint actions inspired by the way primates co-construct joint plans. *Keywords: Cooperation, Vision, Touch, Mobile*

1. INTRODUCTION

The research effort into multi-robot systems is driven by the assumption that multiple agents have the possibility to solve problems more efficiently than a single agent. Agents must therefore cooperate in some way. Before we can begin to consider cooperation between robots we must define exactly what is meant by cooperation. Next we will investigate how cooperation can be achieved within a behaviour-based framework, by implementing cooperative behaviours using mobile robots to perform a concrete task.

The basic control design approaches can be broadly divided into four types. These are defined by [Mataric 92a] and we briefly reiterate them and their shortcomings here. The *purely reactive* approaches use a mapping from sensor sets to associated actions; a set of rules [Brooks 87]. The *planner-based* strategies originated with the symbolic AI community and employ a sense-plan-act cycle. The plan stage uses cognitive techniques to reason about a symbolic world model. There also exist hybrid systems which employ reactive components beneath planner-based systems to provide the benefits of both. Another approach is *behaviour-based*, which uses a set of interacting distributed concurrent behaviours, each of which may incorporate memory and learning of environment representations.

The purely reactive approach achieves robust real-time performance, but the tasks that can be achieved are limited because of the lack of any cognition. The planning approach suffers from a number of problems including slow interaction with the environment due to slow processing, the frame problem and the symbol grounding problem. The hybrid approaches attempt to marry two incompatible philosophies and still suffer from many of the problems of the planning approaches. For more depth on these problems refer to [Pfeifer 95]. The behaviour-based approach has the potential of real-time response and cognitive processing in a uniform manner. We have chosen to implement our cooperative behaviour within the behaviour-based philosophy.

2. OUR AIMS

As research into multi-agent cooperation is still a relatively new field there are no standard formalism's for describing cooperation nor benchmarks for measuring the performance of techniques. The aim of our research is to determine the defining characteristics of cooperative behaviour; examine the effect of these characteristics on performance and to propose a scheme for implementing cooperation within the framework of behaviour-based robotic systems.

In order to assess the performance of different levels of sophistication of cooperative behaviour we have focused on a concrete application. The task we have chosen is for two autonomous mobile robots to clean the floor of our laboratory. The 'Yamabico' robots [Yuta 91] shown in Figure 1 each have different tools and sensors such that neither can accomplish the task alone.

 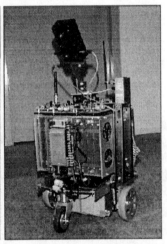

Figure 1 -   (a) Yamabico Flo   (b) Yamabico Joh

The robot 'Joh' navigates by vision, and has a vacuum cleaner that can be turned on and off via software. Joh's task is to vacuum piles of litter from the laboratory floor. It cannot vacuum close to walls or furniture. It has the capability to 'see' piles of litter using its vision system, but not fine particles scattered over the floor.

The other robot 'Flo', has a brush tool that is dragged over the floor to sweep distributed litter into larger piles for Joh to pick up. Flo uses whisker sensors to navigate (see Figure 1a).

The task is to be performed in the unmodified indoor environment of a laboratory. Our laboratory is cluttered and the robots have to contend with furniture, other robots, people, opening doors, changing lighting conditions, equipment with dangling cables and other hazards.

3. WHAT IS COOPERATION?

The word 'cooperation' has been applied to behaviour between robots and also between humans and robots, without much thought having been given to its exact meaning. Although its literal reading - simultaneous operation - is quite general, the word has historically been used primarily to refer to the joint behaviour of humans, and sometimes animals. As such its traditional meaning may be loaded with anthropomorphic assumptions which do not hold in the modern robotics context. Before we consider an appropriate definition for robot cooperation we need to examine cooperative behaviour in humans and animals. As cooperation is a social phenomena we may look to sociology and social psychology for insight into its mechanisms [Tinbergen 53][Wilson 75][Wright 95].

3.1 The origins of Altruism

Given that Charles Darwin's theory of natural selection implies that individuals behave in a completely selfish manner to increase their own fitness, an obvious question is 'why do biological organisms cooperate at all?'. This very question perplexed Darwin who introduced a mechanism he called group selection to account for the cooperation he observed - without any idea as to why group selection should exist. The answer to the question came once the gene was discovered and was identified as the basic selectionist unit, not the individual. Modern sociobiology and evolutionary psychology is based on the application of Darwin's theory at the level of the gene instead of the individual. This selfish-gene approach has been popularised across disciplines in a number of books by Richard Dawkins [Dawkins 96]. It is interesting to note that according to the *social intelligence hypothesis* intelligence in primates "*originally evolved to solve social problems and was only later extended to problems outside the social domain*" [Byrne 95][Dautenhahn 96a].

The evolutionary path from early primates to humans was a long and convoluted one. It was considerably accelerated once performance in the social context became the dominant driving force behind selection rather than just survival in the natural environment. This was largely due to the beginnings of language. A detailed account of this is given in [Wright 95] and [Diamond 91]. What we should keep in mind from this is that the dynamics of cooperation between humans implicitly includes a history all the peculiarities of our particular path through the evolutionary space of possibilities.

3.2 Characterising Cooperation

The robotics community has investigated various characteristics of cooperation. Using the terms defined by [Cao 95], which we can relate to the social organisation of

biological systems, some of the characteristics that define the *group architecture* of a cooperative system are:
- Centralisation/Decentralisation - of planning for cooperative actions
- Differentiation - homogeneous or heterogeneous capabilities
- Communication Structures - none, implicit, explicit, dialogue for planning
- Modeling of Other Agents - no awareness, awareness, modeling of others

A more detailed description of these and a classification of current research into these categories is given in [Jung 96].

Our research focuses on systems which are distributed, allow heterogeneous individuals, explicit communication, and modelling of other agents. Individuals are autobiographical - learn from experience, engage in explicit communication, and dynamically plan cooperative actions. Although cooperation is dynamically planned, the mechanisms are determined at design time - not learned via social or cultural learning. The learning will be limited to learning the environment. The robots will also be given an implicit model of some actions of other agents.

4. REALISING COOPERATION

In the sections to follow we will describe four experiments we have designed around the cleaning task. The first will be a simple scenario not involving any explicit communication or learning, and will serve as a benchmark against which to assess the performance of the other three. They will increase the level of cooperative sophistication. Before we describe the experiments we will briefly describe the robot hardware we are using.

4.1 Hardware

We have the two Yamabico robots mentioned above, each equipped with basic locomotion and four ultrasonic range sensors. The robot Joh also has a CCD camera and video transmitter that sends video to our *Fujitsu MEP tracking vision system*. The vision system does template correlation, and can match about 100 templates at frame rate. The vision system can communicate with the robot, via UNIX host, over a radio modem to close the loop. We have implemented a vision-based navigation system that is capable of landmark based navigation and can operate safely in dynamic environments at speeds up to 600 mm/sec [Cheng 96].

The robot Flo has a brush and some novel proportional whisker sensors we have developed for accurately sweeping close to walls and furniture. Flo has two whisker sensors mounted on its left side for wall following and two whiskers in front for collision detection (see Figure 1a). The basic behaviours we have implemented on Joh and Flo are discussed below. First we will outline the three experiments we have designed to carry out the cooperative cleaning task and assess performance.

4.2 Experiments

4.2.1 Emergent Cooperation

The simplest experimental case, which will serve as a benchmark against which to access the following three, involves no awareness of the other robot and hence no explicit cooperation or communication. Any communication is implicit via interaction through the environment. The cooperation can be described as emergent. We have found that very little needs to be added to the base competency of each robot to achieve this kind of cooperation.

Flo uses wall following and obstacle avoidance behaviours to brush around the perimeter of the laboratory close to the walls. Periodically, Flo decides to deposit the litter collected so far into a pile in clear space farther away from the wall where Joh can vacuum it.

Joh navigates our laboratory using visual landmarks. A foraging behaviour is executed in which it wanders about the laboratory, avoiding obstacles, people and Flo while searching for piles of litter. The piles of litter are identified using the vision system. To conserve power Joh only turns its vacuum on when its sees litter. Occasionally Joh may attempt to vacuum artefacts on the floor that aren't litter, such as a patch of sunlight or a piece of equipment. In the case of sunlight no harm is done, while in the case of a piece of equipment the bump sensor will stop the robot.

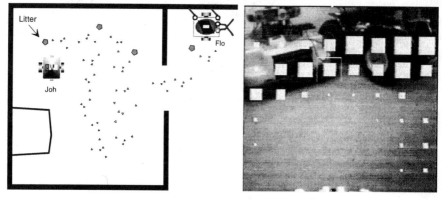

Figure 2 - (a) Cleaning by emergent cooperation   (b) Visual obstacle avoidance

Using these simple behaviours Flo and Joh clean the floor in an inefficient manner. Neither robot requires the capacity to purposively navigate around the laboratory using a learned map in this case.

Since we required a close high speed wall following capability in Flo for sweeping close to the walls, after investigating existing technologies, we decided to develop a unique proportional whisker sensor [Jung 96a]. We utilised a Kalman filter to estimate the robot position relative to the wall by fusing the information from the side whiskers with the wheel odometry. Flo is also fitted with a simple scoop so that wall following at an appropriate distance causes litter to be scooped from the wall. Flo then periodically drives into free space away from the wall and reverses leaving a pile of litter from scoop behind.

The wall following behaviour was extended to follow rough contours around the laboratory by using the front whiskers to detect corners and obstacles and to make an appropriate turn. Flo also has other miscellaneous simple behaviours such as reflex stopping, door traversal, tracking along straight trajectories and others [Jung 96a].

Joh has the ability to visually distinguish between the carpet on our laboratory floor and other obstacles at video frame rate (30Hz). This is accomplished using our template matching vision system and has been discussed in [Cheng 96]. This provides a good free space wandering and obstacle avoidance behaviour as shown in Figure 2b. Joh also needs the ability to detect piles of litter left by Flo. For this we use an 'interest' operator that segments areas of non-carpet surrounded by carpet (see Figure 3a). The 2D array of template correlation values must be normalised to compensate for camera lens distortion first. Once a possible pile of litter has been detected a visual servoing behaviour moves the robot over it. The vacuum can be turned on and off via software control. If the object was an obstacle not litter, then since Joh is fitted with a row of bump sensors on the front, it will be stopped from trying to vacuum it.

 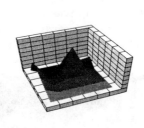 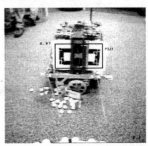

Figure 3 - (a) Pile of litter and Interest operator output   (b) Flo from Joh's camera

The beauty of this approach to litter detection is that it is very simple. It requires little computation, no model, is very robust, and delivers enough information for the task. A scheme that used sophisticated classical computer vision techniques, perhaps using a model of litter, would be computationally expensive, difficult to implement and would still fail some times. Hence almost nothing would be gained since the bump sensors would still be required as backup.

4.2.2 Cooperation by Observation

The second experiment, a more interesting case, utilises explicit cooperation using implicit communication by passive observation (cf. [Kuniyoshi 94]). In this case Joh uses the vision system to identify Flo by matching a unique geometric pattern. By observing Flo's actions Joh can determine the approximate location of the litter deposited by Flo. This new behaviour augmented the existing foraging behaviour and improved the efficiency of the cleaning task. This case requires awareness of Flo's existence by Joh.

In particular Joh needs to be able to identify and track the motion of Flo in the live video. This is accomplished by placing a unique pattern on the sides of Flo and tracking templates from it with geometric constraints (Figure 3b). The match correlation value for each template is fused using a network of Kalman filters. Other work in our laboratory used this technique with the vision system for robust human head tracking and is discussed in [Zelinsky 96]. Because we know the size and shape of the pattern we can also easily estimate the distance and heading of Flo relative to Joh.

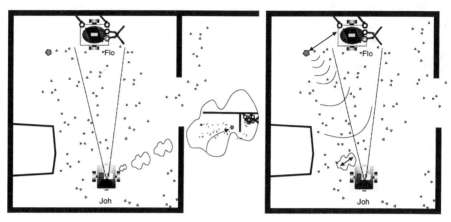

Figure 4 - (a) Cooperation by observation   (b) Cooperation with communication

4.2.3 Cooperation with Communication

The third experiment, by including explicit communication between the robots, can further improve the efficiency of the task. When Joh can see Flo it communicates this fact. In response Flo communicates the location where it next plans to dump a pile of litter, specified relative to its own location. Since Joh knows Flo's location relative to itself, the dump location can also be calculated relative to itself. Joh then navigates to the approximate dump location and initiates a visual search for the litter pile. If it is acquired a visual servo behaviour vacuums over it. This case only requires some simple communicative behaviours over the last experiment. Since it is not always possible to predict Flo's actions from observation alone, the communication removes ambiguity and improves performance. For example, visually determining if, when and where Flo plans to dump litter by observation alone is somewhat uncertain.

The communication between the robots is achieved using two pairs of radio modems between each robot and a UNIX host in combination with some networking software we developed for the custom robot operating system.

4.2.4 Co-construction of Joint Plans

All the capabilities required to implement the first three experiments have been described. The fourth experiment requires a significant increase in the sophistication of the basic behavioural capabilities of the robots. In particular they must be able to learn a map of their environment and purposively navigate by it. A mechanism for this is currently being implemented, and is briefly described below.

The are two possible dialogues in this experiment. When Joh sees Flo it chooses a name for the location where it sees Flo and labels the location in its internal map. It then tells Flo to label the location in its map. Joh and Flo have different representations for their maps. The map is represented in terms of the behavioural and sensory space of the robots, which means Joh will have visual landmarks in its map while Flo will have whisker based landmarks in its. Hence there is no possibility for communicating an absolute location except where the location has been labelled as just described.

The second dialogue occurs when Flo dumps litter, in which case it communicates the litter location in terms of a relative position from the nearest named location. In this case Joh simply navigates to the location using its map and visually searches for the litter to vacuum. Alternatively Joh could just note the location and vacuum the litter pile when it next comes near the location.

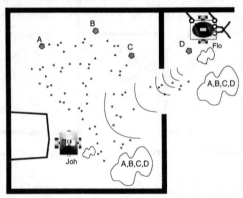

Figure 5 - Joint planning (eg. communication of rendezvous points)

The communication is completely grounded in the behaviour-sensor space of the robots even though each has a different such space. There is no communication of symbolic concepts that have been anthropomorphically designated by the designer. For example, we may consider that Flo can detect 'doors' using its whiskers and Joh can detect 'doors' visually. Hence we might imagine some communication using the concept of a 'door'. However, since a door to Flo is really just a pattern of whisker movements and to Joh a geometric arrangement of matched templates, there will be situations where the implicit identity assumption in our anthropomorphic designation will break down. As Connah and Wavish state *"communication between them (robots) will have to be flexible and natural; growing out of the perceptions that are part of their general behaviour patterns"* [Connah 90].

The mechanism we have developed for purposive navigation uses a spatial and topological map representation that is integrated into our action selection mechanism, which incorporates learning. This mechanism is briefly described below because it is also the basis of our proposed scheme for the dynamic co-construction of joint cooperative action plans, which is currently being implemented.

One major problem to be solved in robotics is, given a robot with a repertoire of basic behaviours - which behaviour should be selected next? Various action selection mechanisms have been proposed, such as Rodney Brooks' subsumption architecture [Brooks 87]. We have developed a mechanism based loosely on Pattie Maes' spreading activation scheme [Maes 90a], extended to add integrated learning and adapted to a behaviour based framework. This is beyond the scope of this paper, please refer to [Jung 97b].

Briefly, in simplified form, behaviours are connected in a network, where each behaviour has a set of preconditions. The preconditions are the outputs of feature detectors that are basically virtual sensors. Only one behaviour is active at once, and it must have all its preconditions satisfied. Behaviours are also connected to those feature detectors that become satisfied as a result of its execution - its postconditions. Behaviours become active according to an activation level - which is a result of a spreading activation algorithm that depends on the connections. For an example see Figure 6. The precondition-behaviour connections are designed, but the behaviour-postcondition connection are learnt according to correlations between behaviours and feature detector outputs. The spreading activation achieves distributed planning of action sequences. It can handle multiple hypotheses and contingencies.

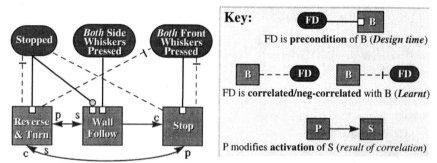

Figure 6 - Behaviour network implementation of a simple perimeter follower

The important point to note is that action sequences are always linked via a condition that is directly sensed from the environment. If a behaviour doesn't have the desired effect on the state of the environment then its successor behaviour will not be blindly activated.

We have designed a scheme for using this same mechanism for representing topological environment maps. These are landmark feature detectors linked by behaviours to take the robot from one landmark to the next. We have also incorporated spatial mapping into the scheme by using feature detectors for locations which self-organise like nodes of a Kohonen self-organising-map.

Primates are very social animals. As Bond writes in reference to vervet monkeys *"They are acutely and sensitively aware of the status and identity of other monkeys, as well as their temperaments and current dispositional states"* [Bond 96]. Humans, as other primates, have the ability to co-construct plans with more that one interacting person, and flexibly adapt and repair them all in real time.

Bond goes on to describe the construction and execution of joint plans in monkeys. He defines a *joint plan* as a conditional sequence of actions and goals involving the subject and others. In order to achieve interlocking coordination each agent needs to adjust its action selection based on the evolution of the ongoing interaction. The cooperative interaction will consist of a series of actions - including communication acts - where each agent attempts different plans, assesses the other agents' goals and plans, and alters the selection of its own actions and goals to achieve a more coordinated interaction where joint goals are satisfied. This model of interaction is similar to that proposed by human conversation theorists [Goodwin 81].

The action selection mechanism outlined above can be effortlessly applied to planning this sort of joint plan. This is because it is irrelevant which robot causes a change in the environment that triggers the precondition of the next action of a sequence. For example a cooperative interaction may consist of a sequence of actions by each robot interleaved (see Figure 7).

Figure 7 - Interleaved cooperative action sequence

This type of cooperative interaction, where the repertoire is essentially fixed at design time, is adequate for realising the fourth experiment. If we consider communicative utterances as behaviours and heard utterances as sensed features then this mechanism is also capable of planning dynamic communication dialogues. We may investigate this in future research.

5. CONCLUSION

We have presented the design and implementation of some experiments in cooperative cleaning behaviour between autonomous mobile robots. In particular we:
- Implemented a novel proportional whisker sensor and data fusion technique for close wall following.
- Implemented basic robot behaviours for emergent robot cleaning, including whisker and vision based navigation capabilities.
- Improved the performance of cleaning by adding, first implicit - by observation, and then explicit, communication between the robots.
- Proposed an action selection mechanism for joint planning of cooperative actions, including the possibility of planning communication dialogue.

Although this was a slightly contrived task one can easily imagine applications where there exists a trade-off between using one or a few very complex robots or a larger number of much simpler and less expensive robots to perform a task. For example, you may design a large complex and expensive automated cleaning robot, but in the event of failure the task cannot be performed. If instead there were numerous smaller, cheaper and simpler robots performing the same task cooperatively, then the reliability of any single robot will be higher. In addition, because they are cheaper to manufacture it may be possible to have redundancy so that the failure of a single unit does not render the whole system useless. In addition the time required to develop a few robots with simple behaviours may be less than that required for a single robot with complex behavioural requirements.

We also believe that the approach to action selection we are proposing will prove effective for implementing high-level behaviour in simple behaviour-based robots with meagre perceptual abilities, including the possibility of distributed planning of cooperative behaviour and dialogue for communication.

Videos demonstrating our experiments are available from our web site.

REFERENCES

[Bond 96]   Bond, Alan H., "**An Architectural Model of the Primate Brain**", *Dept. of Computer Science, University of California*, Los Angeles, CA 90024-1596, Jan 14, 1996.

[Brooks 87]   Brooks, Rodney A. and Connell, Jonathan H., "**Asynchronous Distributed Control System for a Mobile Robot**", *SPIE's Cambridge Symposium on Optical and Opto-Electronic Engineering Proceedings*, Vol. 727, October, 1986, 77-84.

[Byrne 95]   Byrne, R., "**The thinking ape, evolutionary origins of intelligence**", *Oxford University Press*, 1995.

[Cao 95]   Cao, Y. Uny, Fukunaga, Alex S., Kahng, Andrew B. and Meng, Frank, "**Cooperative Mobile Robotics: Antecedents and Directions**", *IEEE 0-8186-7108-4/95*, 1995.

[Cheng 96]   Cheng, Gordon and Zelinsky, Alexander, "**Real-Time Visual Behaviours for Navigating a mobile Robot**", *Proceedings of the IEEE/RSJ International Conference on Intelligent Robots and Systems (IROS)*, vol 2. pp973. November 1996.

[Connah 90]   Connah, David and Wavish, Peter, "**An experiment in Cooperation**", *Decentralized A.I.*, Yves Demazeau & Jean-Pierre Müller (Eds.), Elsevier Science Publishers B.V. (North Holland), 1990.

[Dautenhahn 96a]   Dautenhahn, K., "**Social modelling - on selfish genes, social robots and the role of the individual**", submitted to: *Int Journal in Computer Simulation*, Special issue on the Simulation of Social Behavior, August 1996.

[Dawkins 96]   Dawkins, Richard, "**Climbing Mount Improbable**", *Penguin Books*, 1996. ISBN 0-14-026302-0.

[Diamond 91]   Diamond, Jared, "**The rise and fall of the third chimpanzee**", *Vintage Books*, 1991.

[Goodwin 81]   Goodwin, Charles, "**Conversational Organization : interaction between speakers and hearers**", *Academic Press*, New York and London, 1981.

[Jung 96]   Jung, David and Zelinsky, Alexander, "**Cleaning: An Experiment in Autonomous Robot Cooperation**", *Proceedings of the World Automation Conference (WAC) - International Symposium on Robotics and Manufacturing* (ISRAM), Montipellier, France, March, 1996.

[Jung 96a]   Jung, David and Zelinsky, Alexander, "**Whisker-Based Mobile Robot Navigation**", *Proceedings of the IEEE/RSJ International Conference on Intelligent Robots and Systems (IROS)*, vol 2. pp497. November 1996.

[Jung 97b]   Jung, David, "**An Architecture for Cooperation between Autonomous Mobile Robots**", *PhD Thesis, Intelligent Robotics Laboratory, University of Wollongong*, submitted December 1997. For further information on this research see the web site http://pobox.com/~david.jung.

[Kuniyoshi 94] Kuniyoshi, Yasuo, Rougeaux, Sebastien, Ishii, Makoto, Kita, Nobuyuki, Sakane, Shigeyuki and Kakikura, Masayoshi, "**Cooperation by Observation [The framework and basic task patterns]**", *Proceedings of the 1994 IEEE International Conference on Robotics and Automation*, San Diego, California, May 8-13, 1994.

[Maes 90a] Maes, P., "**Situated Agents Can Have Goals.**", *Designing Autonomous Agents*. Ed: P. Maes. MIT-Bradford Press, 1991. ISBN 0-262-63135-0. Also published as a special issue of the Journal for Robotics and Autonomous Systems, Vol. 6, No 1, North-Holland, June 1990.

[Mataric 92a] Mataric, Maja J., "**Behavior-Based Systems: Key Properties and Implications**" in *Proceedings, IEEE International Conference on Robotics and Automation*, Workshop on Architectures for Intelligent Control Systems, Nice, France - May 1992, 46-54.

[Parker 95] Parker, Lynne E., "**The Effect of Action Recognition and Robot Awareness in Cooperative Robotic Teams**", *IEEE 0-8186-7108-4/95*, 1995.

[Pfeifer 95] Pfeifer, Rolf, "**Cognition - Perspectives from autonomous agents**", *Robotics and Autonomous Systems 15*, pp47-70, 1995.

[Steels 90] Steels, Luc, "**Cooperation between Distributed Agents through Self-Organisation**", *VUB AI Lab*, Pleinlaan 2, 1050 Brussels Belgium, IEEE, *IROS 90*.

[Steels 96a] Steels, Luc, "**Synthesising the origins of language and meaning using co-evolution, self-organisation and level formation**", Artificial Intelligence Laboratory, Vrije Universiteit Brussel. In: Hurford, J. (ed.) *Evolution of Human Language*, Edinburgh University Press. Edinburgh, July 1996.

[Steels 96b] Steels, Luc, "**Self-organizing vocabularies**", In: Langton, C. (ed.) *Proceedings of Alife V*, Nara Japan, 1996.

[Tinbergen 53] Tinbergen, N., "**Social Behaviour in Animals**", *Methuen & Co. Ltd.*, London, 1953.

[Wilson 75] Wilson, E. O., "**Sociobiology: The New Synthesis**", *Harvard*, 1975.

[Wright 95] Wright, Robert, "**The Moral Animal : Why We Are the Way We Are : The New Science of Evolutionary Psychology**", *Vintage Books*, Luann Walther (Ed.), ISBN:0679763996, 1995.

[Yuta 91] S. Yuta, S. Suzuki and S. Iida, "**Implementation of a small size experimental self-contained autonomous robot - sensors, vehicle control, and description of sensor based behavior**", *Proc. Experimental Robotics*, Tolouse, France, LAAS/CNRS (1991).

[Zelinsky 96] Zelinsky, Alexander and Heinzmann Jochen, "**Real-Time Visual Recognition of Facial Gestures for human-Computer Interaction**", *Proceedings of the 2nd International Conference on Automatic Face and Gesture Recognition*, IEEE Computer Society Press, Killington, Vermont, October 1996.

Self-reconfigurable Robots for Navigation and Manipulation

Keith Kotay and Daniela Rus
Dartmouth Robotics Laboratory
Department of Computer Science
Dartmouth
Hanover, NH 03755, USA
{rus,kotay}@cs.dartmouth.edu

Abstract: Self-reconfigurable robots consist of a set of one or more identical autonomous modules that can adapt their shape and function to tasks. We describe a module that can function as a climbing robot, a manipulator, or a leg in a multi-legged walker. This module can make autonomous transitions between these states. We present the control algorithms that enable our robot to be a versatile navigator and manipulator and report on our experimental results.

1. Introduction

A robot designed for a single purpose can perform some specific task very well, but it will perform poorly on a different task, in a different environment. This is acceptable if the environment is structured; however if the task is to autonomously explore and manipulate an unknown environment, then a robot with the ability to change its shape to suit the environment and the required functionality will be more likely to succeed than a fixed-architecture robot. In addition, robots designed to explore remote environments should be very reliable and fault-tolerant.

A self-reconfigurable robot is a set of one or more autonomous modules that has the ability to self-organize, that is, to adapt their shape and functionality according to task autonomously. Self-reconfigurable robots can be viewed as a minimalist approach to designing versatile, extensible, and reliable navigators and manipulators. A single, architecturally-lean autonomous module can aggregate in a variety of structures with other identical modules. If the module is robust and the aggregation protocol is provably correct, the end result is a range of reliable robots. When these modules are autonomous robots, they can independently organize themselves as robust three-dimensional structures whose geometric configurations determine the resulting robot function. Such structures have built-in redundancy and are thus naturally fault-tolerant.

Applications of robots capable of changing their geometric configuration include tasks that require different modalities of locomotion and manipulation. For example, a robotic module such as the Inchworm robot described here, that can propel itself along smooth, arbitrarily oriented surfaces in three dimensions can aggregate with other similar modules to form a six-legged walker.

Such a system could self-assemble for traversing non-smooth terrains, or self-disassemble into a distributed system of robots that can inspect a three dimensional structure. Another application would allow a set of modules to self-reconfigure, enabling interesting locomotion and manipulation gates. For example, the modules could self-organize as a linear structure for traversing a narrow tunnel, and reconfigure as a multi-fingered arm upon exit, to manipulate objects.

We have made progress towards building, analyzing, planning, and using this class of robots. We have designed two types of modules capable of self-reconfiguration: an inchworm robot, and a universal molecule (to be described in our forthcoming work). Inchworm robots are capable of a variety of manipulation and navigation modalities by themselves, or cooperating as a distributed system. Universal molecules are capable of aggregating as arbitrary three dimensional structures.

In this paper we describe the Inchworm robot [KR96], a module that can be an autonomous navigator of three-dimensional structures, a manipulator, a multi-legged walker (where each leg is an inchworm), and a snake (with several inchworms connected together in a linear structure.) More specifically, each Inchworm robot can (1) function as an autonomous navigator of three dimensional structures, regardless of the presence, absence, or direction of gravitation; (2) attach the front end to an object, lift the object, and push the object on a straight line; (3) attach the back end to an object and pull that object as it propels itself; (4) come together with six Inchworms to form a multi-legged walker that can navigate on rough terrain; and (5) transition autonomously between all these tasks, which illustrates the self-reconfigurability of the Inchworm robot. Applications of this work include inspection tasks, where several Inchworms can attach themselves to a platform and carry tools and power supplies to the base of a complicated bridge or tower. At this point, the Inchworms can detach and distributively inspect the structure by climbing.

2. Related Work

Related work in designing modular robots includes [PK95, PK93, NS96, HS96]. In [PK95] a method for designing various robotic arms with different reachability properties out of the same set of 7 modules is proposed. The mechanical design algorithm is implemented as simulated annealing that starts with a random mechanical design and converges to the design with desired reachability properties. The modules get aggregated by hand as the computed shape. Our work is different in that our modules could self-aggregate (without human intervention) and the planning phase is of a task-directed, geometric nature. Our modular robots are aggregated according to task and we can view reachability as a specific kind of task.

Related work in self-organizing robots includes robots in which modules are reconfigurable using external intervention [CLBD92]. In [FK90] a cellular robotic system is proposed to coordinate a set of specialized modules. [CB97] describe a theoretical framework for counting the number of unique configurations realizable from a set of modules and joints, without considering implementation issues. [Yim93] studies multiple modes of locomotion that are achieved by composing a few basic elements in different ways. [Mu94] consider a system of modules that can achieve planar motion by walking over each other due to changes in the polarity of magnetic fields. [PCSC] describes metamorphic robots that can aggregate as stationary two-dimensional structures with varying geometry and implement planar locomotion. Our self-reconfigurable robots are different than metamorphic robots in that each module is autonomous. The resulting structures are three dimensional, can move along any axis in a three dimensional space, and have motion autonomy relative to a three-dimensional world.

Related work on biologically-inspired robots includes [Bro89, HNT91, FS96, DRJ95]. Brooks [Bro89] proposes insect intelligences and six-legged walkers. Hirose [HNT91, NH94] describes a quadruped robot that uses suction cups for attachment and can climb on straight surfaces (vertically and horizontally) and make transitions between surfaces with joint angles of 90 degrees. This robot is much heavier than ours (99 pounds vs. 1.25 pounds). Our Inchworm robot is different than the previous climbing robots [MD92, Neu94, HNT91, GR90, Nis92] in that it is much smaller, lighter, it needs less workspace to operate and it can handle web structures, as well as solid walls.

Related work on the mechanical analysis of Inchworm robots and climbing robots includes [KM94, CG91]. The kinematic analysis of our Inchworm is inspired by the analyses presented in [KM94, CG91].

3. The Inchworm Robot

The Inchworm is a biologically-inspired robot, designed to imitate the movements of the inchworm caterpillar. This functionality was achieved by creating a light, linear structure made of four sections. The sections are linked with three joints providing three degrees of freedom (see Figure 1). These joints allow the Inchworm to extend and flex. The first and fourth sections are the *feet* of the Inchworm. These sections contain the attachment mechanisms, equivalent to the legs and protolegs of the inchworm caterpillar, which allow the Inchworm to adhere to the surface it is traversing and provide the anchoring force needed to support the Inchworm walking motion. A fourth degree of freedom is provided by a pivot joint which allows the body of the Inchworm to rotate relative to the attachment mechanism of the rear foot. This allows the robot to turn. This pivot joint is currently being integrated in the robot architecture.

The weight of the Inchworm is 566 grams. When fully extended, the length of the robot is 330 millimeters and its height is 80 millimeters. When fully contracted, the length of the robot is 175 millimeters and its height is 160 millimeters. The speed of the robot is 0.75 meters/minute.

The configuration space of this robot is

$$SE(2) \times S^1_{(-\frac{\pi}{2},\frac{\pi}{2})} \times S^1_{(-\frac{\pi}{2},\frac{\pi}{2})} \times S^1_{(-\frac{\pi}{2},\frac{\pi}{2})}.$$

Here, $S^1_{(-\frac{\pi}{2},\frac{\pi}{2})}$ denotes a subset of S^1 defined for the range of angles $(-\frac{\pi}{2},\frac{\pi}{2})$.

Figure 1. The Inchworm robot. The body of robot consists of four links. The first and last link of the robot comprise the feet. Each foot has two 1-inch electromagnets attached to it. The robot has one front IR sensor and a pair (IR sensor, contact switch for touch sensing) attached to each electromagnet.

In the absence of inertial effects, the propelling locomotion of this robot will be governed by the equation [KM94]

$$k\dot{x} + (1 - k - 2l)(\dot{x} + \dot{h}) = 0,$$

where x is the distance traveled, k is the length of the foot of the robot, l is the length of each of the two middle links of our Inchworm, and $h = l\cos\theta_1 + l\cos\theta_2$, with θ_1 and θ_2 denoting the angle between the front, respectively the back link of the robot relative to the line of motion. Thus, we can rewrite the locomotion of the robot as $\dot{x} + (1 - k - 2l)(\dot{x} - l\dot{\theta}_1\sin\theta_1 - l\dot{\theta}_2\sin\theta_2) = 0$.

The Inchworm robot can navigate three dimensional structures and it can manipulate objects. The robot can propel itself on arbitrarily oriented surfaces in an on-line fashion, in the absence of geometric models and independent of the direction and magnitude of gravity. The robot can walk, climb, and walk inverted. The robot can also make transitions between surfaces whose joint angle is convex or concave. This robot can also attach its rear foot to an object, which enables the robot to pull an object as it traverses a surface. All motions are compliantly controlled on-line. The control algorithms are simple, compliant, adaptive, and reliable. We have implemented each of these skills as task-level primitives. The rest of this section is devoted to describing these task-level primitives for navigation and manipulation.

3.1. Stepping

The simplest task level primitive is the **step**. A **step** consists of two phases, the extension phase and the contraction phase. During the extension phase, the rear foot is attached to the surface. An extension is accomplished by raising the front foot, extending it forward, and then lowering the foot until it contacts the surface. At this point, the front foot is attached to the surface and the rear foot is detached. The contraction phase then occurs by raising the rear foot, contracting the rear foot to bring it closer to the front foot, and then lowering the foot until it contacts the surface.

3.2. Concave Transitions

concave-transition is a primitive that implements a move from one surface to another surface when the relative orientation between the surfaces ranges from 45 to 90 degrees relative to the first surface. (Transitions to surfaces whose relative orientation is between 1 and 45 degrees are implemented as part of a step as explained in [KR96]). The control algorithm for concave-transition consists of raising the front foot, rotating it until it is parallel to the new surface, extending it far enough so that there is space for the rear foot behind the front foot on the new surface, and then bringing the rear foot from the old surface to the new surface.

3.3. Convex Transitions

convex-transition is a primitive used for moves from one surface to another surface when the relative orientation between the surfaces ranges from -45 to -90 degrees relative to the first surface. (Transitions to surfaces whose relative orientation is between -1 and -45 degrees are implemented as part of a step in the same way as for a concave transition as explained in [KR96]). The control algorithm for convex-transition consists of moving the front foot to the new surface, bringing the back foot close to the corner, extending the front foot further on the surface to make room for the back foot and then bringing the back foot to the new surface (see Figure 2).

Figure 2. Illustration of the movements performed during a convex transition.

3.4. On-line walking in three dimensions

We have implemented an on-line navigation algorithm which implements a greedy strategy where the robot attempts to preserve the initial direction of motion. If the IR sensor pointed forward registers an object (*e.g.* it returns a small value) it follows that an obstacle or a surface with a relative concave orientation has been encountered. The robot uses concave-transition to attach itself to this new surface and continues travelling. If the first IR sensor of the bottom of the front foot registers free space (*e.g.* it returns a large value while the second IR sensor on that foot returns a small value) a surface with a relative convex orientation has been encountered. The robot uses convex-transition to attach itself to this new surface and continues travelling. Otherwise, the robot continues stepping in the current direction.

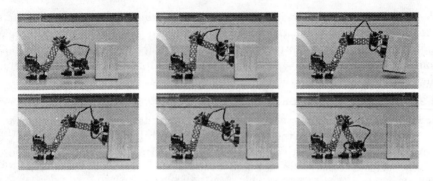

Figure 3. Six snapshots taken from a sweep sequence

3.5. Lifting and placing an object

When the Inchworm is performing a concave transition, it must detect the surface in its path and then attach its front foot to that surface. These are the same skills the Inchworm needs to detect and grasp a movable object. We have implemented a lift-and-place algorithm which grasps an object, lifts the object, moves the object forward, lowers the object to the surface, and returns the Inchworm to its original pose. We have used lift-and-place to develop a higher-level algorithm, sweep, which can move an object in a straight line. sweep repeatedly performs a step until an object is encountered, then positions the robot, invokes lift-and-place, and continues stepping (see Figure 3).

When an object is detected by the forward IR sensor during a step movement, the step is terminated and the Inchworm is positioned to perform a grasp. A raise skill with forward compliance is then used to lift the front foot while maintaining a constant distance to the object. Then a position skill is performed to align the bottom of the front foot with the object, followed by an extend skill which terminates on the detection of free space by the forward, downward-pointing IR sensor on the front foot, signifying that the front foot is near the top of the object. A lower-to-contact skill places the electromagnets on the surface of the object and an attach completes the grasp. Once the object is grasped, the raise skill is used to lift and lower the object, and the extend skill is used to move the object forward. A detach releases the object.

3.6. Pulling an object

In our previous work, we developed control algorithms which allowed the Inchworm to move autonomously in an unstructured environment. We now present three manipulation primitives which allow the Inchworm to attach itself to an object, move itself and the object, and detach itself from the object. These primitives enable the Inchworm to manipulate objects in its environment. Our current experimental platform is a wheeled object with an attached steel surface at an angle of approximately 45 degrees (see Figure 4). This object rests on a level steel surface, or "floor". The robot manipulates the object by attaching one foot to the steel surface of the object, attaching the other foot to the floor, and then generating a propelling motion which moves the entire system.

The manipulation primitives are **connect**, **stride**, and **disconnect**. **connect** allows the Inchworm to attach its rear foot to the object to be manipulated. This is done using a series of **position** movements which place the rear foot in close proximity to the inclined steel surface, followed by a **lower-to-contact** movement which moves the rear foot into contact with the surface. **attach** is then used to activate the attachment mechanism. When the rear foot is attached, the front foot can be raised off the "floor" surface due to the stability of the wheeled platform. This allows the front foot to be placed in the starting point for the **stride** primitive.

The **stride** primitive is used to move the Inchworm and wheeled platform system across the steel floor. The algorithm begins with an extension of the front foot identical to that in the extension phase of the **step** algorithm described in Section 3.1. After the front foot is in contact with the surface, it is attached to the steel floor and three **position** movements are performed which, in turn, push down on the floor raising the small castor ball on the wheeled platform, pull the wheeled platform forward toward the front foot, and lower the castor ball to the floor. This results in a net forward motion of the object.

disconnect is a primitive that is used to detach the Inchworm from the manipulation object. It consists of a **detach** skill which deactivates the attachment mechanism on the rear foot, followed by a **raise** skill which lifts the rear foot off the steel attachment surface. Two **position** movements and a **lower-to-contact** movement return the rear foot to contact with the floor.

Using these primitives the Inchworm can perform manipulation tasks on objects such as the wheeled platform (see Figure 4). Due to the current lack of turning capability on the Inchworm, manipulation is limited to straight-line motion. We have only implemented a pulling motion, but a pushing motion is also feasible.

3.7. Transitions between navigation and manipulation

The design of our control code enables autonomous transitions between manipulation and navigation tasks. This is demonstrated by the use of a navigation primitive, **step**, and a manipulation primitive, **lift-and-place**, within the **sweep** algorithm. **sweep** autonomously transitions between navigation and manipulation primitives in the process of moving an object. In the object pulling task, the Inchworm uses manipulation primitives to move the object but once the robot has detached itself from the object it can switch to navigation primitives to move away from the object as shown in Figure 4. The ability to autonomously transition between navigation and manipulation tasks gives the Inchworm the flexibility to navigate or manipulate as circumstances dictate.

3.8. Multi-legged walker

In Section 3.6 we showed how the Inchworm can manipulate an object by pulling it. Multiple Inchworms can run the same algorithm in synchrony. This would be useful when the weight of the object being pulled is too large to be moved by one Inchworm alone. If the frictional forces between the object being pulled and the surface are high, multiple Inchworms can attach themselves to the object and lift it to suspend the object in air. The Inchworms can then

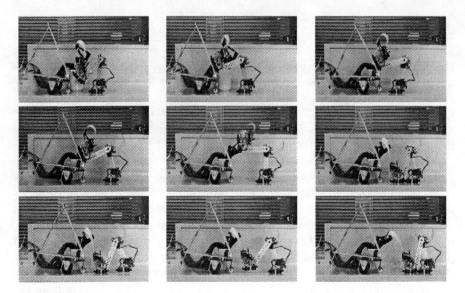

Figure 4. Eight snapshots taken from a pull sequence followed by a step.

propel themselves in synchrony to relocate the object. This resulting structure can be thought of as a multi-legged walker, where each inchworm functions as one leg. At least 4 Inchworms are needed to implement such a structure (see Figure 5). Three Inchworms are needed to be in contact with the surface at all times to stably support the object. A four-legged structure can walk by sequentially allowing each Inchworm to perform the stride described in Section 3.6. A six-legged walker can walk by using the tripod gate. Other interesting gates can be formulated when more than six legs are involved. As described in Section 3.7, the Inchworm can autonomously transition between navigation and manipulation tasks. This capability enables individual Inchworms to independently navigate to the walker structure and then use manipulation primitives to attach to and move the structure. Similarly, Inchworms can detach from the walker structure and switch to navigation primitives in order to perform individual tasks.

3.8.1. Globally-controlled leg

Although we have not yet implemented this algorithm because we have not completed the construction of all six Inchworms, we have implemented the function of one leg. The **stride** movement performed as part of the pulling task executes a motion which propels an attached object forward. This one-leg control can be replicated on five other Inchworms. Here we assume that a global controller coordinates the specific roles of the Inchworms in this structure (that is, tells the Inchworms when to stay fixed to support the object and when to execute the stride). Of course, many interesting challenges arise in a distributed model, where Inchworms have local control and some communication mechanism. We plan to address the control of a distributed set of Inchworms in our future work.

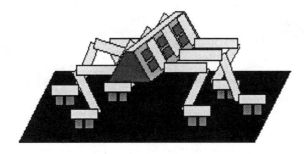

Figure 5. A six-legged walker composed of four Inchworm robots.

4. Experiments

We implemented the **sweep** algorithm to test the **lift-and-place** primitive, and the **pull** algorithm to test the **connect**, **stride**, and **disconnect** primitives. All experiments were conducted on a level steel surface.

The object used in the **sweep** experiments was a balsa wood block with a steel plate attached to one side. The weight of the block is 116 grams. A successful **lift-and-place** means that the object was grasped, lifted, moved forward, and lowered. Each successful **lift-and-place** also implies that the object was successfully detected and the Inchworm was correctly positioned for the **lift-and-place** operation, however object detection and Inchworm positioning are part of the **sweep** algorithm not **lift-and-place**. In this experiment, all objects detected in front of the Inchworm were assumed to be movable objects. As shown in Figure 6, in 115 trials **lift-and-place** was successful over 92 percent of the time.

In the straight-line pulling experiment, the Inchworm uses **connect** to attach itself to a wheeled platform, pulls the platform using two **stride** operations, detaches from the object using **disconnect**, and then performs two **step** operations. The results of 35 runs of this experiment are shown in Figure 6. As the data shows, the primitives are very reliable. It is worth noting, however, that the **connect** primitive is sensitive to the initial positions of the robot and the object. For the above experiment, the initial positions of the robot and the object were set by hand. The reason for this is that the Inchworm does not have a backward facing sensor to detect the position of the object. We intend to add a sensor to overcome this limitation.

Figure 6 also includes data for **step** and the globally-controlled leg algorithm. These tasks were components of the straight-line pulling experiment. Each pulling experiment included two **step** tasks, and this data was added to the data presented in [KR96]. The result is 200 successful steps out of 210 tries. The globally-controlled leg data consists of data for the **stride** component of the pulling experiment. This was done because a **stride** consists of the leg motion which would be used for each Inchworm in the multi-legged walker.

Task	Tries	Success	Reliability
Lifting and Placing	115	107	92.5 %
Straight-line Pulling	35	34	97.1 %
Step	210	200	95.2 %
Concave-transition	57	46	80.7 %
Globally-Controlled Leg	35	35	100.0 %

Figure 6. This table contains reliability data for the lifting and placing, straight-line pulling experiments, and on-line navigation experiments.

5. Discussion

We have described a self-reconfigurable robot that can accomplish complex multi-modal navigation and manipulation tasks in three dimensional environments. The experimental data demonstrates that this is a feasible and robust way of creating versatile robots. At the moment we have hardware experiments with one module only. A second robot will be ready within a month and we expect to have the other four robots ready by the end of the year. We are currently designing a walker frame. Individual Inchworms will be able to "home in" on the frame in order to assemble as a multi-legged walker. The frame will also be used to provide power and communication to the Inchworms. The central challenge that remains is to design the coordination algorithms that allow the robots to cooperate for walking. Much work remains to be done.

Our long term goal is to develop a theoretical and experimental foundation for creating general self-reconfigurable modules. The main challenges here are (1) to design a simple, light-weight universal module capable of the range of motions required to aggregate in arbitrary configurations with other identical modules; (2) to characterize geometrically the class of robots that can be constructed using the universal module; (3) to develop planners for a system of universal modules; and (4) to characterize the locomotion and manipulation gates that can be planned for.

Acknowledgements

This paper describes research done in the Dartmouth Robotics Laboratory. Support for this work was provided through the NSF CAREER award IRI-9624286 and the NSF award IRI-9714332. Support for our research was also provided by Microchip Inc., the Motorola University Support Program, Omron Inc., and RWI Inc.

References

[AB90] C. Angle and R. Brooks, Small Planetary Rovers, in *Proceedings of the IEEE/RSJ International Workshop on Intelligent Robots and Systems*, Ikabara, Japan, 383-388, 1990.

[Bro89] R. Brooks, A robot that walks: emergent behaviors from a carefully evolved network, in *Proceedings of the IEEE Conference on Robotics and Automation*, Scottsdale, 1989.

[CB97] I. Chen and J. Burdick, Enumerating the Non-Isomorphic Assembly Configurations of a Modular Robotic System, to appear in the *International Journal* of Robotics Research.

[Che90] F. Chernousko, On the mechanics of a climbing robot, *Mechatronic systems engineering*, 1:219-224, 1990.

[CLBD92] R. Cohen, M. Lipton, M. Dai, and B. Benhabib, Conceptual design of a modular robot, *Journal of Mechanical Design*, March 1992, pp. 117-125.

[CG91] G. Chirikjian and J. Burdick, Kinematics of a hyper-redundant robot locomotion with applications to grasping, in *Proceedings of the IEEE International Conference on Robotics and Automation*, 1991.

[DRJ95] R. Desai, C. Rosenberg, and J. Jones, Kaa: an autonomous serpentine robot utilizes behavior control, In *Proceedings of the 1995 International Conference on Intelligent Robots and Systems*, Pittsburgh, 1995.

[DJR96] B. Donald, J, Jennings, and D. Rus, Minimalism + Distribution = Supermodularity, in *Journal of Experimental and Theoretical Artificial Intelligence*, 1996. (to appear)

[DJR94a] B. Donald, J. Jennings, and D.Rus. Information invariants for distributed manipulation, *The First Workshop on the Algorithmic Foundations of Robotics, eds. K. Goldberg, D. Halperin, J.-C. Latombe, and R. Wilson*, pages 431-459, 1994.

[FS96] A. Fernworn and D. Stacey, Inchworm Mobility - Stable, Reliable and Inexpensive, In *Proceedings of the Third IASTED Internation Conference on Robotics and Manufacturing*, Cancun, 1995.

[FK90] T. Fukuda and Y. Kawauchi, Cellular robotic system (CEBOT) as one of the realization of self-organizing intelligent universal manipulator, in *Proceedings of the 1990 IEEE Conference on Robotics and Automation*, pp. 662-667.

[GR90] V. Gradetsky and M. Rachkov, Wall climbing robot and its applications for building construction, *Mechatronic Systems Engineering* 1:225-231, Kluwer Academic Press, 1990.

[HS96] G. Hamlin and A. Sanderson, Tetrabot modular robotics: prototype and experiments, in *Proceedings of the IEEE/RSJ International Symposium of Robotics Research*, pp 390-395, Osaka, Japan, 1996.

[HNT91] S. Hirose, A. Nagakubo, and R. Toyama, Machine that can walk and climb on floors, walls, and ceilings, in *Proceedings of the International Conference on Advances in Robotics*, Pisa, 753-758, 1991.

[Ina93] M. Inaba, Remote-Brained Robotics: Interfacing AI with real world behaviors, in *Robotics Research: The Sixth International Symposium*, Hidden Valley, 1993.

[KM94] S. Kelly and R. Murray, Geometric phases and robotic locomotion, CDS Technical Report 94-014, California Institute of Technology, 1994.

[KR96] K. Kotay and D. Rus, Navigating 3d steel web structures with an inchworm robot, in *Proceedings of the 1996 International Conference on Intelligent Robots and Systems*, Osaka, 1996.

[Lat91] J. C. Latombe, *Robot Motion Planning*, Kluwer Academic Publishers 1991.

[MD92] A. Madhani and S. Dubowsky, Motion planning of multi-limb robotic systems subject to force and friction constraints, in *Proceedings of the IEEE Conference on Robotics and Automation*, Nice, 1992.

[Mu94] S. Murata, H. Kurokawa, and Shigeru Kokaji, Self-assembling machine, in *Proceedings of the 1994 IEEE International Conference on Robotics and Automation*, San Diego, 1994.

[Neu94] W. Neubauer, A spider-like robot that climbs vertically in ducts or pipes, in *Proceedings of the 1994 International Conference on Intelligent Robots and Systems*, Munich, 1994.

[NS96] B. Neville and A. Sanderson, Tetrabot family tree: modular synthesis of kinematic structures for parallel robotics, in *Proceedings of the IEEE/RSJ International Symposium of Robotics Research*, pp 382-390, Osaka, Japan, 1996.

[NH94] A. Nagakubo and S. Hirose, Walking and running of the quadruped wall-climbing robot, in *Proceedings of the IEEE Conference on Robotics and Automation*, pp. 1005-1012, San Diego, 1994.

[Nis92] A. Nishi, A biped walking robot capable of moving on a vertical wall, *Mechatronics*, vol2, no. 6, pp. 543-554, 1992.

[PCSC] A. Pamecha, C-J. Chiang, D. Stein, and G. Chirikjian, Design and implementation of metamorphic robots, in *Proceedings of the 1996 ASME Design Engineering Technical Conference and Computers in Engineering Conference*, Irvine, CA 1996.

[PK95] C. Paredis and P. Khosla, Design of Modular Fault Tolerant Manipulators, in *The First Workshop on the Algorithmic Foundations of Robotics*, eds. K. Goldberg, D. Halperin, J.-C. Latombe, and R. Wilson, pp 371-383, 19 95.

[PK93] C. Paredis and P. Khosla, Kinematic Design of Serial Link Manipulators from Task Specifications, in *International Journal of Robotic Research*, Vol. 12, No. 3, pp 274–287, 1993.

[Yim93] M. Yim, A reconfigurable modular robot with multiple modes of locomotion, in *Proceedings of the 1993 JSME Conference on Advanced Mechatronics*, Tokyo, Japan 1993.

Human-Robot Interface System with Robot Group Control for Multiple Mobile Robot Systems

● José **Beltrán Escavy**   Tamio **Arai**   Akio **Nakamura**
Shinjiro **Kakita**   Jun **Ota**

Arai-Ota Lab., Dept. of Precision Machinery Engineering
The University of Tokyo
7-3-1 Hongo, Bunkyo-ku, Tokyo 113, Japan
Tel.: +81-3-38122111 ext.: 6486   Fax: +81-3-38128849

e-mail: {pepe, arai, nakamura, kakita, ota}@prince.pe.u-tokyo.ac.jp

Abstract: A system to allow control of multiple mobile robot systems by a single human operator is proposed. The main feature of the system is that the unit of control is, not the individual robot, but the **group** of robots. This control paradigm is tested by creating a system that allows a human user to control 4 robots by means of a joystick. Experiments show that this kind of control is feasible and useful.

1. Introduction

Nowadays, a great amount of research is being done in the field of multiple mobile robot systems. It is not necessary to stress how useful they can be in many industrial fields, if we manage to create a system that is robust and reliable enough.

However, when we undertake the task of building such a system, we run into a very serious problem: Given the current state of technology, it is impossible for us to give those robots anything even remotely approaching **true** autonomy.

It follows, then, that we must have (at least in the foreseeable future) a human user that will supervise and control the system. But, then, assuming that, we find another serious problem: with many robots working at the same time, it is extremely difficult for a human to understand and follow what is happening (not to mention actually **controlling** it). With, let us say, 20 robots working at the same time, and with information from them all coming back to the user, the system may collapse because the human user is unable to absorb such a flood of information.

Therefore, we must try to find a way of allowing a human user to exert appropriate control and supervision. Also, it is necessary to develop some way of sending information to the user that will not overwhelm him/her with unmanageable amounts of data. If we want to have systems with many robots performing tasks in a real-world environment, we must solve these two problems.

Another factor that affects the ability of a human user to control a system with many robots is that, till now, almost all of the research that has been done regarding control of mobile robots assumed that one user will control one single robot (or a small number of robots) at a given time. If we keep this "control paradigm" in systems with many robots, it quickly becomes impossible to exert an appropriate supervision and control on the part of the human.

However, we can make the following change in the control paradigm: instead of considering that the "unit of control" is the individual robot, we consider that this "unit" is instead formed by a **group of robots** doing some task. In that way, there is less information coming back to the user, and less trouble exerting control over the robots. Of course, if needed, the user could ask for data about and control of a single robot, but this would be the exception, not the rule.

The idea, then, is to allow the user to keep track of **groups** and send commands to **groups**. The computer system takes the command for the group, "translates" it and adapts it for the individual robots of the group, sending those individual commands to the robots in the work area. This process is transparent to the user; he/she must not worry about controlling individual robots.

Some tentative research has been done in the field of controlling and operating groups of robots [2] [3] [4]; however, there has not been much dedication. Among the best research is that of Hasegawa et al. [1], who developed a system that interpreted "intuitive gestures" made by a human operator to control mobile robots either one by one or in groups. But it is very difficult for a computer to interpret complex intuitive gestures from a human, and this limits the applicability of this method.

A related field of study might be the control and teleoperation of cooperative manipulators. Many works have been done in this field; however, their results are not fully applicable to multiple mobile robot systems, as the paradigms are very different. Besides, studies on cooperative manipulators rarely focus on control of more than 2 or 3 manipulators at a time, whereas, in the field of multiple mobile robots, it is not unusual to have to control 4 or more robots simultaneously.

Another concept that we apply in our research is that of **semi-autonomy**. We have said that it is impossible to give real autonomy to the robots in our system. However, it would be highly impractical to require total control on the part of the human user. We have to find a "middle ground". In our case, this "middle ground" consists in having a set of built-in reactive behaviours in our robots that are activated in case of need ('subsumption architecture'-like).

When are those behaviours activated? When the environment dictates so. It can be said that "semi-autonomy" means having the environment help the user when controlling the robots of the system. For instance, we can have a group of two robots that have been told to push an object. They might start moving together; however, their speeds might be slightly different. That means that one robot would reach the object before the other. If the robots just kept moving, we would not be able to push the object efficiently. But, in that case, we have semi-autonomy kicking in: the robot that arrives first to the object will stop and wait for the other. Then, they will start moving together again.

All those concepts are embedded in the system built in our laboratory. Experiments show that they indeed work properly. That is why we propose this system as a suitable method to have multiple mobile robot systems in real-life situations.

2. Concept of the proposed system

The system that we have built is essentially a benchmark to test whether the control method that has been developed is workable. This method, as has been said, relies in commanding the robots as **groups**. The assumptions so far are:

- The robots do not know their environment beforehand
- The robots are very simple and have a low intelligence level

The robots must cooperate in a task. The task is moving an object by pushing it around in the workspace. The fundamental commands available to the human user are:

1. **Grouping** the robots in one or more groups (by default, initially all robots are grouped into a single group).

2. **Moving** a group of robots.

3. **Defining** a formation for the robots (triangle, square, line).

4. **Scaling** a formation.

5. **Rotating** a formation.

All those commands are available, at this moment, by means of joystick and keyboard input in the computer.

3. System architecture

The system on which we are making our research consists of: a *Silicon Graphics Indigo Computer*, which hosts a graphic interface program; an *IBM-PC* connected to the SG machine and the robots; *4 Khepera robots* and a *CCD camera* taking a bird's eye view of the work area. **Figure 1** shows schematically the architecture of our system. In our

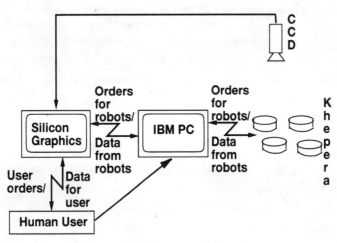

Fig. 1 – System architecture

system, the SG machine receives input from the CCD camera, and uses it to display the work area to the user. The user can make selections and send orders to the robots, via joystick and keyboard. Then, the interface program "translates" these orders into commands for the robots of the group receiving the order.

We want to connect the SG via RS-232 to the IBM-PC. This is for: 1) relaying commands to the robots via the IBM-PC, and 2) relaying data from the robots to the SG via the IBM-PC. The IBM-PC also connects with the 4 Khepera robots that are the experimental subjects. We use an IBM-PC computer to control the Khepera because Khepera drivers were not available for SG machines (we want the SG to be the main computer in our system). The robots receive orders relayed by the IBM-PC and follow them. They can send sensor data back to the IBM-PC, in order for it to relay them to the SG for the interface program.

The Khepera robots used in our system have been fitted with extra contact sensors and an array of LEDs for easy following through the CCD camera image. **Figure 2** shows two of the robots used in our system; **Figure 3** shows one robot and the attachment with the extra contact sensor and LEDs. **Figure 4** shows one robot with the attachment installed. **Figure 5** shows a general view of the system.

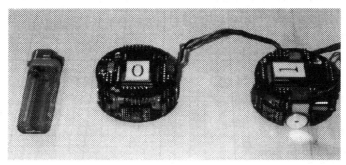

Fig. 2 – Khepera robots

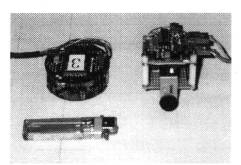

Fig. 3 – Khepera and attachment

Fig. 4 – Attachment mounted

Another thing that we want to have is some kind of graphic interface for the user. This interface would incorporate the CCD camera image in such a way that the user would be able to interact with the robots via joystick and mouse only (keyboard commands are somewhat cumbersome). At this moment, we are starting to develop this graphic interface.

4. Experiments and simulations

For the experiments, we defined a simple task to be fulfilled by the robots. This task was "moving objects by pushing them". Once this decided, we performed two experiments. The first one was designed to test the "semi-autonomy" behaviour, by having two robots pushing an object in such a way that one of them would arrive to the object in advance of the other. As for the second, it involved an "assembly" task, in which the 4 robots would have to push two objects (one concave and one convex) in order to "assemble" them. This second experiment was designed to test the general applicability of the system to "real life" situations.

We also developed some simulations of the graphic interface that we want to have, and tested whether the control of the system was really enhanced by using it. **Figure 5** shows a general view of the whole system taken while performing the experiments.

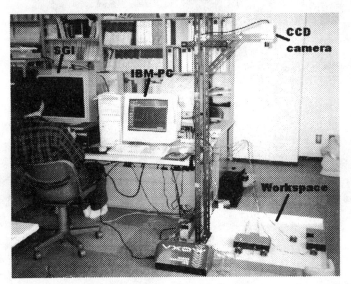

Fig. 5 – *Experimental system*

4.1 "Semi-autonomy" experiment

This experiment was done by using only two of the four robots. The robots were "assigned" to the same group, with the task of pushing an object together. However, from the beginning the starting position was such that one of the robots would surely arrive to the object before the other. **Figures 6 to 9** show this experiment, as seen from the CCD camera. The images shown are screen snapshots taken from the SG machine. In **figure 6**, the robots are starting to move. **Figure 7** shows the left-hand robot that has arrived to the object. But it stops and does not move, as the other robot has not yet arrived. **Figure 8** shows that the right-hand robot has finally arrived; at that point, both robots start moving together. **Figure 9** shows how they push the object.

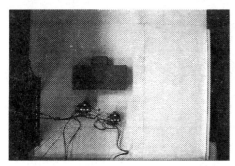

Fig. 6 – Robots start moving

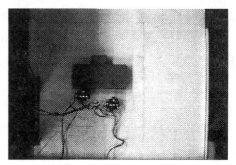

Fig. 7 – Left robot arrives first

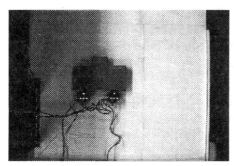

Fig. 8 – Right robot arrives

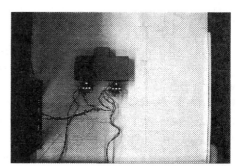

Fig. 9 – Both robots push

This experiment shows clearly that "semi-autonomy" is a great help when controlling groups of robots. The user just had to tell the robots to go in a certain direction to push; the system took care of synchronizing the robots.

4.2 "Assembly" experiment

This experiment was done by using the 4 robots. They were divided in two groups with two robots each, and their task was to avoid an obstacle and manoeuvre a couple of blocks (one concave, one convex) so that they could be assembled. The experiment took some time to complete, but it was successful. **Figures 10 to 21** are screen snapshots taken from the SG machine during the assembly experiment (**Figure 5** was also taken during that experiment).

4.3 Simulations

While the system was being prepared, and the experiments carried on, we also programmed some simulations of the kind of graphic interface we want to have for the human operator to use. We want to be able to give orders to the robots using only mouse and joystick inputs. We defined, again, the task to be done as "pushing" an object in an obstructed workspace by means of 4 robots (divided in 2 groups of two robots each). **Figures 22 to 25** are screen snapshots taken during the execution of the interface simulation.

The way in which we performed this simulation was the following: first, the robots were divided in two groups, which were sent to opposite sides of the object in order to rotate it by pushing. When the object was rotated 90 degrees, one group would be put aside to "rest", while the other one aligned appropriately and pushed the object to its final destination.

The simulation was used to test the efficiency of the "group approach" when controlling multiple mobile robot systems. We repeated the same task, but using 4, 6, 8, 10 and 12 "robots". And, each time, we compared the results obtained with our method to the results obtained by controlling the robots individually. The comparison was made by checking how long it took to complete the task in both situations. **Figures 26 and 27** show the results of the simulation.

It can be easily seen that, when performing the same task, the control system oriented to groups requires only 55% of the time needed by a control system in which robots must be commanded individually. It can also be seen that the ratio tends to become better with a growing number of robots in the system.

This shows that the basic idea of the control paradigm proposed in our research is sound. Commanding robot systems at the level of robot groups not only makes keeping control easier, but also more efficient.

5. Conclusions and further research

5.1 Conclusions

We need some system to allow a human user to keep control of multiple mobile robot systems in an efficient way. The proposed system has been shown to do precisely that. Not only it provides good control, but it also enhances efficiency when compared to "classic" control methods. To sum up, we are confident that this control paradigm will help in developing multiple mobile robot systems into elements used in real-world industrial environments.

5.2 Further research

From now, what do we want? First, we want to develop a GUI that for easier interaction and control of the system. We also want to define metrics to provide the user with information about the general state of the system and perhaps to "fine-tune" robot behaviour in accordance to the environment. Some of these metrics could be:

- Average speed of the robots
- "Cluttering" factor (how many obstacles are being detected at each sampling interval)
- How many robots are stopped
- ......etc......

Also, we are now re-fitting the robots with new, improved contact sensors, so that, instead of having only one, they will have at least four. And we want to realize sensor integration on the robots. Our Khepera robots have infrared and contact sensors, and we want to integrate the data provided by them, so that we can process that information as efficiently as possible. Our final goal is to create an interface system that will be useful, versatile, and practical. And we think that it can be done.

References

[1] Hasegawa K. et al., "Novel Teleoperation System for Multi-Micro Robots based on Intuition Understanding through Operator's Intuitive Behaviours", Journal of the Robotics Society of Japan, 1996, Vol. 14, No. 4, pp 156–573 **(in Japanese)**

[2] Nakamura A. et al., "Teleoperation of Robot Groups (1st. Report: Object Manipulation by a Robot Group)", Proceedings of the 14th Meeting of the Robotics Society of Japan, 1996, Vol. 2, pp 659–660 **(in Japanese)**

[3] Beltrán-Escavy J. et al., "Teleoperation of Robot Groups (2nd. Report: System for Multiple Robot Control)", Proceedings of the 14th Meeting of the Robotics Society of Japan, 1996, Vol. 2, pp 661–662

[4] Mataric M.J. et al., "Cooperative Multi-Robot Box-Pushing", Proceedings of ICRA '95, pp 556-561

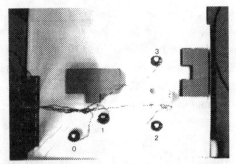

Fig. 10 – Assembly: Initial position

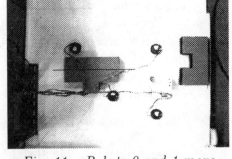

Fig. 11 – Robots 0 and 1 move

Fig. 12 – Robots 0 and 1 rotate object

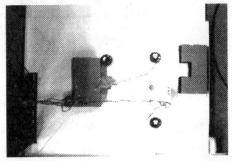

Fig. 13 – Convex object rotated

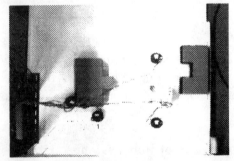

Fig. 14 – Robots 0 and 1 move to push

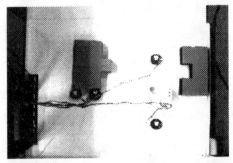

Fig. 15 – Robots 0 and 1 pushing

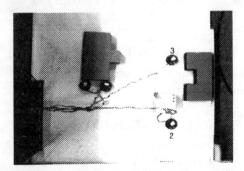

Fig. 16 – Robots 2 and 3 start moving

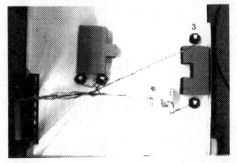

Fig. 17 – Robots 2 and 3 ready to push

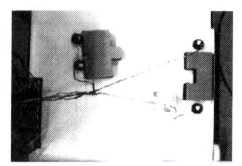
Fig. 18 – Robots 0 and 1 ready to push

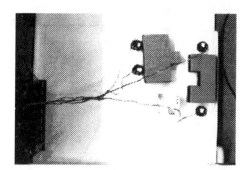
Fig. 19 – Robots 0 and 1 pushing

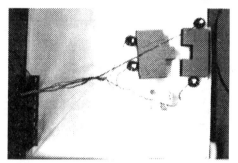
Fig. 20 – Robots 2 and 3 pushing

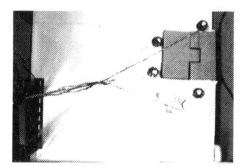
Fig. 21 – Assembly finished

Fig. 22 – Simulation starts

Fig. 23 – Robots move towards object

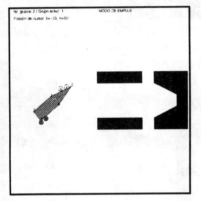

Fig. 24 – Robots rotate object

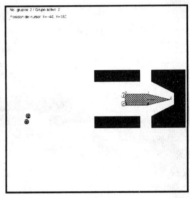

Fig. 25 – Robots push object

	4 robots		6 robots		8 robots		10 robots		12 robots	
	Group	1 by 1	Group	1 by 1	Group	1 by 1	Group	1 by 1	Group	1 by 1
Robots go towards object	25	40	35	60	45	80	65	100	75	120
First coasting (adjusting)	5	20	5	30	7	40	10	50	10	60
Rotate the object	6	6	6	6	6	6	6	6	6	6
Take away all robots except 2	3	6	8	12	10	18	13	24	15	30
Second coasting (only 2 robots)	7	14	7	14	7	14	7	14	7	14
Pushing of the object by 2 robots	11	11	11	11	11	11	11	11	11	11
TOTAL:	57	97	72	133	86	169	112	205	124	241
Total time ratios:	0.59		0.54		0.51		0.55		0.51	

Fig. 26 – Table with times

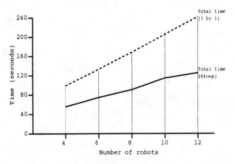

Fig. 27 – Graphic: results of simulation

Chapter 14

Learning & Skill Acquisition

Robot programming and control architectures must be equipped enough to face unstructured environments, which may be partially or totally unknown at programming time, or environments which are time varying and unpredictable. It is often hard to obtain an explicit model of many difficult robot tasks and skills, or to explicitly design a corresponding controller with an algorithm capable of performing the task. Actuation, sensory imprecision, and noise further complicate the problem. On the other hand, a human being has considerable expertise, both in manipulation and in moving himself in complex environments, with no apparent difficulties. Moreover, he is capable of learning by himself, or with the assistance of an instructor or teacher. This motivates the development, improvement, and application of automatic learning and teaching techniques that, taking advantage of sensory information, would enable the acquisition of new robot skills and avoid
some of the difficulties of explicit programming.

The first paper by Araujo and Almeida deal with the problem of mobile robots learning to navigate from a known starting position to a predefined goal region, on an initially unmodelled world. The application of the parti-game self-learning approach is demonstrated. It is based on a multiresolution grid, where the resolution is selectively increased, according to the world that is being faced by the system. The path to the goal is planned using a game-like minimax shortest-path approach. Improvements to the parti-game architecture are proposed.

Laschi, Taddeucci, and Dario propose a model for artificial perception, and sensory-motor coordination. The psychological and neural levels that, according to the involvement of conscious reasoning, can be identified in human learning, inspired the inclusion of two levels on the proposed architecture: low and high levels of processing and learning. An application of the architecture is illustrated with a reinforcement learning implementation applied to a robot edge-following task. It uses a radial geometry variable resolution vision sensor, and an integrated fingertip comprising tactile, dynamic, and thermal sensors.

Lloyd and Pai describe an integrated interactive simulation system, for enabling easier programming of part-mating and contact tasks, by possibly unexperienced operators. For example, barrier potentials may be used for preventing collisions or for maintaining a particular contact. A programmed path may be optimized by removing unwanted motions and unnecessary contacts while preserving feasibility. This optimization is done by having forces applied on the path: attractive forces by the nearest neighbors, and repulsive forces by the unwanted obstacles.

The fourth and final paper, by Liu, Tang, and Khatib, describes an approach for learning and planning robot manipulation strategies, using a dynamical discrete-event system model. In the model each state triggers an associated action and corresponds to a state in the robot task environment, and each arc corresponds to a possible state transition that is induced by an action. A reliable manipulation strategy is achieved by searching for a connected state transition path in this model.

Exploration-Based Path-Learning by a Mobile Robot on an Unknown World

Rui Araújo, and A. T. de Almeida
Institute of Systems and Robotics (ISR),
Electrical Engineering Department,
University of Coimbra
Coimbra, Portugal
emails: {rui,adealmeida}@isr.uc.pt

Abstract: In this paper we are concerned with the problem of navigating an autonomous mobile robot, on an unknown indoor environment, without any external support or human intervention, from an initial location to a specified goal region. The parti-game self-learning approach is used, for simultaneous learning of a multiresolution world model, and learning a motion trajectory to the goal. The presented approach has been validated with both simulation experiments and experiments a with a real mobile robot.

1. Introduction

It is important for an autonomous mobile robot to be able to navigate on unknown environments, where the location, shape and size of obstacles is unknown, and where there in no map or model of the world initially available. In fact, it is difficult to provide the robot control system with a global map model of its world. This, may easily become a tedious and time consuming programming task. In addition, robot programming and control architectures must be equipped to face unstructured environments, which may be partially or totally unknown at programming time.

Many path planning techniques, such as road-map, cell decomposition, and potential field methods (see [1] for an overview and further references), generally assume that a complete model of the robot's environment is available. There are approaches that incorporate graph models of the environment, typically representing the world as a network of free-space regions (e.g. [2]). Other methods use a uniform resolution spatial grid for representing the world (e.g. [3]). However, it is difficult to select a resolution for the grid, that is suitable for representing, and to serve as a basis for reasoning on, the entire world. A very localised feature of the world may impose a very high (constant-)resolution grid over the entire state-space. This implies higher memory requirements, and induces unnatural and excessive detail on world modelling, on reasoning, and on the paths that result from reasoning under such a model. In order to introduce selectivity concerning the nature and clutter of the world, some approaches have been based on a variable resolution state-space partition (e.g. [4]). In [5], it is proposed a variable resolution approach incorporating a learning ability enabling both path planning, and environment modelling to be performed simultaneously. Learning has also been incorporated in reactive approaches. For example [6] proposes a reactive approach that learns to coordinate some predefined behaviours.

In this paper we demonstrate the effectiveness of the parti-game self-learning multiresolution approach [7] to the specific case of learning a mobile robot path from an initial position, to a known goal region in an unknown world. We also propose original algorithm modifications aimed at improving its performance. Results of experiments, both simulated and with a real robot, demonstrate the effectiveness of both, the navigation approach, and the proposed improvements on the algorithm.

2. Learning Architecture

The problem we analyse in this work may be defined as follows: The mobile robot is initially on some position on an unknown environment, and then it must learn to navigate, through an obstacle-free path, to a known goal region in the world. The mobile robot controller architecture used in this work, is based on the application of the parti-game learning algorithm [7]. It is a multiresolution approach that incorporates ideas from both graph-based, and grid-based methods. Spatial resolution is chosen using a game-theoretic cell-splitting criterion. A database of previous experiences is incrementally maintained in real-time and used for planning. The algorithm does not use any predefined behaviour, and does not have any initial internal representation, map, or model, of the world. In particular the system has no initial information regarding the location, the size, or the shape of obstacles. The mobile robot can simultaneously, learn a kind of map of its environment, and learn to navigate to the goal, having only the predefined abilities of doing straight-line motion (between the current position and other specified position), and obstacle detection (not avoidance) using its own distance sensors. The two concurrent learning abilities may be seen as cooperating and enhancing each other in order to improve the overall system performance.

The main ideas of the algorithm are outlined in figure 1, and will be detailed in the rest of this section. The algorithm assumes that a local greedy controller is available, which can be asked to move the system greedily towards a desired state. However, there is no guarantee that a request to the greedy controller will succeed. For example, in this work the greedy controller is the "straight-line mover". Two comments regarding straight-line motion ability are in order. First, the movements may fail because of the presence of an obstacle that is detected, and second those motions require the knowledge of the robot current position. However, in this paper we do not deeply address the problem of mobile robot localisation. We simply use accumulation of encoder information to perform robot localisation. Even though this simple approach induces errors, it was sufficient to validate the learning approach.

The parti-game algorithm is based on partitioning the state-space. It begins with a large partition. Then it increases the resolution by subdividing

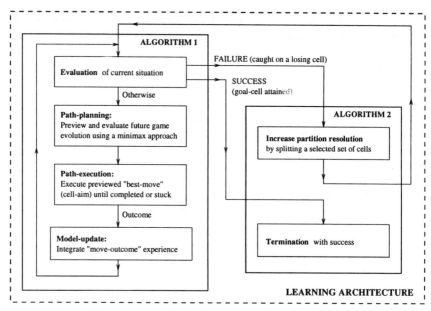

Figure 1. Outline of the learning approach.

the state-space (see figures 5 and 6) where the learner predicts that a higher resolution is needed. As usual a *partitioning* of the state-space is a finite set of disjoint regions, the union of which covers the entire state-space. Those regions, which will be called *cells*, are all axis aligned rectangles, and will be labelled with integers $1, 2, \ldots, N$. We define a *real-valued state*, as a vector $s = [X\ Y]^T$, of real numbers in a 2D space, and whose elements are the two position coordinates of the mobile robot (X and Y). Real-valued states and cells are distinct entities. Each real-valued state is an element of one cell, and each cell is a continuous set of real-valued states. Let us define **NEIGHS**(i) as the set of neighbours, or cells which are adjacent to i. Specifically, two cells are said to be neighbours if their intersection is a segment of a line with more than one point. When we are at a cell i, applying an *action* consists on actuating the local greedy controller "aiming at cell j". A cell i has an associated set of possible actions that is defined as **NEIGHS**(i). Each action is labelled by a neighbouring cell.

The algorithm uses an environmental model, which can be any model (for example, dynamic or geometric) that we can use to tell us for any real-valued state, control action, and time interval, what will be the subsequent real-valued state. In our case the "model" is implemented by the mobile robot (real or simulated), and takes the current position, and position command, to generate the next robot position (a geometric model).

Let us define the **NEXT-PARTITION**(s, j) function that tells us in which cell we end up, if we start at a given real-valued state, s, and using the local greedy controller, keep moving toward the centre of a given cell, j, until either we exit the initial cell or get stuck. Let i be the cell containing the real-valued state s. If we apply the local greedy controller "aim at cell j" until either cell i is exited or we become permanently stuck in i, then

$$\text{NEXT-PARTITION}(s,j) = \begin{cases} i \text{ if we became stuck} \\ \text{the cell containing the exit state otherwise} \end{cases}$$

The test for sticking performs an obstacle detection operation with the mobile robot distance sensors. In general, we may have $\text{NEXT-PARTITION}(s,j) \neq j$, because the local greedy controller is not guaranteed to succeed. In fact, a cell-aim operation may fail because of either a detected obstructing obstacle, or by attaining a cell that was not the selected one. Since there is no guarantee that a request to the local greedy controller will succeed, each action has a set of possible *outcomes*. The particular outcome of an action depends on the real-valued state s, from which the system starts "aiming". The outcomes set, of an action j in cell i, is defined as the set of possible next cells:

$$\text{OUTCOMES}(i,j) = \left\{ k \; \middle| \; \begin{array}{l} \text{exists a real-valued state } s \text{ in cell } i \text{ for which} \\ \text{NEXT-PARTITION }(s,j) = k \end{array} \right\}$$

The algorithm uses a supporting database that organises relevant information. This database is incrementally constructed, and implements a kind of map or model of the world that assists the learning operation. A database of cell-outcomes, observed cell-aim experience, is memorised. Cells have also associated neighbourhood relations. Those two features are conceptually organised on a graph-of-cells data structure, with two two types of edges: edges representing neighbourhood relations, and edges representing cell-outcomes. In addition, cells are simultaneously organised in a kd-tree [8] for easy state-to cell mapping. This reflects the hierarchical structure of the partitioning, that results from the selective and incremental subdivision of cells.

When the system is on the real-valued state s of a cell i, it has to decide the cell at which the system should aim using the local greedy controller. The sequence of cells traversed to reach the goal cell, is planned using a game-like minimax shortest path approach. All next-cell outcomes, that are possible results of a certain cell-aim operation, may be viewed as "response-moves" available to an imaginary adversary that would be working against our objective of reaching the next cell, and ultimately reach the goal. The next cell on the path is chosen taking into account a worst case assumption, i.e. we imagine that for each cell we may aim, the adversary is able to place us on the worst position on the current cell such that the next cell that results from the aim is also the worst. In this way we always aim at the neighbouring cell with the best worst-outcome. In this framework, we can define the minimax shortest path from cell i to the goal, $J_{WC}(i)$, as the minimum number of cell transitions to reach the goal assuming that, when we are in a certain cell i and the intended next cell is j, an adversary is allowed to place us in the worst position within cell i prior to the local controller being activated:

$$J_{WC}(i) = \begin{cases} 0 & , \text{if } i = \text{GOAL} \\ 1 + \min_{k \in \text{NEIGHS}(i)} \max_{j \in \text{OUTCOMES}(i,k)} J_{WC}(j) & , \text{Otherwise} \end{cases} \quad (1)$$

The $J_{WC}(i)$ values can be obtained by by the application of Dynamic Programming methods [9]. The value of $J_{WC}(i)$ can be $+\infty$ if, when we are at cell i, our adversary can permanently prevent us from reaching the goal. By definition such a cell is called a *losing cell*. With this method, the next cell to aim is the neighbour, i, with the lowest $J_{WC}(i)$. Using this approach we are sure that, if $J_{WC}(i) = n$, then we will get n or fewer transitions to get to the goal, starting from cell i. However, the method is too much pessimistic because, regions of a cell that will never be actually visited, are available for the adversary to place us. But those may be precisely the regions that lead to an eventual failure of the process. So although this method guarantees success if it finds a solution, it may often fail on solvable problems.

Algorithm 1, reduces the severity of this problem by considering only all empirically observed outcomes, instead of all possible outcomes for a given cell. Another argument contributing to this solution, is that, as a learning algorithm, it is more important to learn the outcomes set, only from real experience on the behaviour of the system. Besides that, it could be difficult or impossible, to compute all possible outcomes of an action. Whenever an **OUTCOMES**(i, j) set is altered due to a new experience obtained, equation (1) is again solved in order to find the path to the goal. Before an action is experienced, we can not leave the **OUTCOMES**(i, j) set empty. In these situations we use, the default optimistic assumption that we can reach the neighbour that is aimed. **Algorithm 1** of figure 1 keeps applying the local greedy controller, aiming at the next cell, on the "minimax shortest path" to the goal, until either we are caught on a losing cell ($J_{WC} = \infty$), or reach the goal cell. Whenever a new outcome is experienced, the system updates the corresponding **OUTCOMES**(i, j) set, and equation (1) is solved, to obtain the, possibly new, "minimax shortest path". Algorithm 1 has three inputs: (1) The current (on entry) real-valued state s; (2) A partitioning of the state-space, P; (3) A database, the set D, of all previously different cell transitions observed in the lifetime of the partitioning P. This is a set of triplets of the following form: (start-cell, aimed-cell, actually-attained-cell). At the end Algorithm 1 returns three outputs: (1) The updated database of observed outcomes, D, (2) the final, real-valued system-state s, and (3) a boolean variable indicating SUCCESS or FAILURE.

Algorithm 1 gives up when it discovers it is in a losing cell. Assuming that all paths through the state space are continuous, and that a path to the goal actually exists through the state-space, i.e. the problem is solvable, then there must be an *escaping-hole* allowing the transition to a non-losing cell and eventually opening the way to reach the goal. This hole has been missed by Algorithm 1 by the lack of resolution of the partition, but can certainly be found on the cells at the borders between losing and non-losing cells. Taking this comments into account, the top level **Algorithm 2** of figure 2, divides in two the cells in the borders, in order to increase the partition resolution, and to allow the search for the mentioned escaping-hole. This partition subdivision takes place between, each successive calls to Algorithm 1 that keep taking place while the system does not reach the goal region, i.e. while Algorithm 1 fails because of the system being caught on a losing cell.

ALGORITHM 2
WHILE (s is not in the goal cell)
1. Run Algorithm 1 on s and P. Algorithm 1 returns the updated database D, the new real-valued state s, and the success/failure signal.
2. **IF FAILURE** was signalled **THEN**
    2.1 Let $Q :=$ All losing cells in P ($J_{WC} = \infty$).
    2.2 Let $Q' :=$ The members of Q who have any non-losing neighbours.
    2.3 Let $Q'' := Q'$ and all non-losing members of Q'.
    2.4 Split each cell of Q'' in half along its longest axis producing a new set R, of twice the cardinality.
    2.5 $D := D + R - Q''$
    2.6 Recompute all new neighbour relations, and delete from the database D, those triplets that contain a member of Q'' as a start point, an aim-for, or an actual outcome.
LOOP

Figure 2. Top level Algorithm 2.

3. Algorithm Improvement

In this section we discuss some aspects of the behaviour of the algorithm presented in section 2. This discussion will motivate and enable the presentation of algorithm modifications aimed at increasing its effectiveness. Experimental experience has shown that, with this novel contribution, the algorithm behaviour has been able to improve on most world environments. Additionally, in the experiments that were carried with the new approach, we were able to tune the new algorithm, such that it never exhibited a strongly worse, or ever slightly worse, behaviour. It is possible to imagine "worst-case" world environments for which the new approach is somewhat inferior. However if the original algorithm was able to solve those path-planning problems, the new approach will also be successful.

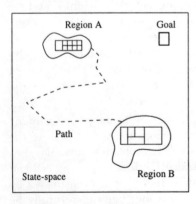

Figure 3. Two cell clusters forming the Q'' set.

As a simplified example of a situation that the algorithm of section 2 often faces, suppose that the set of cells to be split can be decomposed in two subsets (figure 3). The cells of each of these subsets form two separated, closed, and bounded regions on the state-space, Region A and Region B. Region A is composed of "small-sized" cells, and Region B is composed of "big" cells. Here there are various possibilities for defining the *size* of a cell (e.g. the area or the length of the greatest (or the smallest) side of the rectangle). Assuming that a path to the goal actually exists through the state-space, i.e. the problem is solvable, then there must be an escaping-hole, either in Region A or in Region B, allowing the transition to a non-losing cell and eventually opening the way to reach the goal. However, since we don't know where this hole really is, there is no information supporting a preference for concentrating the search in only one of the two regions. But this implies that, there must be at least one system traversal of the possibly long path between the two regions, for every time a set of splits occurs (every cycle of Algorithm 2 - section 2). This path can become quite long, not only because of the distance between the two regions, but also because of the existence of obstacles. When learning with a physical system, the traversal of this path typically implies the use of a certain number of sampling intervals. This fact along with an increase of computational costs that in most of the cases is also associated to the traversal, leads to the increase of the system learning time. Thus we need a method for decreasing those traversals.

The original cell splitting strategy of the algorithm (section 2) is based on splitting of all cells of Q'', resulting on the generation of the set of cells R. For improving the algorithm in a situation like the one we are analysing, we argue that there is no reason for insisting to search for an extremely narrow escaping-hole in Region A before we search for a wider escaping-hole on the bigger cells of Region B. Thus if we change the strategy in order to favour the splitting of the bigger cells, we may expect the system not to take so much of its time travelling to and from Region A, because it is no longer so much "interested" on analysing the smaller cells. Note that a worse case occurs when the cell sizes are balanced between Regions A and B. In this case more travelling occurs between the two regions, but this is no worse than the original problem that we want to solve. Specifically as a first approach we will only split a subset, Q_1'', of cells of Q'', whose size is equal to one of the N_1 biggest sizes of cells of Q''. This splitting produces a new set of cells $R_1 \subset R$. This is the **Rule 1** of the new strategy - see figure 4.

However, there are situations when it may be worth splitting smaller cells from Region A. Suppose the system is close to Region A, and all the cells of Q'' were split. Then the minimax algorithm was used and a resulting shortest path includes some small cells of Region A. In this case it is better also to split the smallest cells of Q'', because in this way we are able to promptly analyse the, possibly feasible, shortest path before we travel all the path to Region B where the biggest cells will be explored.

Algorithm 3 (see figure 4) summarises and precisely specifies the modifications on the cell splitting strategy that were made in Algorithm 2 in order to improve it. Note that, since $Q'' = Q_1'' + Q_2''$, when using Option 1, all the cells are split when the condition of Rule 2 is verified.

4. Simulation Experiments

In this section we present results of simulation experiments obtained with the Khepera mobile robot [10], regarding the application of the algorithm presented in section 2, and including the improvements of section 3. Concerning Algorithm 3 (see figure 4) we used, $N_1 = 1$, $N_2 = N_3 = 2$, and Option 1 of Rule 2. On each trial, the robot started on a given position and tried to reach a goal region. In our example the world has a great number of walls (see figure 5). The robot uses its infrared sensors to perform obstacle detection. The experiments were organised in two separate simulations, also called "runs". Each run is composed by a sequence of trials to navigate from a starting position,

> **ALGORITHM 3** (an alternate splitting strategy to be used in **ALGORITHM 2**)
> 1. Let $Q_1'' :=$ All of cells of Q'', whose size is equal to one of the N_1 biggest sizes of cells of Q''. Call R_1, the set of all the cells resulting from the splitting of cells from Q_1''. Note that $R_1 \subset R$.
> 2. Let $Q_2'' := Q'' - Q_1''$. This implies $Q'' = Q_1'' + Q_2''$. Call R_2, the set of all the cells resulting from the splitting of cells from Q_2''. Note that $R = R_1 \cup R_2$.
> **Rule 1.** Split all the cells of Q_1''. Let $\hat{Q}'' := Q_1''$.
> 3. As a temporary situation for preparing Rule 2, split all the cells of Q'', and use minimax algorithm to obtain the shortest-path from the new actual-cell to the **GOAL** cell.
> **Rule 2.** IF N_2 cells among the N_3 initial cells of the shortest-path from the new current-cell to the **GOAL**, belong to R_2 **THEN**:
>   **R2.1** Let $\hat{Q}'' := \hat{Q}'' + Q_\alpha''$, where the set Q_α'' is composed according to one of the following options:
>   **Option 1.** Q_α'' contains all the cells of Q_2'' (note that $\hat{Q}'' = Q''$ in this case).
>   **Option 2.** Q_α'' contains all the cells of Q_2'' belonging to the shortest-path.
> 4. Change ALGORITHM 2 to split all the cells of $\hat{Q}''$ instead of splitting all cells of Q''.

Figure 4. Algorithm 3: alternate cell-splitting strategy.

to a predefined goal region. The system has no initial knowledge about the location, and shape of obstacles in the world. Thus, the first trial starts with no model of the world. Subsequent trials start with, and build upon, the world model that was learned until the end of the previous trial. From trial to trial, the system accumulates knowledge about the world, as described in section 2.

On figure 5(a) it can be observed the mobile robot trajectory on trial 1. Since the robot starts without any initial knowledge about the world, it performs a considerable amount of exploration on the first trial. However, the its already able to reach the goal. The robot world-exploration effort continues on the succeeding trials. On Figs. 5(b), 5(c), and 5(d), we can observe the robot trajectories, and final state-space partition, on trials 2, 4, and 6 respectively. As can be seen, by inducing environment exploration, the algorithm is able to incrementally construct a world-representation, that is based on a partitioning of the state-space, and is suitable for the simultaneous implementation of its own path-learning capabilities. The robot, in an attempt to find a path to the goal, selectively increases the partition resolution on regions where it faces greater difficulties to navigate. As obviously expected, these difficulties arise on areas that are close to obstacles. In contrast, on free-space areas there is no need to increase the resolution that therefore is kept lower. At the beginning of the sixth trial, the robot has learned all the knowledge required to promptly navigate, through an obstacle-free path to the goal (see figure 5(d)).

For assessing the effectiveness of the algorithm modifications proposed in section 3, we have selected six performance indexes. The first is the *Number of cells*. In some sense, we can state that, for the same environment, it is a measure of the difficulties the algorithm faced for modelling the world. When comparing two algorithms, it is also a measure of the "over-partitioning" effort of one of them. It is in general also true that with a greater number of cells we have higher computational costs. The second index, the *Steps*, is the number of time intervals, the robot was advancing in straight-line to a certain point (the

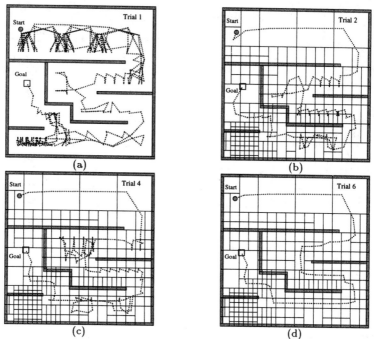

Figure 5. Mobile robot path, final state-space partition, and obstacles perceived at the end of (a) Trial 1, (b) Trial 2, (c) Trial 4, and (d) Trial 6.

centre of an aimed cell). Since we used a constant robot velocity, it is clearly a measure of the robot travelling distance. It does not include the time steps used at the beginning of each aim, in order to point the front of the robot towards the centre of the aimed cell. The third index, the *Aims*, is the number of times the robot aimed a neighbouring cell. It is a measure of both the exploration effort, and the computational costs. At a level more abstract than the steps, it is also a measure, of travelling effort. The fourth index, *Aim Fails*, is the percentage of aims that failed, because either the robot became stuck due to an obstacle, or it attained a cell other than the aimed one. It is a measure of the exploration difficulties. The fifth index, the *Number of minimaxs*, is the number of times the minimax problem of equation (1) was solved. It is a measure of the important proportion of computational cost, that must be used for solving those problems. Finally, the *CPU Time*, strictly represents the time used by the learning algorithm presented in sections 2 and 3. It does not include the robot simulation time, or the interprocess communication time.

In order to analyse the benefits introduced on the algorithm, by the modifications of section 3, we collected the performance indexes on simulations with and without the modifications. Table 2 presents the indexes for the simulation with the modifications. This corresponds to the simulation presented on figure 5, and associated discussion text above. Table 1 presents the indexes for the same simulation (same world, start position, goal region) but without modifications of section 3. In both tables, and for all indexes, both the "trial-cumulative" and "trial-differential" figures are presented. As can be seen, the

Table 1. Simulation Results – without Algorithm Modification

		Trial 1	Trial 2	Trial 3	Trial 4	Trial 5	Trial 6
Number of cells	Σ	324	334	490	490	490	490
	Δ	324	10	156	0	0	0
Steps	Σ	103383	117312	129150	149291	158898	166696
	Δ	103383	13929	11838	20141	9607	7798
Aims	Σ	1022	1174	1274	1641	1722	1767
	Δ	1022	152	100	367	81	45
Aim Fails (%)	Σ	48.34	46.59	45.05	45.03	44.08	43.07
	Δ	48.34	34.87	27.00	44.96	24.69	4.44
Number of minimaxs	Σ	1040	1129	1296	1664	1746	1792
	Δ	1040	154	102	368	82	46
CPU Time (unit)	Σ	39.84	49.82	56.97	84.90	93.72	100.00
	Δ	39.84	9.98	7.15	27.93	8.82	6.28

Table 2. Simulation Results – with Algorithm Modification

		Trial 1	Trial 2	Trial 3	Trial 4	Trial 5	Trial 6
Number of cells	Σ	162	199	199	231	231	231
	Δ	162	37	0	32	0	0
Steps	Σ	59923	73955	88007	104152	121220	128702
	Δ	59923	14032	14052	16145	17068	7482
Aims	Σ	444	574	713	849	991	1033
	Δ	444	130	139	136	142	42
Aim Fails (%)	Σ	51.35	48.95	46.28	44.29	42.48	40.76
	Δ	51.35	40.77	35.25	33.82	31.69	0.00
Number of minimaxs	Σ	487	622	762	903	1046	1089
	Δ	487	135	140	141	143	43
CPU Time (unit)	Σ	14.61	19.23	24.89	31.11	38.56	41.36
	Δ	14.61	4.62	5.66	6.22	7.45	2.80

modifications clearly improved the behaviour of the algorithm. In fact all the indexes have improved with the modifications. The Number of Cells have decreased to less than a half, reflecting lower world modelling difficulties and less computational costs. The Steps also decreased indicating a lower robot travelling distance. The number of Aims also decreased reflecting lower exploration effort and lower computational costs. Additionally, among the lower number of Aims that were performed, the percentage of Aim Fails also decreased, confirming lesser exploration difficulties. The Number of minimaxs problems solved also decreased, indicating again a decrease of computational effort. Finally, the CPU Time was strongly decreased to less than a half, due to the introduction of the modifications. The "quality" of the final solution obtained on both cases, as measured in terms of a final travelling path of approximately 7500 steps on both cases, was quite similar. However, the modifications allowed the decrease of the modelling difficulties, the exploration effort, and the computational costs, to achieve the solution. This same analysis was performed on other different world environments, and the same results were consistently observed, i.e. the algorithm modifications lead to an improvement on the performance.

5. Experiments with a Real Robot

This section presents results obtained with a real Nomad 200 mobile robot. Again, the infrared ranging sensors were used to perform obstacle detection, while learning to navigate to the goal. On the presented example (one run),

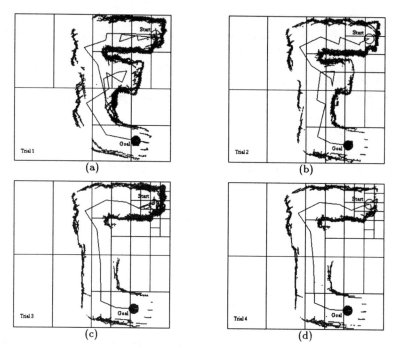

Figure 6. Mobile robot path, final state-space partition, and obstacles perceived at the end of (a) Trial 1, (b) Trial 2, (c) Trial 3, and (d) Trial 4.

the world environment is composed of some obstacles, that make it difficult the task of learning to navigate to the goal (see figure 6). The algorithm of section 2 with the modifications of section 3 was used. Concerning Algorithm 3 of figure 4, we used again $N_1 = 1$, $N_2 = N_3 = 2$, and Option 1 of Rule 2.

In figure 6(a) we can observe the trajectory of the mobile robot, and the final partition of the state-space, on the first trial to reach the goal. As can be seen, since the robot has no initial knowledge about the world, it performs a considerable amount exploration on this first trial. However, the robot is already able to reach the goal on the first trial. The world-model knowledge accumulated, enables on the second trial (see figure 6(b)) a much more direct navigation to the goal. However some (but lower) exploration effort is still performed because of some newly created cells. On the third trial (see figure 6(c)), the robot faces some difficulties in getting out of the neighbourhood of the starting position, with some cells being split and the resolution increased there. This leads to some additional exploration on the starting area. However, on the remaining part of the path, the navigation to the goal is already straightforward. On trial 4 (see figure 6(d)) we see that the robot has concurrently learned a world model and a path to the goal, both of sufficient quality to enable it to promptly navigate, without exploration effort.

On figure 6(b), but mainly on figure 6(a), we observe that some obstacles are slanted when compared to figures 6(c) and 6(d). The reason for this is that, we used a simple encoder accumulation method for robot localisation. In this way, since the robot performs greater amounts of motion on the first two trials, the corresponding localisation errors are also greater. Errors on robot self-localisation, are reflected on errors regarding the position where the it "thinks" the obstacles are. From the figures we can roughly say that the most important errors are on the orientation component. The location accumulators were set to correct values at the beginning of each trial. In spite of these localisation errors, the robot was still able to learn to navigate to the goal.

6. Conclusion

We have demonstrated the application of the parti-game learning approach, for learning to navigate a mobile robot on an unknown world. Original modifications for improving the algorithm operation, were proposed. Results of both simulation, and real robot experiments were presented demonstrating the effectiveness of the algorithm and the proposed modifications. An interesting line of further work with the learning method is predictive on-line trajectory filtering for enabling lower exploration efforts. The learning method has some ability to cope with changing environments. As easily seen, it is able to overcome the introduction of new obstacles. On the other hand, the free-space created by the removal of an obstacle, is not considered for exploring a better path than an already existing solution, until/unless a newly created object obstructs the previous solution path. Further work on this method, has the potential to be able to provide it with, the ability to cope with even more general dynamic environments.

References

[1] Latombe J C 1991 *Robot Motion Planning*. Kluwer Academic Publishers, Boston, NJ, USA

[2] Crowley J L 1985 Navigation for an intelligent mobile robot. *IEEE Journal of Robotics and Automat*, RA-1:31–41

[3] Elfes A 1987 Sonar-based real-world mapping and navigation. *IEEE Journal of Robotics and Automat*, RA-3:249–265

[4] Kambhampati S, Davis L S 1986 Multiresolution Path Planning for Mobile Robots. *IEEE Journal of Robotics and Automat*, RA-2:135–145

[5] Zelinsky A 1992 A mobile robot exploration algorithm. *IEEE Trans Robotics and Automat*, 8:707–717

[6] Beom H R, Cho H S 1995 A sensor-based navigation for a mobile robot using fuzzy logic and reinforcement learning. *IEEE Trans Syst Man Cybern*, 25:464–477

[7] Moore A W, Atkeson C G 1995 The parti-game algorithm for variable resolution reinforcement learning in multidimensional state-spaces. *Machine Learning*, 21:199–233

[8] Friedman J H, Bentley J L, Finkel R A 1977 An algorithm for finding best matches in logarithmic expected time. *ACM Trans on Mathematical Software*, 3:209–226

[9] Bertsekas D P 1987 *Dynamic Programming: Deterministic and Stochastic Models*. Prentice-Hall, Inc, Englewood Cliffs, NJ

[10] Floreano D, Mondada F 1996 Evolution of Homing Navigation in a Real Mobile Robot. *IEEE Trans Syst Man Cybern Part B Cybernetics*, 26:396–407

An Anthropomorphic Model of Sensory-Motor Co-Ordination of Manipulation for Robots

Cecilia Laschi Davide Taddeucci Paolo Dario
Scuola Superiore Sant'Anna
Pisa, Italy
E-mail: {cecilia, davide, dario}@arts.sssup.it

Abstract. This paper investigates the problem of artificial perception related to manipulation tasks in robotics.

The proposed approach is based on biological models of perception and sensory-motor co-ordination in humans and aims at devising anthropomorphic solutions to the problems of perception, learning and control in robotics. In particular, our approach involves the integration of different sensory modalities and the interpretation of sensory data aimed at the control of robot behaviour.

We consider as sensory modalities, in relation to manipulation tasks, vision and haptic perception, i.e. the integration of tactile proprioceptive and exteroceptive data.

The experimental part of this work is aimed at investigating some aspects of the proposed anthropomorphic model of perception in manipulation by means of anthropomorphic visual and tactile sensors on a robotic manipulator and a pan-tilt head, and a processing module based on neural network computational models integrated with the reinforcement learning paradigm.

1 Introduction

Recent applications of robotics, such as service robotics and robotics for personal assistance [1], require high level of flexibility and adaptability. For this reason, the problems of perception, interpretation of surrounding environments, learning and adaptive behaviour are becoming increasingly important. Since, especially in the field of 'personal' robotics, robots are introduced into environments modelled on humans' needs (that cannot in general be structured for robots) and operate in collaboration with humans, anthropomorphic solutions are emerging as the most suitable for the robot in order to achieve proper behaviour in human environments.

The growing interest towards anthropomorphic solutions in robotics is also demonstrated by the increasing research activity for the development of humanoid robots. In this framework, in 1996 the Waseda University of Tokyo has promoted the First International Symposium on Humanoid Robots [2], during which works on many different aspects of the design and development of anthropomorphic robots were presented [3] [4] [5].

In humans, the capability of performing a variety of motor actions by interacting with the unpredictable real environment and without any particular conscious efforts, is the result of the combination of several sophisticated sub-systems, which have been refined in the course of evolution. Human behaviour is controlled on the basis of what is perceived by human sensors, but the modalities of sensory-motor co-ordination have never been exhaustively described by a quantitative model, though noticeable attempts have been made in the fields of psychology and philosophy, and also, more recently, in the areas of automatic control and high level programming [6].

How different sensory data are integrated into one perception and how this is co-ordinated with motor actions still remain important issues of investigation for neuroscience, both for improving the knowledge on human brain and for the application of such knowledge to the design and development of anthropomorphic robotic sensory-motor co-ordination systems.

It is widely accepted that *thinking* and *motor behaviour* in the human brain can be identified as two different entities able to co-ordinate, when necessary. It has also been demonstrated that such entities physically reside in separate regions of the brain and the split brain experiment by R. W. Sperry showed that, when surgically disconnected, they function as two separate brains [7].

More precisely, *thinking* is the part of reasoning involving conscious cognitive processing and including sophisticated pattern recognition, emotional evaluators and behavioural situation/action rules. Instead, the processes of *motor behaviour* are mostly instinctive and perform low-level pattern recognition and simple behavioural reactions, with no involvement of conscious reasoning.

For both aspects of human behaviour, it has been established that learning is one of the basic issues for the acquisition of motor co-ordination schemes [8]. Learning is widely practised in humans all along childhood, during which, along different phases, motor schemes are defined for a variety of tasks, from simple co-ordination of walking to more sophisticated actions such as speaking, reading and writing. Learning is however always present in humans, for the definition of motor schemes related to new tasks.

Also for learning, two different levels can be identified (the psychological and the neural), according to the involvement of conscious reasoning. Walking, eating, balancing on a rocky terrain are learnt instinctively, without any conscious reasoning efforts; very different is, instead, learning reading and writing, sailing or playing piano.

2 Proposed model of artificial perception and sensory-motor co-ordination

Based on the general model of human brain processes described before, we propose a model of artificial perception and sensory-motor co-ordination for the specific case of manipulation. The conceptual scheme of the proposed model is depicted in Fig.1.

We take into account vision and touch, as the most relevant perceptual modalities in humans and the most important to be replicated in robots for artificial sensory-motor co-ordination. Tactile exteroceptive perception (including pressure distribution, object temperature and texture, contact force and torque) and proprioception of arm and hand configurations are all "fused" as "haptic perception".

According to the proposed anthropomorphic approach, the paradigm of active perception [9] [10] is assumed in the proposed model of artificial perception.

The proposed approach for the processing module starts from the assumption that two different kinds of processing are replicable from the human model: low level processing and high level processing.

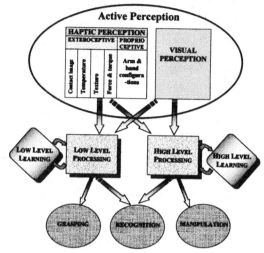

Fig.1. The conceptual scheme of the proposed anthropomorphic model of perception

As described in Section 1, low level processing is the fusion and interpretation of perceptual data with no involvement of conscious cognitive processes. On the contrary, high level processing is just the involvement of conscious cognitive processes in the understanding of perceived data and planning of appropriate behaviour learning is a fundamental feature at both levels, allowing proper functioning of each module.

Different tasks which include sensory integration, data processing and sensory-motor co-ordination, such as grasping, manipulation and recognition, were identified as specific cases for the implementation of robotic models. Whereas grasping can be considered as essentially a low-level task, and manipulation as a high level task, recognition has a two-fold nature. When grasping an object, the hand is configured so as to fit the object shape, and grasping force is tuned on the object physical features. All this is instinctively achieved through low level processing with no particular conscious effort in humans. On the contrary, manipulative actions, such as handwriting, opening a door, or pouring water from a bottle, require a varying involvement of conscious attention. Recognition, instead, is achieved at different levels: the sensation coming from the contact with an object surface gives us the perception of the material the object is composed of, without conscious reasoning on thermal properties, texture or stiffness. On the other hand, unknown objects require a deeper analysis to be recognised, involving high level reasoning.

3 Implementation: neural networks and reinforcement learning

The implementation of the proposed concept of artificial perception and sensory-motor co-ordination still follows an anthropomorphic approach, which is materialised in the structure of the visual and tactile sensors and of the actuators for arm/hand motor actions, and in the architecture of the modules implementing processing and learning models.

We focused on the implementation of low level processing and learning, based on a *neural* approach integrated with the *reinforcement learning* paradigm.

The neural approach to low level processing is well suited for an anthropomorphic solution, for a number of reasons [11] [12]:

1. the processing performed by neural networks is achieved thanks to the connections in the nodes of the network, just as low level processing is performed in humans by physical connections among neurones;
2. information is processed by neural networks in parallel, as sensory data are perceived simultaneously and fused in human multi-sensory perception;
3. one of the main features of neural networks are the plasticity and adaptability of the structure, where connections among nodes can be modified dynamically, as they are reinforced or cancelled in the human brain;
4. neural networks are generally redundant and guarantee the production of output values even if some units or connections are missing, by applying the same solution applied by Nature in animals, based on redundancy;
5. finally, the concept of learning is intrinsically related to artificial neural networks, just as learning is intrinsic in the human mental processing.

Encouraging results have been achieved in previous works of Dario and Rucci [13][14] with the implementation of such an anthropomorphic sensory-motor and computational system.

Human beings can learn through a number of different paradigms; at present, though, a unified understanding of the problem has not been reached yet. Computational models of learning have been proposed which try to replicate some aspects of animal learning process. Essentially, a rough taxonomy of computational learning comprises of Supervised Learning, Unsupervised Learning and Reinforcement Learning. Supervised learning is achieved when a teacher is available that specifies the input and the expected output the system must learn; during the Unsupervised Learning, instead, a system discovers regularities in the training phase producing clusters of similar data in a self organised fashion; finally, in the Reinforcement Learning the system has to emulate an expected behaviour without a direct specification of the output but having as feedback from the environment only a scalar value representing how good was the action performed in response to a particular input.

We selected reinforcement learning as the paradigm for the implementation of the proposed model of low level learning, because of its strong analogy with the biological (and anthropomorphic) mechanism of correction or enhancement behaviours.

Reinforcement Learning is learning about, from, and while interacting with an environment in order to achieve a goal [15]; in other words, it is a relatively direct model of the learning of people and animals plausible both from a neurophysiological (Hebbian rule [11]) and psychological (Pavlovian conditioning) point of view.

In robotics, problems are typically defined in terms of abstract goals rather than specific pair <perception, action>. Reinforcement procedures usually fit this type of learning since they require only scalar values measuring the relative desirability of the state goodness [16]. The dynamic achieving of a robotic goal could be seldom attainable via an immediate reward (positive value), when the system has a good behaviour, and a punishment (negative value), in case of wrong actions.

While many algorithms such as Q-Learning [17] and Time Delay [18] have been developed for the problem of maximising a long term reward, in the common case of immediate reward a proven algorithm does not exist, but rather some empirical learning rules chosen for the particular problem to solve. Some work on reinforcement learning-based models of robotic controllers have been published recently. Shaal and Atkenson [19] built a two-arm robot that learns to juggle a particular device in a six-dimensional state space after about 40 initial attempts. Mataric [20] describes a robotic experiment in a very high complex dimensional space to control four mobile robots that collect and transport small disks in a particular area. Crites and Barto [21] developed a very robust system for an elevator-dispatching task. Morgan, Patterson and Klopf [22] solved the well-know balancing pole problem with an adaptive controllers based on drive-reinforcement neurons. Using a miniature "Khepera" robot controlled via neural networks, Claude Touzet [23] implemented obstacle avoidance behaviour using an immediate reinforcement learning paradigm. Experiments on active perception based on reinforcement learning are carried out by Shibata et al. [24], by training a neural network system to perform recognition of simple patterns by moving a visual sensor controller with reinforcement signals.

4 Experimental Validation

In order to validate our anthropomorphic model of a computational and anthropomorphic sensory system, we implemented a system able to perform a visual and tactile edge tracking, considered as the first step of sensory-motor co-ordination in grasping actions.

The experimental scenario, depicted in Fig.2, is composed of: a robotic arm (possessing proprioceptive perception of current configuration) equipped with a finger supporting a multi-sensory integrated fingertip, and a pan-tilt head incorporating a retina-like sensor. The arm and the head are subsystems of a more complex mobile robotic system developed in the framework of the TIDE-MOVAID project [25] [26] for the assistance to disabled and elderly people.

Aim of the experiment is to co-ordinate the movement of the fingertip along an object edge, by integrating visual information on the edge, proprioceptive information on the arm configuration, and tactile information on the contact, and by processing this information in a neural framework based on the reinforcement learning paradigm. The different components of the experimental system (integrated fingertip, retina-like sensor, early visual and tactile data processing, robotic arm, robotic head, computational model based on reinforcement learning) are described in detail in the following sub-sections.

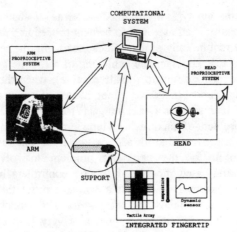

Fig.2. Experimental scenario

4.1 The Integrated Fingertip

The integrated fingertip is a multi-functional miniature sensor comprising of ae "Tactile Array sensor", a "Dynamic sensor" and a "Thermal sensor".

The Tactile Array sensor, based on force sensing resistor technology, has been described by the author elsewhere[28] The Tactile Array sensor possesses a space-variant disposition of the sensing sites, aiming at increasing the size of the tactile sensing area and reproducing the concept of tactile "focus of attention" at the fingertip (very similar to the analogous well known concept of foveal vision [27]).

The Dynamic sensor is based on a bimorph piezo-ceramic element, generating a signal related to the applied stress.

The Thermal sensor is composed of two miniature resistors embedded in thermally conductive rubber: one is for heating the sensor and the other (a thermistor) is for detecting temperature variations, so that the sensor can measure the thermal flow from the fingertip to the explored material. Proper interface electronics acquires the signals from the sensor and put them in numerical format. The integrated fingertip is contained in a compact and lightweight frame of 27.4 x 25 x 52 mm that is intended to be mounted on a commercial hand (Barrett Hand, by Barrett Technology Inc., Cambridge, MA); the Tactile Sensor and the Thermal Sensor are placed on the contact surface, while the Dynamic Sensor slightly protrudes from the fingertip surface. In order to limit possible damages to the dynamic sensor during grasping operations, the sensor stick can glide into the fingertip structure. The electronic boards for data acquisition are embedded inside the fingertip structure, too. The tactile acquisition electronics is connected to the Tactile Sensor through a custom connector thus optimizing compactness. Fig.3 shows the integrated fingertip.

Further technical details on the design and development of the sensor can be found in [28]. The main characteristics of the integrated fingertip are summarised in the following tables.

Tactile Array Sensor Physical Characteristics	
External Dimensions	24.36mm x 34.9 mm.
Overall Area	850 mm²
Sensitive Area	432 mm²
Number of sensitive sites	64
Maximum resolution (in the centre)	1 mm.
Minimum resolution (at the periphery)	5 mm.
Number of wires	16
Signal Pad Area	2.25 mm² (each)

Table 1

Tactile Sensor Performance	
Force resolution	0.1 N (approx.)
Linearity & Hysteresis	± 12% Fs
Uniformity	± 15% Fs @ 3N

Table 2

Fig.3. The integrated fingertip

4.2 The retina-like sensor

The vision system used in the experiment is based on a retina-like sensor [27] with a circular physical layout and a space-variant resolution of active sites, with consequent space-variance in the resulting image. This feature drastically reduces the amount of data of a single image, corresponding to the degradation of peripheral areas of the image. The sensor is composed of photocells located on the radii of concentric circles; the same number of cells is located on each circle, so that they result in a higher density in the central area. Photocells are scanned from centre to periphery sequentially on each radius. The first version of this retina-like sensor was developed on CCD technology, while an improved recent version exploits C-MOS technology. The chip incorporating the C-MOS retina-like sensor is shown in Fig.4 (a), while Fig.4 (b) shows the sensor completed with lens and electronic board for digital data acquisition mounted on the head of the MOVAID robotic system.

 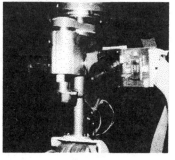

(a) *(b)*

Fig.4. The C-MOS retina-like sensor (a), completed with lens and electronics mounted on the MOVAID head (b)

4.3 Robotic Arm & Head

The robotic actuator used in the experiment is the 8 d.o.f. Dexter arm, produced by S. M. Scienzia Machinale srl, Pisa, Italy, for incorporation in the URMAD service robotic system and then integrated in the MOVAID mobile robotic system [25] [26].

The Dexter arm has a highly anthropomorphic physical structure. Cable transmission has been realised for joint movements, but the link structure reproduces human body from trunk to wrist. The reason of such an anthropomorphic choice mainly relies on the application the arm is addressed to, i.e. service tasks in unstructured environment involving strict interaction with humans. The anthropomorphic structure allows instinctive approaches in driving tele-operation tasks, thus reducing the training phase of operators. Furthermore, it also increases the acceptability from the user's side and enhances the possibility of integration into a domestic environment, conceived for human limbs.

The redundancy in its cinematic structure facilitates dextrous manipulation, enables the arm to be configured for work at various heights above the ground, and allows the arm to fold so that its work envelope can be minimised. Cable transmission allows for achieving a low weight and inertia of the distal joints; all motors of the 6 d.o.f. distal part are located in the second link which represents the trunk. The first two proximal joints are aimed at pre-positioning the distal 6 d.o.f. manipulator so as to increase the overall workspace.

The MOVAID robotic head is mounted on the first link of the arm, so as to reduce the possibility of image obstruction by the arm. It is a quite simple mechanical structure allowing pan-tilt movements of two cameras, which can be installed on two side supports.

A view of the arm and head sub-system (also supporting an auto-localisation system for MOVAID) is shown in Fig.5.

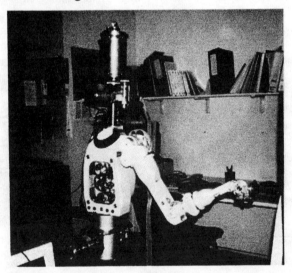

Fig.5. A view of the MOVAID robotic arm and head

4.4 Reinforcement Learning Implementation

As already mentioned, the aim of the experiment is to implement an edge following behaviour by integrating visual and tactile information and using a reinforcement learning paradigm. Our approach is to pursue this strategy starting essentially from a totally random policy and enhancing it via rewards and punishments.

The system in the state s_t senses visuo-tactile patterns and performs an action **a** carrying it in the state s_{t+1}; two heuristic modules evaluate the goodness of the new state with respect to the desired behaviour producing a reinforcement value **r**. The value **r** is positive if the state s_{t+1} is good, and negative otherwise; **r** is then backpropagated to the learning system to adjust the policy. In order to evaluate the 'goodness' of the new state an evaluation module is required.

Fig.6 shows an overall scheme of the processing and learning modules.

The visual information is fed to Neural Network 1 (NN1) which computes the next movement of the fingertip; a visual heuristic module (Heuristic 1) compares this movement to the direction suggested by an early vision processing module and determines a reinforcement value to adjust the policy of NN1.

Analogously, the tactile image is used as input to Neural Network 2 (NN2) in order to compute the required rotation for the fingertip. An early tactile processing module computes the centre of mass used by a tactile heuristic (Heuristic 2) for determining the reinforcement value for NN2 policy adjustment.

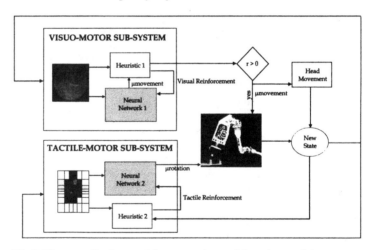

Fig.6. The overall scheme of the processing and learning modules

In our experiment the expected behaviour is obtained by decomposing the problem in two main neural networks performing 1) μmovements and 2) μrotations. We define **μmovement** a short movement of the finger of 1 cm in the direction illustrated in Fig.7 (a), along a straight line, and **μrotation** a finger rotation of different amplitude (±5°,±10°,±15°,±20°) in the two directions, as illustrated in Fig.7 (b). By composing these actions, the edge following is performed.

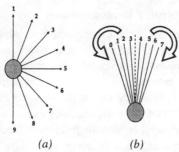

Fig.7. (a) μmovements and (b) μrotations for the fingertip.

4.4.1 Early Tactile Processing

For the purposes of the experiment, data from the tactile array sensor are early processed in order to provide input data to the processing module.

The early tactile processing of the experiment aims at detecting the centre of mass of tactile image, as a relevant feature for the evaluation of the fingertip contact of the fingertip on the explored surface.

In Fig.8 two different tactile images, with the corresponding detected centres of mass, are shown.

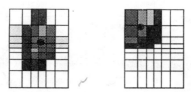

Fig.8. Two different tactile images, with the corresponding centres of mass

4.4.2 Early Vision Processing

Visual information is processed at an early stage in order to detect relevant features for further processing and integration with tactile data for learning.

Early vision processing can be divided into four phases:
1. space-variant image acquisition;
2. extraction of the foveal part of the image;
3. filtering and edge detection;
4. identification of the line approximating the edge segment focused in the fovea.

An example of the grey level image acquired with the retina-like sensor is given in Fig.9 (a), while Fig.9 (b) shows the foveal part of the same image. Fig 9 (c) shows the edge image and the line approximating the edge, considered as an indication of the next fingertip movement along the edge.

The grey scale image is filtered in order to reduce possible random noise with a smoothing algorithm and edge detection is then performed by means of a gradient-based algorithm.

The line approximating the edge is detected by applying the Hough method [27] on the edge image. The Hough method is able to detect the parametric curves that cross the image points, and creates an accumulation array in which such curves are stored together with a scoring value allowing the detection of the one best fitting the edge. A modified method has been applied which identifies lines, by attributing a higher score to those fitting the right end of the edge, considered as the most relevant edge segment for the detection of the fingertip movement along the edge.

An analysis of the accumulator array allows determining one line out of the set of detected lines, whose parameters provide an indication of the next movement direction.

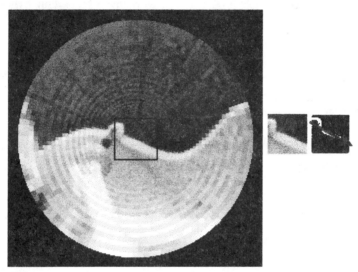

Fig.9 (a) The space-variant grey level image; (b) foveal area of the image; (c) edge image; and (d) identified movement direction

4.4.3 μmovement NN

As illustrated in Fig.6, the visuo-motor system determines the direction of the μmovement by looking at the next segment of the edge. NN1 receives as input a 40x40 pixel window centred on the fovea of the retina-like sensor and produces the 9 possible actions as output (**μmovement s**).

The rough algorithm is the following:
1) position the head towards the fingertip: the fingertip is in the fovea of the retina-like sensor;
2) extract the following target movement **t** along the segment on the curve in the left-to-right direction;
3) apply the 40x40 pixel window centred on the fovea to the NN1 and compute the relative output **a**;
4) compare **t** with **a** and compute the reinforcement **r**;
5) backpropagate **r**,
6) IF the reinforcement is positive THEN perform the action **a** and go to step 1 ELSE go directly to step 1.

In order to allow an exploration on the search space, at each iteration a small noise is added to the output of the neural network so that, coming back to step 3, it is possible to obtain a different action.

The architecture chosen for the neural network is the classic BackPropagation [7] modified according to the following reinforcement learning algorithm [20][28]:

1) $a_i = o_i + rnd_i$, i=1 ... p; p=number of possible actions, o_i real output of the network on unit i, rnd_i small random value, a_i new value of the output node i;

2) Performance $\phi(t) = -\sum_{i=1}^{9}(t_i - a_i)^2$. How much the output is near to the desired output.

3) Reinforcement $r = \dfrac{d\phi}{dt} = \phi(t+1) - \phi(t)$. How much the performance is enhanced.

4) Error $e_i = (r*rnd_i)^2$, i=1 ... p;

5) Backpropagate the error vector to the BP network.

4.4.4 μrotation NN

The scheme of the μrotation neural network (NN2) is also illustrated in Fig.6. The aim of this subsystem is to train a neural network to maintain the "fovea" of the fingertip perpendicularly in contact with the surface of the shape. Dynamically, during the training, the network receives the tactile image as input and activates a unit indicating the size of the rotation and its direction as output.

The rough algorithm is the following:

1) calculate the distance of the centre of mass from the fovea on the tactile image at time t d(t);
2) calculate the output of the NN2;
3) execute the rotation;
4) calculate the distance of the centre of mass from the fovea of the new centre of mass on the tactile image, d(t+1);
5) calculate the reinforcement;
6) backpropagate r;
7) IF r>0 THEN go to step 2 ELSE rotate back to the initial position and go to step 1.

The implementation of NN2 is similar to that of NN1 but the performance is not computed and the reinforcement

$$r = d(t) - d(t+1)$$

The error vector is computed with the same formula.
An example of state transition is shown in Fig.10.

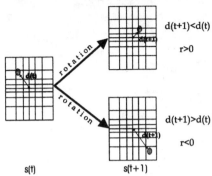

Fig.10. An example of state transition through a μrotation

5 Experimental Results and Conclusions

The implemented system has been validated through experimental trials organised as follows: a) the edge to be tracked is located in front of the visual sensor so as to foveate the edge; b) the fingertip is put into contact with the edge surface, according to the spatial position detected by vision; c) according to the reinforcement learning paradigm, the fingertip is moved along the edge through μmovements and μrotations.

At the very beginning such movements are basically randomly executed, but the reinforcement calculated on the visual and tactile data tunes them on the edge following; d) the visual sensor is moved, too, according to the current position of the fingertip on the basis of proprioceptive information from the arm.

Two phases of the experimental trials are shown in Fig.11.

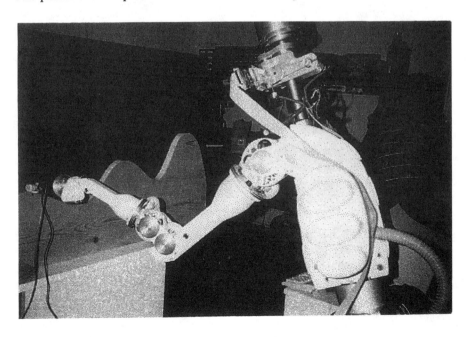

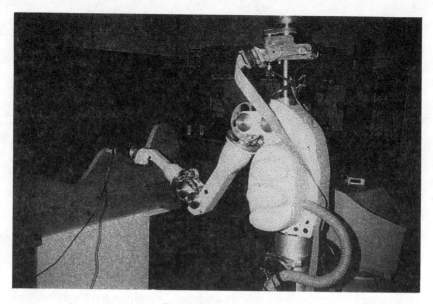

Fig.11. Experimental trials

The overall performance of the system has matched rather effectively the expected results. This represents an encouraging step towards the full validation of the proposed approach.

The anthropomorphic structure of the tactile and visual sensors has improved the computational properties of the system, by exploiting the concept of fovea which encodes relevant information in a reduced and compact amount of data. A sequence of retina images acquired during an experimental trial is shown in Fig.12.

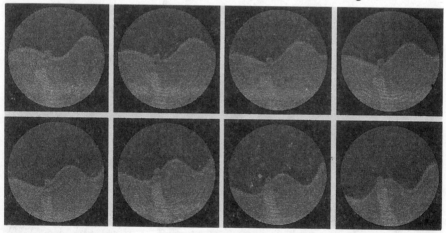

Fig.12. A sequence of images acquired by the retina-like sensor during an experimental trial

The neural approach has demonstrated the expected flexibility, adaptability and learning capability. The reinforcement learning paradigm has produced the expected natural-like behaviour in robot movements. In few experimental trials, it was possible to observe the development of not optimal local policies (decreasing oscillations of the fingertip around the optimal fovealisation of the centre of mass), due to the simple heuristic implemented in the experimental system for the calculation of the reinforcement.

6 Future Work

As outlined before, the experimental work was aimed at validating a specific aspect of the proposed general model of anthropomorphic sensory-motor co-ordination, namely the low level processing and learning.

Ongoing and future work is aimed at implementing the overall model and applying the system to additional robotics tasks. Future experiments will be carried out on grasping tasks, by using an anthropomorphic three-finger hand on which one fingertip sensor is mounted at each finger. The vision system will be improved by the use of a pair of cameras (still based on the retina/like sensor) so as to obtain full three-dimensional information on the scene.

7 Acknowledgement

The integrated fingertip sensor was developed in collaboration with the Korean Institute of Science and Technology in the framework of the Project "CENTAUR" [4].

8 Bibliography

[1] Bekey G., Crisman J. 1996 The 'grand challenge' for robotics and automation, Special Panel discussion, *IEEE Conference on Robotics and Automation*, Minneapolis, Minnesota, USA, April 22-28

[2] *Proc. of the First International Symposium on HUmanoid RObots*, 1996, sponsored by Waseda University, Tokyo, Japan, October 30-31

[3] Brooks R. 1996 Prospects for human level intelligence for humanoid robots, *Proc. of the First International Symposium on HUmanoid RObots*, Waseda University, Tokyo, Japan, October 30-31, 17-24

[4] Lee C.W 1996 Project "CENTAUR" in mid-entry strategy – KIST 2000 human robot system project, *Proc. of the First International Symposium on HUmanoid Robots*, Waseda University, Tokyo, Japan, October 30-31, 73-82

[5] Yamaguchi J. 1996 Development of a humanoid robot – Design of a biped walking robot having antagonist driven joints using nonlinear spring mechanism, *Proc. of the First International Symposium on HUmanoid RObots*, Waseda University, Tokyo, Japan, October 30-31, 102-110

[6] Albus J.S. 1981 *Brains, Behaviour & Robotics*, BYTE Books, Subsidiary of Mc Graw Hill, Peterborough, N.H., USA

[7] Sperry R.W. 1970 Perception in the absence of the Neocortical Commisures, in *Perception and its Disorders*, Research Publication of the Association for Research in Nervous and Mental Diseases 48.

[8] Piaget J. 1976 *The grasp of consciousness: action and concept in the young child*, Harvard University Press, Cambridge, MA, USA

[9] Aloimonos J., Weiss I., Bandyopadhay A. 1988 Active Vision, *International Journal of Computer Vision*, 1:333-356

[10] Bajcsy R. 1988 Active Perception, in *Proc. of the IEEE*, 76, 8:996-1005

[11] Rumhelhart D., Mc Clelland J. 1986 *Parallel Distributed Processing*, Cambridge, Massachusetts, MIT Press

[12] Haykin S. 1994 *Neural Networks: A Comprehensive Foundation*, IEEE Comp. Spc. Press, McMillan
[13] Dario P., Rucci M. 1993 A neural network- based robotic system implementing recent biological theories on tactile perception, in *Proc. of 3rd International Symposium on Experimental Robotics*, Kyoto, Japan, 162-167
[14] Rucci M., Dario P. 1994 Development of cutaneo-motor co-ordination in an autonomous robotic system, *Autonomous Robots*, 1:93-106
[15] Sutton R.S. 1996 Reinforcement Learning, *NIPS Tutorial*, Dec. 2
[16] Meeden L.A. 1994, An incremental Approach to developing intelligent neural network controllers for robots
[17] Watkins C.J.C.H. 1989 Learning from delayed rewards, Ph.D. Thesis, University of Cambridge, UK
[18] Sutton R.S. 1988 Learning to predict by the methods of temporal differences, *Machine Learning* 3, 9-44
[19] Shaal S., Atkenson C. 1994 Robot Juggling: an implementation of memory based learning, *Control System Magazine*, 14
[20] Mataric M.J. 1994 Reward functions for accelerated learning, in *Proc. of the 11th International Conference on Machine Learning*, Morgan Kaufmann
[21] Crites R.H., Barto A.G. 1996 Improving elevator performance using reinforcement learning, in Touretzky D., Mozer M., Hasselmo M. editors *Neural Information Processing Systems 8*
[22] Morgan J.S., Patterson E.C., Klopf A.H. 1990 Drive-reinforcement learning: a self-supervised model for adaptive control, *Network: Computation in Neural Systems*, 1:439-448
[23] Touzet C. 1994 *Neural Implementations of immediate reinforcement learning for an obstacle avoidance behaviour*, Technical Report NM.94.6, EERIE, Nimes, France
[24] Shibata K., Nishino T., Okabe Y. 1995 Active perception based on reinforcement learning, in *Proc. of WCNN'95*, Washington D.C., 2:170-173
[25] Dario P., Guglielmelli E., Laschi C., Guadagnini C., Pasquarelli G., Morana G. 1995 MOVAID: a new European joint project in the field of rehabilitation robotics, in *Proc. of the 7th International Conference on Advanced Robotics (ICAR '95)*, Sant Feliu de Guíxols, Spain, September 20-22, 51-59
[26] Dario P., Guglielmelli E., Laschi C., Teti G. 1997 MOVAID: a mobile robotic system for residential care to disabled and elderly people, *Proc. of the 1st MobiNet Symposium*, Athens, Greece, May 15-16, 1997, pp.45-68
[27] Sandini G., Dario P., De Micheli M., Tistarelli M. 1993 Retina-like CCD sensor for active vision, in *Robots and Biological Systems*, Dario P., Sandini G., Aebischer P. Ed.s, Berlin-Heidelberg: Springer-Verlag, 553-570
[28] Dario P., Lazzarini R., Magni R., Oh S.R., 1996 An integrated miniature fingertip sensor, in *Proc. of the 7th International Symposium on Micro Machine and Human Science (MHS '96)*, Nagoya, Japan, October 2-4
[29] Allotta B., Bosio L., Chiaverini S., Guglielmelli E. 1993 A redundant arm for the URMAD robot unit, in *Proc. of the 6th International Conference of Advanced Robotics (ICAR '93)*, Tokyo, Japan, November 1-2, 655-660
[30] Allotta B., Bioli G., Colla V. 1996 A repeatable control scheme for redundant manipulators observing joint mechanical limits, in *Proc. of Robotics towards 2000: 27th International Symposium on Industrial Robots*, Milan, Italy, October 6-8, 521-526
[31] Hough P.V.C. 1962 Method and means for recognising complex patterns, U.S. Patent 3069654, December 18th
[32] Williams R.J. 1992 Simple statistical gradient-following algorithm for connectionist reinforcement learning, *Machine Learning*, 8:229-256

Extracting Robotic Part-mating Programs from Operator Interaction with a Simulated Environment

John E. Lloyd and Dinesh K. Pai

Department of Computer Science
University of British Columbia

{lloyd,pai}@cs.ubc.ca

Abstract

We describe an integrated system for programming part-mating and contact tasks using simulation. A principle goal of this work is to make robotic programming easy and intuitive for non-trained users. Using simulation, an operator can specify part placement and contact motions by simply "putting things where they belong", without resorting to textual descriptions. We describe the simulation system, which models objects as polyhedra and emulates collision and contact interactions in combination with a simplified "operator-friendly" dynamics. The combination of contact simulation and graphical fixtures enables the operator to easily manipulate this virtual environment using a simple 2D mouse. We also describe the program generation system, which takes the operator's action sequence and transforms it into a set of robot motion commands which realize the prescribed tasks.

1 Introduction

This paper describes work at the University of British Columbia (UBC) on the problem of using simulation as a tool for robotic programming. Our ultimate objective is to make programming a robot extremely easy and natural, so that it can be done intuitively by non-specialists using simple desk-top computer systems.

Particular emphasis is directed at tasks involving contact, since, at the time of this writing, there are virtually no commercially available robot programming systems which support contact-based tasks. While the reasons for this are complex, we suspect that programming difficulties are a significant part of the problem.

We believe that an effective way to program contact-based tasks involves having the operator perform the required task in a simulated virtual environment. The operator's actions are then interpreted and used to generate a robot program which replicates the indicated task at the actual work site.

Using a simulated environment for programming allows the operator to directly "show the system" where an object is to be placed or moved in relation to other objects. Textual descriptions of such tasks, can, by contrast, be very tedious. Also, within a simulated environment, we can adopt a "task-centric" view of programming, letting the operator manipulate workpieces directly, with less emphasis on the robot which will ultimately realize the task.

1.1 System requirements and overview

A simulation programming system requires several key subsystems:

1. *Model Generator:* builds and maintains the work site model used by the simulator;

2. *Task Simulator:* provides a simulation of the work environment, along with operator-friendly dynamics that makes task-specification easy and intuitive.

3. *Program Generator:* takes the motions specified in the virtual environment and creates a set of robot motion commands capable of realizing the prescribed task;

4. *Execution monitor:* verifies task execution at the robot site, and provides correction information to the model generator.

The second and third items are the main focus of this paper.

At UBC, we have built a test platform in accordance with the above structure. As a challenging but simplified example, we chose a task domain consisting of a puzzle of wooden blocks that can be assembled within a rigid frame. Tasks which the operator can specify include placing or moving a particular block in contact with other objects in the work site.

The model generator, described in [1], uses a gray-scale vision system that can rapidly recognize the location of objects characterized by straight-line edge features [2]. The simulation system (described in detail in Section 4) lets the operator manipulate blocks using inputs from a simple 2D mouse. We chose to use a mouse as an input device because it is cheap and ubiquitous, and matches with our above-stated desire for a system that can be used by non-specialists using simple desk-top computer systems. The program generator (Section 5) uses the motions specified by the operator to create a set of robot motion commands capable of realizing the specified task. Execution monitoring is the subject of ongoing investigation.

1.2 Paper outline

The remainder of this paper is organized as follows: Related work is discussed in Section 2, and a summary of the test platform components and hardware is given in Section 3. Task simulation and program generation are described in Sections 4 and 5, and experimental results are presented in Section 6.

2 Related Work

Our work closely follows the *teleprogramming* work of Funda, Sayers, and others [3, 4], in which operator interaction with a simulated environment is used to overcome problems which can arise in telerobotic systems due to time delays between the remote site and the operator station. Our work differs in that the simulation is somewhat more complete, and the system is more "task oriented": the sequence of operator motions may be modified radically before being sent to the robot as motion commands.

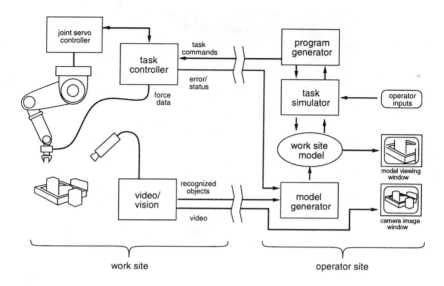

Figure 1: System architecture.

Teleprogramming was predated by the use of graphical simulation as an operator aid in time-delayed telerobotic applications [5, 6]. The introduction of synthetic "fixtures" into the operator's display to assist in task specification has also been considered [7, 8]. While commands sent to a remote site in teleprogramming systems tend to be at the level of "guarded moves", the idea of sending more robust commands, which can allow the manipulator system to plan for and recover from unanticipated contact states, is investigated in [9], using a Petri-Net-based contact state model.

Virtual reality simulation has also been used as a platform on which fine motion task skills can be learned; a good example in the context of this paper is given in [10].

It should also be mentioned that our general goal of making robotic systems usable by non-expert operators has recently been explored in the context of position-based robotic applications accessible on the World Wide Web [11, 12, 13].

3 System Description

A somewhat simplified block diagram of the UBC system is given in Figure 1. The main components will be summarized very briefly here; a more detailed description is given in [1].

3.1 Work site

The work site (Figure 2) contains a 6 DOF CRS A460 robot (Puma-type geometry), controlled at the lowest level by 1 KHz joint servos (supplied by the manufacturer), which are in turn driven by a *task controller*. The task controller is implemented using the Robot Control C Library (RCCL) [14] running in real-time on a Sun Sparc 5. It

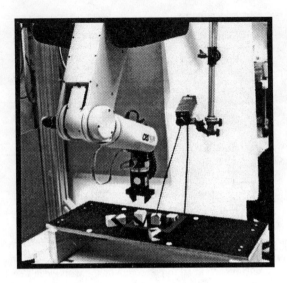

Figure 2: Remote site, showing the robot, camera, and work area.

accepts Cartesian motion commands from the operator site, and generates the required trajectories at 100 Hz. The trajectory generator also receives input from a force sensor, allowing it to implement both guarded moves and a position-based impedance control similar to that described in [15].

A *video/vision* module continuously collects images from a camera and processes them using a model-based vision algorithm [2] to locate objects in the scene. The objects and their positions are continuously sent back to the operator site where they are incorporated into the work site model under the control of the operator. The camera image itself is also transmitted back to the operator site, where it is displayed in a separate window.

As mentioned above, the work site's task domain consists of a puzzle of wooden blocks that can be assembled within a rigid frame.

3.2 Operator site

The operator site consists of an SGI Indy with a 15 Mflop CPU. It hosts a model of the work site environment, with which the operator interacts, using a mouse, via the task simulator (Section 4). Model data includes polyhedral representations of the work space objects, plus kinematic and geometric information about the robot manipulator (dynamical information is not necessary for the low-speed contact operations presently being investigated). Other information about work site objects, such as friction and stiffness models, may be added later if required. Model information is updated, based on recognition data received from the video/vision module, by the *model generator* (which at present is under operator control [1]). The operator views the model through the *model viewing window*, implemented using the SGI 3D modeling package Open Inventor [16]. A *program generator* (Section 5) creates the robot motion commands

 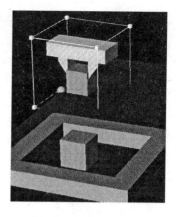

Figure 3: Dragger fixtures. After a part is selected by "clicking" on it with a mouse, a rendering of the manipulator's gripper appears, along with a graphical "dragger box" (right figure). Dragging the mouse cursor along one of the planes of the box causes a displacement parallel to the plane, which is converted into a "virtual force" acting on the workpiece's center.

required to realize specific tasks and sends them to the task controller at the work site.

4 Task Simulation

The task simulator moves a selected object, or *workpiece*, around inside the work site model, in response to operator inputs. Contacts and collision dynamics are modeled, allowing the workpiece to bump into, slide along, and realign itself with other objects. This, in turn, makes it very easy for the operator to place the workpiece into some desired contact state with respect to the rest of the environment.

The operator can also request that the workpiece specifically maintains contact with certain selected *capture* objects. Once the workpiece makes contact with a capture object, its motion is constrained so as to maintain that contact (this is implemented using barrier functions, described in Section 4.2).

The simulated environment is visible to the operator, from any angle, through the model viewing window. Using the mouse, the operator selects a workpiece to be moved by "clicking on it". A simulated gripper then appears, showing how the workpiece will be grasped, along with a graphical "dragger fixture" (at present, a box) that maps 2D mouse inputs into 3D spatial motions and permits the workpiece to be moved about (Figure 3). Rendering only the gripper preserves the "task-centric" focus of the operator's actions; more proximal parts of the robot could be rendered if necessary.

Displacements between the dragger box and the workpiece are used to create a virtual force f_a acting on the workpiece. When the workpiece is brought into contact with other objects, normal forces arise in reaction to the applied force. The normal forces plus the applied force create a net total force on the workpiece, from which a simple first-order dynamics is used to compute the workpiece velocity. A first order dynamics is used deliberately, because it (a) is inexpensive to implement, (b) is stable, and (c) produces results which are simple and intuitive for the operator.

4.1 Implementing the contact model

The simulator keeps track of the distances between objects using I-COLLIDE [17]. Objects closer than ϵ_c are assumed to be in contact, in which case information provided by I-COLLIDE is used to determine a suitable finite set of contact points $\mathbf{p}_i$ and normals $\mathbf{n}_i$ modeling all the contacts[1]. Reaction forces $\mathbf{f}_i$ acting along the contact normals, in response to the applied force $\mathbf{f}_a$, are determined using Baraff's algorithm [19]. The net force $\mathbf{f}$ and moment $\mathbf{m}$ acting on the workpiece are then given by

$$\mathbf{f} = \sum_i \mathbf{f}_i + \mathbf{f}_a, \quad \mathbf{m} = \sum_i \mathbf{p}_i \times \mathbf{f}_i.$$

The first order dynamics is then used to determine the workpiece's spatial velocity $(\mathbf{v}^T \omega^T)^T$, according to

$$\mathbf{v} = d_t \mathbf{f} \quad \text{and} \quad \omega = d_r \mathbf{m} \tag{1}$$

where d_t and d_r are suitable constants.

The task simulator computes and applies the workpiece's velocity once per time step (currently every 50 msec) and uses this to update the workpiece's position, as described in the next section.

4.2 Preventing Collision using Barrier Potentials

When the workpiece is extremely close to other objects, second order constraints or numerical errors can make collision-free motion impossible, even when a feasible velocity exists. The effect of this is to have the workpiece appear to "stick", unreasonably, at certain configurations. Moreover, the information returned by I-COLLIDE becomes unreliable when objects are very close together.

Good simulator performance thus requires trying to keep the workpiece a minimum distance ϵ_b away from other objects, where typically $\epsilon_b = \epsilon_c/2$. This, in turn, is accomplished using a potential barrier.

Let d_i be the distance between the workpiece and another object i, and let $\delta \equiv d_i/\epsilon_b$. Then define the potential $U_i(d_i)$ by

$$U_i(d_i) = \begin{cases} K[\delta - 1 - \ln(\delta)] & \text{if } 0 < \delta < 1, \\ 0 & \text{if } \delta \geq 1. \end{cases} \tag{2}$$

where K is a suitable constant.

To create motion, the desired spatial velocity from equation (1) is in turn associated with an attractive potential U_v that decreases uniformly along the velocity direction in proportion to the work done by $\mathbf{f}$ and $\mathbf{m}$. If motion in the direction of $(\mathbf{v}^T \omega^T)^T$ is parameterized by s, such that $s \in [0, 1]$ corresponds to one simulator time step of Δt, then

$$U_v(s) = -\left[\frac{\|\mathbf{v}\|^2}{d_t} + \frac{\|\omega\|^2}{d_r}\right] \Delta t \, s.$$

---

[1] Face-face or edge-face contacts can be reasonably simulated using a finite set of point contacts; see [18].

Summing U_v and the U_i for all appropriate objects yields a net potential $U(s)$ that varies along the direction of motion. During each simulation step, the workpiece is moved so as to minimize $U(s)$. If there are no obstacles nearby, all $U_i = 0$ and this minimum will occur at $s = 1$, corresponding simply to the application of **v** and ω for time Δt. For purposes of performing the minimization, $U_i(d_i)$ is taken to be $+\infty$ for $d_i < 0$. Because $U_i(s)$ is not smooth (see below), the minimization is done using a golden section search.

To help keep each $d_i \geq \epsilon_b$, the velocity **v** in equation (1) is modified to include, for any object i for which $d_i < \epsilon_b$, a repulsive component computed from the gradient of $U_i(d_i)$ with respect to the workpiece's translational position.

Why not treat the entire problem in terms of potential minimization and calculate both **v** and ω from the gradient of U with respect to the workpiece's overall spatial position? The problem is that this gradient is not simple to calculate. Even though $U_i(d_i)$ is smooth, d_i itself is not smooth in the configuration space of a polyhedral object, and so U_i is not smooth with respect to the configuration space either. Hence in many cases a formal gradient doesn't exist. While non-smooth optimization techniques exist that don't require an explicit gradient, the very thin size of the barrier means that convergence could be quite slow without a good estimate of initial direction. Indeed, the Baraff calculation (Section 4.1) can be thought of as simply a good way of estimating this direction.

The potential method described here can also be used to implement motions in which the workpiece is constrained to *maintain* a particular contact. This is done by modifying $U_i(d_i)$ so that in addition to approaching infinity at $d_i = 0$, it also approaches infinity as d_i nears some small outer boundary value.

5 Program Generation

Once a workpiece has been satisfactorily positioned within the work site model, the operator indicates this to the system using a keystroke. The program generator then sets about creating robot motion commands to realize the requested operation and replicate any specifically requested contact motions.

The system records the workpiece's position and contact state after each simulation step. Each of these positions and contact states defines a *node* along a sequence called the *workpiece path*.

One way to achieve a required task would be to generate a separate robot motion command for each of the nodes on the workpiece path; in other words, closely replicate the operator's actions (similar to what is done in teleprogramming environments [3, 4]). However, in the context of our system, there are problems with this:

1. The workpiece path may contain many unwanted or unnecessary motions, such as those induced by the operator "feeling" her way around;

2. Because motions are constrained to directions permitted by the dragger fixtures, the resulting path may have superfluous kinks and bends;

3. The path may contain unwanted contacts, caused by the operator dragging the

workpiece across or along obstacles, or specifically using obstacles for alignment.

It should be noted, however, that the workpiece path *is* collision free (within the resolution limits imposed by the simulator's step size). What we need to do is modify the workpiece path so as to remove unwanted motions and unnecessary contacts, while preserving its feasibility.

5.1 Stretching and Shrinking the Path

Such a modification can be done by treating the path as a deformable spatial curve which can be bent, stretched, or shrunk in order to move it away from objects or compress its length. To achieve this, we apply to each of the path nodes

1. A constant tension force attracting it to each of its two nearest neighbors;

2. A spring-like repulsive force that pushes it away from unwanted obstacles.

Path endpoints are kept fixed, as well as the endpoints of any required contact motions. The tension force (item 1) is calculated with respect to both position and orientation. A constant, rather than variable, tension is used to keep the path from becoming overly stiff when stretched. The repulsive force (item 2) is calculated with respect to translation only, due to difficulties in computing a repulsive gradient with respect to orientation (as mentioned in Section 4.2). To keep nodes from processing along the path, we eliminate any repulsive force component which is tangential to the path.

It is important that the deformed path remains collision free and preserves any required contacts. This is achieved by moving each node using the contact simulation software, with the combined tension and repulsive forces assuming the role of the applied force $\mathbf{f}_a$. Each node in the path is moved in succession, with the whole procedure being repeated until the path stablizes. After the deformation is complete, the number of nodes is reduced using a simplification scheme similar to a polygonal path approximation algorithm.

Our path deformation approach closely follows the work of Quinlan [20], who originated it to simplify and smooth collision-free paths in configuration space produced by a motion planner. Our situation differs in that we specifically include contact states, and nodes are moved using contact simulation software. In Quinlan's work, collision avoidance was guaranteed by surrounding path points with free space "bubbles", which would be difficult to calculate here because of our use of spatial coordinates and the proximity of obstacles in contact, which would require computing bubbles at extremely fine resolutions.

5.2 Robot Motion Commands

The robot command sequence for a task begins with a "guarded grasp" with which the manipulator grasps the workpiece at the location corresponding to the start of the path. Force sensor data is used to help execute and verify this command.

A motion command is then created for each of the succeeding path nodes, specifying a spatial position which the robot should move to (using linear spatial interpolation). For nodes which involve contact or pass near objects, the robot's speed is lowered, and its impedance is controlled to emulate a spring-damper system with low stiffness. Contact is ensured by adding to the target position a small translational bias (currently around 3-4 mm) in the directions of contact, and contact is verified by checking for observed forces along the directions of the contact normals. Contact instabilities are minimized by clipping the output velocity of the impedance controller to a magnitude not exceeding the current robot speed (on the principle that this should be large enough to remove observed forces within one control cycle).

6 Demonstrations and Observations

Figure 4 illustrates the behavior of the system for two tasks: cornering a block, and dragging a block around the outside of a corner while maintaining contact. Actual execution of the later task is shown in Figure 5.

The task simulator runs easily in real-time (20 Hz) on a 15 MFlop SGI Indy. A barrier size of $\epsilon_b = 1$ mm was used (with a work area diameter ≈ 350 mm and a typical object dimension of ≈ 30 mm). The golden section search (Section 4.2) was performed to an accuracy of about 1.0^{-5}, requiring about 25 I-COLLIDE calls per simulation step, or 500 per second. I-COLLIDE's collision and distance computations can sometimes fail in non-generic situations (*e.g.*, when two object faces are very close to parallel) and so improvement is needed here. Overall performance and smoothness of the contact simulation is greatly enhanced by the use of barrier functions (Section 4.2). However, when the workpiece is constrained to maintain contact with an object, sticking will occasionally occur in some configurations, such as going around a corner.

The discrete-time nature of the simulation means that in certain pathological situations the simulator can "tunnel" through an object without detecting a collision (similar observations were made in [21]). With a maximum speed of 400 mm/sec, a sample rate of 20 Hz, and 4 tests per sample, we currently need to be wary of objects thinner that 5 mm.

The path modification algorithm (Section 5.1) produces paths that make good intuitive sense. In particular, it is particularly good at placing node points so as to ensure reliable and "snag free" approaches and departures from contact situations. The only caveat is that path nodes must be placed fairly close together (currently on the order of 15 mm) for this to work. Algorithm convergence is not a problem, except that the constant tension force can cause minor instabilities to arise when nodes are very close together. This was corrected by using a tension proportional to distance for nodes closer than a certain minimum distance. In general, the whole problem of maintaining proper node spacing requires additional work.

Finally, while not the primary focus of this paper, actual path execution turned out to be quite robust, using the simple methods described in Section 5.2.

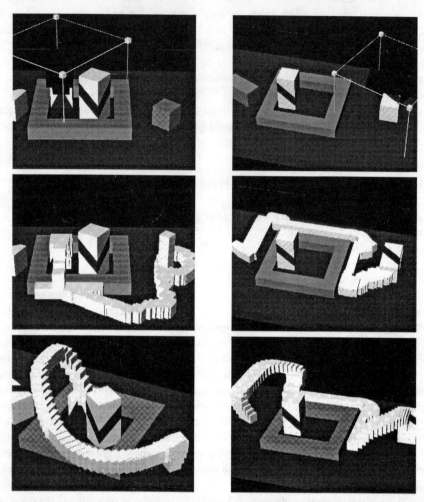

Figure 4: **Left column:** Cornering. The operator has placed the workpiece in a corner of the frame (top); its original location is shown by a gray "shadow block" at the right. The outside of the frame was used to help align the workpiece, as can be seen from the initial workpiece path (middle), where the tightly spaced nodes are shown as half-sized white blocks. After modification (bottom), the nodes have been pushed away from the table and have kept their distance from the tall block, and the initial and final positions are approached at angles that prevent collision with the frame. **Right column:** Here, the operator has moved the workpiece around the outside of the frame (top), while requesting that contact with the frame and table be maintained while going around the corner. The resulting initial workpiece path is shown in the middle. After the path is modified (bottom), frame/table contact is maintained near the corner, while nodes are pushed away from the table elsewhere.

Figure 5: Execution of the corner-rounding task of Figure 4.

7 Conclusion

We have developed an integrated system linking together vision, contact simulation, localized planning, and manipulator control, with the intention of creating an environment in which programming contact and part mating tasks is extremely easy.

The use of a simulated environment appears well suited to programming contact-based tasks, since the work space objects themselves can then serve as "virtual fixtures" that help guide the operator's actions. Indeed, the combination of dragger fixtures plus contact constraints tended to make part placement quite easy, even when using an ordinary mouse as an input device. In contrast, we discovered during earlier experiments that the manual selection of free space motion via-points between goals turns out to be fairly tedious for an operator. However, such points were easily produced by the system using the methods of Section 5.1.

For future work, we wish to add additional local planning to make the contact motion sequences more robust, and to improve the system's abilities so that it can handle contact operations involving tight fits.

References

[1] J. E. Lloyd, J. S. Beis, D. K. Pai, and D. G. Lowe, "Model-based Telerobotics with Vision". *Proceedings of the 1997 IEEE International Conference on Robotics and Automation*, Albuquerque, New Mexico, April 23-25, 1997.

[2] J. S. Beis and D. G. Lowe, "Learning indexing functions for 3-D model-based object recognition," *IEEE Conference on Computer Vision and Pattern Recognition,* Seattle (June 1994), pp. 275-280.

[3] J. Funda, T. S. Lindsay, and R. P. Paul, "Teleprogramming: Toward delay-invariant remote manipulation". *Presence*, Winter 1992, pp. 29–44 (Vol. 1, No. 1).

[4] C. R. Sayers, "Operator Control of Telerobotic Systems for Real World Intervention". Ph. D. thesis, Department of Computer and Information Science, University of Pennsylvania, Philadelphia, PA 19104 USA, 1995.

[5] G. Hirzinger, B. Brunner, J. Dietrich, and J. Heindl, "Sensor-Based Space Robotics – ROTEX and Its Telerobotic Features". *IEEE Transactions on Robotics and Automation*, October 1993, pp. 649–663 (Vol. RA-9, No. 5).

[6] A. K. Bejczy, W. S. Kim, and S. C. Venema, "The Phantom Robot: Predictive Displays for Teleoperation with Time Delay". *Proceedings of the 1990 IEEE International Conference on Robotics and Automation*, Cincinnati, Ohio, May 1990, pp. 546–551.

[7] C. R. Sayers and R. P. Paul, "An Operator Interface for Teleprogramming Employing Synthetic Fixtures". *Presence*, Fall 1994, pp. 309–320, (Vol. 3, No. 4).

[8] National Research Council (U.S.A.), *Virtual Reality. Scientific and Technological Challenges*. National Academy Press, Washington, D.C. 1995.

[9] Y. J. Cho, T. Kotoku, and K. Tanie, "Discrete-Event-Based Planning and Control of Telerobotic Part-Mating Process with Communication Delay and Geometric Uncertainty". *Proceedings of the 1995 IEEE/RSJ International Conference on Intelligent Robots and Systems (IROS)*, Pittsburgh, Pennsylvania, August 1995, pp. 1–6 (Vol. 2).

[10] R. Koeppe and G. Hirzinger, "Learning Compliant Motions by Task-Demonstrations in Virtual Environments". *Experimental Robotics IV (Lecture Notes in Control and Information Sciences No. 223)*, Springer, London, pp. 299–307.

[11] E. Paulos and J. Canny, "Delivering Real Reality to the World Wide Web via Telerobotics". *Proceedings of the 1996 IEEE International Conference on Robotics and Automation*, Minneapolis, Minnesota, April 1996, pp. 1694–1699.

[12] http://cwis.usc.edu/dept/garden/.

[13] http://telerobot.mech.uwa.edu.au/.

[14] J. E. Lloyd and V. Hayward, *Multi-RCCL User's Guide*. Technical Report, Center for Intelligent Machines, McGill University, April 1992.

[15] M. Pelletier and M. Doyon, "On the Implementation and Performance of Impedance Control on Position Controlled Robots". *Proceedings of the 1994 IEEE International Conference on Robotics and Automation*, San Diego, California May 8-13, 1994, pp. 1228–1233 (Vol. 2).

[16] J. Wernecke, *The Inventor Mentor*. Addison-Wesley, Reading, Massachusetts, 1994.

[17] J. Cohen, M. Lin, D. Manocha and K. Ponamgi, "I-COLLIDE: An Interactive and Exact Collision Detection System for Large-Scaled Environments". *Proceedings of ACM Int. 3D Graphics Conference*, 1995, pp. 189–196.

[18] S. Goyal, E. N. Pinson, and F. W. Sinden, "Simulation of Dynamics of Interacting Rigid Bodies Including Friction I: General Problems and Contact Model". *Engineering with Computers*, Springer-Verlag, London, 1994, pp. 162–174 (Vol. 10).

[19] D. Baraff, "Fast Contact Force Computation for Nonpentrating Rigid Bodies", *SIGGRAPH 94 Conference Proceedings*, July 1994, pp. 23–34.

[20] S. Quinlan, *Real-time Modification of Collision-Free Paths*. Ph. D. thesis, Computer Science Department, Stanford University, December 1994.

[21] D. Baraff, "Interactive Simulation of Solid Rigid Bodies", *IEEE Computer Graphics and Applications*, May 1995, pp. 63-75.

Modeling and Learning Robot Manipulation Strategies

Jiming Liu   Y. Y. Tang
Department of Computing Studies, Baptist University
224 Waterloo Road, Kowloon Tong, Hong Kong
{jiming,yytang}@comp.hkbu.edu.hk

Oussama Khatib
Robotics Laboratory, Department of Computer Science
Stanford University, CA, USA
ok@robotics.stanford.edu

Abstract: This paper describes a general approach to learning and planning robot manipulation strategies. Here, the strategies are represented using a discrete-event dynamical systems model where each node corresponds to a state in the robot task environment that triggers certain action schemata and each arc corresponds to a plausible action that brings the task environment into a new state. With such a representation, a manipulation strategy plan can be derived by searching a connected state transition path that is the most reliable. Here, we define the notion of reliability in terms of the estimated chance of success in reaching a desirable state. In the paper, we first present the formalism of discrete-event dynamical system in the context of robot manipulation tasks. Throughout the paper, we provide both illustrative and experimental examples to demonstrate the proposed approach.

1. Introduction

This paper is concerned with the problem of serial patterning of robot manipulation operations at a finer grain granularity as compared to the classical assembly planning. The operations planned at this level are called manipulation strategies. The motivation behind this approach comes from the fact that many robot operations can share similar manipulation strategies. They are selected and executed as long as certain conditions are observed from the robot environment.

1.1. Previous Work

Hasegawa *et al.* developed a model-based robot manipulation system which utilizes some pre-defined *manipulation skills* [7]. Each skill is implemented as motion primitives of hybrid position-force-velocity control strategies in making contact state transitions. Miura *et al.* [12] developed a system in which the uncertainty about the effects of a certain operator was considered, and a learning mechanism was implemented that could record the observations of failure or success. Others have addressed this issue using a traditional probabilistic network based approach [5]. In [9], a teaching system was demonstrated in some manipulation tasks where a robot acquired reusable task plans by recognizing human action sequences. The hierarchical task plans were represented in terms of temporally ordered operations. Kang and Ikeuchi [8] considered the problem of mapping human functional grasps into equivalent manipulator grasps.

In our present work, we are not concerned with the recognition and recording of the action sequences as demonstrated by human operators [1]. Instead, what we are interested in is to acquire some manipulation strategies by learning general state transition functions of a discrete-event system. Such learning takes place during robot task execution in which robot sensory states are monitored.

Whitney [15] addressed the issue of autonomous robot manipulation with dynamically planned adaptive motions. In his model-based approach, the models and evaluators can be probabilistic and empirical in nature, since robot tasks that we deal with in real life are often repeated but in somewhat slightly different ways. Our approach shares this probabilistic view of modeling. In our work, we represent the manipulation strategies with a Markov chain [14].

1.2. Organization of the Paper

In this paper, we first describe how a manipulation strategy can be represented with the Markov chain model [4]. Following such a modeling, we then show how the strategies can be algorithmically derived from empirical data. Finally, we present the experiments that validate the proposed approach.

2. A Markov Chain Model of Manipulation Strategies

In the state-oriented representation of robot task manipulation [2, 3, 10, 11], a series of robot operations may be considered as a sequence of input tokens that changes the states of a robot task environment from an initial configuration to a desirable goal configuration. The states that the task environment can be in are described with a restricted representation as derived from robot sensory input, and represented by means of a state transition network showing the transitions depending on the input tokens. Various geometric and physical properties of objects may be represented, depending on the specific manipulation task.

In our present framework, the robot manipulation strategies are modeled in terms of state transitions. In other words, if the set of all possible states in the robot environment is $\mathcal{S} = \{S_1, S_2, ..., S_n\}$, then

$$\forall O \in \mathcal{O}, \ \exists a \in \mathcal{A} \subseteq \{\{S_i, S_j\} \mid S_i, S_j \in \mathcal{S}\} \subseteq \mathcal{S} \times \mathcal{S} \tag{1}$$

where $\mathcal{O}$ and $\mathcal{A}$ are sets of manipulation strategies and state transition arcs also called *action schemata*, respectively.

2.1. Transition Function Φ

In order to account for the uncertainty inherent in the outcome of a manipulation strategy due to some errors, such as those in control and in feedback, here we model the action schema as a Markov chain whose state transition function defines a probability mass. Formally, our model can be written as the following triple:

$$\mathcal{T} = \langle \mathcal{S}, \mathcal{A}, \Phi \rangle \tag{2}$$

where $\mathcal{S}$, $\mathcal{A}$, and Φ are defined as follows:

$$\begin{aligned}
\mathcal{S} &\triangleq \{S_1, S_2, \ldots, S_n\} \\
\mathcal{A} &\triangleq \{a_1, a_2, \ldots, a_m\} \\
\Phi &\triangleq \{\phi(S^{t+1} = S_j \mid S^t = S_i, a_{ij}) \mid S_i, S_j \in \mathcal{S}, a_{ij} \in \mathcal{A}, t \geq 0\}
\end{aligned} \tag{3}$$

In our Markov chain model of manipulation strategies, actions at state S_i^t are represented as a collection of directed arcs, each of which, a_{ij}, connects two states in the robot task environment; namely, S_i^t and one of the possible S_j^{t+1}. The action schema oriented state transition is associated with a probability, $\phi(S^{t+1} = S_j \mid S^t = S_i, a_{ij})$, that signifies the probabilistic nature of the manipulation outcome. The set of all possible outcomes of an action is denoted as follows:

$$\Theta_{S_i^t, a_{ij}} \triangleq \{S_1^{t+1}, S_2^{t+1}, \ldots, S_l^{t+1}\} \tag{4}$$

Similar to the notion of frame-of-discern as in the Dempster-Shafer theory of evidence [6, 13], here we further distinguish the notion of *ignorance* about a next state. The ignorance is signified by the probability mass assigned to $\Theta_{S_i^t, a_{ij}}$, as denoted by $\phi(\Theta_{S_i^t, a_{ij}} \mid S^t = S_i, a_{ij})$.

And $\phi(S^{t+1} = S_j \mid S^t = S_i, a_{ij})$ satisfies the following properties:

$$\phi(\emptyset) = 0$$

$$\sum_{S_j^{t+1} \subseteq \Theta_{S_i^t, a_{ij}}} \phi(S_j^{t+1} \mid S_i^t, a_{ij}) + \phi(\Theta_{S_j^t, a_{ij}} \mid S_i^t, a_{ij}) = 1.0 \tag{5}$$

In addition, the probability of S_j^{t+1} is dependent of the probability of S_i^t, which can be written as follows:

$$\begin{aligned}
p(S_j^1) &= \sum_i \phi(S_j \mid S_i, a_{ij}) p(S_i^0) \\
p(S_j^2) &= \sum_i \phi(S_j \mid S_i, a_{ij}) p(S_i^1) \\
&\vdots \\
p(S_j^{t+1}) &= \sum_i \phi(S_j \mid S_i, a_{ij}) p(S_i^t)
\end{aligned} \tag{6}$$

In other words, the probability of the task environment at $t+1$ is dependent of all the probabilities of the states at t that have direct state transitions to the state at $t+1$.

In the present task planning case, we assume that the plan to be derived will be sequential and can be written as a chain of state transitions of length N, denoted as a_chain^N. Therefore, Eq. 6 can be rewritten as:

$$p(S_j^{t+1}) = \phi(S_j \mid S_i, a_{ij}) p(S_i^t), i, j \in \{1, 2, \cdots, l\}. \tag{7}$$

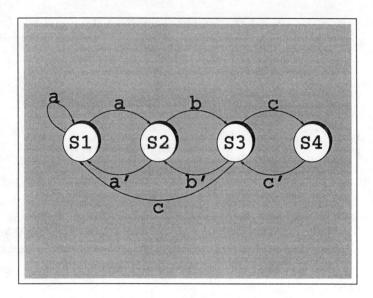

Figure 1. A Markov chain model of the action schema network.

Thus, the probability of a desirable final state can be expressed as follows:

$$p(S^N) = \left(\prod_{t=0}^{N-1} \phi(S_j^{t+1} \mid S_i^t, a_{ij})\right) p(S^0) \qquad (8)$$

where $S_i \longrightarrow S_j$ and $S_i, S_j \in a\_chain^N$.

2.2. Task Planning Based on Action Schemata

The Markov chain model of manipulation strategies can be viewed as a generating function that gives all the combinations and permutations, which can also be called enumerator. The permutation is an ordered arrangement of strategies. It should be noted that the network is a recursive network, since some strategies may yield non-deterministic state transitions.

Without loss of generality, let us consider the Markov chain model of a set of manipulation actions as depicted in Figure 1. Suppose that we want to derive a task plan that will change the states from S_1 to S_4. In this example, it is straightforward to see that there exists a single connected path from S_1 to S_4, that is, $a_a a_b a_c$. However, this sequence of operations expresses only one possible situation where the outcome of a_a causes a state transition to S_2. The complete task plan from this model would be as follows:

$$\begin{aligned} P &\longrightarrow a_a\ P \\ P &\longrightarrow a_a\ P' \\ P' &\longrightarrow a_b\ P'' \\ P'' &\longrightarrow a_c\ P \\ P'' &\longrightarrow a_c \end{aligned} \qquad (9)$$

In other words, the above plan expresses the following manipulation strategy:

	S_j	$\neg S_j$
$S_i \wedge a_{ij}$	$N_{(S_i \wedge a_{ij}) \wedge S_j}$	$N_{(S_i \wedge a_{ij}) \wedge \neg S_j}$
$\neg(S_i \wedge a_{ij})$	$N_{\neg(S_i \wedge a_{ij}) \wedge S_j}$	$N_{\neg(S_i \wedge a_{ij}) \wedge \neg S_j}$

Figure 2. A contingency table where cells indicate the number of co-occurrences.

1. perform a_a until S_2 is reached;
2. when in S_2, perform a_b;
3. when in S_3, perform a_c; and
4. if S_4 is reached stop, otherwise goto Step 1.

3. Learning Manipulation Strategies by Estimating State Transition Functions

In this section, we describe how action schema oriented state transitions can be derived in an algorithmic fashion. Our proposed algorithm induces state transition networks using a relatively small number of empirical data samples.

3.1. Automatic Induction of Manipulation Strategies

The basic idea behind the empirical induction of state transition functions is that in an ideal case, if there is a state transition relation $S_i \xrightarrow{a_{ij}} S_j$, then we would never expect to find the co-occurrences that S_j is TRUE but not $S_i \wedge a_{ij}$, from the empirical data samples. This translates into the following two conditions:

$$P(S_j \mid (S_i \wedge a_{ij})) = 1 \qquad (10)$$

$$P(\neg(S_i \wedge a_{ij}) \mid \neg S_j) = 1 \qquad (11)$$

In reality, however, due to sensory and control uncertainty, Conditions 10 and 11 may not be satisfied. Our action schema induction algorithm takes into account the imprecise/inexact nature of manipulation strategies and verifies the above conditions by computing the lower bound of a $(1-\alpha_c)$ confidence interval around the measured conditional probabilities. If the verification succeeds, a state transition between the two nodes is asserted. A probability mass is associated with the transition. In the present state transition based representation, the weight is estimated based on conditional probabilities $P(S_j \mid (S_i \wedge a_{ij}))$ and $P(\neg(S_i \wedge a_{ij}) \mid \neg S_j)$. The induction algorithm is given in Figure 3.

Here it is assumed that the conditional probability is p in each sample, and all n samples are independent. If X is the frequency of the occurrence, then X satisfies a binomial distribution, i.e., $X \sim Bin(n,p)$, whose probability function $p_X(k)$ and distribution function $F_X(k)$ are given below:

```
begin
   set a significance level α_c and a minimal conditional probability p_min
   for node_i, i ∈ [0, n_max − 1] and node_j, j ∈ [i + 1, n_max] do
      for all empirical case samples do
         compute a contingency table T_ij = | N_11  N_12 |
                                           | N_21  N_22 |
         where N_11, N_12, N_21, N_22 are the numbers of
         occurrences with respect to the following combinations:
            N_11 :   S_i ∧ a_ij = TRUE   ∧   S_j = TRUE
            N_12 :   S_i ∧ a_ij = TRUE   ∧   S_j = FALSE
            N_21 :   S_i ∧ a_ij = FALSE  ∧   S_j = TRUE
            N_22 :   S_i ∧ a_ij = FALSE  ∧   S_j = FALSE
         test the following inequality:
         P(x ≤ N_error_cell) < α_c
         based on the lower tails of binomial distributions Bin(N, p_min) and Bin(Ñ, p_min)
         where N and Ñ denote the occurrences of antecedent satisfaction in the inferences
         using the state transition as an inference rule. α_c is the alpha error of the
         conditional probability test
         if the test succeeds then
            assert a state transition
         endif
      endfor
   endfor
end
```

Figure 3. The state transition induction algorithm.

$$p_X(k) = \binom{n}{k} p^k q^{n-k} \tag{12}$$

$$F_X(k) = p(X \le k) = \sum_{j=0}^{k} \binom{n}{k} p^k q^{n-k} \tag{13}$$

where $p = 1 - q$.

4. Action Selection

In the preceding section, we have described how to acquire manipulation strategies by means of learning action schema oriented state transitions. As can be noted from the algorithm, the availability of empirical state observations is essential to the binomial statistical testing. Figure 4 presents a schematic diagram that outlines the process of empirical sample generation.

At a given time t, the robot first evaluates its current state to see whether a predefined final state has been reached. If the final state is not reached, it starts to search the schema network in an attempt to find a state transition that can bring the robot closer to its final state in the most effective and reliable way. If no state transition can be matched with the current state, the robot finds a state that is close to the current state and has been associated with a state transition arc. The action that corresponds to this state transition will become a trial action performed by the robot. Next, the new state that the

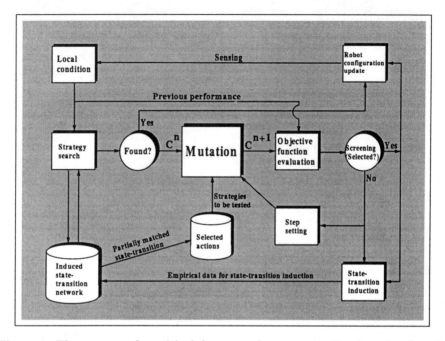

Figure 4. The process of empirical data sample generation by the robot based on a randomized search mechanism called mutation.

robot task environment is in will be evaluated. If the final state is still not reached and the resulting state does not match any of those transition states, the robot will decide whether to repeat the previously selected trial action or to *randomly* select other actions as trial actions (i.e., mutate to other actions). The decision will be based on how the previously selected trial action affects the optimality evaluation results.

While the robot executes a series of actions in a trial-and-error fashion, three kinds of information will be observed and recorded by the robot; namely,

1. the previous sensory readings that correspond to the previous state of the robot environment,

2. a specific action, as selected from a set of primitive actions, that the robot has executed, and

3. the current sensory readings that correspond to the current state of the robot task environment.

Such information will be collected as empirical data samples for manipulation strategy learning.

5. Experiments

The experimental setup that was designed to validate the manipulation strategy learning approach is depicted in Figure 5. More specifically, we investigated a

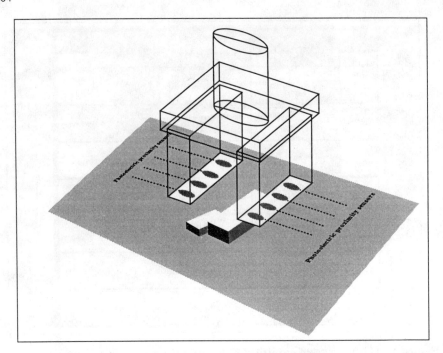

Figure 5. The configuraton of a two-fingered parallel jaw electric gripper equipped with eight downward facing photoelectric sensors.

robot manipulation task in which a two-fingered parallel jaw electric gripper was used for picking small objects. This task is essential in handling delicate devices, such as IC chips. The width of the parallel jaws is $3.5cm$. The gripper was brought by the robot to a downward facing position about $3 - 4cm$ away from the objects. Eight downward facing photoelectric sensors (i.e., optical fiber sensors of $0.7cm$ diameters) were mounted at the tip of each finger. From these sensors, the robot could determine whether its parallel jaws were properly positioned and aligned with respect to the dimensions of the object to be picked up. If the gripper was found not appropriately positioned or oriented, the robot would perform a certain action with its gripper, as selected from those in Figure 6, in order to find the right grasping position and orientation. The action that the robot selected and executed constitutes a manipulation strategy of the robot in response to a particular state in its task environment.

5.1. Empirical Sample Generation

In order to use the above task for validating our learning algorithm, we first applied the randomized search based technique, as described in Section 4, to generate and record a set of empirical observations. This entails the definition of an appropriate grasping position and orientation, i.e., the final optimal state of the task environment. In our experiment, we defined the final optimal state in terms of the readings from the eight sensors as such that no sensor could receive the reflected light from the object. Such sensory feedback directly

Action symbol	Description
7	Approaching
6	Randomized motion
5	Roll in counterclockwise direction
4	Roll in clockwise direction
3	Displacement in y- direction
2	Displacement in y+ direction
1	Displacement in x- direction
0	Displacement in x+ direction

Figure 6. A set of actions that an end-effector can perform. Each action indicates the coupled joint movements of the robot that realize the precise translatioal/rotational increment of the robot end-effector in Cartesian space.

corresponded to a state in which the object was aligned in between the two jaws. It should be pointed out that here we assumed the robot gripper would not be operated in free space but instead always be placed above the region where the objects were located.

The predefined final optimal state of the gripper would be used to evaluate the effectiveness of a trial action, that is, to calculate the difference between the resulting state of the action and the final optimal state. If the trial action reduced the difference, it would become the default action which would be used if the same task state was observed again.

5.2. State Transition Induction

Using the empirical data samples as obtained from the randomized optimal state search, we then applied the state transition induction algorithm as given in Figure 3. To induce one single transition $S_i \xrightarrow{a_{ij}} S_j$ using the contingency table, we required the total number of occurrences in four situations to be at least 30, and the α_c value for the statistical testing to be 0.15.

In theory, there could exist as many as 256 states of the robot task environment with respect to the readings from the eight photoelectric sensors. However, in reality, since many of them are redundant, the actual number of

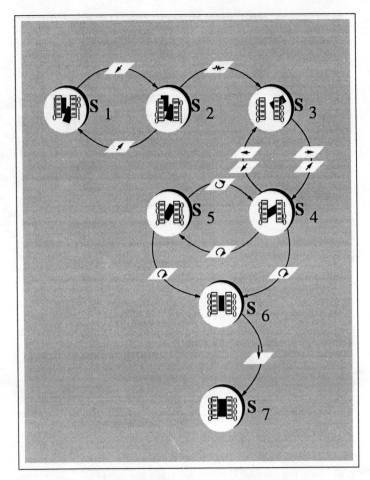

Figure 7. Examples of induced state transitions.

the possible states was here reduced to 29. The actions that compose the manipulation strategies were given in Figure 6, where the increments for the translation Δx and Δy were both set to $0.3cm$, and $\Delta \theta$ was set to $2°$. As a result of the induction, a total number of 34 state transitions were derived, and some of them were depicted in Figure 7.

5.3. Orientation Correction

Given an induced schema network, we investigated the validity of the induced state transitions in modifying the orientation of the gripper in order to align itself with the object to be picked up. The width of the object was $3cm$. The two jaws of the gripper were in an open position with a distance of $4cm$ apart from each other. The initial orientation of the parallel jaws relative to the long sides of the object was set to $25°$. The robot was to search appropriate manipulation strategies from the induced state transitions, such as those presented in Figure 7. Each time when a specific action was selected and performed as

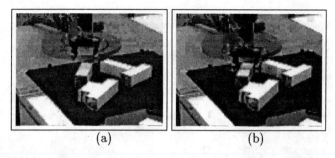

Figure 8. The end-effector modifies its orientation with respect to the objects to be picked up.

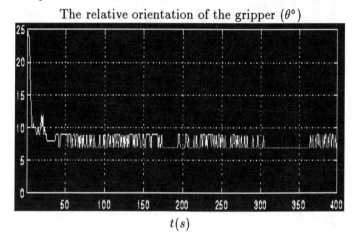

Figure 9. The change of the gripper orientation angle with respect to the long sides of the object, when the parallel jaws of the gripper were in a 4cm open position.

shown in Figure 8(a), the relative orientation was recorded and updated.

Figure 9 shows the history of the relative orientation angle change over a period of time. It can be observed from the figure that after a few seconds, the gripper was capable of reaching its final state, i.e., all the sensors received no reflected light. When this occurred, as shown in Figure 8(b), the actual relative orientation recorded was about 8°.

From our experiment, it was also evident that the position/orientation precision could be sensitive to the configuration of the sensors that were to extract the feedback about the robot task environment. Figure 10 presents the history of the gripper orientation change when the two jaws were kept in a 3.5cm open position. In such a case, the actually recorded relative orientation of the gripper could be brought down to as lower as ±2°.

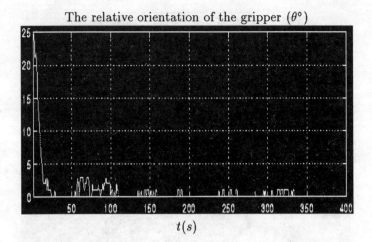

Figure 10. The change of the gripper orientation angle with respect to the long sides of the object, when the parallel jaws of the gripper were in a 3.5cm open position.

5.4. Randomized Motion

One of the useful manipulation strategies as induced by the robot was that of randomized motion. In our present experimental robot task, this strategy was found to be effective when more than four photoelectric sensors detected the presence of cluttered objects. This is illustrated in Figure 7 as the transition from state S_2 to state S_3. Figure 11 presents the snapshots of our experimental robot that selected and performed a randomized motion strategy in separating the objects.

6. Conclusion

In this paper, we have described an empirical learning based approach to robot manipulation task planning. The manipulation strategies are represented within the Markov chain model where the state transitions are automatically induced based on statistical testing. The significance of the presented approach lies in that it enables:

- the relaxation of requirements on explicit task modeling, and
- the selection of near-optimal actions and the incremental construction of new action schemata.

References

[1] Haruhiko Asada and Haruo Izumi. Automatic program generation from teaching data for the hybrid control of robots. *IEEE Transactions on Robotics and Automation*, 5(2):166–173, 1989.

[2] M. Brady. The problems of robotics. In M. Brady, editor, *Robotics Science*, pages 1–35. MIT Press, Cambridge, MA, 1989.

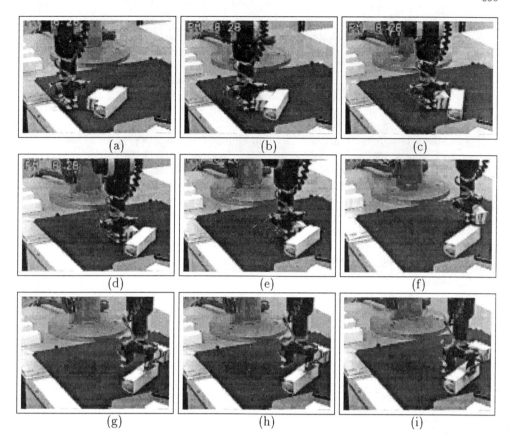

Figure 11. The end-effector separated the objects by executing a randomized motion strategy.

[3] S. J. Buckley. Planning compliant motion strategies. *The International Journal of Robotics Research*, 8(5):28–44, Oct. 1989.

[4] Christos G. Cassandras. *Discrete Event Systems: Modeling and Performance Analysis*. Aksen Associates Incorporated Publishers and IRWIN, Homewood, IL, 1993.

[5] Thomas L. Dean and Michael P. Wellman. *Planning and Control*. Morgan Kaufmann Publishers, San Mateo, CA, 1991.

[6] A. P. Dempster. A generalization of Bayesian inference. *Journal of the Royal Statistical Society*, 30:205–247, 1968.

[7] Tsutomu Hasegawa, Takashi Suehiro, and Kunikatsu Takase. A robot system for unstructured environments based on an environment model and manipulation skills. In *Proceedings of the 1991 IEEE International Conference on Robotics and Automation*, pages 916–932, Sacramento, CA, April 1991.

[8] Sing Bing Kang and Katsushi Ikeuchi. Robot task programming by human demonstration: Mapping human grasps to manipulator grasps. In *Proceedings of the 1994 IEEE/RSJ/GI International Conference on Intelligent Robots and Systems*, pages 97–104, 1994.

[9] Yasuo Kuniyoshi, Masayuki Inaba, and Hirochika Inoue. Learning by watching: Extracting reusable task knowledge from visual observation of human performance. *IEEE Transactions on Robotics and Automation*, 10(6):799–822, 1994.

[10] Jean-Claude Latombe. *Robot Motion Planning*. Kluwer Academic Publishers, Norwell, MA, 1991.

[11] T. Lozano-Pérez and R. H. Taylor. Geometric issues in planning robot tasks. In M. Brady, editor, *Robotics Science*, pages 227–261. MIT Press, Cambridge, MA, 1989.

[12] Jun Miura, Isao Shimoyama, and Hirofumi Miura. Integration of problem-solving and learning in intelligent robots. *Advanced Robotics*, 7(4):309–328, 1993.

[13] G. Shafer. *A Mathematical Theory of Evidence*. Princeton University Press, Princeton, NJ, 1976.

[14] Antonio Tornambe. *Discrete-Event System Theory*. World Scientific Publishing Co., Singapore, 1995.

[15] Daniel E. Whitney. A survey of manipulation and assembly: Development of the field and open research issues. In Michael Brady, editor, *Robotics Science*, System Development Foundation Benchmark Series, pages 291–348. The MIT Press, Cambridge, MA, 1989.

Lecture Notes in Control and Information Sciences

Edited by M. Thoma

1993–1998 Published Titles:

Vol. 186: Sreenath, N.
Systems Representation of Global Climate Change Models. Foundation for a Systems Science Approach.
288 pp. 1993 [3-540-19824-5]

Vol. 187: Morecki, A.; Bianchi, G.; Jaworeck, K. (Eds)
RoManSy 9: Proceedings of the Ninth CISM-IFToMM Symposium on Theory and Practice of Robots and Manipulators.
476 pp. 1993 [3-540-19834-2]

Vol. 188: Naidu, D. Subbaram
Aeroassisted Orbital Transfer: Guidance and Control Strategies
192 pp. 1993 [3-540-19819-9]

Vol. 189: Ilchmann, A.
Non-Identifier-Based High-Gain Adaptive Control
220 pp. 1993 [3-540-19845-8]

Vol. 190: Chatila, R.; Hirzinger, G. (Eds)
Experimental Robotics II: The 2nd International Symposium, Toulouse, France, June 25-27 1991
580 pp. 1993 [3-540-19851-2]

Vol. 191: Blondel, V.
Simultaneous Stabilization of Linear Systems
212 pp. 1993 [3-540-19862-8]

Vol. 192: Smith, R.S.; Dahleh, M. (Eds)
The Modeling of Uncertainty in Control Systems
412 pp. 1993 [3-540-19870-9]

Vol. 193: Zinober, A.S.I. (Ed.)
Variable Structure and Lyapunov Control
428 pp. 1993 [3-540-19869-5]

Vol. 194: Cao, Xi-Ren
Realization Probabilities: The Dynamics of Queuing Systems
336 pp. 1993 [3-540-19872-5]

Vol. 195: Liu, D.; Michel, A.N.
Dynamical Systems with Saturation Nonlinearities: Analysis and Design
212 pp. 1994 [3-540-19888-1]

Vol. 196: Battilotti, S.
Noninteracting Control with Stability for Nonlinear Systems
196 pp. 1994 [3-540-19891-1]

Vol. 197: Henry, J.; Yvon, J.P. (Eds)
System Modelling and Optimization
975 pp approx. 1994 [3-540-19893-8]

Vol. 198: Winter, H.; Nüßer, H.-G. (Eds)
Advanced Technologies for Air Traffic Flow Management
225 pp approx. 1994 [3-540-19895-4]

Vol. 199: Cohen, G.; Quadrat, J.-P. (Eds)
11th International Conference on Analysis and Optimization of Systems – Discrete Event Systems: Sophia-Antipolis, June 15–16–17, 1994
648 pp. 1994 [3-540-19896-2]

Vol. 200: Yoshikawa, T.; Miyazaki, F. (Eds)
Experimental Robotics III: The 3rd International Symposium, Kyoto, Japan, October 28-30, 1993
624 pp. 1994 [3-540-19905-5]

Vol. 201: Kogan, J.
Robust Stability and Convexity
192 pp. 1994 [3-540-19919-5]

Vol. 202: Francis, B.A.; Tannenbaum, A.R. (Eds)
Feedback Control, Nonlinear Systems, and Complexity
288 pp. 1995 [3-540-19943-8]

Vol. 203: Popkov, Y.S.
Macrosystems Theory and its Applications: Equilibrium Models
344 pp. 1995 [3-540-19955-1]

Vol. 204: Takahashi, S.; Takahara, Y.
Logical Approach to Systems Theory
192 pp. 1995 [3-540-19956-X]

Vol. 205: Kotta, U.
Inversion Method in the Discrete-time Nonlinear Control Systems Synthesis Problems
168 pp. 1995 [3-540-19966-7]

Vol. 206: Aganovic, Z.;.Gajic, Z.
Linear Optimal Control of Bilinear Systems with Applications to Singular Perturbations and Weak Coupling
133 pp. 1995 [3-540-19976-4]

Vol. 207: Gabasov, R.; Kirillova, F.M.; Prischepova, S.V.
Optimal Feedback Control
224 pp. 1995 [3-540-19991-8]

Vol. 208: Khalil, H.K.; Chow, J.H.; Ioannou, P.A. (Eds)
Proceedings of Workshop on Advances inControl and its Applications
300 pp. 1995 [3-540-19993-4]

Vol. 209: Foias, C.; Özbay, H.; Tannenbaum, A.
Robust Control of Infinite Dimensional Systems: Frequency Domain Methods
230 pp. 1995 [3-540-19994-2]

Vol. 210: De Wilde, P.
Neural Network Models: An Analysis
164 pp. 1996 [3-540-19995-0]

Vol. 211: Gawronski, W.
Balanced Control of Flexible Structures
280 pp. 1996 [3-540-76017-2]

Vol. 212: Sanchez, A.
Formal Specification and Synthesis of Procedural Controllers for Process Systems
248 pp. 1996 [3-540-76021-0]

Vol. 213: Patra, A.; Rao, G.P.
General Hybrid Orthogonal Functions and their Applications in Systems and Control
144 pp. 1996 [3-540-76039-3]

Vol. 214: Yin, G.; Zhang, Q. (Eds)
Recent Advances in Control and Optimization of Manufacturing Systems
240 pp. 1996 [3-540-76055-5]

Vol. 215: Bonivento, C.; Marro, G.; Zanasi, R. (Eds)
Colloquium on Automatic Control
240 pp. 1996 [3-540-76060-1]

Vol. 216: Kulhavý, R.
Recursive Nonlinear Estimation: A Geometric Approach
244 pp. 1996 [3-540-76063-6]

Vol. 217: Garofalo, F.; Glielmo, L. (Eds)
Robust Control via Variable Structure and Lyapunov Techniques
336 pp. 1996 [3-540-76067-9]

Vol. 218: van der Schaft, A.
L_2 Gain and Passivity Techniques in Nonlinear Control
176 pp. 1996 [3-540-76074-1]

Vol. 219: Berger, M.-O.; Deriche, R.; Herlin, I.; Jaffré, J.; Morel, J.-M. (Eds)
ICAOS '96: 12th International Conference on Analysis and Optimization of Systems - Images, Wavelets and PDEs:
Paris, June 26-28 1996
378 pp. 1996 [3-540-76076-8]

Vol. 220: Brogliato, B.
Nonsmooth Impact Mechanics: Models, Dynamics and Control
420 pp. 1996 [3-540-76079-2]

Vol. 221: Kelkar, A.; Joshi, S.
Control of Nonlinear Multibody Flexible Space Structures
160 pp. 1996 [3-540-76093-8]

Vol. 222: Morse, A.S.
Control Using Logic-Based Switching
288 pp. 1997 [3-540-76097-0]

Vol. 223: Khatib, O.; Salisbury, J.K.
Experimental Robotics IV: The 4th International Symposium, Stanford, California, June 30 - July 2, 1995
596 pp. 1997 [3-540-76133-0]

Vol. 224: Magni, J.-F.; Bennani, S.; Terlouw, J. (Eds)
Robust Flight Control: A Design Challenge
664 pp. 1997 [3-540-76151-9]

Vol. 225: Poznyak, A.S.; Najim, K.
Learning Automata and Stochastic Optimization
219 pp. 1997 [3-540-76154-3]

Vol. 226: Cooperman, G.; Michler, G.;
Vinck, H. (Eds)
Workshop on High Performance Computing
and Gigabit Local Area Networks
248 pp. 1997 [3-540-76169-1]

Vol. 227: Tarbouriech, S.; Garcia, G. (Eds)
Control of Uncertain Systems with Bounded
Inputs
203 pp. 1997 [3-540-76183-7]

Vol. 228: Dugard, L.; Verriest, E.I. (Eds)
Stability and Control of Time-delay Systems
344 pp. 1998 [3-540-76193-4]

Vol. 229: Laumond, J.-P. (Ed.)
Robot Motion Planning and Control
360 pp. 1998 [3-540-76219-1]

Vol. 230: Siciliano, B.; Valavanis, K.P. (Eds)
Control Problems in Robotics and Automation
328 pp. 1998 [3-540-76220-5]

Vol. 231: Emel'yanov, S.V.; Burovoi, I.A.;
Levada, F.Yu.
Control of Indefinite Nonlinear Dynamic
Systems
216 pp. 1998 [3-540-76245-0]